K. Voss   H. J. Genrich   G. Rozenberg   (Eds.)

Concurrency and Nets

Advances in Petri Nets

Springer-Verlag

Berlin Heidelberg New York
London Paris Tokyo

Dr. Klaus Voss
Dr. Hartmann J. Genrich
Gesellschaft für Mathematik und Datenverarbeitung
Schloss Birlinghoven
D-5205 Sankt Augustin 1

Professor Dr. Grzegorz Rozenberg
Department of Mathematics and Computer Science
University of Leiden
Niels-Bohr-Weg 1
NL-2300 RA Leiden

ISBN-13: 978-3-642-72824-2   e-ISBN-13: 978-3-642-72822-8

DOI: 10.1007/978-3-642-72822-8

© Springer-Verlag Berlin Heidelberg 1987

Softcover reprint of the hardcover 1st edition 1987

2145/3140-543210

Preface

The present special volume of *Advances in Petri Nets* has been prepared in tribute to Carl Adam Petri on the occasion of his 60th birthday. It is devoted to an outstanding personality and his pioneering and fruitful scientific work. Meeting Carl Adam Petri means to experience a warmhearted, modest, far-sighted person with high technical and human standards. Getting acquainted with his scientific work means to plunge into a well-founded systems theory of broad scope, strongly influenced by, and deeply concerned with, fundamental practical problems.

During 25 years since his doctoral dissertation appeared, Petri has exerted a great scientific influence. His ideas were conveyed as much through private conversations (usually intended for a couple of minutes but lasting for several hours) as through printed documents providing new insights into the nature of distributed systems and opening important lines of research. The contributions in this volume express great respect and affection for Carl Adam Petri who, like nobody else, has foreseen the development of computers – not so much in technological terms but rather in terms of its function as a novel communication medium – and the central role *concurrency* was therefore to play in computer science.

This volume consists of two parts.

Part I presents the congratulatory addresses and invited talks that were delivered at a Colloquium organised by Gesellschaft für Mathematik und Datenverarbeitung (GMD) to celebrate the birthday (unfortunately one of the talks given during the Colloquium – on theoretical aspects of Petri's work – could not be included in this volume). The Colloquium was held in Schloß Birlinghoven on the afternoon of September 12, 1986, during the 2nd Advanced Course on Petri Nets that was taking place in Bad Honnef at this time. The contributions of this part honour Carl Adam Petri and his work from many different perspectives.

Part II is a collection of papers discussing various aspects of the theme *Concurrency and Nets*. About half of these papers are contributed by researchers that were or still are associated, one way or the other, with the Petri institute at GMD while the other half is contributed by authors that through the years had rather intense contacts either with Petri himself or with the Petri net community.

We are very honoured by having had the opportunity to edit this volume and we express our gratitude to all contributors, to our colleagues Eike Best, César Fernández, Kurt Lautenbach, Wolfgang Reisig and Gernot Richter, and to Springer-Verlag for their cooperation in this task.

Klaus Voss   Hartmann J. Genrich   Grzegorz Rozenberg

Sankt Augustin, March 1987

TABLE OF CONTENTS

PUBLICATIONS BY CARL ADAM PETRI

Petri, C.A.: *Kommunikation mit Automaten.* Rheinisch-Westfälisches Institut für Instrumentelle Mathematik an der Universität Bonn, Schrift Nr. 2, 1962
Also: *Communication with Automata.* Griffiss Air Force Base, New York, Technical Report RADC-TR-65-377, Vol. 1, Suppl. 1, 1966 [English translation]

Petri, C.A.: *Fundamentals of a Theory of Asynchronous Information Flow.* Proc. IFIP Congress 62, München. Amsterdam: North Holland 1963, pp. 386–390

Petri, C.A.: *Grundsätzliches zur Beschreibung diskreter Prozesse.* In: Händler, W.; Peschl, E.; Unger, H. (Hrsg.): 3. Colloquium über Automatentheorie, Hannover, 1965. Basel, Stuttgart: Birkhäuser 1967, pp. 121–140
Also: *Fundamentals of the Representation of Discrete Processes.* Gesellschaft für Mathematik und Datenverarbeitung, St. Augustin, ISF-Report 82.04, 1982 [English translation]

Petri, C.A.: *Concepts of Net Theory.* Mathematical Foundations of Computer Science: Proceedings of Symposium and Summer School, High Tatras, Sep. 1973. Mathematical Institute of the Slovak Academy of Sciences, Bratislava, 1973, pp. 137–146

Petri, C.A.: *Interpretations of Net Theory.* Gesellschaft für Mathematik und Datenverarbeitung, St. Augustin, Interner Bericht ISF-75-07, Second Edition, 1976

Petri, C.A.: *Nicht-sequentielle Prozesse.* Universität Erlangen-Nürnberg, Arbeitsberichte des IMMD, Vol. 9, Nr. 8, 1976, pp. 57–82
Also: Gesellschaft für Mathematik und Datenverarbeitung, St. Augustin, Interner Bericht ISF-76-06, 3., revidierte und ergänzte Auflage, 1977
Also: *Non-Sequential Processes: Translation of a Lecture given at the IMMD Jubilee Colloquium on 'Parallelism in Computer Science'.* Gesellschaft für Mathematik und Datenverarbeitung, St. Augustin, Interner Bericht ISF-77-05, 1977 [English translation]

Petri, C.A.: *General Net Theory.* In: Shaw, B. (ed.): Computing System Design: Proceedings of the Joint IBM University of Newcastle upon Tyne Seminar, Sep. 1976. University of Newcastle upon Tyne 1977, pp. 131–169

Petri, C.A.: *Communication Disciplines.* In: Shaw, B. (ed.): Computing System Design: Proceedings of the Joint IBM University of Newcastle upon Tyne Seminar, Sep. 1976. University of Newcastle upon Tyne 1977, pp. 171–183

Petri, C.A.: *Modelling as a Communication Discipline.* In: Beilner, H.; Gelenbe, E. (eds.): Measuring, Modelling and Evaluating Computer Systems. Amsterdam: North-Holland 1977, pp. 435–449

Petri, C.A.: *Kommunikationsdisziplinen.* Wechselwirkungen zwischen informationstechnischen Entwicklungen und Organisationsstrukturen, ASA-Seminarbericht 76/02. Köln: Selbstverlag AGF 1977, pp. 24–45

Petri, C.A.: *Concurrency as a Basis of Systems Thinking.* Gesellschaft für Mathematik und Datenverarbeitung, St. Augustin, Interner Bericht ISF-78-06, 1978
Also in: Jensen, F.V.; Mayoh, B.H.; Møller, K.K. (eds.): Proceedings of 5th Scandinavian Logic Symposium, Aalborg, Jan. 1979. Aalborg: Universitetsforlag 1979, pp. 143–162

Petri, C.A.: *Über einige Anwendungen der Netztheorie.* In: Böhling, K.H., Spies, P.P. (Hrsg.): GI - 9. Jahrestagung. Informatik-Fachberichte 19. Berlin, Heidelberg, New York: Springer 1979, pp. 81–87

Petri, C.A. (Hrsg.): *Ansätze zur Organisationstheorie rechnergestützter Informationssysteme.* Berichte der Gesellschaft für Mathematik und Datenverarbeitung 111. München, Wien: R. Oldenbourg 1979

Petri, C.A.: *Kommunikationsdisziplinen.* In: Petri, C.A. (Hrsg.): Ansätze zur Organisationstheorie rechnergestützter Informationssysteme. Berichte der Gesellschaft für Mathematik und Datenverarbeitung 111. München, Wien: R. Oldenbourg 1979, pp. 63–76

Petri, C.A.: *Introduction to General Net Theory.* In: Brauer, W. (ed.): Net Theory and Applications. Lecture Notes in Computer Science 84. Berlin, Heidelberg, New York: Springer 1980, pp. 1–19

Petri, C.A.: *Concurrency.* In: Brauer, W. (ed.): Net Theory and Applications. Lecture Notes in Computer Science 84. Berlin, Heidelberg, New York: Springer 1980, pp. 251–260

Petri, C.A.: *State-Transition Structures in Physics and in Computation.* International Journal of Theoretical Physics, Vol. 21, No. 12, pp. 979–992 (1982)

Petri, C.A.: *Some Personal Views of Net Theory.* In: Pagnoni, A.; Rozenberg, G. (eds.): Application and Theory of Petri Nets. Informatik-Fachberichte 66. Berlin, Heidelberg, New York: Springer 1983, pp. 1–13

Petri, C.A.; Smith, E.: *The Pragmatic Dimension of Net Theory.* One-Day Seminar at the Bocconi University of Milan on Applicability of Petri Nets to Operations Research. Bocconi University, Milan, 1986, pp. 44–58

Petri, C.A.: *Concurrency Theory.* In: Brauer, W.; Reisig, W.; Rozenberg, G. (eds.): Petri Nets: Central Models and Their Properties. Advances in Petri Nets 1986, Part I. Lecture Notes in Computer Science 254. Berlin, Heidelberg, New York: Springer 1987, pp. 4–24

Petri, C.A.: *"Forgotten" Topics of Net Theory.* In: Brauer, W.; Reisig, W.; Rozenberg, G. (eds.): Petri Nets: Applications and Relationships to Other Concurrency Models. Advances in Petri Nets 1986, Part II. Lecture Notes in Computer Science 255. Berlin, Heidelberg, New York: Springer 1987, pp. 500–514

Petri, C.A.: *Concurrency and Continuity.* In: Rozenberg, G. (ed.): Advances in Petri Nets 1987. Lecture Notes in Computer Science 266. Berlin, Heidelberg, New York: Springer 1987, pp. 273–292

Remark: The above list of publications has not been checked by Carl Adam Petri.

The editors

K o m m u n i k a t i o n

m i t

A u t o m a t e n

Von der Fakultät für Mathematik und Physik

der Technischen Hochschule Darmstadt

zur Erlangung des Grades eines

Doktors der Naturwissenschaften

(Dr. rer.nat.)

genehmigte

Dissertation

vorgelegt von

C a r l   A d a m   P e t r i

aus Leipzig

Referent: Prof.Dr.rer.techn.A.Walther

Korreferent: Prof.Dr.Ing.H.Unger

Tag der Einreichung: 27.7.1961

Tag der mündlichen Prüfung: 20.6.1962

D 17

Bonn   1962

Facsimile of the title page of Carl Adam Petri's dissertation

CONGRATULATORY ADDRESS

Friedrich Winkelhage
Gesellschaft für Mathematik und Datenverarbeitung (GMD)
St. Augustin, Federal Republic of Germany

On behalf of the Board of Directors of GMD

Dear Mrs. Petri, dear Dr. Petri,
Ladies and Gentlemen,

It is a special pleasure for me to welcome you all at the occasion of this colloquium here at the Gesellschaft für Mathematik und Datenverarbeitung at Birlinghoven Castle. We have planned and organized this colloquium on the special occasion of the 60th birthday of Carl Adam Petri and in connection with the 2nd Advanced Course on Petri Nets. We are very glad to welcome so many friends and partners, representatives of international and national scientific organizations, of our Supervisory Board, our Scientific-Technical Advisory Board, and the participants of the 2nd Advanced Course on Petri Nets. Welcome again!

Carl Adam Petri was born in 1926. When he was 18 years old, in 1944, he had to enter the army and returned home only in 1948. He then studied mathematics at the Technical University of Hannover till 1956, and in 1962 he got his doctor degree from the Technical University of Darmstadt for his fundamental work and his thesis on "Kommunikation mit Automaten" (Communication with automata). Till 1968 he worked as a scientist at the University of Bonn. When GMD was established in 1968, Carl Adam Petri was appointed as the head of the "Institut für Informationssystemforschung", and now he is one of the heads of the "Institut für Methodische Grundlagen" at GMD.

During the following years the universities of Karlsruhe and Dortmund offered him a chair - he decided to stay at GMD. He was invited by and worked as a guest researcher at well-known international research institutes and universities in Europe, the United States of America, Japan and China. There is no doubt: in the international scientific world of informatics Carl Adam Petri is an outstanding personality and scientist.

The 2nd international Advanced Course on Petri Nets gives vivid evidence for this statement. The lectures given at this course are examples for the strong impact of his ideas, of his work on the development in the theoretical and the application-oriented field of informatics. This also is especially true for the research and development activities of GMD. Petri-nets play a major role in many projects where GMD is cooperating internationally with research institutes and industrial partners. This, for example, is the case in the areas of software technology, software specification, higher level protocols and office automation, and I could tell many things in addition.

In this context, and regarding the development of GMD as a national research center we highly appreciate the well-balanced advice of Carl Adam Petri and his ability to give a fundamental orientation for long-range planning and design of research and development activities.

Dear Dr. Petri, on the behalf of GMD I would like to thank you very much for the long time of scientific engagement and contributions to the development of GMD, for the very fruitful cooperation and for your very valuable advice. We are looking forward to continue this work in the future. Thank you very much again, and as a sign for this thank I have a special award for you and your wife.

Ladies and gentlemen, let me say just a few words to the afternoon here. I have to announce that Professor Kotov unfortunately cannot attend this colloquium. He regrets it very much and gives you his best greetings.

The congratulation address from IFIP will be given by the West German representative to IFIP who is also chairman of IFIP Technical Committee for education, Professor Wilfried Brauer. Then we will have the congratulation address by Professor Grzegorz Rozenberg who is president of the European Association of Theoretical Computer Science. Professor Krückeberg will talk to us as president of the Gesellschaft für Informatik, and Dr. Genrich as one of the scientists who have been working with Carl Adam Petri from the very beginning.

CONGRATULATORY ADDRESS

Wilfried Brauer
Technische Universität München
München, Federal Republic of Germany

On behalf of the International Federation for Information Processing (IFIP)

Dear Mrs. Petri, dear Dr. Petri,
Ladies and Gentlemen!

It is a great pleasure and a special honour for me to contribute to the congratulations to Dr. Carl Adam Petri in two different ways: first, as the IFIP representative, and later personally by my lecture.

Most of you surely know something about IFIP, the International Federation for Information Processing. It is a world-wide umbrella organization of more than 40 national informatics societies and several other international societies. Two years ago, as you may know, IFIP celebrated its 25th anniversary in Munich. The Petri net community probably knows IFIP mainly because of the activities of its technical committees and working groups, notably of the technical committee on programming, the TC2. But also other technical committees, particular those on communications and on information systems take Petri nets into their considerations.

You have possibly also heard of TC3, the technical committee on education, which has its own world congresses and, in 1988, will organize also a large European congress on computers in education in Lausanne. By the way, the informatics competition for secondary school students which is now rather popular in Germany and which is just now run by GMD and the Gesellschaft für Informatik (GI) goes back to an initiative of TC3. And, as the chairman of TC3, I would suggest here that we all should carefully think about ways, methods, techniques to teach already in schools Petri nets and the concepts behind and related to them, since teaching children only how to program and how to handle computers might be too narrow-minded.

Another very important activity of IFIP are the large international congresses held every three years. The tenth IFIP congress has just been held last week in Dublin, and at that congress due to its program committee chairman, Professor Dines Bjørner from Copenhagen and the responsible sub-committee chairman, Professor Kotov from Novosibirsk, IFIP's appreciation of net theory became known to a world-wide audience by the invited lecture on "Net Theory and its Applications" given by Dr. Hartmann Genrich from GMD. Since Petri nets are now so much estimated also in IFIP, the president of IFIP, Dr. Kaoru Ando, Fujitsu,

Tokio, personally and on behalf of IFIP, congratulates you, Dr. Petri, to your 60th birthday, and Professor Bjørner joins these greetings heartily.

I now have the great honour to read the congratulation letter from the IFIP president:

"To Carl Adam Petri.

The international Federation for Information Processing (IFIP) hereby wishes to recognize the outstanding profound and epoche-making contributions you have made, both to the basic fundamental research in our science and to the wide-ranging, clarifying and simplifying applications of net theory. IFIP therefore expresses its deepest appreciation.

You have truly created a long ranging school of thought and science. IFIP wishes you all the very best both to you and for the future of your contributions.

Dr. Kaoru Ando,

President of IFIP."

CONGRATULATORY ADDRESS

Grzegorz Rozenberg
Rijksuniversiteit te Leiden
Leiden, The Netherlands

On behalf of the European Association for Theoretical Computer Science
(EATCS)

Dear Mrs. Petri, dear Dr. Petri,
Ladies and Gentlemen!

In my function as the president of the European Association for Theoretical Computer Science I would like to say the following on the occasion of the 60th birthday of Dr. Petri.

Dr. Carl Adam Petri has made a deep, broad and lasting contribution to the field of distributed information processing systems. The field of research founded by him, called Petri nets, is currently one of the most active areas of research in the field of distributed information processing. In Europe alone teams of researchers pursuing both theoretical and practical lines of work are currently active in West Germany, France, Italy, the Netherlands, Denmark, Great Britain, Spain and Finland among others. Apart from numerous articles presented at computer science conferences and published in reputed journals and about ten books on Petri nets that have appeared in recent years, there is an annual conference under the heading "European Workshop on Applications and Theory of Petri Nets". This year the 7th of such meetings was held in June in Oxford. Significant advances in theory and practice are published on a regular basis in a special book series by Springer Verlag under the title "Advances in Petri Nets". In the industry, Petri nets have been applied in a wide range of problem areas.

Dr. Carl Adam Petri achieved his initial conceptual breakthrough in the early sixties while managing one of the first academic computer centers created in West Germany at Bonn University. His main idea was to recognize even at that very early stage that the significant future of the computer was not just its ability to carry out quickly and reliably intricate mathematical computations, but rather its ability to serve as a sophisticated communication medium, a medium which establishes a desired pattern of information flow between a number of independent agents. Dr. Petri founded his field by creating mathematical objects called nets, later called Petri nets by the scientific community. This nets captured simply, precisely and elegantly the intuition that digital computers may be profitably viewed as information processing instruments serving the community of users rather than as just computational devices serving a single user. The initial and fundamental results concerning nets due to Dr. Petri appeared as a doctoral thesis in 1962 under the eminently suitable title "Kommunikation mit Automaten". The sixties and the major portion of seventies were a difficult period for Dr. Petri in terms of finding broad support and acceptance of his ideas. A major reason was that during this period the academic community was still grabbling with the deep problems thrown out by the classic Turing machine model on the theoretical side and its realization in the form of von Neumann architecture on the practical side. A second reason was that even though the paradigm of communication media suggested by Petri was rapidly becoming reality in commercial applications, it was still not transparent enough due to the high cost of hardware.

Fortunately, Dr. Petri was able to gather a small group of dedicated workers around him at the newly founded GMD, and further developed his theory of distributed information processing systems. His major achievements during this period were to construct the mathematical tools required for specifying distributed systems at varying levels of detail, to create the notion of a non-sequential process, to capture the formal semantics of systems capable of concurrent activities and to establish an elegant and powerful link between his theory and classical logic. During the last decade, a period of intense activity in the field of distributed computing, the theory and the tools created by Petri and his group have assumed a dominant position. His work has directly or indirectly influenced a variety of alternative approaches to the study of distributed information processing. A number of his co-workers strongly influenced by Petri's insistence that conceptual elegance ought to be combined with the pragmatics of applications have become recognized leaders in their own right in the area of distributed computing.

It is difficult to give a complete account of the creative achievements of Dr. Petri and the resulting impact of his work in the field of information technology. His contributions range from beautiful theorems on the Dedekind continuity of non-sequential processes to piercing remarks concerning the construction of asynchronous switching circuits. Perhaps his most significant contribution is this: with his uncompromising scientific standards, deep sense of culture and a mature understanding of the possibilities of harnessing information technology to meet the need of society as a whole, Dr. Petri has for more than two decades served as a shining example and a source of inspiration to his fellow scientists. I consider this to be an honour for our organization, the European Association for Theoretical Computer Science, and for me personally, to be invited today to take part in this celebration of Dr. Petri's 60th birthday.

Dear Dr. Petri,

I wish you still many, many years of fruitful work. I also wish you that – and actually I am convinced that this will happen – in the years to come you be witnessing with great satisfaction how the field you have founded will develop as a most thriving area of both intellectual challenge and practical applications.

Happy birthday, Dr. Petri!

CONGRATULATORY ADDRESS

Fritz Krückeberg
Gesellschaft für Mathematik und Datenverarbeitung (GMD)
St. Augustin, Federal Republic of Germany

On behalf of the Gesellschaft für Informatik (GI)

Dear Mrs. Petri, dear Carl Adam,

As the President of the Gesellschaft für Informatik, and as a longtime friend, it gives me great pleasure to congratulate you on the occasion of your 60th birthday. You are an outstanding member of our association, and we consider it a great honour and privilege to have you as a member. The coming generations of computer scientist will be strongly influenced by your scientific work, by your basic ideas and perspectives. We wish you every success in developing and communicating your present and future ideas, and we are looking forward to a mutually fruitful and profitable cooperation with you for a long time to come.

Again on behalf of the Gesellschaft für Informatik, I would like to present to you a copy of the first volume of the complete work of Werner Heisenberg. The choice of this book is no accident, since it is evident from the subtitles – i.e., order of reality, interpretation of quantum mechanics, causality, and uncertainty relations – that its concepts correspond closely to some of your basic ideas concerning the foundations of computer science and their close relationship with the foundations of physics.

Once more, I extend the most cordial congratulations from the Gesellschaft für Informatik and from me personally.

CONGRATULATORY ADDRESS

Hartmann J. Genrich
Gesellschaft für Mathematik und Datenverarbeitung (GMD)
St. Augustin, Federal Republic of Germany

On behalf of the Collaborators of Carl Adam Petri

Dear Carl Adam,

I have been asked to say a few words on behalf of your past and present collaborators – all those who have had an opportunity to work at your institute for a shorter
or longer period of time. So the following is said on behalf of the community that
was – in due modesty – most important for the development of your ideas in the
past.

I am not going to explain what it meant to me personally to be your collabo-
rator for almost two and a half decades; that might become just too personal a
confession. Rather, I shall try to explain to an imaginary new colleague what it
could mean for him or her working with us for some time.

In doing so I wish to distinguish two aspects, a technical one and a personal
one.

The technical aspect of our question *"What can somebody expect when coming to
work with us?"* is deeply influenced by a vision, an utopian idea you have outlined
many years ago. It concerned the future role of computers in society and the task
computer science had to accomplish in this context – a computer science that had
just started to develop.

The role of computers or, in modern terms, information technology was in your
view that of a communication medium that serves people, that connects rather
than alienates them. Thus, the task of computer science was to lay the foundations
that make it possible for computers to serve that purpose. And for you this
meant, above all, to establish computer science following the standard of the exact
sciences.

Since the first days of its planning, GMD has experienced about twenty years of
changeful history. In all these years, your vision has been apparent in our work,
in our plans and reports. Under changing internal and external conditions, vogue-
words have come and gone, but your ideas about the role of computers and the
task of computer science remained valid and served as a guideline.

Your predictions that computers would become an integral part of human society
have become true. The hope that computers bring together rather than alienate
people still remains. I cannot yet say this hope has been realized.

To make the development and usage of information technology transparent and
socially acceptable was the aim and justification of our work. It was the ongoing

task that we, your collaborators, had to undertake. In this way we were able to contribute a little to establishing computer science in the direction you pointed out. The results of our work are part of what is known today as net theory and is presented during these two weeks of the Advanced Course to a large audience. It is our birthday present for you.

The other, personal aspect I was mentioning in the beginning is best character-ized by stating: *Carl Adam Petri is a modest man.*

He never tries to impose his views upon others. He has no contempt for those who do not share his views or goals. However, he tries over and over again to find, in the work of others, traces that relate to his ideas. He tries to sow his ideas into our minds that they may grow within us.

From his own experience he knows about the woes and fears of those who have undertaken the risk of scientific research. He has always shown tolerance for others; he has never dropped somebody because of a private or professional low. Rather, he has generously given his time and effort to help us emerge.

He acknowledges the need for a certain amount of organization and bureaucracy but he suffers from it very much.

He is loyal to everyone who does not misuse this loyalty. He showed us often enough how much he needs us, how much he depends in his work and his well-being on a stable well-intentioned group of collaborators.

We are glad that Carl Adam Petri did not retire on his 60th birthday, and we thank the directors of GMD for making it possible for him to stay with us for some while. We hope that he will work with us for many years to continue to make his visions – which also have become ours – to become true.

Dear Carl Adam, our best wishes and many thanks to you!

Carl Adam Petri and Informatics *)

Wilfried Brauer
Technische Universität München

Dear Mrs Petri, dear Dr Petri, Ladies and Gentlemen!

It is a particular distinction and a great honour as well as an extraordinary pleasure for me to be invited to lecture on the occasion of this colloquium in your honour, Mr Petri, even though, as you know, I am not belonging to the circle of your disciples.

Be that as it may, our acquaintance nevertheless stretches back more than 22 years. When I first came to Bonn in 1964, I got to know you as an experienced practitioner in data processing, oriented towards physics and hardware and dealing with the quite mundane problems of running a computing centre.

I can still clearly recall a seminar given by you in which we studied the latest computer concepts (e.g. IBM/360 and CDC 6600) - with special attention being paid to the possibilities of parallel processing - in a quite down-to-earth way and oriented to practical use.

I was therefore more than a little surprised when I was asked by you and Mr Böhling to give the first course starting the new Computer Science curriculum in the summer semester of 1967 in Bonn just on the theory of algorithms and computability. What is more, you recommended me to use Smullyan´s "Formal Systems" as a basis for the course, a book which is so formal and that far located in the abstract heights of theory that I would not inflict this neither upon me nor my students.

This exceptional quality of yours, of being a down-to-earth practitioner working with very simple examples and supporting your arguments with vivid illustrations while at the same time also being a high-level theoretician working and philosophising with and in formal systems, is reflected in all your works and lectures.

---

*) A translated and thereby slightly changed version of the original German lecture.

I should like to give just two examples of this:

- In your doctoral dissertation written in 1960/61 , the formal theoretical way of thinking is predominant (for obvious reasons); nevertheless, the motivation and argumentation are based on the practitioner´s way of thinking.

  Your dissertation is aimed at a general philosophical-mathematical axiomatic theory of communication which is closely related to modern theoretical physics (particularly quantum physics and the theory of relativity). It is for this purpose that you discuss completely down-to-earth practical questions and reflect on their technical realisation.

- The second example is your course given at GMD´s 1972 Summer Seminar.

  In these lectures, your practical way of thinking was predominant: You dealt with concrete problems relating to the practical development and organisation of real systems such as computer systems (i.e. computer nets or man-machine systems as in information and communication systems or even human/social systems such as administrations). You proposed totally abstract tools of higher mathematics such as linear algebra, relational calculus, category theory and, in particular, general net theory (in several formal variants) as aids in overcoming these practical problems.

For the uninitiated, this vast range of thoughts gives rise to considerable problems, not only when attending your lectures but also when reading your papers. Initially, the distance between the two aspects of earthly practice and high-level theory, concrete examples and abstract formalism is overwhelming. The uninitiated is lacking the ladder required to climb up from the low to the high. Indeed, he does not even have the tools with which to build himself such a ladder.

Only after long conversations and deep discussions with you, Mr Petri, does it become clear (and this was not only the case with me but certainly with many others) that you were in possession of this ladder from the very beginning and that you always had an intuitive awareness of the connection between the low and the high.

Moreover, from the very outset, your intention was not only to shape the heights by net theory, but also to make available both the ladder and even the tools to build this ladder. In your lecture given in 1978 at the MFCS conference in Zakopane, you stressed this explicitly as the program for the future development of general net theory - a program which has since led to a multibranched and intertwined hierarchy of net types and related concepts, and to methods for moving from one step to another. And this without any end in sight to the range of possibilities offered.

Let us now take a look at the new subfield of informatics created by Carl Adam Petri; to simplify matters, I shall simply refer to it as the field of Petri Nets, although the nets - as graphic objects - are only an illustrative and practical means of expression - what Carl Adam Petri has developed and promoted extends far beyond this. When attempting to categorise the field of Petri Nets within informatics, we are faced with a similar difficulty to that faced when attempting to categorise Mr Petri himself as a theorist or practitioner.

For this purpose, let us consider the following practical question: To whom should I offer a course on Petri Nets to be given in Munich?
- Future theorists?
- Future software engineers?
- Future users?

Opinions are divided on this point: Many practitioners consider this field to be pure theory, while many theorists believe that it is primarily for practitioners.

Or should I teach Petri Nets to beginners as a fundamental informatics area? In other words, Petri Nets as a broad-spectrum language for philosophical, logical and mathematical fundamentals for the formal theory and for the user-oriented requirement specification, program specification, programming and implementation.

You, dear Mr Petri, you may well say that I should try to be brave and make the field of Petri Nets the main subject of the introductory informatics course and build up the rest of informatics upon this. After all, this is the aim intimated at in your doctoral thesis, i.e. that of giving new foundations to informatics.

Two questions arise in this context:
1. What arguments are there in favour of making such an attempt?
2. What arguments are there against this (even if only at the present moment)?

Net theory is an extension of a fundamental subfield of informatics - automata theory.

Automata theory arose from the efforts to formalise and automate the activity of an individual human being as a computist or, more generally, as a processor of information. Think of the reasoning of A. M. Turing in the course of the derivation and development of the concept of the Turing machine.

In my opinion, the aim of net theory is more comprehensive: It is aimed at creating the basis for the formalisation and automation of cooperation amongst several people, amongst groups of people. Its prime concern is not what the individual does, but rather the communication and cooperation between members of the group. If we start from the popular assumption that informatics is the science of the systematic processing of information, then Petri´s approach of 1960 appears almost self-evident today:

What is the point of information, indeed what is information, if it is not used in the social interaction between several human beings?

Petri´s approach was foresighted.

The technique of information processing has in fact developed along the lines envisaged by Petri towards distributed processing, computer nets and communication systems. What is more, the problems foreseen by Petri at that time in theory are today emerging in practice.

As mentioned above, the starting points for Petri´s ideas were automata theory and questions related to hardware. At the outset, therefore, net theory primarily developed on the basis of the theory of automata and formal languages. It was here that net theory had its first major successes; profound and interesting problems such as the reachability and the liveness problem arose. A number of important theorems and methods which are today closely related to other parts of theoretical informatics and mathematics were discovered.

However, the real aim pursued by Petri was quite different and was hardly understood at that time. Petri was not concerned with extending an existing theory and developing it in accordance with the traditional sequential pattern. He simply saw informatics differently. For Petri, concurrency forms the central fundamental concept - both in the philosophic-logical and the formal mathematical sense - as reiterated in his opening lecture at the Advanced Course on Petri Nets held in Bad Honnef in September 1986. An abstract formal axiom system with a formalisation of the concept of concurrency and, of course, of the concept of sequentiality or causality forms the basis for Petri´s construction of the theory of informatics.

Net theory is based on a system concept which is more general than the customary one. Firstly, net theory assumes that the components of an information processing system are only able to use local information for their work and for their communication with each other, and that they only have contact with a limited number of adjacent components. Secondly - and this is the decisive point - it is assumed that there is generally no central clock, no common clock pulse for all components. It is therefore assumed that it is also not generally possible to speak of global states of the system, i.e. that the work or the behavior of the whole system must function without recourse to global status information. This clearly shows the influence of the theory of relativity.

The classical system concept, particularly that of the finite automaton, then emerges as the special case in which a common clock exists.

These basic considerations are augmented by an important tool for both the theory and in particular the applications: The graphic representation of this type of distributed systems, also developed by Petri - in other words: the representation as Petri nets. As early as 1965 I saw graphs with round nodes representing places and square nodes representing transitions in your lecture Mr Petri given at the Conference on Automata Theory in Hannover.

Petri nets give an explicit representation of the structure of the system, i.e. the dependencies and concurrencies of the components.

It was a major step forward that on the basis of this representation and with the aid of new mathematical techniques many characteristics of systems can now be systematically and formally investigated. This applies to, for example, deadlock freeness, liveness, invariance, fairness, etc.

Moreover, together with algebraic formalisms, the graphic aid provided by the nets makes it possible to represent concrete systems at various levels of abstraction, i.e. with varying degrees of detail. This variety in the possible types of representation constitutes the main attraction of Petri nets for users.

The reason why formal methods of analysis are not as popular is because the computer-aided tools for this are being developed only slowly. A large number of such tools could be seen at the Advanced Course and it is likely that these will now develop quickly.

However, there is one thing which has not been achieved to date: The Petri-net-based higher programming language.

This ought to be very easy to develop. Carl Adam Petri already mentioned this in his doctoral dissertation, even if only briefly. Is it not possible to simply use one of the higher net types being used so successfully by practitioners for specifying distributed systems as the basis for such a programming language?

It is apparently only necessary to construct compilers which are able to correctly perform the transitions from the higher level (specification-oriented) to the lower level (hardware-related) net types.

Naturally, there are still a number of problems to be solved, for example:
- the transition techniques between the levels are not available or are not sufficiently sophisticated,
- the question as to how a higher net can and should be implemented on a computer system has not been solved to a satisfactory degree.

However work is underway on these problems and solutions will be found.

Why, then, do those involved in the development of programming languages for communicating systems and for distributed processing not use Petri nets?

This has finally brought us to the second of our two questions: What arguments are there against building the introductory course for informaticians on the basis of net theory (even if only at the present moment)?

The answer I should like to give here is of fundamental nature. I believe that Petri nets and programming languages are based on two different ways of thinking. In net theory, one starts with the distributed system and describes its internal behaviour using local conditions, events, actions and information flows.

In the field of programming one adopts the opposite viewpoint: Here, one starts with the result required; everything that the desired system should do is described in the most problem-oriented way, not taking into account the actual realisation of the system. The aim is then to develop a solution systematically (automatically if possible) and design a system which functions as required but whose internal structure is not of great interest.

The basic concept of net theory is the concurrency within the system - great importance is attached to giving an exact description of everything which is sequentially dependent and everything which is not dependent in this way, i.e. which is concurrent (or using another usual way of saying, everything which can occur in parallel).

In programming, abstraction is performed at the description level of the concrete temporal interlinking of concurrent actions, whereby nondeterminism of the detailed behaviour is permitted - only globally or at the end everything must be determined.

I also see a third difference: Programming is imperative (even if an "applicative language" is used); commands are given in order that something has to be done, and one must reckon on these commands being executed immediately.

Net theory is descriptive; dependencies and concurrencies are described and possible means of behaviour represented. No assumption is made that a unit of action executes an action immediately even if the conditions for this are fulfilled - it may even be that the unit of action waits until the conditions are no longer fulfilled as a result of the actions of other units.

In order to further clarify the differences in approaches between net theory and programming, I should now like to ask you to imagine the distributed systems under discussion here not as systems of machines as it is usual, but as human organisations - as groups of human beings working together on a major project and communicating and cooperating with one another on this. For example, consider a working group in an institute of the GMD carrying out a large research or programming project.

The members of the group will work partly on tasks which depend on each other and partly on unrelated tasks, they will exchange results and put questions to each other etc. What is the basic approach used by net theory to describe such a system?

The feature which, in my opinion, characterises net theory is its view of such a system as being "dominance-free", as being a community of human beings having equal rights, in which there is no central administration and no supreme chief, and in which the only things specified are the means and ways of communication and cooperation, the conditions under which actions are possible, and which consequences these actions have.

Nothing is forbidden; anything which is possible is allowed. There are no constraints; everyone is free to decide when and what he does - the only general obligation which exists, is to carry out the activity at some point in time (after a finite period) if the conditions for executing an action are fulfilled and do not lapse.

In order to enable the system to achieve the performance expected of it the structure of the system - i.e. the distribution of the tasks and the communication paths - must be carefully chosen. We are therefore dealing with an extremely free, "totally democratic" system model.

In my view, the chief programmer looks at the world in a completely different way: He primarily expects performance from the system, from the working group and from every individual.

In order to achieve this efficiently,

- strict leadership is necessary,
- the working group must be hierarchically structured wherever possible.
- passing of certain informations on to some members of the group must be forbidden.

In order to ensure that communication and cooperation function efficiently, for example, the following demand is made: If a member of the group extends his hand for a handshake, the other member should follow suit as soon as possible and the two should then hold on to each other until each one has made his contribution to the cooperation (similar to the handshakes in ADA and CSP).

Alternatively, it may be stipulated (e.g. in the programming language LOGLAN) that each member is obliged to become active as soon as possible, such that a maximal number of concurrent tasks is executed.

Basically, the approach of the programmer does not differ from the normal way in which work is organised for groups of human beings given the task of achieving goals as efficiently as possible (e.g. the organisation of administration and companies).

This programming approach as I have outlined it in simplified form is the approach which today predominates in informatics. It has proven itself in practice. A budding informatician must acquire this approach in order to be successful in his job. Therefore I will - at least for the time being - build my introductory informatics course on this programming approach and not use net theory as my basis.

However - and this is the point I would like to make: The prime contribution made by Petri is to have shown us that it is possible to see informatics in a completely different light and to make us reconsider our approach to programming (which most of us spontaneously believe to be the only type of approach possible).

Whether Petri´s net theory approach will win through against the programming approach in the future I don´t know - but it is definitely worth pursuing the path trodden by Petri - particularly if we do not think predominantly of the construction of individual isolated systems as is the case today, but look at the computerised society as a whole, i.e. consider the whole system of human beings with their computer-aided communications and processing systems.

The major problem faced by informaticians of the future will probably be the functionability of this entire system, and not the development of efficient individual solutions. Net theory and its associated approach may well form a suitable basis for this extended task of the informaticians. All informaticians would be well advised to start thinking about these points already today.

And finally, I should like to turn once again to Carl Adam Petri himself: I believe that it is not only pure coincidence that you, Mr Petri, designed Petri nets as systems free of dominance and constraint.

For, even if we are still unsure as to the significance the new approach of net theory will have in the informatics of the future, it has already been shown today that this liberal approach functions in a system which exists purely of human beings, and that such a system can achieve significant results - Carl Adam Petri has proven this with his working group at the GMD.

APPLYING

PETRI NET BASED MODELS

IN

THE DESIGN OF SYSTEMS

Michel DIAZ

LAAS du CNRS
7, avenue du Colonel-Roche
31077 Toulouse
FRANCE

ABSTRACT.   Many   approaches   are   being   developed   for
handling   the   different   phases   in   the   design   of   complex
information   systems,   namely   specification,   verification,
evaluation,   implementation and testing.   These approaches are
more or less applicable to the various specific aspects of the
design phases and partially supported by efficient tools. This
paper deals with one of the most important approaches used   in
the design of information systems, i. e., the Petri net based
models approach,   which proves particularly interesting. It is
indicated   how   Petri net based models are used   to   represent
systems   behavior   and then some of the resulting   advantages
are given.   Furthermore,   it is emphasized why and how a Petri
net based approach can support all the design phases and it is
made clear for which purposes and in which applications   Petri
net   based designs are important.   A few illustrative examples
are also given.

Keywords: Information Systems,   Computer   Systems,   Design,
Specification,   Validation,   Implementation,   Testing,   Petri
nets, Petri net based models, Formal Techniques, Verification,
Simulation, Performance, Dependability

CONTENTS

I - INTRODUCTION

It is a fact that a great deal of information is presently available to people and that it usually has to be communicated, shared and processed by organizations, humans and machines; hence the need of complex information systems.

Owing to the great number of parties involved, it is important to represent the complex resulting interactions which have been defined in order to fully understand the way in which the parties act, communicate and solve some subtle global problems such as parallelism, synchronization, and competition. As a result, careful modelling of such systems is needed to thoroughly understand their behavior and then if required, to define their internal structure.

One major area of information systems is computer systems, of course; because the latter are playing an ever increasing part and since machines are concerned, it is evident that all possible behaviors and interactions must be fully understood.

As a matter of fact, whatever the system, either when designing it before implementation or only in order to fully

understand it, a formal and non ambiguous description of the
global behavior is mandatory. This formal description has to
be based on the definition and use of a formal model which can
depict all aspects, including necessities and constraints, of
the system under consideration.

One important parameter, prior to choosing a formal
model, is to select the properties required by such a model.
Hence the importance of the list given in Table 1 which shows,
in the case of complex systems, the various aspects which have
to be covered when using a model in a global system approach:
in order for the model to be used by all design teams in all
phases of the design, it must meet the requirements given in
Table 1.

> The model must:
> -Support a methodology for
> > Specification
> > Validation
> > Implementation
> > Testing
> -Allow expression of
> > Functional
> > Performance and
> > Dependability requirements
> -Cover different fields of Application

TABLE 1

The necessity of such a global support for handling all
aspects of a complex information processing system is, in the
author's opinion, of fundamental and definitive importance.
It must be understood that as soon as a formal model is
selected it is possible to define a related methodology, to
automatically handle the selected formalism and to define and
build support tools which will greatly assist the providers
and users of the considered information system.
The selected methodology and the actual support tools
must therefore be related as follows:
1. The tools must support the given design methodology.
In other words they must aid the designers and be usable by
possibly various design teams during the design phases.

Thus, in the case of a classical 4-phase design methodology, i.e., specification, validation, implementation, testing, the model and its related tools or set of tools must be capable of:

- representing the system specification, including all its important aspects.

- handling the corresponding validations because the system is generally very complex

- facilitating the actual implementation in order to decrease costs and delays

- helping the generation of related testing procedures by means of the already existing information.

Note that since the system is usually highly complex, a careful validation of the specification and a careful definition of the testing sequences will be required.

2. The methodology and the tools must permit the definition and handling of all major necessities and constraints in order to specify and design any system of interest. Of course, the usual functional requirements must be expressed. However usually this is not the only data of interest to systems and for instance performance and dependability constraints may equally be of special importance to some systems.

Consequently, when performance requirements have to be taken into account the model must equally be able to specify and handle all performance aspects. Similarly, when probabilities play a major part in the behavior or in the properties of the system, the model and the tools must allow for probability specification and evaluation.

It is shown in this paper how Petri net based models can express all the requirements needed for the design of complex systems and how they can be used in the different phases of the classical 4-phase design methodology; this will highlight their importance as a basic and global support tool in the design of complex interacting systems. Of course, due to their nature and importance, Computer Systems will be taken as the most illustrative domain and the examples herein will be derived from it.

Some general comments regarding the global design methodology are first given in the next paragraph. Then some Petri net based models which have been developed for complex systems are defined in the third paragraph. To illustrate the possibilities resulting from the use of these models, some simple but hopefully significant applications are described.

II. DESIGN METHODOLOGY AND NETS

It is now clear that one of the major advantages of Petri nets is that they enable to understand and express simply and naturally all problems and properties related to parallelism and concurrency /PET/ /GEN2/. Therefore, it appears that the resulting models use this characteristic and are rather powerful when dealing with parallel and distributed sytems. Let us now consider the four classical phases of design and the way in which Petri net based models apply to them.

II.1. Specification

Depending on the system considered, either for a thorough understanding or prior to attempting an implementation, a specification of the considered system must first be derived from the informal requirements. Since in general complex systems are generally distributed in nature, the specification model must therefore handle parallelism and synchronization, as allowed by net based models .

Another aspect is worth noting because of its high practical value. Because complex systems are difficult to understand and describe it is useful to define and consider separately a well structured set of sub-systems. In such a case, the different sub-systems are first defined and specified one after the other; second, the resulting sub-systems are connected together in order to give the global representation, and third, all system level analyses are performed on this global specification. One interesting property of net based models is that they can easily be interconnected. Consequently, global specifications, in which a global net represents the behavior of the whole system, can be obtained by interconnecting subnets, where subnets represent the models of the different sub-systems: subnets can be interconnected and lead to a global model by transition or place mergings, shared places, or more complex mechanisms depending on the system specified. As an example, when transition merging is used, it means that two given transitions, one in each of the two considered subnets, will be merged in the resulting global net in which they will constitute only one transition.

For instance, consider a simple example, the mutual exclusion mechanism, expressed by the Petri net in Figure 1.

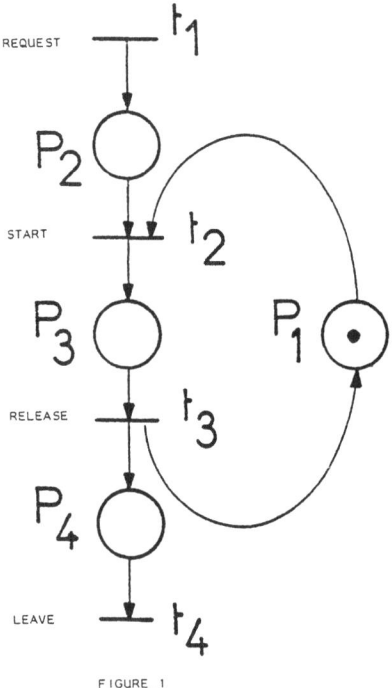

FIGURE 1

In this basic example, the synchronization problem is that of mutual exclusion where different processes request one of a set of resources; only one of the processes at the most is allowed to use the resource at any time; it starts operating with the resource as soon as it has acquired the right to use it and when it ceases to use it, the resource is released. In Figure 1, when a process (either one or many) requests the resource, it fires transition t1 and goes into Place P2; Place P1 initially contains one token which represents the right to access the resource and the other places are initially empty. Place P2 may contain a given number of tokens, one for each firing of transition t1. When

there is at least one token in Place P2 and the token in Place
P1, then t2 may fire, the corresponding process gets the right
to access the resource and one token is put into Place P3
which means that the resource is being used. The corresponding
process is in the mutual exclusion section and no other
process can utilize the resource as there is no more token in
Place P1 and transition t2 cannot be fired until the resource
has been released by firing transition t3 : this is done by
the process after having used the resource, i.e., when
leaving the exclusion section, by firing transition t3 which
releases the resource.

This exclusion problem is one of the basic
synchronization mechanisms occurring in the presence of
competing processes.

Two consequences may be derived from this simple example:

1. the net given in Figure 1 is very easily defined and
understood. As a result, Petri net based models allow to
nicely express the basic synchronization mechanisms which then
appear easily accepted by people; hence, their importance in
Education and Teaching.

2. Composition can also be illustrated in this example.
Figure 1 can be seen as composed of two different parts
existing in the system: one for the processes, the other for
the access right. Then, the part on the left of Figure 1
stands for the processes and the one on the right represents
the access to the resource. In such a case, transitions t2 and
t3 of Figure 1 can be seen as obtained by merging on the one
hand a transition t2P "the process starts using the resource"
in the process part and a transition t2R "the resource is
allocated" in the resource part; and on the other hand a
transition t3P "the process releases the resource" in the
process part and a transition t3R "the resource is released"
in the resource part.

It clearly appears that this type of composition can be
extended to building blocks and used to compose nets, as for
instance in /BAU1/.

II.2. Validation

The second phase in the design is validation and usually
it constitutes a difficult problem. When using net based
models, designers rely on a imperative state based approach

for specification; consequently, in this case, a natural solution to validation and verification consists on using the set of all possible states of the system, i.e., all possible reachable states; this set of states can be depicted by the so-called reachability graph which gives the set of all possible reachable states and can be used to check the correctness of the global system behavior.

Naturally, this set of states may be difficult to obtain, in terms of space and time. In order to handle the resulting complexity, some methods which reduce the graph complexity have been proposed while other methods are still being investigated /JEN2/. For example, the size of the graph may be reduced by searching for duplicate, equivalent or covering markings: the graph construction is stopped after a given new marking when the newly obtained marking does not provide any new information with respect to a previously considered marking; when this is the case, the construction of the graph is stopped at this marking and continued from another pending marking.

Another interesting property of Petri nets is that they allow to automatically obtain place and transition invariants: given a net, place and transition invariants can be derived by solving a set of integer linear equations. Another valuable property of nets is that they easily permit invariance checking; if some invariants are given by the designer, it is rather simple to check whether they are real invariants for the considered net.

Consequently, major properties can be checked when using Petri net based models, namely:
- general properties such as boundedness,
                                liveness,
                                cyclic behavior
- specific properties such as mutual exclusion,
                                robustness.

These properties can be checked directly on the reachability graph or by using invariants.

For example, let us consider bounded nets, boundedness meaning that , for all possible markings, the number of tokens in any place of the net is bounded. One possible interpretation of this property is obtained when the places correspond to real objects to be implemented, as buffers for instance; in this case it appears that nets must be bounded, because in the opposite case, there exists a behavior for which the real capacity can be overflowed. Thus, if the place represents one buffer, and if this place is not bounded, an owerflow may occur in the corresponding buffer of the implemented system. Liveness can also be defined: if a net is live, it means that there will not be any dead transitions after any behavior; in the implemented system, this property implies that no part of it will die during the computational process. Cyclic behavior means that the system behavior is repetitive and, as a specific case, that it can be re-initialized.

Note that, on the other hand, invariants are dependent on the specific properties of the system; they can be expressed by the objects which exist in the specification, such as places, transitions, variables, etc.

II.3. Implementation

Quite often, the aim of the design is the implementation of the sytem. Then, this step in the design must start with the specification which should have been validated and it must be able to derive an adequate implementation from the specification.

Now another advantage of the formal approach is clear: since the model is perfectly defined, all that is required to implement any system is to give a well defined implementation of all the constructions of the model or of the set of important primitives of the model.

For Petri net based models, it is worth noting that firing a transition is a very basic mechanism:

- before a transition is fired all its input places must be tested to check whether they hold one or more tokens; in other words the corresponding places must be accessed in an atomic way

- if the firing conditions are fulfilled, when a transition is fired, the tokens must indivisibly be removed from the input places and located in the output places.

- furthermore, if the net based model allows to handle variables, then the corresponding variables must also be tested and updated in an indivisible way.

Thus firing concepts are closely related to indivisibility and it must be well understood that they are applied at the net level, or in other words at the given level of abstraction, depending on the system representation which has been selected.

For nets, as usual, the implementation can be first direct or indirect, and second centralized, parallel or distributed.

Centralized solutions are the simplest. Parallel solutions, i.e. using parallel processors or processes with a common memory are more complex and make use, for instance, of interrupts, semaphores or more sophisticated abstractions like monitors. This aspect will be developed in the next chapter. In the case of distributed systems, the problem is much more difficult to solve because the implementation has to cope with the use of a computer network, the visibility of such networks depending on the services they provide; consequently, the distributed implementation of a given net can be either easy or very difficult to achieve depending on the functions required, the network considered and the services it provides. Note that direct implementations are used when possible, either in hardware for efficiency as it will be seen later or in software where numerous solutions exist, as for esxample the one in Hitachi /KOM/.

II.4. System testing

The fourth part in the design is testing. Given a net representation of the system, the user has defined a functional representation of the behavior of the system; then the designer intends to check wheter or not behavior of the implemented system is correct.

The basic idea is to work directly at the specification level, i.e., the net level, looking at some possible faults, instead of working at the implementation level, i.e., the real system. Then, two different views are possible, depending on the type of test, either by stopping the tested part during its test or by having the considered system part actually running while testing it.

A - Let us first consider classical testing, where the functional operation of the tested part is stopped. The easiest way to obtain the test sequences, which is for complex systems a difficult testing problem, is to directly use the reachability graph and subsequently to go through some of its arcs. The ideal solution would be to have tests which would go through all the paths of the reachability graph. But clearly this is not feasible in most cases. Consequently, a subset of the possible paths is selected in the complete reachability graph. Then some test sequences may derived by,

- going at least once through all the arcs of the graph,

- traversing at least once all the states of the graph,

- selecting the next step, when being in a state, either automatically, or randomly or in some user-defined bases, - etc.

Another more complex possibility is to consider the effect of some faults at the net level. For example, let us suppose that a possible fault leads to the deletion of an arc in the net representation; note that nothing is said about the physical origin of the fault. Such a choice is made because if this type of error occurs, then synchronization is highly affected and it is necessary to test that sort of fault. In this case a test is required to compare the behaviors of the initial net and of the modified net. This is an interesting but difficult way of tackling the problem.

B - Let us now consider run time checking. For this kind of test an approach relying on nets and on an observer concept can be used.

As usual, a specification is first selected and subsequently implemented in some way. A worker is obtained: the worker is the name of the real implementation of the system.

Then, the specification is used again: it is generally simplified by extracting its major mechanisms and it is this simplified specification which is used in the observer. In other words, the simplified specification is in fact a model which, in turn, enables one to obtain the observer: the observer is an implementation, as direct as possible, of the simplified model.

With the implementation of the worker on the one hand and the implementation of the observer on the other, it is possible to compare during run time the corresponding behaviors of the two implementations. This comparison, which detects any discrepancy between the two behaviors, permits to detect a fault in the system. It appears that the observer and the worker must be compared during the system run time.

Note that the two major concepts of redundancy and distinctness in dependability and testing are present in this approach. The first concept relies on the fact that there are two different and in some sense redundant systems, the worker and the observer. The second concept originates from the real difference between worker and observer: the worker being an actual implementation on a given system with given constraints; the observer being the simplest possible implementation which sometimes takes the form of a model simulation. Hence the design of the observer must rely on a model which is both formal (because the observer is a reference whose form must be well known) and easily executable (because it has to be easily implemented).

Petri nets proved to be well suited to this kind of run time testing: they can be validated and then give a reliable reference for observation; they are very simple to implement by direct simulation of the net behavior. They have been used successfully for defining observers and an industrial example of their use is given in the next chapter.

From the previous paragraphs, it becomes evident that nets can be used in the different methodological phases and that they are particularly useful to support the various steps and the different teams involved in the design.

III - MODELS AND SYSTEM SPECIFICATION

This chapter is concerned with the different aspects of systems which have to be specified and hence must be specified by using net based models. Of course, distributed systems can be specified since parallelism and distribution are naturally and well modelled by Petri nets.

Furthermore, using Petri nets leads to a state based approach which presents the advantage of being well accepted and well known to system users and providers and in particular to computer science users. To specify all aspects of systems, two types of information must be added to the usual functional specification, namely :

time values because in some systems explicit time values play a preeminent part

and probabilities since it is often the only way of knowing the average behavior of systems.

III- 1- High Level Petri Nets

One set of approaches for functional specification mainly uses natural extensions of classical Petri nets or Place-Transition nets leading to Predicate-Transition nets /GEN/ and Coloured Petri nets /JEN/; those two extensions are used to differentiate tokens by attributing a name or identity, when needed, to each token, for example when it is necessary to have some form of token individuality. One of the important advantages of such nets is that they lead to highly compact descriptions. Such approaches allow for instance the use of Logic Programming for data description and manipulation /AZE/, the association of Abstract Data Types and Petri nets /VAU/. For example, if coloured tokens are used in Figure 1 then the processes can have an identity and can be distinguished.

The second set of approaches considers that systems consist of two parts, control and data; it is based on the use of predicates and actions on markings and on program variables. In this case, a label of the form "if Predicate do Action" is added to each transition. The firing rule is now slightly different because such a transition is fired only if the input places are correctly marked and if the predicate is evaluated to be true; when the transition is fired, the action is executed. In the case where the logical predicate and the action are general, it is possible to represent parallel programming (for instance parallel programs) in a general way. Such an approach was first proposed in /KEL/ where named Predicate-Action transition systems were introduced. The transitions represent atomic and indivisible computations; their names are a pair of the form (Pt,Ft) where Pt is a predicate and Ft a partial function defined when Pt is true. One example of such a transition is given in Figure 2.

One of this set of approaches has led to the definition of Numerical Petri Nets (NPN) /SYM/ /BIL/. In NPN, a coloured Petri net defines the control and to this net are added some extensions:

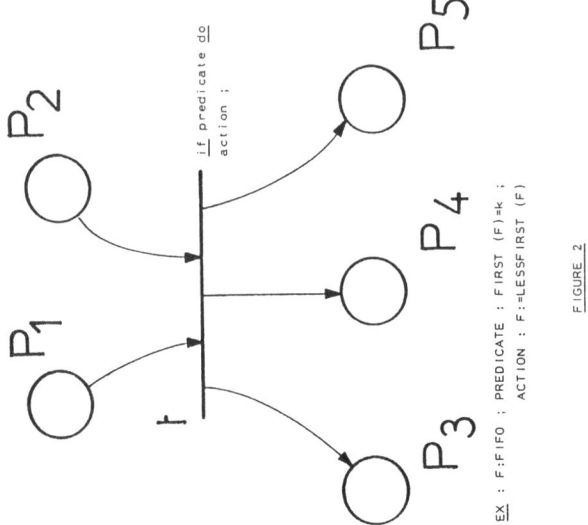

if predicate do
   action ;

P1   P2

t

P3   P4   P5

FIGURE 2

EX : F:FIFO ; PREDICATE : FIRST (F)=k ;
     ACTION : F:=LESSFIRST (F)

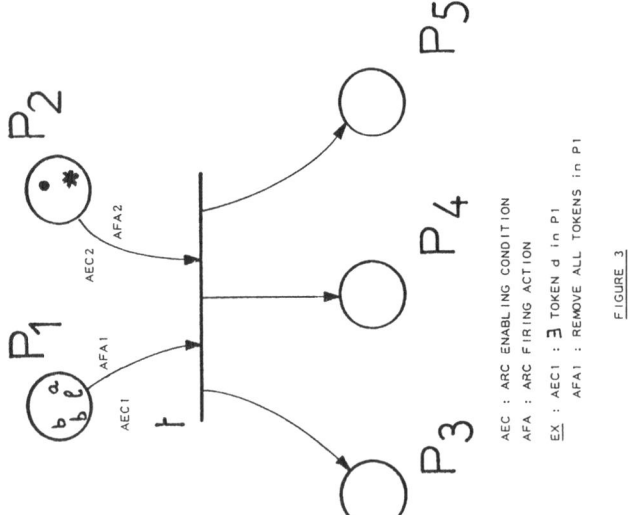

P1   P2

AEC1   AFA1   AEC2   AFA2

t

P3   P4   P5

AEC : ARC ENABLING CONDITION
AFA : ARC FIRING ACTION

EX : AEC1 : ∃ TOKEN d in P1
     AFA1 : REMOVE ALL TOKENS in P1

FIGURE 3

- to each transition is associated a label of the form "
when Pt(x) do x'= Ft(x) " as in /KEL/ in which the predicate
Pt and the action Ft are functions of the set of program
variables x
- for the transitions, for each ingoing arc to a
transition, a distinction is made between the arc enabling
condition AEC related to its input place and the arc firing
action AFA which removes the tokens from this place. In
order for a transition to be fired,

       * its transition predicate Pt must be true
       * its arc enabling conditions AEC must be
fulfilled.
    When the transition fires,
       * the marking is updated by the transition firing
rules by using the -local- AFAs corresponding to the ingoing
arcs of this transition,
       * the -global- action Ft is executed.
Note that what is introduced is the possibility of having
separate specifications for AEC and AFA. /WHE/ gives a formal
definition of NPN by using: a directed Petri net; two
functions that assign to each place the maximum number of
tokens and the maximum number of copies of a given type of
token; a set of program variables which have a scope defined
as a set of transitions; operators and predicates on this set
of variables; an initial marking and a firing rule. The
notation uses bags and a few predefined inscriptions which
define the input conditions, the creation and removal of
tokens, the transition names conditions and operations, the
place names and capacities, and the initial marking.

    Figure 3 gives an example of a NPN transition; note that
AEC and AFA can be implicitly " >=N " and " =N "
respectively, which leads to the firing rule in Place-
Transition nets of arcs with a weight of N. From some examples
/BIL/ /BIL2/ /SYM/ /WHE/, it appears that NPNs seem rather
well adapted for deriving concise and powerful specifications
of distributed systems behaviors; they also allow the design
of efficient simulators.

III- 2- Time Petri Nets

Another major advantage of Petri net based models for the specification of real time and distributed systems is that they allow one to deal with <u>explicit values</u> of time.

Until now, two main approaches have been proposed to handle explicit value of time /RAM/ /MER/ /SIF2/, respectively known as Timed Petri nets and Time Petri nets. They are illustrated in Figure 4.

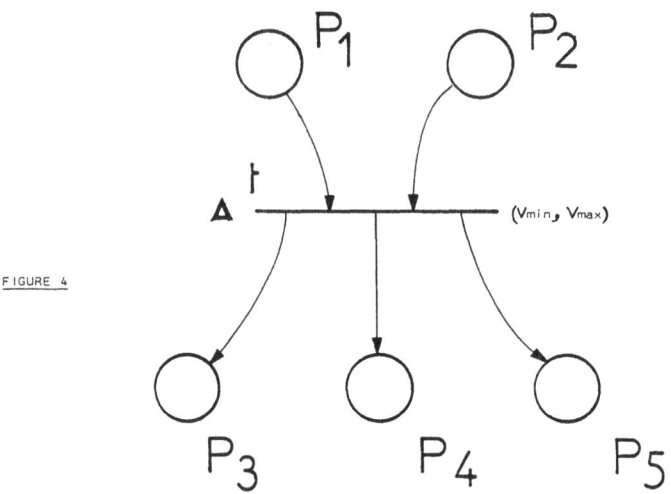

FIGURE 4

Δ : TIMED PN, DURATION IN THE TRANSITION
(Vmin, Vmax) : TIME PN, MIN AND MAX VALUES

Ramchandani's Timed Petri nets are obtained from Petri nets by associating a firing time with each transition of the net. The firing rule is modified by considering that firing a transition is initiated at the same moment it is enabled and that firing the transition takes the value of the firing time associated with the transition: in other words, this means that, when firing a transition, the token(s) remains in the transition during the time associated with the transition. This type of model has mainly been used for performance evaluation, after having been extended in different directions -see for instance /SIF2/ /ZUB/.

The most general approach, and the only one developed here, related to the specification of time is obtained by using Time Petri Nets (TPN), as given by Merlin /MER/. TPN are more general than Timed Petri nets and are important because they constitute a well adapted formalism allowing to handle time verification for explicit values of the time parameter. The TPN approach attaches a pair of non negative numbers, (vmin,vmax) to every transition. The first value gives the minimal time the transition must be enabled before its firing or occurrence. The second value gives the maximal time at which the transition will fire or occur if it has continuously been enabled during the interval (0,vmax). Thus, if a transition remains enabled, it will in any case be fired at vmax; for example, if a global time μ is considered and if a transition has been enabled at time μ, then it may not fire before time μ+vmin and must fire before or at time μ+vmax unless it has been disabled before μ+vmax by the firing of another transition.

In the case where the two values are zero and infinity, standing for the firing time initial and final values respectively, the Time Petri net offers the advantage of working as in the corresponding underlying Place-Transition net, i.e., the net without time, where all vmin and vmax values have been deleted. This model allows easy representation of some important behaviors of a system, such as delays, because the same value can be given to vmin and vmax. It must be emphasized that it is rather difficult to understand these nets and to formalize their behavior.

For example, if some natural numbers are attributed to the two explicit values of time, the instant of firing of the transition can then be chosen as one of the infinite number values in the corresponding interval; there are in general an infinite number of "one step" or "one firing" reachable states. It appears that the condition to fire is given by two subconditions: first the place must be marked and second, firing instant must fulfil the inequality vmin<=firinginstant<=vmax .

Note that such a behavior of the firing rule allows perfect description of time-outs: it permits to model the waiting time defined by a time-out when it is triggered and it also models -with the min and max values - the possible uncertainty of the actual time at which the time-out will time out due to always possible drifts.

Such nets are difficult to analyze, due to:

*the infinite number of possible times for a transition to fire between vmin and vmax,

*the fact that, when a transition fires, its firing induces some subtle constraints on the transitions that remain enabled.

It appears that Time Petri nets induce additional constraints on the set of the reachable markings in the underlying Place-Transition net – the TPN in which all vmin values are zero and vmax values are infinity: some markings reachable in the underlying Place-Transition net may not be reachable in the time Petri net.

The first complete solution for analysing nets with explicit time values has been given in /BER/ : it allows one to build the set of a forward reachable graph of Time Petri nets. In order to solve the problems related to the infinite number of markings /BER/ introduces the concept of state classes: a finite representation of the behavior of an important family of TPNs can be computed in terms of state classes and of reachability relations between them. The resulting enumerative representation is rather similar in its spirit to the reachability method used for Place-Transition nets. A state Class is a pair (M,D), where M is a marking and D is the set of all the solutions of a system of inequalities representing the firing constraints and two state Classes are equal iff they have the same marking and the same set of solutions for their systems of inequalities. A virtual ring protocol has also been analysed and a software package has been written in APL /ROU/.

As usual the state graph can be very large but it is currently the only way of checking the complete behavior of systems with specified explicit time constraints.

III- 3- Stochastic Petri nets

Another type of extension of Petri nets, called Stochastic Petri nets or SPNs, is aimed to cover the important aspects of system performance analysis and dependability evaluation.

SPNs /FLO1/ /FLO2/ /FLO3/ /MOL/ /BETO/ /MAR/ /RAMHO/ consist of extending Petri nets by assigning a random variable to each transition, this variable representing a random firing delay of the corresponding transition; these variables can be defined by some adequate distributions such as exponential or geometric distributions.

There are two main assumptions in the definition of Stochastic Petri nets:

a) for any SPN there is an underlying Place-Transition or Predicate-Transition net which is obtained by removing all the random variables; then, the underlying net can be used if needed. Hence the firing rules and the possibilities of verification are fully maintained.

b) there is an isomorphism between any SPN and a related Markov chain by using the reachability graph of the underlying net.

Two specific aspects of SPNs are worth noting:

The first one concerns the introduction of stochastic Petri nets in order to solve conflicts in transition firings; this is done by associating a probability value to all transitions in conflict in order to value the corresponding firings for all the transitions in conflict.

As a simple example, a place having two output transitions in conflict will lead to a probability assignment for the system under consideration such that one of the two transitions will be fired with a probability P and the other transition with a probability (1-P). Such a possibility, which can be extended to any set of mutually exclusive transitions, is very often defined by a constant value for P and so for (1-P). It generally corresponds to a specific behavior of the system under condideration.

A second possible utilization of such nets is obtained by considering the instant at which a transition fires after being enabled as a continuous random variable associated to a given distribution function, the time of transition firing being given by this probability. Once again, the main assumption is that the net without the probabilities of firing is the same as the behavior of the underlying Place-Transition net. This sort of behavioral knowledge is often used for performance evaluation when one needs to know how the system probabilistically operates in some given states.

As it may be understood, SPNs allow to get performance evaluation /RAMHO/ /FLO3/ /MAR/ and dependability evaluation /BEO/ /BEY/ /FLO1/ about the considered system and the corresponding analyses are performed by using the Markov models equivalent to the SPNs. Given a system described by a

SPN, the equivalent Markov model can be generated and solved by using the set of the reachable markings. In fact, some different SPNs have been proposed /PAG/ by modifying the firing rule of the underlying Place-Transition net; generally, the proposed modifications are such that the time between the enabling and the firing of the transition could be considered as a continuous random variable. Now, if the distribution functions are adequately restricted, then the marking graph, i.e., the reachability graph, forms a homogeneous Markov chain /BETO/ /FL04/ . Then, ergodic SPNs are defined in a way similar to ergodic Markov chains. Figure 5 illustrates the relationship between SPNs and Markov processes.

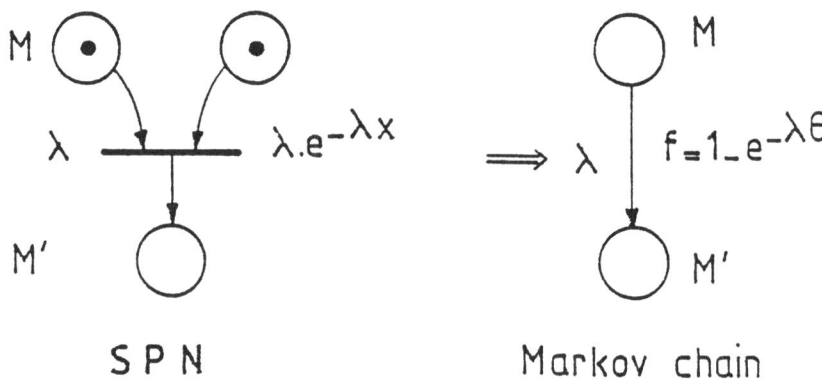

FIGURE 5

In the general case, the analysis will consist of two steps:

1. Check for correctness the reachability graph, which gives the Markov chain when associating the probabilities,

2. Compute the probability of each state in this graph, as usual in Markov processes.

Given the correct behavior of the system and the state probabilities, the figures for delays and throughputs are deduced.

Firing the transitions allow to obtain the marking graph in which are added the given probabilities or rates; thus, it appears that this extended marking graph constitutes a Markov chain from which computation is done. For example it is possible to compute the probability of being in a given marking. Once this is done the probability of a given place to be marked is the sum of the probabilities of all markings in which this place is marked.

This rather recent type of extension is of great interest since it allows to handle uniformly analysis of correctness and performance or dependability evaluation.

Such an approach has two principal advantages:

1. It allows performance analyses when the considered system includes a tight synchronization, something which is rather difficult to achieve when using Queueing Theory.

2. On the other hand, because Petri nets may consist of subnets connected by place or transition mergings, it is much easier to use a SPN description for complex systems than to directly build the Markov chain which often turns out to be manually impractical.

Hence, the undeniable advantage of stochastic Petri nets.

From the previous points, it appears that Petri net based models are able to specify and handle functionalities, times and probabilities of interest in system specification.

IV - EXAMPLE OF APPLICATIONS

In order to illustrate the possibility of using nets to gather all the aspects of system design, some examples will be now given. They have been selected in order to present a few interesting characteristics which have to be handled when considering all aspects of complex systems.

Many fields of application have been tackled by using net based models, not only within the framework of Computer Science but also in different and important related fields such as Office Automation, Computer Integrated Manufacturing, etc. Some other research work also applies nets to more general systems such as for instance Man-Machine Interaction, Pragmatism, etc.

Of course, due to space limitation and illustrative interest, a few examples only can be selected; hence, only those areas which have already been well developed and

covered in previous work will be presented here because in
such areas nets have already proved to be of real interest.
The following areas have therefore been selected;

* Operating systems,
* Distributed cooperation protocols,
* Flexible automation,
* Maintenance and spare management.

It must first be noticed that the first important field
of application of net based models has probably been the
digital control of electronic systems and the design of logic
programmable controllers. In the corresponding work, hardware
realizations played a preeminent part; they can be classified
into two groups: implementations using a transformation of the
nets and direct implementations of nets. The first set of
implementations is based on the transformation of a safe Petri
net into a state machine. One possibility is to compute the
reachability graph and then to obtain the logic equations of
the state machine; finally, these equations are implemented by
using a Programmable Logic Array which gives the synchronous
circuit /LEU/. Another possibility is to decompose the net
into state machines and then synchronize the resulting
machines. Of course, such realizations are of limited
complexity but when applicable, lead to a very efficient
circuitry. The second set of implementations has led to the
definitions of asynchronous modules /AUG/, asynchronous
optimized flip-flops /COVS2/, and asynchronous matrix
implementations /PAT/, /COVS3/. Asynchronous matrices for the
synchronization of parallel multiprocessor architectures have
also been investigated /COVS4/. All the previous
implementations deal with safe nets. Weighted Petri nets have
also been considered in /COVS/ /COVS1/ by using modular and
matrix solutions. As before, such realizations, when
applicable, lead to a very efficient circuitry.

IV.1. Operating Systems

In the area of operating systems, let us consider for
instance the classical example of readers and writers given in
Figure 6 : a set of readers and a set of writers share some
data. Place P1 characterizes the maximum possible number of

parallel accesses, given by N, the initial number of tokens in P1. The system operates as follows.

A reader comes and makes a request for reading by entering place P2. Then it takes one of the tokens in Place P1 and starts reading. In the case where another reader makes another request, (N-1) tokens are still available. This second reader will then take another "reading resource" and starts reading. Parallel reading is therefore feasible. Upon leaving, when finishing reading, the reader releases the resource it had been allocated for reading by firing transition t3.

Now when a writer steps in and requests to write by entering place P5, due to the weight of the output arc of place P1, the writer needs all N available resources to fire transition t6 and write. Hence, when at least one reader performs reading, a writer will not be able to carry out writing. Only when any reader does not read will a writer be able to write. When allowed to write, a writer will remove all the resource tokens by firing transition t6. Then, when a reader enters place P2, the reader has to wait for the writer to complete its work and put all N tokens back in P1 by transition t7 before starting to read.

A more sophisticated solution is given in Figure 7: it consists of granting priority to the writers. This can be easily achieved by using Predicate Action Nets. In this case the number of writers requesting to write is counted by W, initially set to 0 and updated when a request to write occurs by firing transition t5. When at least one writer has requested to write, the value of W becomes greater than zero; the predicate "if X<=0" in transition t2, becoming false, gives the priority to the writers: after a writer has asked for writing, readers can no longer access the critical section.

Note that the reader processes are in the left part of the net, the writer processes are in the other part, and two variables are shared by the reader and writer processes, namely place P1 and W. Since this data must be shared, one possible implementation is to use Monitors.

In such a case, it is possible to use K- monitors /KES/ to directly derive from the Predicate-Action net a monitor which is a direct implementation of the net. The monitor contains two variables, W and Pex (corresponding to place P1)

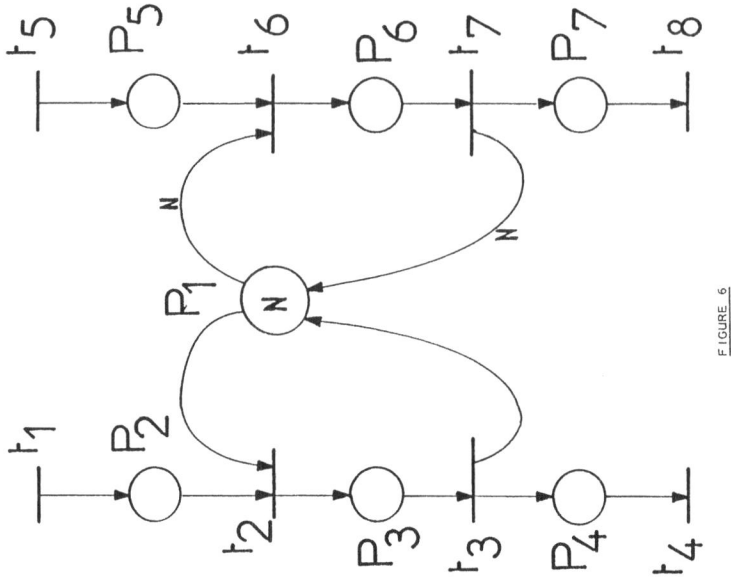

FIGURE 6

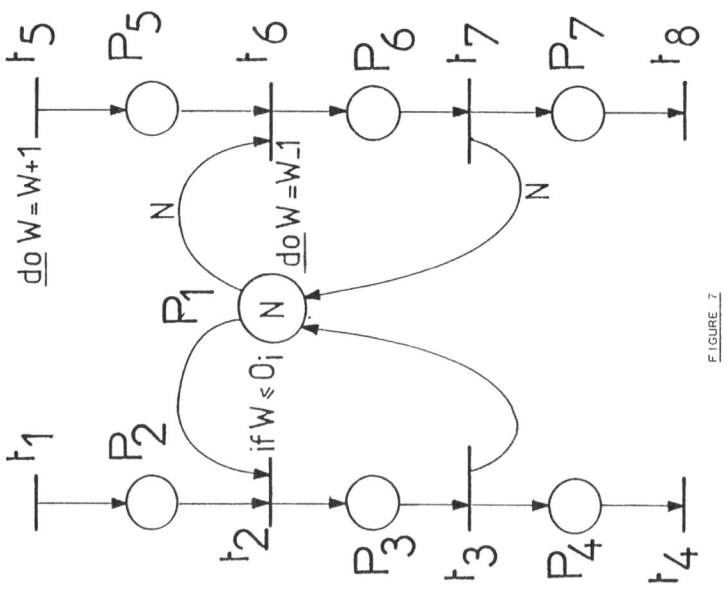

FIGURE 7

and two conditions, one for the writers and one for the readers, which have to be fulfilled to fire the corresponding "authorization to write" and "authorization to read" transitions in the net. The first condition of the monitor, "Pex=N", concerns firing transition t6 and the second condition "W<=0 and Pex>0" concerns firing transition t2.

```
RW: monitor
begin W:integer; Pex:integer;
readallowed: condition ( W<=0 and Pex>0 );
writeallowed:condition ( Pex=N );
      procedure:startread
      begin  wait readallowed;
                  Pex := Pex-1 end
      procedure:endread
      begin  Pex := Pex+1 end
      procedure:startwrite
      begin  W := W+1;
                  wait writeallowed;
                  Pex := Pex-N end
      procedure:endwrite
      begin  Pex := Pex+N end
W:=0; Pex:=N;
endmonitor
```

Table 2

Four procedures, startread, endread, startwrite, endwrite are needed in the monitor to manage the synchronization of the accesses. The four procedure and the monitor can be directly derived from the net, as given in Table 2. Note that startread concerns transitions t1 and t2, endread t3, startwrite t5 and t6, endwrite t7.

Let us notice the similarity between the behavior of the monitor and the behavior of the net. Such an approach can be extended to other synchronization mechanisms based on the same specification principle.

IV.2. Communication and cooperation protocols

The second selected area of application deals with communication and cooperation protocols in distributed systems.

In distributed systems, interactions between distant processors must obey well defined rules called protocols: a protocol is a set of rules defining the messages which have to be exchanged and the implications of possible exchanges on processor state, sequencing, actions, etc; a protocol defines the actual exchange of messages used to implement a distributed function.

It is now clear to protocol designers that, in complex distributed systems, protocols must be organized as a set of layers, leading to a layered architecture; such a structuring must be carried out by using principles such as those given by the OSI architecture of the ISO and CCITT Reference Model /ZIM/. As usual, layering is applied in distributed systems to structure complexity. The service concept, see /VIS/ for instance, defines how the lower layers act when they are used by a considered layer which is located on top of them. The service must allow hiding of everything -in the lower layers- unneeded for the design of the considered layer.

In other words, on the one hand, a service defines the functions offered to a given layer by the layers under it and how these layers are seen from the considered layer; on the other hand, a protocol defines how the functions of a given service are actually realized, i.e. by which real exchanges of messages. In some sense, a service defines a global function, a protocol defines a possible distributed realization.

The concept of service ensures a real independence between layers. Hence, when a designer has to specify a distributed system, a layered approach allows him to use a service in place of the description of all the layers located under the considered layer. It can be easily understood that such a possibility will lead to a much simpler specification with all related advantages; this is because it will be sufficient to model the service instead of modelling the behaviors of all lower layers.

When the architecture, its services and its protocols have been selected, the designer will have to deal with each layers.

For each layer, the designer will have to understand the layer functionalities and to select the level of detail of the modelling. Petri nets ,Place-Transition Nets with infinite capacity have now been used for some years to specify and validate protocols because they allow the user to state clearly the interactions existing between a sender and a receiver /DAN/ /DIA1/ /DIA2/. More recently, Predicate-Transition nets have also been used for protocols /GEN/ /VOS/. One of the advantages of Petri nets stems from the important use of state machines based specifications for protocols and services in distributed systems, including standards in ISO and CCITT /BOC1/ /BOC2/ /see standards/.

Usually, the communicating entities, i.e. those participating in the cooperation, are represented by a set of communicating processes: such communicating processes can be described by Petri net based models which model the processes sending and receiving messages.

Hence, to specify a system one has to:

1. Identify each one of the processes and describe its behavior by a net model

2. Connect EXPLICITLY the previous models to obtain the global model for the distributed system.

Note that the selected connection model must provide a formal view of the ACTUAL (implemented) service /DIA1/ /DIA4/.

The advantage of Petri nets is that they allow an explicit modelling of the needed interactions and that in fact the complexity of distributed system modelling is often that of modelling its interactions. General approaches are under development to define building blocks for their use in the specification of OSI services and protocols /BAU/.

Of course, using net based models enables the user to apply the usual validation methods developed for these models. Another point which is of major importance in layering is the service concept. As already stated, services of a layer (N-1) are used by layer N to implement the N-protocols and the latter layer N provides its N-services to the layer (N+1). Then, it appears rather interesting to show that the set "N-protocols,(N-1)-services" meets the requirements of the N-services /BUR/, /JUA/. One possibility has been given in /LAM/ for the general case. In /COU/ a projection is performed on the reachability graph.

Petri nets based models have been applied to real and significant protocols such as the CCITT no 7 signaling protocol /AYA1/, the X21 interface /WES/ /RAZ/, X25 /RAZ2/, a virtual ring protocol /AYA2/ /ROU/, Transport Service /BIL/ and protocol /BERTR/, TCP /ROS/, remote servers /AZE1/, two step commitment /BAE/, the CCR algorithm /BAU2/, cache coherence /CHA/, CSMA/CD performance /MAR2/, etc.

Let us take a specific example, a virtual ring to access the bus in a local area network used for real time industrial control. A set of functional processors is connected via a set of two buses. A virtual ring is used as the layer 2 protocol access method: each processor is granted, in turn, a privilege giving the right to send messages on the bus. Of course only one processor is granted this right at a time because otherwise the communication would not be possible electronically. This is a very basic part of a local area network. One aspect of the system design is to design this part and then to define a certain protocol between the communicating processors.

After selecting a given protocol, a model of it has to be built; one possibility stated in /AYA2/ is given in Figure 8. This figure shows the model for one processor; and if there are n processors there are n times this model, one for each processor. The global model is then derived from those n models by connecting them adequately depending on the underlying service. In /AYA2/ the underlying service is a datagram which means that the sending and the reception of a message are connected by a shared place. It is then shown that the resulting global model is bounded, safe and cyclic and that the protocol ensures exclusion, fairness and robustness. By place invariants it has been shown that the exclusion mechanism operates correctly: there is, at the most, one processor which is granted the privilege. Fairness and robustness have been proved by transition invariants. For fairness, the invariants show that, whatever the communication processor, it will be granted the privilege. For robustness, assumptions must first be made about faults which are about to occur. The transition invariants must then be asked for and it can be shown that for any fault of a set of faults, any processor is granted the privilege: when the fault occurs,

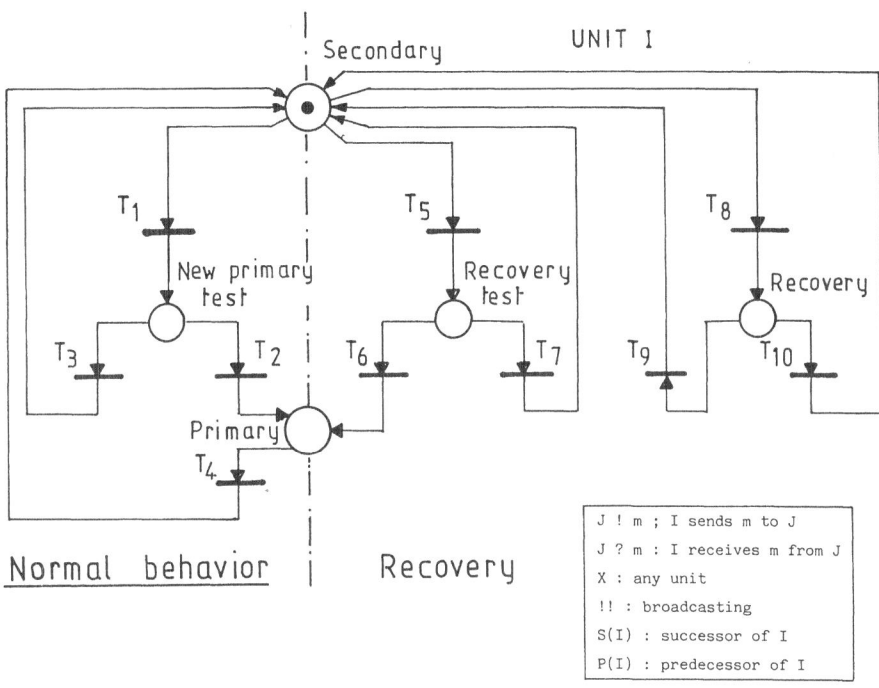

J ! m ; I sends m to J
J ? m : I receives m from J
X : any unit
!! : broadcasting
S(I) : successor of I
P(I) : predecessor of I

TRANSITION		PREDICATE	ACTION
NORMAL BEHAVIOR	T_1	X ? primary status (J)	timer-set (TPS,TO) ; N P = J ; OK.RECOVERY = false
	T_2	J = I	
	T_3	J ≠ I	
	T_4		!!primary status (S(I)) ; timer-set (TPS,TO) ; NP = S(I);
RECOVERY	T_5	X ? become-primary	
	T_6	OK.RECOVERY = true	OK.RECOVERY = false ;
	T_7	OK.RECOVERY = false	
	T_8	clock ? time-out (TPS)	NP=S(NP); OK.RECOVERY = true ; timer-set (TPS,TO) ;
	T_9	I = S(NP)	delay(t);NP ! become-primary ; OK.RECOVERY = false ;
	T_{10}	I ≠ S(NP)	

Figure 8

this right, if lost, is recovered and passed through the
processors.

Once this model is validated, it can be used as a basis
for an observer, as already said. The complexity of the model
is first reduced after selecting the events to be observed and
the simplified model is located in a specific processor,
connected to the bus and generaly devoted to network
management: one processor is dedicated to testing and
observing.

This approach led to the creation of a genuine commercial
product for the industrial local area network FACTOR of the
APTOR company. Inside the observer there is a Petri net model
of the Link Control protocol as well as the Transport
protocol. The use of such a dedicated processor allows to
obtain a great deal of valuable information from the
distributed system.

IV.3. Flexible Manufacturing

The third area selected for an illustrative purposes
concerns flexible manufacturing. In the latter, /BRU/ /NAR/
/VAL1//VAL2//ALA//MUR1//MUR2/, Petri net based models are
used at different levels of system design, both at the highest
level, the global description of the manufacturing system and
at the lowest level, the control of the mechanical parts of
the machines, for example for machine tools.

A - Machine tool control.

Major companies in logic control currently utilize a
Petri net based model for the specification of logic
controllers; this is because the needed logic control
consists in fact of extensions of state machines
/MUR1//MUR2//VAL1//KOM//BRU/. Two slightly different models
are used in this area: Predicate-Action nets and the Grafcet
model; basically, the two models are similar but the Grafcet
is more oriented towards low level logic control /VAL1/. In
the two models, inputs and outputs are the logical values of
the wires which control the behavior of the machine, i.e. of
sensors and actuators; also the predicates and the actions
often handle the values of the sensors and actuators. Thus it
is possible to represent the machine behavior and its
interactions with its environment. It is worth noting that the

Grafcet model is currently considered by the International Electrotechnical Commission with a view to achieving international standardization. Furthermore, these major companies utilizing these Petri net based models also provide automatic solutions or aids for the resulting implementation: hardware software implementations are avalaible; software implementation generally consists of a direct programming of the net in a microprocessor based programmable controller.

B - The second level of interest is flexible automation where nets are used for the specification of a global synchronization mechanism, i.e. a synchronization between different machines or sub-parts of the factory /NAR/ /VAL2/ /ALA/. A very important example is that of the transport system in a factory. In a factory, transportation is ensured by self-propelled trucks handling parts of the equipment to be built as well as some parts of the tools required for constructing it. A set of trucks is therefore in motion in the factory, depending on cost and performance constraints. The trucks follow a set of wires or are wire-guided; in addition, there is a number of dedicated equipment or plots where the trucks stop and exchange information about their position and the next working sequence while stopping at the contacts of the plots.

Such a valuable example is given in /ALA/. The trucks have to follow a wire net consisting of wires which have to be followed by the trucks tuned to a given frequency from the line. After a first possible definition of the wire net by the designer, the system has to be represented and evaluated. /ALA/ shows that such a transportation system can be specified by means of cells and sections (Figure 9) . A section can be simply defined: it is made of the same wire and of its contacts; at any time, one truck at the most is present in the section and is either moving along the wire or halting at a contact. Depending on the wire net, the cells must then be defined. A simple example is that of a crossing between two wires. Naturally two trucks cannot be present on the cell at the same time because once they have left the contacts the system no longer has any information about their position until they reach the next contact and they may collide. It is therefore necessary to define a behavior model of such a cell. Generally, cells consist of several sections and contacts and

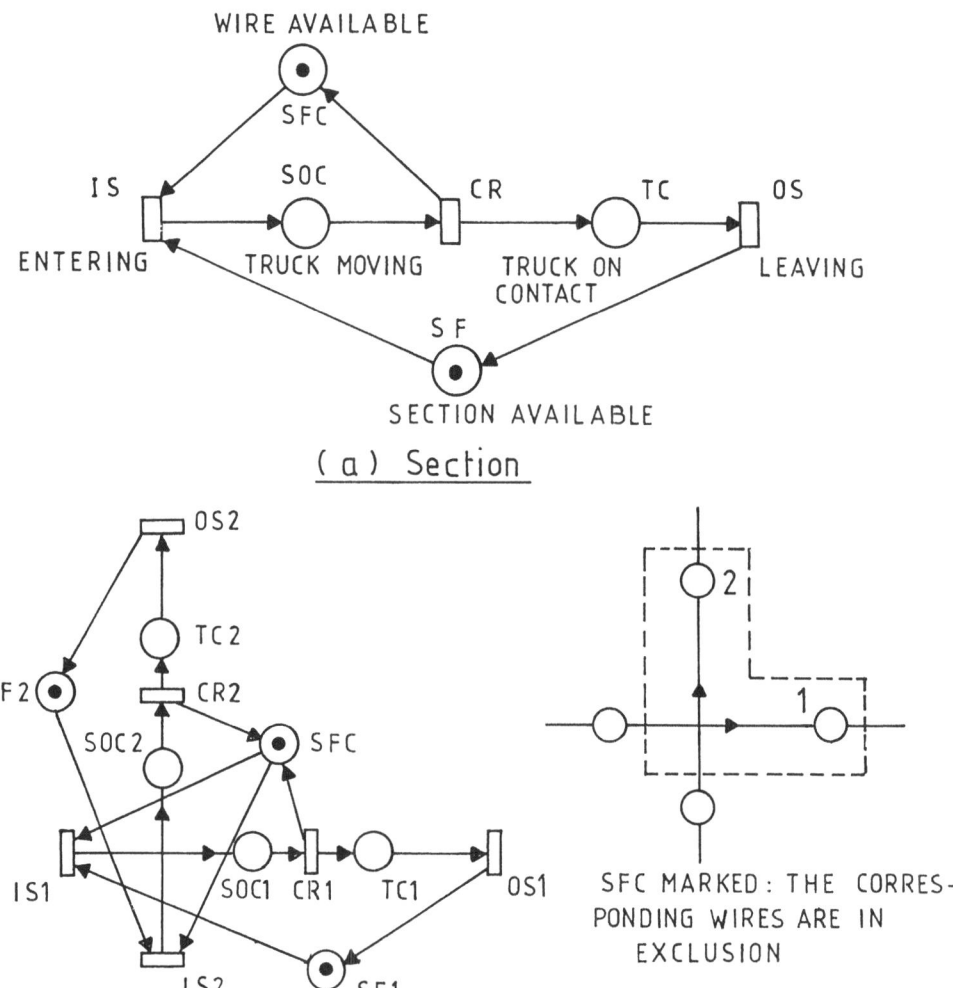

WIRE AVAILABLE

SFC

SOC

IS   CR   TC   OS

ENTERING   TRUCK MOVING   TRUCK ON   LEAVING
                              CONTACT

SF

SECTION AVAILABLE

(a) Section

OS2

TC2

SF2   CR2

SOC2   SFC

IS1   SOC1 CR1   TC1   OS1

IS2   SF1

(b) Crossing cell

2

1

SFC MARKED: THE CORRES-
PONDING WIRES ARE IN
EXCLUSION

Figure 9

the representation of their behavior has to be modelled in a similar way. Note that cells tend to be rather simple and that the specification of the global wire net will be derived from the structuring of the models of the different cells constituting the wire net.

A section model and the model of the cell previously discussed is given in Figure 9; in the section model, Place SFC means, when it is marked, that the wire is available. On the crossing model, the 2 wires cross each other. There must therefore be an exclusion. The latter is given by merging together the two places SFC of the two sections giving the place SFC of the cell model. Thus a cell model is obtained for this kind of crossing. If there are different cells, as in Figure 10, the different corresponding models are defined and the global model of the transportation system is obtained by connecting the different cell models.

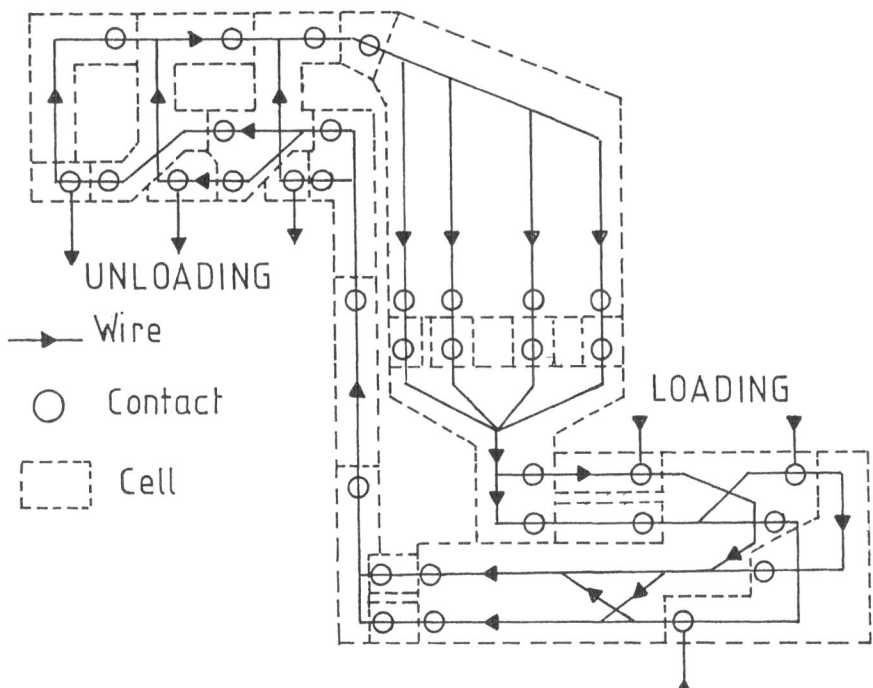

Figure 10

It is also interesting to know the performance of the transport, say for instance, the number of items of equipment carried by time unit. This can be done by using explicit values for time, for instance by adding to the previous nets the time spent by the truck at a plot or to move in the section from one plot to another. Thus a global timed model of the transport system is obtained. Handling of the timed model allows to compute mean time values in order to get some insight into the general behavior of a transportation system. Of course, simulation can also be used. Simulation and mean time evaluations have been conducted in /ALA/.

IV.4. Maintenance and Spare Management

Let us now consider as the last area of application, spare management in dependability and maintenance; some items of equipment or units are operating and, in the case of a faulty unit, the latter has to be replaced and repaired. A net associated with this example /FLO1/ is given in Figure 11 where a set of equipment is considered. In this net, the tokens in place P1 represent the set of the non faulty units, the tokens in place P2 the set of the failed units, the tokens in place P3 the number of available spare units and the tokens in place P4 the units under repair.

It is then possible to use this net to think out a maintenance policy based on some significant rates, e.g., failure rate, replacement rate and repair rate. The failure rate controls the firing of transition t1, which models the transition of a unit from the fault-free state to the failed state, the replacement rate controls the firing of transition t2 and the repair rate controls the firing of transition t3.

Let us suppose that in the initial state there are five operating units and three spares. If the failure rate of one unit is lambda, the correct units may fail with a failure rate of "(marking-of-P1)*(lambda)", depending on the number of fault free units: this is why the marking M(P1) appears in transition t1. When a unit fails, the corresponding token goes into place P2 and if one spare is available, it is used as a replacement unit and the faulty unit goes under repair; this is done by firing transition t2 with the replacement rate of

mu1; once repaired according to the repair rate of mu2 the
previously faulty unit becomes a spare. The net is a
probabilistic Petri net, and more precisely a markovian Petri
net as the distributions are selected to be exponential.

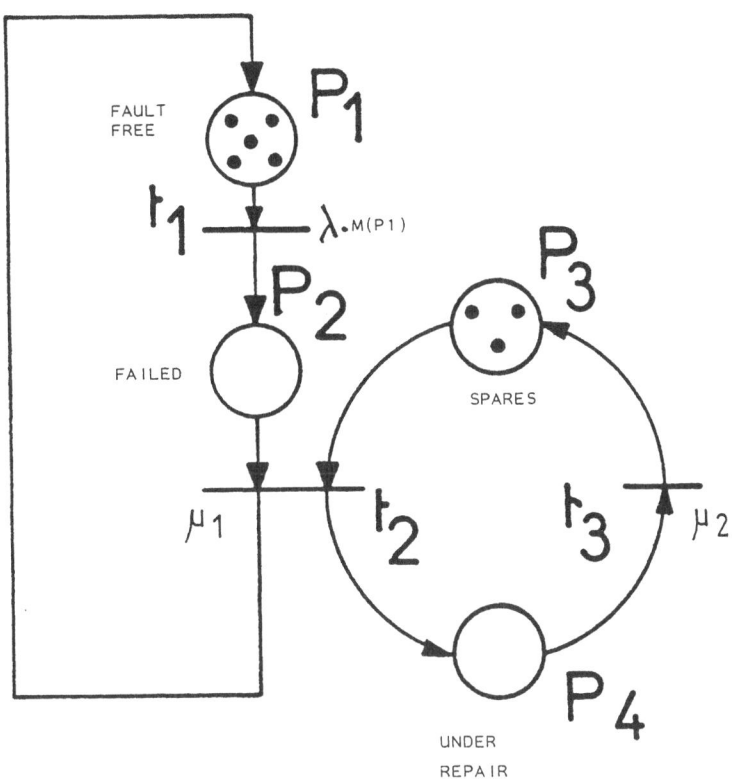

FIGURE 11

With this model, /FL01/ shows for example that the
probability of system breakdown can be computed. If the system
is such that as least one non faulty unit is needed for the
system to operate smoothly, then the probability of breakdown
is that of having no token in place P1: if there is no token
in place P1, no unit can operate and this causes a breakdown.
Hence if the probability of having place P1 not marked is
known, then it represents the probability of system breakdown.
Similarly, the probability of having the system ready to

repair is that of having at least one token in place P3. It is also possible to estimate the needed value for lambda if the other figures are known, i.e. given mu1, mu2 and the necessary probability of the system breakdown.

Of course, such values can be obtained by computing from the Markov chain related to this stochastic Petri net. Hence the interest of such an approach.

V - SUPPORT TOOLS

As highlighted by the references, real systems lead to complex nets. It is therefore mandatory to use software tools to cope with the resulting complexity. Sophisticated packages exist for Petri net based models as indicated by questionnaires in the European Workshops on Theory and Application of Petri Nets, 1980-1986... and during the Advanced Course in Bad Honnef in September 1986. These tools are mainly dedicated to verification or simulation. Most of them are still being developed and based on the use of a reachability graph.

Probably the first tool dedicated to the proof of Petri nets then extended with a particular emphasis on protocols was OGIVE /CHE/ /DUF/, industralized by the SYSECA company under the name of OVIDE. The firt release was delivered in 1979 it was developed by LAAS-CNRS from 1977 to 1984.

For simulation purposes, which seem of high importance, an interesting tool /RAZ/ /EST/, SARA, developed at UCLA, explicitly includes a data part. Analyses are made on the reachability graph and an elaborated simulation is possible. Other packages, such as GALILEO /LOP/ at ITT Spain, and PROTEAN, a NPN simulator /BIL2/ at the Australian Telecoms, are under development to simulate complex systems. Other tools are being developed for simulation, but also for graphics, verification, performance and dependability.

VI - CONCLUSION

Other models have been used to represent parallel and distributed systems. The most important ones, in addition to Petri nets, are abstract data types /GUT/, temporal logic /MAN//PNU/, functional programming /MIL/, and many dedicated or general languages.

Some attempts have been made to link some of these models as in the case of temporal logic and Petri nets /DIA3/ /GEN2/ /QUE/ and of abstract data types and Petri nets /VAU/.

Whatever the specification method, the final aim is to realize actual systems and the selected method must therefore deal with all aspects of system design.

The first difficulty lies in that the selected approach must simultaneously and coherently allow: - a high level specification, very far from implementation details and choices; - and, of course, an efficient implementation. The second major difficulty results from the need for an approach encompassing almost all fields of application, and its scope must be sufficiently wide as to be used in various contexts and to express various system constraints.

As a matter of fact, from the previous points and examples developed so far, it appears that Petri nets and Petri net based models are of great interest. They lead, in particular, to specifications which are both powerful and easily accepted by learning people and industrial designers. Finally, some supported tools and a lot of prototype tools exist and allow users to cope with complex and real problems and systems.

ACKNOWLEDGEMENTS The author would like to thank all the researchers at LAAS who contributed to the development and application of Petri nets and to the writing of this paper, especially P. AZEMA, CH. BEOUNES, JP. COURTIAT, M. COURVOISIER, G. JUANOLE, and R. VALETTE.

BIBLIOGRAPHY

It goes without saying that all papers of interest cannot be given here and only a few of them are listed hereafter. For a thorough list the interested reader has to consult the following publications:

Papers presented at the

First European Workshop on Applications and Theory of Petri nets

Strasbourg, F, September 1980.

Second European Workshop on Applications and Theory of Petri nets

Bad Honnef, D, September 1981.

Third   European Workshop on Applications and Theory of Petri nets
        Varenna, I, September 1982.
Fourth   European Workshop on Applications and Theory of Petri nets
        Toulouse, F, September 1983.
Fifth   European Workshop on Applications and Theory of Petri nets
        Aarhus, DK, June 1984.
Sixth   European Workshop on Applications and Theory of Petri nets
        Espoo, Finland, June 1985.
Seventh European Workshop on Applications and Theory of Petri nets
        Oxford, GB, June 1986.
    Applications and Theory of Petri nets, Selected papers from the 3rd European Workshop, Varenna, Springler-Verlag, Informatik 66, A. Pagnoni and G. Rosenberg Editors, 1983.
    Advances in Petri nets 1984, LNCS 188, Springer-Verlag, G. Rozenberg Editor, 1985.
    Advances in Petri nets 1985, LNCS 222, Springer Verlag, G. Rozenberg Editor, 1986.
    -within this Avances series, selected papers from the European Workshops on Application and Theory of Petri Nets appear togheter with other papers-
    IEEE International Workshop on Timed Petri nets
        Torino, I, July 1985.
    IEEE Transactions on Software Engineering,
    IEEE Transactions on Computers,
    IEEE Transactions on Communications,
    Computer Networks

/ALA/ P. ALANCHE, K. BENZAKOUR, F. DOLLE, F. GILLET, P. RODRIGUES, R. VALETTE, "PSI, a Petri net based simulator for flexible manufacturing systems", 5th Eur Workshop on Petri nets, Aarhus, and LNCS 188, Springer-Verlag, 1984, G. Rozenberg Editor, 1984.
/AUG/ M. AUGUIN, F. BOERI, C. ANDRE, "New design using PLAs and Petri nets", MECO 78, Athens, June 1978.
/AYA1/   JM. AYACHE, M. DIAZ, H. KONBER, "Specification and

Verification of Signaling Protocols", Int. Switching Symposium, ISS81, Verdun, CANADA, September 1981.

/AYA2/ JM. AYACHE, JP. COURTIAT, M. DIAZ, "REBUS: a fault-tolerant distributed system for industrial real time control", IEEE Tr on Computers, Special Issue on Fault-Tolerant Computing, July 1982.

/AZE/ P. AZEMA, G. PAPAPANAGIOTAKIS, "Protocol Analysis by using Predicate Nets", Protocol Specification, Testing and Verification , M. DIAZ Editor, North Holland 1986.

/AZE1/ P. AZEMA, B. BERTHOMIEU, P. DECITRE, "The design and validation by Petri nets of a mechanism for the invocation of remote servers", Proc IFIP Congress, Melbourne, October 1980.

/BAE/ JL. BAER, G. GARDARIN, C. GIRAULT, G. ROUCAIROL, "The two step commitment protocol: modelling, specification and proof methodology", 5th Conf on Software Engineering, San Diego, March 1981.

/BAL/ G. BALBO, S.C. BRUELL, S. GHANTA, "Combining queueing network and generalized stochastic Petri nets models for the analysis of some software blocking phenomena", IEEE Tr on Software Engineering, Vol SE12, N 4, April 1986.

/BAU1/ B. BAUMGARTEN, P. OCHSENSCHLAGER, R. PRINOTH, "Building blocks for distributed system design", Protocol Specification, Testing and Verification, Toulouse-Moissac, June 1985, North Holland, 1986, M. Diaz Editor.

/BAU2/ B. BAUMGARTEN, P. OCHSENSCHLAGER, R. PRINOTH, "A formal model of the CCR algorithm", GMD report 186, December 1985, also IFIP Protocol Specification, Validation and Testing, Montreal, June 1986.

/BEO/ C. BEOUNES, JC. LAPRIE, "Dependability evaluation of complex computer systems: stochastic Petri net modelling", IEEE Int Symp on Fault Tolerant Computing, FTCS15, Ann Arbor, June 1985.

/BER/ B.BERTHOMIEU., M. MENASCHE, "An enumerative approach for analysing time Petri nets", Proc. of the IFIP Congress, Paris, September 1983.

/BERTR/ G. BERTHELOT, R. TERRAT, "Petri nets theory for the correctness of protocols", 2nd. Europ. Workshop on Appl. & Theory of Petri nets, Bad Honnef (F.R.G.), September 1981, pp.31-58, also IEEE Trans. on Communications, Vol COM-30, no.12, December 1982.

/BETO/ A. BERTONI, M. TORELLI, "Probabilistic Petri nets and semi Markov processes", 2nd Eur Workshop on Application and Theory of Petri nets, Bad Honnef, September 1981.

/BEY/ B. BEYAERT, G. FLORIN, S. NATKIN, P. LONC, "Evaluation of computer system dependability using stochastic Petri nets", IEEE Int Symp on Fault Tolerant Computing, FTCS11, Portland, June 1981.

/BIL/ J. BILLINGTON, "Specification of the Transport service using numerical Petri nets", 2nd Int. Workshop on Protocol Specification, Testing and Verification, Idyllwild Los Angeles, May 1982, North-Holland, 1982, C. Sunshine Editor. also "Abstract specification of the ISO transport service definition using labeled numerical Petri nets", North Holland, 3rd Int. Workshop on Protocol Specification, Verification and Testing, 1983, H. Rudin C. West Editors.

/BIL2/ J. BILLINGTON, M.C. WILBUR-HAM, M.Y. BEARMAN, "Automated Protocol Verification", Proc. of the 5th Int. Workshop on Protocol Specification, Testing and Verification, M. Diaz Editor, North Holland, 1986.

/BOC1/ G.V. BOCHMANN, "Finite state description of communication protocols", Conf. Computer Network Protocols, Liège, 1978, also in Computer Networks 2, 1978, pp. 361-372.

/BOC2/ G.V. BOCHMANN, "A general transition model for protocols and communication services", IEEE Trans. on Communications, vol. COM-28, no. 4, April 1980, pp.643-650.

/BOC3/ G.V. BOCHMANN, C.A. SUNSHINE, "Formal methods in communication protocol design", IEEE Trans. on Communications, vol. COM-28, no. 4, April 1980, pp. 624-631.

/BUR/ H.J. BURKHART, H. ECKERT, R. PRINOTH, "Modelling of OSI services and protocols using Predicate-Transition Nets", Protocol Specification, Testing and Verification, Skytop, June 1984, North Holland, 1985, Y. YEMINI ET AL Editors.

/BRU/ G. BRUNO, G. MARCHETTO, "Process-translatable Petri nets for the rapid prototyping of process control systems", IEEE Tr on Software Eng, Vol SE12, N 2, February 1986.

/BRU2/ G. BRUNO, P. BIGLIA, "Performance evaluation and validation of tool handling in flexible manufacturing systems using Petri nets", IEEE Proc of the Int Workshop on Timed Petri nets, Torino, July 1985.

/CHA/ C. CHATELIN, C. GIRAULT, S. HADDAD, "Specification and properties of a cache coherence protocol model", 7th Eur Worshop on the Application and Theory of Petri nets, Oxford, June 1986.

/CHE/ B. CHEZALVIEL-PRADIN, "Un outil graphique interactif pour la verification des systemes decrits par reseaux de Petri", These de Dr Ingenieur, UPS, 1979.

/COUT/ J.P. COURTIAT, J.M. AYACHE, B. ALGAYRES, "Petri Nets are Good for Protocols", SIGCOMM 84 Symposium; also in Computer Communications Review, 14, No.2, 1984.

/COVS/ M. COURVOISIER, JP. SECK, "Hardware implementation of generalized Petri nets", Electronics Letters, Vol 15, N24, November 1979.

/COVS1/ M. COURVOISIER, "A matrix based implementation of generalized Petri nets", 3rd Eur Workshop on Petri nets, Varenna, also Springer-Verlag, Informatik 66, Applications and Theory of Petri nets, 1983.

/COVS2/ M. COURVOISIER, "Description et realisation de systemes de commande a evolutions simultanees, revue RAIRO, Fevrier 1985.

/COVS3/ M. COURVOISIER, "An asynchronous logic array for the realization of logic systems with concurrency", Electronics Letters, Vol 14, N 4, December 1977.

/COVS4/ M.COURVOISIER, R. VALETTE, "Description and realization of parallel systems", COMPCON Fall, Washington, September 1977.

/DAN/ A.S. DANTHINE, "Protocol representation with finite-state models", IEEE Trans. on Communications, vol. COM-28, no. 4, pp. 632-643, April 1980.

/DIA1/ M. DIAZ, "Modelling and analysis of communication and cooperation protocols using Petri net based models", Tutorial paper, Computer Networks, December 1982.

/DIA2/ M. DIAZ, J.P. COURTIAT, B. BERTHOMIEU, J.M. AYACHE, "Status of Petri net based models for protocols", IEEE Int. Conf. on Communications, ICC 83, Boston, June 1983.

/DIA3/ M. DIAZ, G. GUIDACCI DA SILVEIRA, "On the specification and validation of protocols by temporal logic and nets", Proceedings of the IFIP 83 Congress, Paris, September 1983.

/DIA4/ M. DIAZ, "Nets in the specification and verification of protocols", Advanced Course on Petri nets, Bad Honnef, September 1986.

/EST/ G. ESTRIN, R.S. FENCHEL, R.R. RAZOUK, M.K. VERNON, "SARA: Modelling, analysis and simulation support for design of concurrent systems", IEEE Tr on Software Eng, Vol SE12, N 2, February 1986.

/DUF/ J. DUFAU, "OGIVED, Un outil pour la verification des protocoles décrits par Réseaux de Petri", Thèse de Docteur-Ingénieur, Univ. Paul Sabatier, Toulouse, Janvier 1984.

/FLO1/ G. FLORIN, S. NATKIN, "Les reseaux de Petri stochastiques", revue AFCET-TSI, Vol 4, N 1, 1985.

/FLO2/ G. FLORIN, S. NATKIN, B. LONC, An evaluation CAD tool based on stochastic Petri nets", IFIP Working Conf on "Reliable computing in the 1980's, London, September 1979.

/FLO3/ G. FLORIN, S. NATKIN, "Evaluation based on stochastic Petri nets of the maximum throughput of a full duplex protocol", 2nd Eur Workshop on Applications and Theory of Petri nets, Informatik 52, Springer-Verlag, C. Giraud and W. Reisig Editors, 1982.

/FLO4/ G.FLORIN, S. NATKIN, "Ergodicity criteria for stochastic Petri nets", 5th EUR Workshop on Application and Theory of Petri nets", Aarhus, June 1984.

/GEN/ H.J. GENRICH, K. LAUTENBACH, "System modelling with high-level Petri nets", Theoretical Computer Science 13, North Holland, 1981.

/GEN1/ H.J. GENRICH, K. LAUTENBACH, "The analysis of distributed systems by means of Predicate-Transition nets", Semantics of Concurrent Computation, Evian 1979, G. Kahn Editor, Lect. Notes in Computer Sciences, vol. 70, Springer Verlag 1979, pp. 123-146.

/GEN2/ H.J. GENRICH, K. LAUTENBACH, P.S. THIAGARAJAN, "Elements of Net Theory", LNCS, 84, 1980.

/GUT/ J. GUTTAG, "Notes on type abstractions", Proc of the Conf on Reliable Software, 1979.

/HUR/ G.S. HURA, H. SINGH, N.K. NANDA, "Some design aspects of databases through Petri net modelling", IEEE Tr on Software Eng, Vol SE12, N 4, April 1986.

/JEN/ K. JENSEN, "Coloured Petri nets and the invariant method", Theor. Comp. Science, 14, 1981.

/JEN2/ K. JENSEN, "High level Petri nets", Advanced Course on Petri nets, Bad Honnef, September 1986.

/JUA/ G. JUANOLE, B. ALGAYRES, J. DUFAU, "On Communications Protocol Modelling and Design" LN in CS, Springer Verlag, 188, Advances in Petri nets 1984, G. Rozenberg Editor.

/KES/ J.L.W. KESSELS, "An alternative to event queue for synchronization in monitors", Comm of the ACM, Vol 20, N 7, July 1977.

/KEL/ R.M. KELLER, "Formal verification of parallel programs", Com. ACM 19-7, July 1976, pp. 371-384, vol. 19, no. 7.

/KOM/ N. KOMODA, T. MURATA, K. MATSUMO, "Petri net based

controller: SCR and its application in factory automation", IEEE Int Symp on Circuits and Systems, ISCAS85, Kyoto, June 1985.

/LAM/ S.S. LAM, A.U. SHANKAR, "Protocol Verification via Projections", IEEE Tr on Software Engineering, Vol SE10, No4, July 1984.

/LEU/ K.C. LEUNG, C. MICHEL, P. LEBEUX, "Logical system design using PLAs and Petri nets programmable hardwired systems", IFIP Congress, Toronto, 1977.

/LOP/ I. LOPEZ, "The use of GALILEO to represent and analyse telecommunications protocols", 2nd Europ. Workshop on Applications and Theory of Petri nets, Bad Honnef, FRG, September 1981.

/MAN/ Z. MANNA, "Logics of programs", IFIP80, North Holland, 1980.

/MAR/ A. M. MARSAN, G. BALBO, G. CONTE, "A class of generalized stochastic Petri nets for the performance evaluation of multiprocessors systems", ACM Tr on Computer Systems, Vol 2, N 2, May 1984.

/MAR2/ M.A. MARSAN, G. CHIOLA, A. FUMAGALLI, "An accurate model of CSMA/CD bus LAN", 7th Eur Worshop on the Application and Theory of Petri nets, Oxford, June 1986.

/MER/ P.M. MERLIN, D.J. FARBER, "Recoverability of Communications protocols", IEEE Trans. on Communications, September 1976.

/MIL/ R. MILNER, "A calculus of communicating systems", LNCS, N 92, Springer Verlag, 1980.

/MOL/ M.K. MOLLOY, "Performance analysis using stochastic Petri nets", IEEE Tr on Computers, Vol 31, N 9, September 1982.

/MUR1/ T. MURATA, N. KOMODA, K. MATSUMOTO, "A Petri net based factory automation controller for flexible and maintainable control specifications", IECON 1984.

/MUR2/ T. MURATA, N. KOMODA, K. MATSUMOTO, K. HARUNA, "A Petri net based controller for flexible and maintainable sequence control and its applications in factory automation", IEEE Tr on Industrial Electronics, Vol IE 33, N 1, February 1986.

/NAR/ Y. NARAHARI, N. VISWANADHAM, "Coloured Petri net models for generalized flexible manufacturing systems", 7th Eur Workshop on Application and Theory of Petri nets, Oxford, June 1986.

/PAT/ S.S. PATIL, "An asynchronous logic array", MIT report, May 1975.

/PAG/ A. PAGNONI, "Stochastic nets and performance evaluation", Advanced Course on Petri nets, Bad Honnef, September 1986.

/PET/ C.A. PETRI, "Introduction to general net theory", LNCS 84, Proc of the Advanced Course on General Net Theory 1979, Springer Verlag, 1980, W. Brauer Editor.

/PNU/ A. PNUELI, "The temporal logic of programs", IEEE Symp on Foundations of Computer Science", 1977.

/QUE/ J.P. QUEILLE, J. SIFAKIS, "Specification and verification of concurrent systems in Cesar", 2nd Europ Workshop on Application and Theory of Petri nets, Bad Honnef, September 1981.

/RAM/ C. RAMCHANDANI, "Analysis of Asynchronous Concurrent Systems by Timed Petri Nets", Research Report MAC-TR 120, MIT, February 1974.

/RAMHO/ C.V. RAMAMOORTHY, G.S. HO, "Performance evaluation of asynchronous concurrent systems using Petri nets", IEEE Tr on Software Eng, Vol SE6, N 5, September 1980.

/RAZ/ R.R. RAZOUK, G. ESTRIN, "Modelling and verification of communication protocols in SARA: the X.21 interface", IEEE Tr on Computers, Vol C-29, no.12, December 1980, pp.1038-1051.

/RAZ2/ R.R. RAZOUK, "Modelling X.25 using the graph model of behavior", Protocol Specification, Testing and Verification, Idyllwild-CA, May 1982, North-Holland, 1982, C. Sunshine Editor.

/ROS/ M.T. ROSE, "Modelling of initial connection handling in TCP using Contour-Transition Nets", Protocol Specification, Testing and Verification, Skytop, June 1984, North Holland, 1985, Y. Yemini et al Editors.

/ROU/ JL. ROUX, B. BERTHOMIEU, "Verification of a local area network with TINA, a software package for Time Petri Nets", 7th European Workshop on Application and Theory of Petri Nets, Oxford, June 1986.

/SHA/ A.U. SHANKAR, S.S. LAM, "An HDLC protocol specification and its verification using image protocols", ACM Trans Computer Systems, Vol 4, Nov 1983.

/SIF/ J. SIFAKIS, "A unified approach for studying the properties of transition systems", Theoretical Computer Science, Vol 18, 1982.

/SIF2/ J. SIFAKIS, "Performance Evaluation of Systems using Nets", LNCS 84, Proc of the Advanced Course on General Net Theory, Springer Verlag, 1980, W. Brauer Editor.

/SYM/ F.J.W. SYMONS, "Representation, analysis and verification of communication protocols", Research Report 7380, Telecom. Australia, 1980.

/THO/ D.T. THOMPSON, C.A. SUNSHINE, R.W. ERICKSON, S.L. GERHART, D. SCHWABE, "Specification and verification of communication protocols in AFFIRM using state transition models", Research Report ISI-RR-81-88, USC, Inf. Sc. Institute, March 1981.

/VAL1/ R. VALETTE, "Nets in production systems", Advanced Course on Petri nets, Bad Honnef, September 1986, Springer-Verlag, 1987.

/VAL2/ R. VALETTE, M. COURVOISIER, H. DEMOU, JM. BIGOU, C. DESCLAUX, "Putting Petri nets to work for controlling flexible manufacturing systems", IEEE 1985 Int Symp on Circuits and Systems, Kyoto, June 1985.

/VAU/ J. VAUTHERIN, "Parallel system specifications with coloured Petri nets and algebraic abstract data types", 7th Eur Workshop on Application and Theory of Petri Nets", Oxford, June 1986.

/VID/ F. VIDONDO, "GALILEO, experiences in the design of a Petri net based language for real time systems", 2nd Europ. Workshop on Applications and Theory of Petri nets, Bad Honnef, FRG, September 1981.

/VIS/ Ch. VISSERS, L. LOGRIPPO, "The importance of the concept of service", 5th Int. Workshop on Protocol Specification,Testing and Verification, Toulouse, June 1985, North Holland, M. Diaz Editor, 1986.

/VOS/ K. VOSS, "Using Predicate-Transition Nets to model and analyse distributed database systems", IEEE Tr on Software Engineering, Vol SE6, No6, November 1980.

/WHE/ G.R. WHEELER, M.C. WILBUR-HAM, J. BILLINGTON, J.A. GILMOUR, "Protocol analysis using Numerical Petri Nets", LNCS 222, Advances in Petri Nets, G. Rozenberg Editor, 1986.

/ZIM/ H. ZIMMERMAN, "OSI reference model. The ISO model of architecture for open systems interconnection", IEEE Trans. on Communications, vol. COM-28, April 1980.

/ZUB/ W.M. ZUBEREK, "Timed Petri nets and Performance Evaluation", 7th Ann Symp on Computer Architecture, May 1980.

Part II

CONTRIBUTED PAPERS

SOME CLASSES OF
LIVE AND SAFE PETRI NETS

Eike Best
Institut für Methodische Grundlagen
Gesellschaft für Mathematik und Datenverarbeitung
5205 St.Augustin 1
Fed. Rep. Germany

Pazhamaneri S. Thiagarajan
The Institute of Mathematical Sciences
Madras 600 113
India

ABSTRACT

We study a series of structural restrictions which gives rise to a set of classes of marked nets: S-systems, T-systems, free choice systems and asymmetric choice systems. This series of restrictions corresponds to a series of behavioural properties of marked nets first introduced by C.A.Petri — sequentiality, determinacy and confusion-freeness.

For each one of these restrictions, we address the question to what extent the behaviour of a marked net is determined by the structure of the underlying net. In particular, we study the characterisation of liveness and the characterisation of safeness in the presence of liveness. The paper presents strengthened, streamlined and elementary proofs of some important results.

1   Introduction

A large portion of the theory of marked nets can be viewed as an attempt to answer the question: To what extent does the underlying net of a marked net determine its behaviour? Since marked nets constitute one (among many) net-based model of distributed systems, the question posed above is actually a more concrete form of the basic question: To what extent does the structure of a distributed system determine its behaviour?

One pleasant aspect of net theory is that the study of the interplay between structure and behaviour can be carried out in systematic manner. One can impose a series of restrictions on the structure of the underlying nets (of marked nets) and study the resulting classes of marked nets. The surprise is that the structural restrictions one imposes (which are meaningful in their own right) correspond to fundamental behavioural phenomena such as conflict, concurrency and confusion — phenomena that have first been identified by C.A.Petri in his seminal papers [13] and [14]. The restrictions one imposes on the underlying nets yield —in a sufficient sense— deterministic (no conflict), sequential (no concurrency) and confusion-free (well, no confusion) classes of marked nets[1].

---

[1]Strictly speaking, we assume safeness at this point.

Our aim here is to reemphasise this very appealing aspect of net theory. We do so by presenting results concerning various subclasses of marked nets. Almost all of the results we mention are known. What is perhaps novel is the choice of the material and the order in which the results have been arranged. In addition, proofs of certain results have been streamlined and, where possible, strengthened. This paper results from combining the framework for presenting the theory of marked nets developed in [16] and the results presented in [1].

In the next section we introduce the basic concepts and relations we require. We then proceed in section 3 to put down some observations that hold for the whole class of marked nets. The major concerns here, as also in the subsequent sections, are to characterise liveness and to characterise safeness in the presence of liveness. In section 4 we deal with S-systems (based on S-graphs) and in section 5 we deal with T-systems (based on T-graphs). Safe S-systems are sequential (if one assumes that the underlying S-graphs are connected, which we do) and safe T-systems are deterministic in their behaviours. Section 6 is the major section of the paper. Here we turn to free choice systems which —when safe— are guaranteed to be confusion-free. We expose part of the substantial theory of this class of marked nets. In particular, we present an elementary proof of an important structural result. Our proof is a strengthened version of the one presented in [1]. In section 7 we point out some results concerning asymmetric choice systems. Safe systems of this type can exhibit only a limited kind of confusion called asymmetric confusion [16].

2 Basic Concepts

Our object of study are place/transition systems [2] without capacity constraints (that is, we will assume all capacities to be infinite) and with a trivial weight function (i.e. the weight equals the flow relation). Thus, in the following we will use the notation $\Sigma = (S, T; F, M_0)$ to denote a P/T-system. As usual, S, T, F and M_0 are the set of places, the set of transitions, the flow relation and the initial marking, respectively.

Remark 2.1 *The underlying net of a system*

> The net $N = (S, T; F)$ will be called the underlying net of the system $\Sigma = (S, T; F, M_0)$. Sometimes we write $N = N_\Sigma$ to emphasize the fact that N belongs to Σ.
> Structural definitions that apply to N only (and do not depend on the initial marking M_0) will be stated in terms of N rather than in terms of Σ. Such definitions are understood to carry over to Σ by the rule that Σ enjoys some property iff its underlying net N enjoys it. ■ 2.1

Without repeating their definitions, we shall use the following concepts [2]: the transition rule, the set of occurrence sequences of Σ, the set of transition sequences of Σ and the set $[M_0\rangle$ of forward reachable markings of Σ.

We will demand two restrictions:

Restriction 2.2 *Finiteness of N and of Σ*

> From now on, we will always assume N (and hence Σ) to be finite, i.e., $S \cup T$ to be a finite set. ■ 2.2

Definition 2.3 *Weak connectedness*

> $N = (S, T; F)$ is weakly connected *iff* $(F \cup F^{-1})^* = (S \cup T) \times (S \cup T)$.
> (We denote by ρ^* the reflexive and transitive closure of a relation ρ.) ■ 2.3

Restriction 2.4

From now on, we will always assume N (and hence Σ) to be weakly connected. ∎ 2.4

In a non-weakly connected system, it is usually possible to study the connected parts in isolation. The existence of *directed* paths is captured by the next definition.

Definition 2.5 *Strong connectedness*

$N = (S, T; F)$ is strongly connected *iff* $F^* = (S \cup T) \times (S \cup T)$. ∎ 2.5

We will not require strong connectedness universally because it is not easily possible to split a non-strongly connected system into strongly connected components without disrupting its behaviour. The F^*-paths of a net $(S, T; F)$ are important, so we introduce them explicitly by means of a definition. The paths that start and end at the same element of $S \cup T$ are particularly important; they will be called the cycles of N.

Definition 2.6 *(Simple) paths and cycles*

Let $N = (S, T; F)$ be a net.

(i)   A path of N is a sequence $(x_0, f_1, x_1, \ldots, x_{m-1}, f_m, x_m)$ such that $x_j \in S \cup T$ $(0 \leq j \leq m)$, $f_j \in F$ $(1 \leq j \leq m)$ and $f_j = (x_{j-1}, x_j)$ for $1 \leq j \leq m$; we shall say that the path leads from x_0 to x_m.

(ii) A path $(x_0, f_1, \ldots, f_m, x_m)$ is called simple iff no element x_j (except perhaps $x_0 = x_m$) appears twice in it, i.e., iff

$$\forall k, j: \ (0 \leq k \leq m \wedge 1 \leq j \leq m - 1 \wedge k \neq j) \ \Rightarrow \ x_k \neq x_j.$$

(iii) A cycle of N is a path with $x_0 = x_m$.

(iv) A cycle is called simple iff it is a simple path. ∎ 2.6

Every path from x_0 to x_m can be shortened (by cutting out cycles) to a simple path also leading from x_0 to x_m. When no confusion can arise, we often write a path in the form $x_0 x_1 \ldots x_m$, with the understanding that $f_j = (x_{j-1}, x_j)$ for $1 \leq j \leq m$.

We shall say that a net $(S, T; F)$ is covered by (simple) cycles iff every $f \in F$ lies on some (simple) cycle. As is easy to see, strong connectedness is equivalent to weak connectedness and the property of being covered by cycles.

Besides paths and cycles, we need the notion of a subnet.

Definition 2.7 *Subnet*

Let $N = (S, T; F)$ and $N_1 = (S_1, T_1; F_1)$ be two nets.
N_1 is a subnet of N *iff* $S_1 \subseteq S$, $T_1 \subseteq T$ and $F_1 = F \cap ((S_1 \times T_1) \cup (T_1 \times S_1))$. ∎ 2.7

The requirement $F_1 = F \cap ((S_1 \times T_1) \cup (T_1 \times S_1))$ means that *all* F-arrows between elements of the subnet (and *only* such F-arrows) lie in the subnet. Hence F_1 is completely defined by F, S_1 and T_1. By contrast, paths or cycles according to definition 2.6 are not necessarily also subnets since there may be more F-arrows between the elements of a path or a cycle than are included in it.

Because F_1 is defined by S_1 and T_1, it makes sense to speak of the subnet generated by a set of elements $X_1 \subseteq S \cup T$:

Definition 2.8 *Subnet generated by a set of elements*

Let $N = (S, T; F)$ be a net and $X_1 \subseteq S \cup T$. Then $N_1 = (S_1, T_1; F_1)$ is the subnet generated by X_1 *iff* [2]

$$S_1 = (X_1 \cap S) \cup {}^\bullet(X_1 \cap T) \cup (X_1 \cap T)^\bullet,$$
$$T_1 = (X_1 \cap T) \cup {}^\bullet(X_1 \cap S) \cup (X_1 \cap S)^\bullet \text{ and}$$
$$F_1 = F \cap ((S_1 \times T_1) \cup (T_1 \times S_1)). \qquad \blacksquare 2.8$$

The special cases that X_1 consists only of places or only of transitions are important. If $X_1 \subseteq S$ then 2.8 simplifies to $S_1 = X_1$, $T_1 = {}^\bullet S_1 \cup S_1^\bullet$ and F_1 as before. If $X_1 \subseteq T$ then 2.8 simplifies to $T_1 = X_1$, $S_1 = {}^\bullet T_1 \cup T_1^\bullet$ and F_1 as before.

It is often necessary to restrict a marking to a part of a net; the next definition introduces the necessary concepts.

Definition 2.9 *Restrictions of markings*

Let $N = (S, T; F)$ be a net, $S_1 \subseteq S$ and M a marking of N.

(i) $M|_{S_1}: S_1 \to \mathbf{N}$ is the marking restricted to S_1, defined by the rule that $M|_{S_1}(s) = M(s)$ for all $s \in S_1$.

(ii) $M(S_1) = \sum_{s \in S_1} M(s)$ is the token load of M on S_1. $\qquad \blacksquare 2.9$

Concurrency and conflict are basic phenomena in place/transition systems. We will introduce them in the next few definitions. A number of simple facts will be stated also.

Definition 2.10 *Concurrent enabling*

For $\Sigma = (S, T; F, M_0)$, let $M \in [M_0\rangle$ be a marking and $t_1, t_2 \in T$ two transitions. t_1 and t_2 are concurrently enabled by M *iff* for all $s \in S$:

$$M(s) \geq \begin{cases} 2 & \text{if } s \in ({}^\bullet t_1 \cap {}^\bullet t_2) \\ 1 & \text{if } s \in (({}^\bullet t_1 \backslash {}^\bullet t_2) \cup ({}^\bullet t_2 \backslash {}^\bullet t_1)) \\ 0 & \text{otherwise.} \end{cases}$$

$$\blacksquare 2.10$$

Fact 2.11 *Exchanging concurrently enabled transitions in an occurrence sequence*

If t_1 and t_2 are concurrently enabled by M then both $M t_1 M' t_2 M''$ and $M t_2 \hat{M} t_1 M''$ are occurrence sequences starting with M and ending with M'' (but not necessarily $M' = \hat{M}$).

Proof: Immediate from the transition rule and from 2.10. $\qquad \blacksquare 2.11$

If M enables both t_1 and t_2 then it is by no means necessarily true that t_1 and t_2 are concurrently enabled by M; failing this, the situation is called a conflict (see Figure 1).

---

[2] As usual, for $x \in S \cup T$, ${}^\bullet x = \{y \in S \cup T \mid (y, x) \in F\}$ denotes the set of immediate predecessors of x and $x^\bullet = \{z \in S \cup T \mid (x, z) \in F\}$ denotes the set of immediate successors of x; for $X_1 \subseteq S \cup T$, ${}^\bullet X_1 = \bigcup_{x \in X_1} {}^\bullet x$ and $X_1^\bullet = \bigcup_{x \in X_1} x^\bullet$.

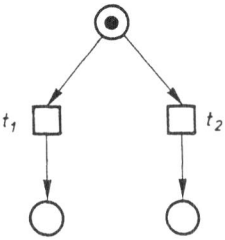

Figure 1: A conflict

Definition 2.12 *Conflict at a marking*

For $\Sigma = (S, T; F, M_0)$, let $M \in [M_0\rangle$ be a marking and t_1, t_2 two transitions.
Then t_1 and t_2 are in conflict at M *iff* M enables both t_1 and t_2 but does not concurrently enable t_1 and t_2. ∎ 2.12

Fact 2.13 *Characterisation of conflict*

t_1 and t_2 are in conflict at M *iff both are enabled at M and $M(s) = 1$ for some $s \in {}^{\bullet}t_1 \cap {}^{\bullet}t_2$.*

Proof: Obvious from the definitions. ∎ 2.13

If t_1 and t_2 do not share an input place then their combined enabling implies their concurrent enabling, as shown by part (a) of the next simple fact.

Fact 2.14 *Sufficient conditions for concurrent enabling*

(a) *If ${}^{\bullet}t_1 \cap {}^{\bullet}t_2 = \emptyset$ and M enables both t_1, t_2 then M concurrently enables t_1, t_2.*

(b) *If $t_1^{\bullet} \cap {}^{\bullet}t_2 = \emptyset$ and $M t_1 M' t_2 M''$ is an occurrence sequence then M concurrently enables t_1, t_2.*

Proof: Again an easy consequence of the definitions. ∎ 2.14

We will now define a number of simple but important behavioural properties of Σ called boundedness, safeness and liveness.

Definition 2.15 *Boundedness and safeness*

Let $\Sigma = (S, T; F, M_0)$ be a P/T-system.

(a) $s \in S$ is n-bounded ($n \in \mathbf{N}$) *iff* $\forall M \in [M_0\rangle : M(s) \leq n$.

(b) Σ is n-bounded ($n \in \mathbf{N}$) *iff* $\forall s \in S : s$ is n-bounded.

(c) Σ is safe *iff* Σ is 1-bounded. ∎ 2.15

Safeness is an important property which holds in many practical situations (e.g.: a variable always has exactly one value; in a sequential program, control is always at exactly one location).

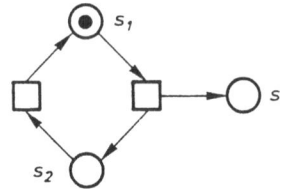

Figure 2: An unbounded system

Definition 2.16 *Liveness*

Let $\Sigma = (S, T; F, M_0)$ be a P/T-system.

(a) $t \in T$ is live *iff* for all $M \in [M_0\rangle$ there is some $M' \in [M\rangle$ such that M' enables t.

(b) Σ is live *iff* $\forall t \in T$: t is live. ∎ 2.16

Liveness means that no transition may become 'dead' in the sense that it can never again be enabled. This corresponds to the absence of (partial) deadlock.

3 Some General Results

We start with two simple observations on liveness and safeness. These two results, at the same time, model the pattern of many of the results presented in this paper. We will first have a result of the form

"A system Σ is live iff (or if, or implies) ..."

and then a result of the form

"A live system Σ is safe iff (or if, or implies) ...".

Proposition 3.1 *A necessary condition for liveness*

If $\Sigma = (S, T; F, M_0)$ is live then $\forall s \in S$: $^\bullet s \neq \emptyset$.

Proof: (Indirect.) Suppose $^\bullet s = \emptyset$ for some $s \in S$. By the definition of a net, $s^\bullet \neq \emptyset$. Because of the liveness of Σ, there is a reachable marking $M \in [M_0\rangle$ (possibly $M = M_0$) at which $M(s) = 0$ holds true. Because $^\bullet s = \emptyset$, we have $M'(s) = 0$ for all $M' \in [M\rangle$, so none of the transitions in $s^\bullet$ can ever be enabled again, contradicting liveness. ∎ 3.1

Proposition 3.2 *A necessary condition for the safeness of live systems*

If a live system $\Sigma = (S, T; F, M_0)$ is safe then $\forall s \in S$: $s^\bullet \neq \emptyset$.

Proof: (Indirect.) Suppose Σ is live and safe, and $s^\bullet = \emptyset$ for some $s \in S$. Again by the definition of a net, $^\bullet s \neq \emptyset$. By the liveness of Σ, there is a reachable marking $M \in [M_0\rangle$ such that $M(s) = 1$ (possibly $M = M_0$). Because $s^\bullet = \emptyset$ and because of the safeness of Σ, we have $M'(s) = 1$ for all $M' \in [M\rangle$, so none of the transitions in $^\bullet s$ can ever occur again, contradicting liveness. ∎ 3.2

The next two results achieve a converse of the previous two.

Proposition 3.3 *A necessary condition for safeness*

> If $\Sigma = (S, T; F, M_0)$ *is safe then* $\forall t \in T : {}^\bullet t \neq \emptyset$.

Proof: (Indirect.) Suppose that Σ is safe and that $t \in T$ such that ${}^\bullet t = \emptyset$. Then $t^\bullet \neq \emptyset$ by the definition of a net. Let $s \in t^\bullet$. It is easy to check that for some $M \in [M_0\rangle$, $M_0[tt\rangle M$ and moreover $M(s) > 1$. This contradicts the safeness of Σ.   ■ 3.3

Proposition 3.4 *A necessary condition for the liveness of safe systems*

> If a safe system $\Sigma = (S, T; F, M_0)$ *is live then* $\forall t \in T : t^\bullet \neq \emptyset$.

Proof: We start with a general observation. Let $N = (S, T; F)$ be a net and M, M' two markings on N such that $M' \geq M$; in other words, $\forall s \in S : M'(s) \geq M(s)$. Suppose that $M[\tau\rangle M_1$ where τ is a transition sequence. Then $M'[\tau\rangle M_1'$ where M_1' satisfies:

$$\forall s \in S : \quad M_1'(s) \ = \ M_1(s) + (M'(s) - M(s)).$$

In other words, if τ is a transition sequence at M then it is also a transition sequence at M'. Moreover, the resulting markings M_1 and M_1' are related to each other exactly as M and M' are related to each other. This follows easily from the definitions.

Turning now to the result to be proved, assume that the safe system $\Sigma = (S, T; F, M_0)$ is live. Suppose that $t \in T$ such that $t^\bullet = \emptyset$. Let $M_0[\tau\rangle M'$ such that t is enabled at M'. The existence of τ and M' are assured because Σ is assumed to be live. Let $M'[t\rangle M$. Then clearly $M' \geq M$ because $t^\bullet = \emptyset$. In fact, for each $s \in {}^\bullet t$, $M'(s) > M(s)$. Since Σ is live and $M \in [M_0\rangle$ we can now find a transition sequence τ_1 and a marking M_1 such that $M[\tau_1\rangle M_1$ and t is enabled at M_1. By the preceding observation, we can find a marking M_1' such that $M'[\tau_1\rangle M_1'$ where M_1' satisfies:

$$\forall s \in S : \quad M_1'(s) \ = \ M_1(s) + (M'(s) - M(s)).$$

Let $s \in {}^\bullet t$. Such an s exists because from $t^\bullet = \emptyset$ it follows that ${}^\bullet t \neq \emptyset$. Then $M_1(s) > 0$ because t is enabled at M_1. As noted earlier, from $M'[t\rangle M$, it follows that $M'(s) > M(s)$. Hence $M_1'(s) > 1$ which contradicts the safeness of Σ.   ■ 3.4

The following is an easy consequence of the previous results:

Corollary 3.5 *A necessary condition for liveness and safeness*

> If $\Sigma = (S, T; F, M_0)$ *is live and safe then* $\forall x \in S \cup T : {}^\bullet x \neq \emptyset \neq x^\bullet$.   ■ 3.5

These results hold good even for non-weakly connected systems, as can easily be checked. For weakly connected systems, the last corollary can be strengthened by relating safeness, liveness and strong connectedness to each other. On the one hand, the previous examples show that a system could be either live or safe and non-strongly connected: The system shown in Figure 1 is not strongly connected but safe (but it is not live), while the system shown in Figure 2 is not strongly connected but live (but it is not safe). On the other hand, we do have the following:

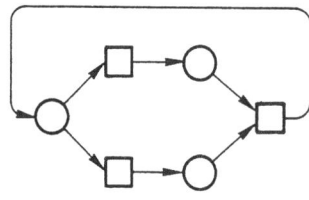 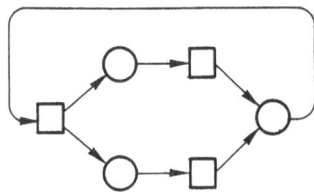

(i) No live marking   (ii) No non-zero safe marking

Figure 3: Two strongly connected nets that have no live and safe markings

Theorem 3.6 *Liveness and safeness imply strong connectedness*

  If the (finite, weakly connected) net N has a live and safe marking then it is strongly connected.

Proof: See [1].   ■ 3.6

The converse of theorem 3.6 is not true, as shown in Figure 3.

4   S-systems

Starting with this section, we will introduce and study four important classes of nets of increasing generality. The first class, called S-graphs (or S-systems if a marking is implied) admits conflict but no synchronisation (and concurrency only in the case of non-safeness or non-connectedness).

Definition 4.1 *S-graphs and S-systems*

   (i)   An S-net is a net $N = (S, T; F)$ such that $\forall t \in T: |{}^\bullet t| \leq 1 \land |t^\bullet| \leq 1$.
   (ii) An S-graph is a net $N = (S, T; F)$ such that $\forall t \in T: |{}^\bullet t| = 1 = |t^\bullet|$.
   (iii) $\Sigma = (S, T; F, M_0)$ is an S-system *iff* its underlying net is an S-graph.   ■ 4.1

Note that every strongly connected S-net is an S-graph. However, a net can be an S-graph without being strongly connected.

Since we are ultimately interested in characterising liveness and safeness and since corollary 3.5 tells us that we should then have $|{}^\bullet x| \geq 1 \leq |x^\bullet|$ for all $x \in S \cup T$, we have based the notion of an S-system on S-graphs rather than on S-nets.

In an S-system there is never any synchronisation, simply because there are no (backward) branched transitions. A transition is enabled as soon as its unique input place has a token. A safe connected S-system is sequential, i.e., at no reachable marking are there any concurrently enabled transitions.

The first result about S-systems characterises liveness.

Proposition 4.2 *Liveness in S-systems*

An S-system $\Sigma = (S, T; F, M_0)$ is live iff it is strongly connected and $M_0(S) > 0$, i.e. there is at least one token.

Proof: To prove ($\Rightarrow$), we first note that in the empty initial marking (i.e. $M_0(S) = 0$), none of the transitions of Σ is live; hence Σ is also not live since $T \neq \emptyset$ by the definition of a net. Next we show that liveness implies strong connectedness (using weak connectedness). To this end, we consider the strongly connected components of Σ, i.e. all sets $R \subseteq S \cup T$ such that $\forall x, y \in R \colon (x, y) \in F^* \wedge (y, x) \in F^*$. This defines a partitioning of $S \cup T$. We define $R_1 \sqsubseteq R_2$ (where R_1 and R_2 are two strongly connected components) if $\exists x \in R_1 \, \exists y \in R_2 \colon (x, y) \in F^*$; then $\sqsubseteq$ is a partial ordering on the set of strongly connected components of Σ. Since Σ is finite, $\sqsubseteq$ must have maximal elements. Because an F-path leads from every place of Σ into some maximal element of $\sqsubseteq$, every token of M_0 can be moved into one of the maximal elements of $\sqsubseteq$ where it will stay. Hence by the liveness of Σ, $\sqsubseteq$ cannot have any non-maximal elements. But then by the weak connectedness of Σ, $\sqsubseteq$ cannot have more than one maximal element. Hence the unique maximal element of $\sqsubseteq$ is $S \cup T$ itself, which implies that Σ is strongly connected.

To prove ($\Leftarrow$), we consider any marking $M \in [M_0\rangle$ and any transition $t \in T$. We have $M(S) = M_0(S)$; this follows directly from the definition of an S-system and the transition rule. Because $M(S) = M_0(S) > 0$, there is a place $s \in S$ which is marked under M, i.e. $M(s) > 0$. By strong connectedness, a directed F-path can be found which leads from s to t; this F-path can be used to enable t. ∎ 4.2

The next result characterises safeness in live S-systems.

Proposition 4.3 *Safeness in live S-systems*

A live S-system $\Sigma = (S, T; F, M_0)$ is safe iff $M_0(S) = 1$, i.e. there is exactly one token.

Proof: To prove ($\Rightarrow$), we use the preceding result to show that $M_0(S) > 0$ and Σ is strongly connected. If $M_0(S) > 1$ then either there is a place $s \in S$ such that $M_0(s) > 1$ (in which case the system is not safe in the initial marking), or there are places $s_1, s_2 \in S$ such that $s_1 \neq s_2$ and $M_0(s_1) + M_0(s_2) > 1$. By strong connectedness, an F-path leads from s_1 to s_2, and the system is not safe. Hence $M_0(S) = 1$.

To prove ($\Leftarrow$), it suffices to notice $M(S) = M_0(S)$ for all $M \in [M_0\rangle$. ∎ 4.3

5 T-systems

The second class of nets that will be studied is called the class of T-graphs (or T-systems if a marking is implied). T-systems admit concurrency and synchronisation, but no conflict.

Definition 5.1 *T-graphs and T-systems*

(i) A T-net is a net $N = (S, T; F)$ such that $\forall s \in S \colon |{}^\bullet s| \leq 1 \wedge |s^\bullet| \leq 1$.

(ii) A T-graph is a net $N = (S, T; F)$ such that $\forall s \in S \colon |{}^\bullet s| = 1 = |s^\bullet|$.

(iii) $\Sigma = (S, T; F, M_0)$ is a T-system *iff* its underlying net is a T-graph. ∎ 5.1

In a T-system there is never any conflict, simply because there are no (forward) branched places. A token can be taken away from a place only by its unique output transition. T-systems are very well understood. The basic references are [4,7], and [12,8] may be consulted for further reading. As for S-systems, we have based the notion of a T-system on T-graphs rather than on T-nets. This is because we are ultimately interested in characterising liveness and safeness and because of corollary 3.5.

The next result cites a well known characterisation of the liveness of T-systems.

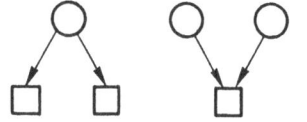 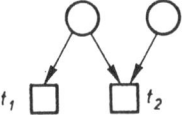

(i) Allowed   (ii) Excluded

Figure 4: Illustrating the free choice structure

Theorem 5.2 *Liveness in T-systems*

A T-system $\Sigma = (S, T; F, M_0)$ is live iff all of its simple cycles carry at least one token at M_0.

Proof: See [4], theorem 1 and [7], theorem (8S).   ∎ 5.2

The next result characterises the safeness of live T-systems.

Theorem 5.3 *Safeness in live T-systems*

A live T-system $\Sigma = (S, T; F, M_0)$ is safe iff it is covered by simple cycles which carry at most one token at M_0.

Proof: See [4], theorem 2 and [7], theorem (28S).   ∎ 5.3

As a matter of fact, strong connectedness nicely characterises the existence of a live and safe marking in a T-graph, so that we have the following converse of theorem 3.6:

Theorem 5.4 *Existence of live and safe markings*

If the T-graph N is strongly connected then it has a live and safe marking.

Proof: See [4], theorem 4, and [7], theorem (32S).   ∎ 5.4

6 Free Choice Systems

Free choice nets have been invented as a common generalisation of S-graphs and T-graphs, with the aim of retaining as much as possible of the nice theory of these classes. They allow concurrency (but only in the 'T-graph way') and conflicts (but only in the 'S-graph way'). The former is to say that if two places share a common output transition then they may not have any further output transitions, and the latter is to say that if two transitions share a common input place then they may not have any further input places. But these two properties are equivalent! They allow the structures shown in Figure 4(i) but exclude the structure shown in Figure 4(ii).

Definition 6.1 *Free choice nets and free choice systems*

(i) A free choice net is a net $N = (S, T; F)$ such that
for all $t_1, t_2 \in T, t_1 \neq t_2$: $^\bullet t_1 \cap {}^\bullet t_2 \neq \emptyset \Rightarrow |^\bullet t_1| = 1 = |^\bullet t_2|$, or, equivalently,
for all $s_1, s_2 \in S, s_1 \neq s_2$: $s_1^\bullet \cap s_2^\bullet \neq \emptyset \Rightarrow |s_1^\bullet| = 1 = |s_2^\bullet|$.

(ii) $\Sigma = (S, T; F, M_0)$ is a free choice system (abbreviated FC system) *iff* its underlying net
is a free choice net. ■ 6.1

It is clear that every T-graph, as well as every S-graph, is free choice; that is, free choice nets are
indeed a common generalisation.

An essential consequence of the free choice property is that if t_1 and t_2 share a common input
place then it can never be the case that one of them is enabled while the other is not. That is,
every marking enables either both of them or none of them. In the (excluded) case of Figure 4(ii),
by contrast, a marking can be found which enables t_1 but not t_2.

Furthermore, safe free choice systems are free of both types of confusion defined in [16].

We will take a closer look at liveness in FC systems. Perhaps the most well known result is the
following structural characterisation of liveness which is due to F.Commoner [3] and M.Hack [9].
The theorem is a generalisation of theorem 5.2. To formulate it, we first need a definition.

Definition 6.2 *Deadlocks and traps*

Let $N = (S, T; F)$ be a net.

(i) A nonempty set $D \subseteq S$ is called a deadlock[3] *iff* $^\bullet D \subseteq D^\bullet$.

(ii) A nonempty set $Q \subseteq S$ is called a trap *iff* $Q^\bullet \subseteq {}^\bullet Q$. ■ 6.2

Deadlocks and traps are both special sets of places. Their important properties are that if a
deadlock is empty under some marking then it remains empty under each successor marking, while
if a trap is marked under some marking then it remains marked under each successor marking.

It is easy to see that the intersection of two deadlocks is again a deadlock; hence it makes sense
to speak of minimal deadlocks. Similarly, the union of two traps is again a trap.

Theorem 6.3 *Liveness in FC systems*

*A free choice system $\Sigma = (S, T; F, M_0)$ is live iff every deadlock of Σ contains a trap which is
marked at M_0.*

Proof: See [9] and [15]. The direction ($\Leftarrow$) of this theorem also follows from theorem 7.5 below;
the direction ($\Rightarrow$) is harder to prove. ■ 6.3

In T-systems, (the set of places on) every simple cycle is a deadlock and a trap. Hence theorem 6.3
indeed reduces to theorem 5.2.

The next theorem characterises the safeness of live free choice systems. This theorem is due to
M.Hack [9]. We will reproduce his proof here with a few formal amendments. The theorem states
that a live and safe free choice system is covered by special kinds of S-graphs called S-components
which, moreover, carry exactly one token each. In order to understand the theorem and the proof,
we need to define S-components first of all.

---

[3]This is established but not very agreeable terminology.

Definition 6.4 *S-components*

$N_1 = (S_1, T_1; F_1)$ is called an S-component of $N = (S, T; F)$ *iff* N_1 is the subnet generated by S_1 and, in addition, $\forall t \in T_1 : |{}^\bullet t \cap S_1| \leq 1 \wedge |t^\bullet \cap S_1| \leq 1$ (where the preset and the postset are taken w.r.t. F, but it would come to the same if they were taken w.r.t. F_1).
N_1 will be called strongly connected iff it is strongly connected as an S-net by itself, that is, if there is a directed F_1-path between any two distinct elements of N_1.   ∎ 6.4

Theorem 6.5 *Safeness in live FC systems*

A live free choice system $\Sigma = (S, T; F, M_0)$ is safe iff
it is covered by strongly connected S-components which have exactly one token·each at M_0.

Proof: ($\Rightarrow$:) Using the sequence of lemmata given below, one can establish two useful claims:

Claim 1: Every minimal deadlock D of Σ is itself a trap, and it contains no other traps (lemma 6.9).

Claim 2: The subnet generated by a minimal deadlock D of Σ is a strongly connected S-graph and hence a strongly connected S-component of Σ (lemma 6.10).

Since an S-component is the subnet generated by its set of places, it suffices to show that every place of Σ is contained in an S-component of Σ that carries exactly one token. So consider any $s \in S$ and a marking $M \in [M_0\rangle$ with $M(s) = 1$; such a marking exists because Σ is assumed to be live ($M(s) \geq 1$) and safe ($M(s) \leq 1$). Define a different marking $\hat{M}$ by

$$\hat{M}(x) = \begin{cases} M(x) - 1 & \text{if } x = s \\ M(x) & \text{if } x \neq s. \end{cases}$$

Then $\Sigma' = (S, T; F, \hat{M})$ is not live because otherwise, Σ would not be safe! This follows from the observation made in the initial part of the proof of proposition 3.4. From theorem 6.3 it follows that there exists a deadlock D of Σ' (and hence of Σ) such that for every trap Q contained in D, $\hat{M}(Q) = 0$. Clearly, we can assume without loss of generality that D is a minimal deadlock. By claim 1, D is itself a trap. Hence $\hat{M}(D) = 0$. On the other hand, $\Sigma = (S, T; F, M)$ is live. Again from theorem 6.3, we then have that $M(D) > 0$. From the definition of $\hat{M}$, it is clear that $s \in D$ and that $M(D) = 1$. The required result now follows from claim 2.

($\Leftarrow$:) Let $N_1 = (S_1, T_1; F_1)$ be an S-component of Σ which carries exactly one token. Then it is easy to see that $\forall M \in [M_0\rangle : M(S_1) = 1$. Hence regardless of the liveness of Σ, the fact that Σ is covered by its set of S-components carrying one token at once implies that Σ is safe.   ∎ 6.5

For lemmata 6.6—6.10, assume globally that $\Sigma = (S, T; F, M_0)$ is a live and safe FC system.

Lemma 6.6

If D is a minimal deadlock of Σ then $\forall t \in D^\bullet : |{}^\bullet t \cap D| = 1$.

Proof: (Indirect.) Suppose $s_1, s_2 \in {}^\bullet t \cap D$ and $s_1 \neq s_2$. By the free choice property, $s_1^\bullet = \{t\} = s_2^\bullet$. Then $D \setminus \{s_1\}$ is again a deadlock, contradicting the minimality of D.   ∎ 6.6

Lemma 6.7

> Let $Q \subseteq D$ be a trap contained in a minimal deadlock D of Σ.
> Then $\forall t \in Q^\bullet$: $|{}^\bullet t \cap Q| = 1 \leq |t^\bullet \cap Q|$.

Proof: We have $|{}^\bullet t \cap Q| \leq 1$ because $Q \subseteq D$ and $|{}^\bullet t \cap D| = 1$ by lemma 6.6; also, we have ${}^\bullet t \cap Q \neq \emptyset$ because $t \in Q^\bullet$. Hence $|{}^\bullet t \cap Q| = 1$. On the other hand, $|t^\bullet \cap Q| \geq 1$ holds because of the trap property $Q^\bullet \subseteq {}^\bullet Q$. ∎ 6.7

As a corollary of this lemma, we have that $M' \in [M\rangle$ implies $M'(Q) \geq M(Q)$, where Q is as in the lemma. That is, the number of tokens on Q can never properly decrease. Lemma 6.7 also implies that $|{}^\bullet t \cap Q| \leq 1$ for any $t \in T$.

Lemma 6.8

> Let $Q \subseteq D$ be a trap contained in a minimal deadlock D of Σ.
> Then $\forall t \in {}^\bullet Q$: $|t^\bullet \cap Q| = 1$.

Proof: We have $t^\bullet \cap Q \neq \emptyset$ because of $t \in {}^\bullet Q$. We prove that $|t^\bullet \cap Q| \leq 1$ indirectly. Assume otherwise, i.e. $|t^\bullet \cap Q| \geq 2$; then $M'(Q) > M(Q)$ whenever $M[t\rangle M'$, since $|{}^\bullet t \cap Q| \leq 1$ by lemma 6.7. Hence the number of tokens on Q properly increases by the occurrence of t which, together with the facts that it can never decrease and that Q is a finite set, contradicts either safeness or liveness (of t). ∎ 6.8

Lemma 6.9

> Let D be a minimal deadlock of Σ.
> Then D is a trap and for all traps $Q \subseteq D$: $Q = D$ (i.e. D is the only trap contained in itself).

Proof: D must contain some trap Q because of theorem 6.3 and the fact that Σ is live. To prove the required result, it suffices to show that $Q = D$. To show that $Q = D$, it suffices to show that ${}^\bullet Q \subseteq Q^\bullet$, since D is minimal. Assume that $t \in {}^\bullet Q$ and suppose that ${}^\bullet t \cap Q = \emptyset$; then we have again that $M'(Q) > M(Q)$ whenever $M[t\rangle M'$, giving the same contradiction as in lemma 6.8. Hence ${}^\bullet t \cap Q \neq \emptyset$, i.e. $t \in Q^\bullet$. ∎ 6.9

Lemma 6.10

> The subnet $N_1 = (S_1, T_1; F_1)$ generated by a minimal deadlock D is a strongly connected
> S-component of Σ.

Proof: From lemma 6.9 we have ${}^\bullet D = D^\bullet$. From lemma 6.6 we have that $|{}^\bullet t \cap D| = 1$ for every $t \in {}^\bullet D$. From lemma 6.8 we have that $|t^\bullet \cap D| = 1$ for every $t \in {}^\bullet D$. Hence N_1 is an S-graph and thus an S-component of Σ. Moreover, it is easy to show that the set of places contained in each strongly connected component of N_1 is a deadlock of Σ. Since D is minimal, N_1 is a strongly connected S-component of Σ. ∎ 6.10

M.Hack, in [9], has described the behaviour of free choice systems in more detail, and in particular, he has stated a kind of dual of the last theorem, namely that a live and safe free choice system is covered not only by S-graphs but also by T-graphs. Unfortunately his proof contains mistakes which have subsequently been corrected in [10] and in [6]. As a whole, the proof is not easily accessible and we undertake here to provide a new proof in elementary terms. This proof is based on the one given in [1]; however, we will modify it by moving parts of the arguments into separate lemmata and strengthen it in a sense to be made precise.

The theorem states that a live and safe free choice system is covered by special T-graphs called (strongly connected) T-components. The notion of a T-component is defined analogously to that of an S-component:

Definition 6.11 *T-components*

$N_1 = (S_1, T_1; F_1)$ is called a T-component of $N = (S, T; F)$ *iff* N_1 is the subnet generated by T_1 and, in addition, $\forall s \in S_1 : |{}^\bullet s \cap T_1| \leq 1 \wedge |s^\bullet \cap T_1| \leq 1$ (where the preset and the postset are taken w.r.t. F, but it would come to the same if they were taken w.r.t. F_1).
N_1 will be called strongly connected iff it is strongly connected as a T-net by itself, that is, if there is a directed F_1-path between any two distinct elements of N_1. ∎ 6.11

Theorem 6.12 *Live and safe FC systems are covered by strongly connected T-components*

Let $\Sigma = (S, T; F, M_0)$ be a live and safe free choice system and let $\hat{x} \in S \cup T$. Then there is a strongly connected T-component $(S_1, T_1; F_1)$ of Σ which contains $\hat{x}$, i.e. $\hat{x} \in S_1 \cup T_1$.

The proof of 6.12 will be given after a sequence of lemmata (6.13—6.17). Lemmata 6.13—6.15 may be of general interest. We will also use them to prove lemma 6.16 and lemma 6.17 which are technical results needed in the proof of 6.12.
In what follows we let $\#_x(\tau)$ denote the number of times the symbol x appears in the sequence of symbols τ.

Lemma 6.13

Let $\Sigma = (S, T; F, M_0)$ be a system which satisfies the following conditions:

(i) $N_\Sigma = (S, T; F)$ is strongly connected;

(ii) N_Σ is not a T-graph but for every $s \in S : |s^\bullet| = 1$.

Then $\forall M \in [M_0\rangle \forall \tau \in T^*: M[\tau\rangle M \Rightarrow |\tau| = 0$
  (that is, no marking is nontrivially reproducible).

Proof: (Indirect.) Assume that $M \in [M_0\rangle$ and $\tau \in T^*$ such that $M[\tau\rangle M$ and $|\tau| > 0$.
First we shall show that every transition of Σ must appear at least once in τ.
We observe that since $|\tau| > 0$ some transition, say $\hat{t}$, must appear in τ. Now consider an arbitrary transition $t \in T$. Since N_Σ is strongly connected, we can find a path of the form

$$t_0 s_0 t_1 s_1 \ldots t_m s_m t_{m+1} \quad (m \geq 0)$$

such that $t_0 = \hat{t}$ and $t_{m+1} = t$.
We will verify that t must appear at least once in τ by induction on m. To this end it is useful to observe that for every $s \in S$,

$$\sum_{x \in {}^\bullet s} \#_x(\tau) = \sum_{y \in s^\bullet} \#_y(\tau).$$

This follows from the fact that $M[\tau\rangle M$.
So, consider the case $m = 0$. Then $t_1 = t$ and

$$\sum_{x \in {}^\bullet s_0} \#_x(\tau) = \sum_{y \in s_0^\bullet} \#_y(\tau).$$

Now $\hat{t} \in {}^\bullet s_0$ and $\#_{\hat{t}}(\tau) > 0$. On the other hand, $s_0^\bullet = \{t\}$ (since $t = t_1$ and $\forall s \in S: |s^\bullet| = 1$). Hence $\#_t(\tau) > 0$.
For the case $m > 0$, the required result ($\#_t(\tau) > 0$) follows easily from the induction hypothesis and the proof of the basis step.
Since N_Σ is not a T-graph and since N_Σ is strongly connected, there exists a place $s_0 \in S$ such that $|{}^\bullet s_0| > 1$. Let $t_0, t_0' \in {}^\bullet s_0$ with $t_0 \neq t_0'$. We can find a simple cycle in N_Σ —because it is strongly connected— that passes through t_0 and s_0, as shown in Figure 5.
Now $s_0^\bullet = \{t_1\}$ and we know that $\#_{t_0}(\tau) > 0$ and $\#_{t_0'}(\tau) > 0$. Therefore $\#_{t_1}(\tau) > \#_{t_0}(\tau)$. But $s_1^\bullet = \{t_2\}$. Hence $\#_{t_2}(\tau) \geq \#_{t_1}(\tau) > \#_{t_0}(\tau)$. Applying this argument m times along the chosen simple cycle, we arrive at the contradiction $\#_{t_0}(\tau) > \#_{t_0}(\tau)$. ∎ 6.13

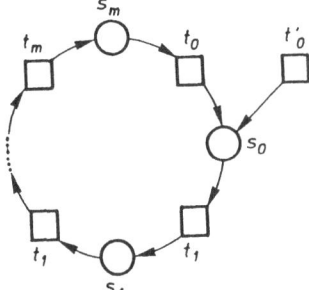

Figure 5: A simple cycle through t_0 and s_0

Lemma 6.14

Let $\Sigma = (S, T; F, M_0)$ be a T-system such that $N_\Sigma = (S, T; F)$ is strongly connected. Suppose $M \in [M_0\rangle$ and $\tau \in T^*$ such that $M[\tau\rangle M$ and $|\tau| > 0$. Then $\forall t \in T$: $\#_t(\tau) > 0$.

Proof: Follows easily from the first part of the proof of the previous lemma (in which premise 6.13(ii) has not been used). ∎ 6.14

For proving the next result it will be convenient to adopt a piece of terminology. Let $\Sigma = (S, T; F, M_0)$ be a system, $M \in [M_0\rangle$ and $\emptyset \neq T' \subseteq T$. Then $\tau \in T^*$ is called a minimal T'-enabling sequence at M *iff* the following conditions are satisfied:

(i)   For some marking $M' \in [M_0\rangle$, $M[\tau\rangle M'$.

(ii) Some transition in T' is enabled at M'.

(iii) τ is a sequence of minimal length which satisfies (i) and (ii).

Lemma 6.15

Let $\Sigma = (S, T; F, M_0)$ be an FC system which is live and safe and let $N' = (S', T'; F')$ be a subnet of $N_\Sigma = (S, T; F)$ which satisfies the following conditions:

(i) N' is strongly connected (with respect to F');

(ii) For every $s \in S'$: $|s^\bullet \cap T'| = 1$.

Then there is a marking $M \in [M_0\rangle$ and a sequence $\tau \in T'^*$ such that $M[\tau\rangle M$ and $|\tau| > 0$.

Proof: Set $M^0 = M_0$.

Now assume inductively that M^{i-1} ($i \geq 1$) has been defined. Then M^i is defined as follows. Let τ^i be a minimal T'-enabling sequence at M^{i-1}, so that $M^{i-1}[\tau^i\rangle M'_i$ and $x^i \in T'$ is enabled at M'_i. The existence of τ^i is assured by the liveness of Σ. Let $M'_i[x^i\rangle M^i$, so that

$$M^{i-1}[\tau^i x^i\rangle M^i.$$

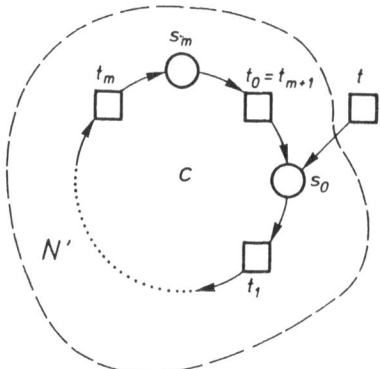

Figure 6: Illustrating the proof that T^{in}-transitions do not occur in τ_1

Since Σ is safe, in the sequence of markings $M^0, M^1, \ldots$ thus defined, we can find two markings M^k and M^{k+l} with $l > 0$ such that $M^k = M^{k+l}$ (in fact, $2^{|S|}$ is an upper bound for l); thus we have:

$$M^k \left[\tau_1 x_1\right) M^{k+1} \left[\tau_2 x_2\right) M^{k+2} \ldots M^{k+l-1} \left[\tau_l x_l\right) M^{k+l},$$

such that $l > 0$, τ_i is a minimal T'-enabling sequence at M^{k+i-1}, $x_i \in T'$ for $1 \le i \le l$, and $M^k = M^{k+l}$.

Define $T^{out} = \{t \in T \backslash T' \mid {}^\bullet t \cap S' \neq \emptyset\}$.
We claim that

$$\boxed{1} \qquad \forall t \in T^{out} \ \forall i \in \{1, 2, \ldots, l\}: \ \#_t(\tau_i) = 0.$$

To see this, suppose that some $t \in T^{out}$ appears in τ_1. Then τ_1 can be expressed as $\tau_1 = \tau_1' t \tau_1''$. Let $M^k[\tau_1')M$. Then for some $s \in {}^\bullet t \cap S'$, $M(s) > 0$. But $s \in S'$ implies that for some $t' \in T'$, $s^\bullet \cap T' = \{t'\}$. Since $t \notin T'$, we have $t \neq t'$ and hence, by the free choice property, that ${}^\bullet t' = {}^\bullet t = \{s\}$. Hence t' is also enabled at M which implies that τ_1' is itself a minimal T'-enabling sequence at M^k, and this contradicts the definition of τ_1. A similar argument is valid for $\tau_2, \ldots, \tau_l$, establishing the claim that no $t \in T^{out}$ occurs in any of the τ_i, for $i \in \{1, \ldots, l\}$.

Next define $T^{in} = \{t \in T \backslash T' \mid t^\bullet \cap S' \neq \emptyset\}$.
We now claim that

$$\boxed{2} \qquad \forall t \in T^{in} \ \forall i \in \{1, 2, \ldots, l\}: \ \#_t(\tau_i) = 0.$$

To see this, once again assume that some $t \in T^{in}$ appears in τ_1. Then τ_1 can be expressed as $\tau_1 = \tau_1' t \tau_1''$. Let $M^k[\tau_1' t)M$ and $s_0 \in t^\bullet \cap S'$. Consider a simple cycle of the form

$$c = t_0 \, s_0 \, t_1 \, s_1 \ldots t_m \, s_m \, t_{m+1}$$

in N' as shown in Figure 6 (c exists because N' is strongly connected). For any $M' \in [M_0\rangle$, define $M'(c)$ as

$$M'(c) = M'(\{s_0, \ldots, s_m\}) = M'(s_0) + \ldots + M'(s_m).$$

Now we already know that ${}^\bullet t \cap S' = \emptyset$ because no transition in T^{out} appears in τ_1. Moreover, we have that for every $s \in S'$, $|s^\bullet \cap T'| = 1$. Consequently $M(c) > M^k(c)$ and $M^{k+1}(c) \geq M(c)$. It is easy to verify, by repeated applications of this argument, that

$$M^{k+l}(c) \geq M^{k+l-1}(c) \geq \ldots \geq M^{k+1}(c).$$

We now have the contradiction

$$M^k(c) = M^{k+l}(c) \geq M(c) > M^k(c).$$

Thus, indeed, no transition in T^{in} can appear in τ_1 and, by similar reasoning, in $\tau_2, \tau_3, \ldots, \tau_l$. Furthermore, the minimality of $\tau_1, \ldots, \tau_l$ immediately implies that the τ_i cannot contain any T'-transition, i.e.:

$$\boxed{3} \qquad \forall t \in T' \; \forall i \in \{1,2,\ldots,l\}: \; \#_t(\tau_i) = 0.$$

Further, we claim that

$$\boxed{4} \qquad \forall t \in T \backslash (T' \cup T^{out} \cup T^{in}) \; \forall i \in \{1,\ldots,l\}: \; \#_t(\tau_i) = 0.$$

To see this, recall from $\boxed{1} - \boxed{3}$ that τ_i must be of the form $t_1 \ldots t_n$, with $n \geq 0$ and $t_j \in T \backslash (T' \cup T^{out} \cup T^{in})$ for $1 \leq j \leq n$. But then we have ${}^\bullet t_j \cap {}^\bullet x_i = \emptyset$, which, together with fact 2.14(a) and fact 2.11 implies that M^{k+i-1} enables x_i for all $i \in \{1,\ldots,l\}$. The minimality of τ_i, once again, implies that $n = 0$; hence, finally, we may deduce that $|\tau_i| = 0$ for all $i \in \{1,2,\ldots,l\}$.
Thus, by $\boxed{1} - \boxed{4}$, we have:

$$M^k \, [x_1\rangle \, M^{k+1} \, [x_2\rangle \, M^{k+2} \, \ldots \, M^{k+l-1} \, [x_l\rangle \, M^{k+l}$$

with $x_i \in T'$ for $i \in \{1,\ldots,l\}$, $l > 0$ and $M^k = M^{k+l}$. Hence with $M = M^k$ and $\tau = x_1 x_2 \ldots x_l$, the claim of the lemma is established. ∎ 6.15

We may contrapose lemma 6.15 and lemma 6.13 to get the following.

Lemma 6.16

Let $\Sigma = (S, T; F, M_0)$ be an FC system and $N' = (S', T'; F')$ be a subnet of $N_\Sigma = (S, T; F)$ which satisfies the following conditions:

(i) N' is strongly connected (with respect to F');

(ii) N' is not a T-graph but for every $s \in S'$: $|s^\bullet \cap T'| = 1$.

Then Σ cannot be both live and safe.

Proof: (Indirect.) Assume that Σ is live and safe.
Then according to lemma 6.15,

$$\exists M \in [M_0\rangle \; \exists \tau \in T'^*: \; M[\tau\rangle M \text{ and } |\tau| > 0.$$

This contradicts lemma 6.13, applied to $\Sigma' = (S', T'; F', M_0|_{S'})$. ∎ 6.16

Finally, in combination with lemma 6.14, lemma 6.15 yields the following:

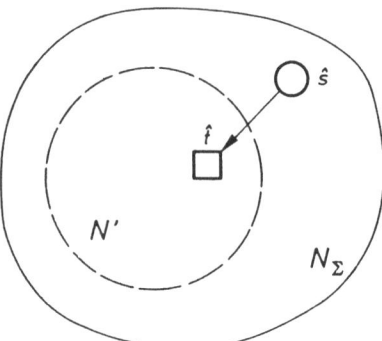

Figure 7: Illustrating the proof that N' is generated by its transitions

Lemma 6.17

Let $\Sigma = (S, T; F, M_0)$ be an FC system and $N' = (S', T'; F')$ be a subnet of $N_\Sigma = (S, T; F)$ which satisfies the following conditions:

(i) N' is a strongly connected (with respect to F') T-graph;

(ii) $(T')^\bullet \subseteq S'$ but $^\bullet(T') \nsubseteq S'$ (where pre-sets and post-sets are taken with respect to F).

Then Σ cannot be both live and safe.

Proof: (*Indirect.*) Assume that Σ is live and safe.

Let $\hat{t} \in T'$ and $\hat{s} \in S \setminus S'$ such that $(\hat{s}, \hat{t}) \in F$; such $\hat{s}, \hat{t}$ exist by $^\bullet(T') \nsubseteq S'$ (see Figure 7).
By lemma 6.15, there are $M \in [M_0\rangle$ and $\tau \in T'^*$, $|\tau| > 0$, such that $M[\tau\rangle M[\tau\rangle M$.
By lemma 6.14, $\#_{\hat{t}}(\tau) > 0$.
Hence the sequence $\tau\tau$ is of the form

$$\tau\tau \ = \ \tau' \hat{t} \tau'' \hat{t} \tau'''.$$

Let $M', M'', M''' \in [M_0\rangle$ be such that $M[\tau'\rangle M'$, $M'[\hat{t}\rangle M''$ and $M''[\tau''\rangle M'''$. Then $M'(\hat{s}) = 1$ because M' enables $\hat{t}$, $M''(\hat{s}) = 0$ because $\hat{s} \notin \hat{t}^\bullet$ (by the assumptions $(T')^\bullet \subseteq S'$ and $\hat{s} \notin S'$), and $M'''(\hat{s}) = 0$ because τ'' contains only transitions from T' and $\hat{s} \notin (T')^\bullet$. However, $M'''(\hat{s}) = 0$ contradicts the fact that M''' enables $\hat{t}$. ■ 6.17

Proof of theorem 6.12:

Assume now that $\Sigma = (S, T; F, M_0)$ is a live and safe FC system.
Since a T-component is generated by its transitions, it suffices to show that every transition of Σ is contained in a strongly connected T-component.
To this end fix $\hat{t} \in T$ and fix a simple cycle

$$c \ = \ t_0 s_0 t_1 s_1 \ldots t_m s_m t_{m+1}$$

with $\hat{t} = t_0 = t_{m+1}$. Since N_Σ is strongly connected by theorem 3.6, the existence of c is assured. Define $N^0 = (S^0, T^0; F^0)$ as follows:

$$
\begin{aligned}
S^0 \ &= \ \{s_0, s_1, \ldots, s_m\} \\
T^0 \ &= \ \{t_0, t_1, \ldots, t_m\} \\
F^0 \ &= \ F \cap ((S^0 \times T^0) \cup (T^0 \times S^0)).
\end{aligned}
$$

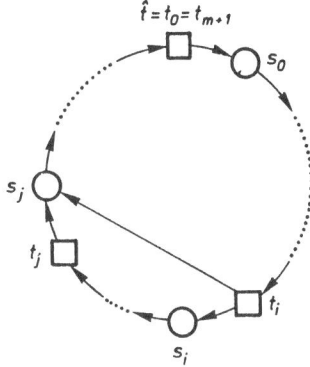

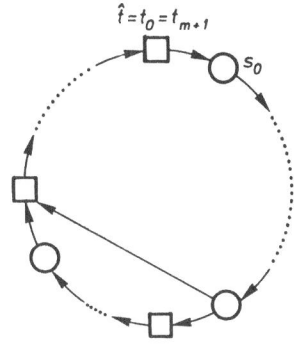

Case 1   Case 2

Figure 8: Illustrating the proof that N^0 is a T-graph

We shall verify that N^0 is a strongly connected T-graph. Thus, we wish to rule out the two cases shown in Figure 8. Case 2 is ruled out by the free choice property and the fact that c is a simple cycle. Case 1 is then ruled out by lemma 6.16.

Now assume inductively that $N^i = (S^i, T^i; F^i)$ $(i \geq 0)$ is a strongly connected T-graph which is a subnet of N_Σ and which contains $\hat{t}$.

If $(T^i)^\bullet$ (in N_Σ) $\subseteq$ S^i then by lemma 6.17, ${}^\bullet(T^i)$ (in N_Σ) $\subseteq$ S^i. Then clearly, N^i is a T-component of Σ and the proof is done.

If, on the other hand, $(T^i)^\bullet$ (in N_Σ) $\not\subseteq$ S^i then the construction can be continued as follows. Let $t \in T^i$ and $(t,s) \in F$ such that $s \notin S^i$. Let $t_0 s_0 t_1 s_1 \ldots t_m s_m t_{m+1}$ be a simple cycle in N_Σ such that $t = t_0 = t_{m+1}$ and $s_0 = s$ (see Figure 9). Put for $0 \leq j \leq 2m$,

$$ x_j \;=\; \begin{cases} s_{(j-1)/2} & \text{if } j \text{ is odd} \\ t_{j/2} & \text{if } j \text{ is even} \end{cases} $$

Put $x_{2m+1} = t_0 = t$. Now let k be the least integer in $\{0, \ldots, 2m+1\}$ such that $x_k \in S^i \cup T^i$ and $x_{k-1} \notin S^i \cup T^i$; k must exist because $x_1 = s_0 \notin S^i \cup T^i$ and $x_{2m+1} = t \in S^i \cup T^i$. With the help of lemma 6.16, it is easy to verify that k is an even number, that is, that $x_k \in T^i$. Now N^i will be extended to N^{i+1} as follows:

$$
\begin{aligned}
S^{i+1} &= S^i \cup (\{x_0, x_1, \ldots, x_k\} \cap S) \\
T^{i+1} &= T^i \cup (\{x_0, x_1, \ldots, x_k\} \cap T) \\
F^{i+1} &= F \cap ((S^{i+1} \times T^{i+1}) \cup (T^{i+1} \times S^{i+1})).
\end{aligned}
$$

As before for N^0, it can be verified that N^{i+1} is a strongly connected T-graph containing $\hat{t}$ and that $S^i \subset S^{i+1}$ and $T^i \subseteq T^{i+1}$.

We can repeat the argument for N^{i+1}. The required result now follows from the finiteness of N_Σ. ∎ 6.12

Remark 6.18

The above proof shows that the place s_0 which lies outside of N^i can be connected back to N^i at *any* output transition of s_0, to yield N^{i+1}. In [1], the weaker statement has been proved

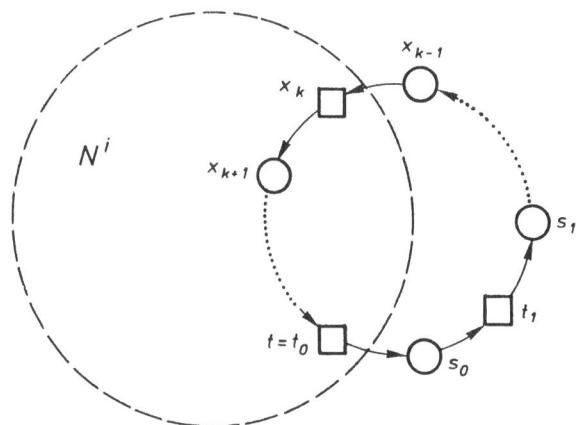

Figure 9: Illustrating the construction of a T-component containing $\hat{t}$

that there *exists* an output transition of s_0 which is suitable to extend N^i.

Another distinction between the two proofs is the fact that the concept of a maximal marking introduced in [17] and used in [1] has been replaced here by the concept of the reproduction of a marking which has played a rôle in lemma 6.15. More precisely, the reproducible markings M in lemma 6.15 are exactly the maximal markings in the sense of [17] and [1].

Lemma 6.15 resembles a key result proved in [17] (namely lemma 3.2, in the enumeration of [17]), but the latter has a slightly different premise.

Lemma 6.16 implies the result proved independently in [5].   ∎ 6.18

7   Asymmetric Choice Systems

F.Commoner [3] has shown that part of theorem 6.3 holds for a class of systems which properly includes the free choice systems. He has called this the class of 'simple' systems. We will rechristen it to be called asymmetric choice systems:

Definition 7.1 *Asymmetric choice systems*

$\Sigma = (S, T; F, M_0)$ is called asymmetric choice (AC) system *iff*
$\forall s_1, s_2 \in S: s_1^\bullet \cap s_2^\bullet \neq \emptyset \Rightarrow (s_1^\bullet \subseteq s_2^\bullet \vee s_2^\bullet \subseteq s_1^\bullet)$.   ∎ 7.1

Figure 10(i) shows some allowed structures while Figure 10(ii) shows the typical structure which is excluded by the asymmetric choice property. It is immediate that every free choice net is also asymmetric choice.

In the terminology of [16], the asymmetric choice property allows asymmetric confusion but disallows symetric confusion.

A useful fact about AC systems is that the conflict relation is transitive. To see that this need not always be true, consider the typical non-AC net of Figure 10(ii) with the marking shown in Figure 11. In this marking, both t_1 and t_2 are in conflict with t_0, but they are concurrently enabled, i.e. not in conflict with each other.

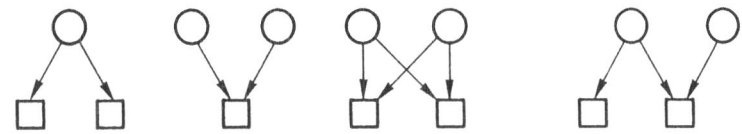

(i) Allowed

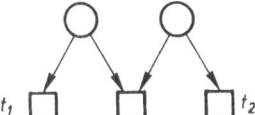

(ii) Excluded

Figure 10: Illustrating the AC structure

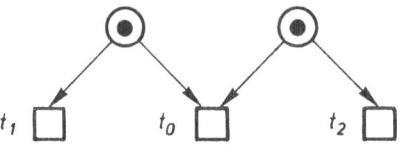

Figure 11: A non-transitive conflict situation

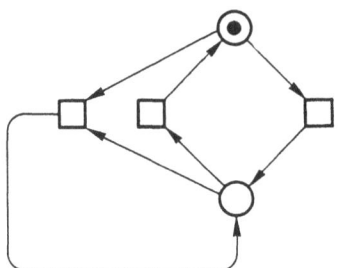

Figure 12: A system which is place-live but not live

Fact 7.2 *Conflict is transitive in AC systems*

Let $\Sigma = (S, T; F, M_0)$ be an AC system, let $M \in [M_0\rangle$ be a marking and let t_0, t_1, t_2 in T be such that both t_1 and t_2 are in conflict with t_0 at M.
Then t_1 is in conflict with t_2 at M.

Proof: By fact 2.13($\Rightarrow$), we may pick $p \in {}^\bullet t_1 \cap {}^\bullet t_0$ and $q \in {}^\bullet t_0 \cap {}^\bullet t_2$ such that $M(p) = 1$ and $M(q) = 1$. By the AC property, either $p^\bullet \subseteq q^\bullet$ or $q^\bullet \subseteq p^\bullet$. If $p^\bullet \subseteq q^\bullet$ then $q \in {}^\bullet t_1 \cap {}^\bullet t_2$ and hence, by fact 2.13($\Leftarrow$), t_1 and t_2 are in conflict at M. If $q^\bullet \subseteq p^\bullet$ then $p \in {}^\bullet t_1 \cap {}^\bullet t_2$, yielding the same conclusion. ∎ 7.2

We will state another typical property of AC nets.

Definition 7.3 *Place-liveness*

$\Sigma = (S, T; F, M_0)$ is place-live *iff* $\forall M_1 \in [M_0\rangle \; \forall s \in S \; \exists M \in [M_1\rangle : M(s) > 0$. ∎ 7.3

Place-liveness means that no place can become and remain unmarked. Figure 12 shows the typical example of a (non-AC) system which is place-live but not live.

Theorem 7.4 *Equivalence of liveness and place-liveness*

Let $\Sigma = (S, T; F, M_0)$ be an AC system. Then Σ is live iff Σ is place-live.

Proof: See [1]. ∎ 7.4

A very similar result is lemma 4.3 of [11] which states that every dead transition in an AC system has an input place which remains unmarked.

The last result can be used to show that the structural criterion for liveness in free choice systems given by theorem 6.3 is sufficient for liveness in AC systems.

Theorem 7.5 *Liveness in AC systems*

Let $\Sigma = (S, T; F, M_0)$ be an asymmetric choice system such that every deadlock of Σ contains a trap which is marked at M_0. Then Σ is live.

Proof: (Sketch.) Assume that some transition t_0 is not live in Σ. Then by the proof of the preceding theorem, t_0 has some input place s_0 which is and remains unmarked. This input place can be used in a backtrack argument to construct a deadlock which is empty at the marking in which t_0 is dead. Since a trap can never become empty, this deadlock cannot have contained a marked trap in the initial marking. ■ 7.5

8 Conclusion

We have considered four classes of Petri nets and their behavioural properties:

S-systems which admit no synchronisation and —in the safe connected case— no concurrency (section 4).

T-systems which admit no conflicts, i.e., are deterministic (section 5).

Free choice systems which admit no confusion (section 6).

Asymmetric choice systems which admit only one kind of confusion (section 7).

Our goal was to investigate, for each of these classes, the extent to which the behaviour of a marked net is determined by the structure of the underlying unmarked net. We have concentrated on characterising liveness and characterising safeness in the presence of liveness. We have collected such characterisations from the literature, and we have also added some small propositions — particularly regarding the general case (section 3)— to complete the picture.

We have attempted to unify the theory and to simplify it by giving elementary proofs for some of the important basic (and often used) facts of the theory of free choice systems: the coverability of live and safe free choice systems by 1-token strongly connected S-components and by strongly connected T-components. We hope that our efforts in trying to make these proofs understandable —by dividing them into sequences of lemmata— will turn out to be of use to other researchers in proving other interesting facts. The interested reader may be referred to [17] which describes more of the theory of free choice systems, and to [1] where a list of further references can be found.

References

[1] E.Best: Structure Theory of Petri Nets: the Free Choice Hiatus. Advanced Course on Petri Nets, Bad Honnef (September 1986), to appear in Springer Lecture Notes in Computer Science.

[2] E.Best and C.Fernández: Notations and Terminology on Petri Net Theory. Arbeitspapiere der GMD No.195 (1986).

[3] F.Commoner: Deadlocks in Petri Nets. Report, Applied Data Inc., CA-7206-2311 (1972).

[4] F.Commoner, A.W Holt, S.Even and A.Pnueli: Marked Directed Graphs. JCSS Vol.5, 511-523 (1971).

[5] J.Desel: A Structural Property of Free Choice Systems. Petri Net Newsletters No.25, 16-20 (December 1986).

[6] K.Döpp: Zum Hack'schen Wohlformungssatz für Free-Choice-Petrinetze. EIK 19/1-2, 3-15 (1983).

[7] H.J.Genrich and K.Lautenbach: Synchronisationsgraphen. Acta Informatica Vol.2, 143-161 (1973).

[8] H.J.Genrich and P.S.Thiagarajan: A Theory of Bipolar Synchronisation Schemes. TCS Vol.30, 241-318 (1984).

[9] M.Hack: Analysis of Production Schemata by Petri Nets. TR-94, MIT-MAC (1972).

[10] M.Hack: Corrections to MAC-TR-94. Computation Structure Notes 17, MIT-MAC (1974).

[11] M.Jantzen and R.Valk: Formal Properties of Place/Transition-Nets. Springer Lecture Notes in Computer Science Vol.84, 165-212 (1980).

[12] R.Johnsonbaugh and T.Murata: Additional Methods for Reduction and Expansion of Marked Graphs. IEEE Tr. on Circuits and Systems, Vol.28/10, 1009-1014 (1981).

[13] C.A.Petri: Concepts of Net Theory. Mathematical Foundations of Computer Science, Proceedings of Symposium and Summer School, High Tatras, 137-148 (1973).

[14] C.A.Petri: Interpretations of Net Theory. GMD-ISF Report 75-07 (1975).

[15] W.Reisig: Petri Nets — an Introduction. Springer EATCS Monographs (1985).

[16] G.Rozenberg and P.S.Thiagarajan: Petri Nets: Basic Notions, Structure and Behaviour. Springer Lecture Notes in Computer Science Vol.224, 585-668 (1986).

[17] P.S.Thiagarajan and K.Voss: A Fresh Look at Free Choice Nets. Information and Control, Vol.61/2, 85-113 (1984).

A model of cooperation and its specification with nets

Burkhardt H.-J., Eckert H., Prinoth R., Raubold E.
Gesellschaft für Mathematik und Datenverarbeitung mbH
Rheinstraße 75, D - 6100 Darmstadt

1. Abstract and Introduction

Cooperation is an essential element in human social life. More and more, traditional forms of human cooperation are supported or even substituted by cooperating computer systems. Nevertheless, the realization of cooperating computer systems as well as their embedding in existing organisational structures is to a far extent a not understood process.

To move the realization of cooperating computer systems towards a well understood engineering discipline, a constructive approach is urgently needed.

This approach has to cover the design of cooperating systems - comprising the global modelling and formal specification of their cooperation - as well as their realization - comprising the derivation of implementations and tests from global specifications of cooperation.

Modelling of cooperation and the method for its formal description determine the practicability and usefulness of such an approach. In this paper, some basic concepts of the modelling of cooperation and its formal specification with a tailored form of PETRI nets are outlined, which have proven within the work of the GMD Institut for System Technology to be well suited.

Acknowledgements: Our thanks are due to our colleagues Dr. B. Baumgarten and Dr. P. Ochsenschläger for helpful discussions, and to D. Kunert and U. Vollhardt for typing the manuscript.

2. Basic concepts of modelling cooperation

In this paper, we deal with communication involving technical systems. Communication is understood as a form of cooperation which is based on the exchange of information.

To become able to support human communication by communicating computer systems, one has to model communication at a level of abstraction where the differences between human communication and computer communication are no longer visible.

To this behalf and at this level of abstraction we regard first the field of human communication. Here, we find natural or legal persons cooperating. Persons are uniquely identified, and have at any point in time a unique location. They are distinguished by consciousness, freedom of will, and the ability to contact other persons and to negotiate cooperation. Furthermore, they own resources.
Each cooperation is directed towards one objective. The range of objectives for cooperations is wide. The objective of a cooperation may be fixed in detail at its very beginning or may be derived - as a partial result of the cooperation itself - from a rough intention. At the level of abstraction regarded, the objec-tive of a cooperation is expressible as a consistent change of consciousness of the persons concerned, eventually accompanied by a consistent alteration of the distri-bution of their resources.

Every cooperation can be structured into five phases:
In the first phase, persons get into touch with other persons they want to coope-rate with using either an acustic channel in cases when they are face to face, or other communication media when they are remote to each other.
In the second phase, persons negotiate the objective of an intended cooperation. The essential precondition to perform this phase is a common initial understanding. This common initial understanding is formed by
- a set of objectives in common
- mechanisms to select a specific objective out of this set for the
   intended cooperation; these mechanisms must guarantee freedom of decision
   (to cooperate or not) for each contacted person
- a texture of roles which is associated with each cooperation objective. Each
   role in this texture will be played by a person. This texture of roles
   determines the transitions from the set of initial states of cooperating
   persons to the set of their final states.
The second phase is necessary in cases, where persons can select the objective of a cooperation and can decide whether they want to cooperate or not.
In the third phase, persons cooperate by playing the roles declared in the texture, which is associated with the agreed upon objective. Roles comprise the production and consumption of information as the operations changing consciousness as well as the sending and receiving of information. The interrelation of roles in the texture is called a protocol.

In the fourth phase, the special roles persons have played during their cooperation are deactivated in a consistent manner.
In the fifth phase, persons release their contacts.

Persons exchange information using an information-transport-system (ITS), e. g. notepaper, the air of an acustic channel, the electrical current of an electric channel. As a consequence, a protocol always comprises conventions concerning the usage of an ITS.
An ITS tranfers signals. Signals are defined as changes of physical states, to which specific meanings are assigned. These meanings are known only to the users of the ITS, and not to the ITS itself.
An ITS is - in contrast to the acting persons - passive, which means that processing of information is completely performed by persons. Especially, receivers of information are not selected by an ITS.

At the same level of abstraction we now consider cooperating technical systems. Here, we find again the concepts of persons, roles, set of roles, consciousness, freedom of choice etc. and their interrelationships - named different only.

The concept 'person' corresponds to a processing unit, the concept 'role' to a program, a set of roles to a program library. A person playing a role corresponds to a process. Consciousness of a person is equivalent to the states of a process and a person's freedom of choice is equivalent to the alternative behaviour of a process, not triggered by external events.

From these considerations we conclude that - at the level of abstraction chosen - human beings and computer systems can be regarded as different "extensions" of abstract communicating instances.

Note: By no means we want to postulate equality between human beings and machines. Nevertheless, we believe that - to a certain degree - observable interactions in human communication and in computer communication can be described by a set of common communication patterns.

The modelling approach presented subsequently
- allows on the one hand, to describe cooperation objectives separately from the methods of realizing them
and
- guarantees, on the other hand, their realizability in a constructive manner.

To model cooperation, we define community to be the set of all instances (persons and/or machines) which are capable to communicate. These instances are called Transport Service users (TS - user).

To model the five phases described above we introduce
- a permanent basis role, which is active in all phases and which is the only role activated in the first and the fifth phase,
- a temporary basis role, which is activated/deactivated by the permanent basis role as result of the first/fifth phase and is active inbetween,
and
- temporary specific roles, which are activated/deactivated by the permanent basis role as result of negotiations in the second and fourth phase and are active in the third phase.

For the purpose of contacting each other and transmitting messages, the members of the community (TS-users) use the services of a Transport Service Provider (TS-provider). This TS-provider is a special ITS.
The TS-provider is a system distributed in space (e. g. telephone network) with discrete interaction points called Transport Service Access Points (TSAPs). TSAPs are the only interaction points at which users may access the services provided.
The TS-provider does only know the members of the community with respect to their basis roles.
The Transport Service Provider treats all Transport Service Users in a uniform way (in their temporary roles) as potential sources and sinks of flows of data. He does not know which roles members are able to play in addition to its basis roles and which roles are played in an actual running cooperation.

Fig. 2.1 is intended to visualize the members of the community c, d, b and m at a point in time where no connection between them is established, i. e. it shows the corresponding TS-users in their permanent basis roles.
Inside the boxes representing users c, d, b and m in their permanent basis role temporary specific roles, these instances are able to play, are delineated.

Fig. 2.2 shows the Transport Service Users c and d at a point in time, where they have successfully passed phase 1 and are now in phase 2, which we call a connection, i. e. it shows the corresponding TS-user instances c and d playing beside their permanent basis roles their temporary basis roles, represented by the small boxes named by c.1 and d.3.

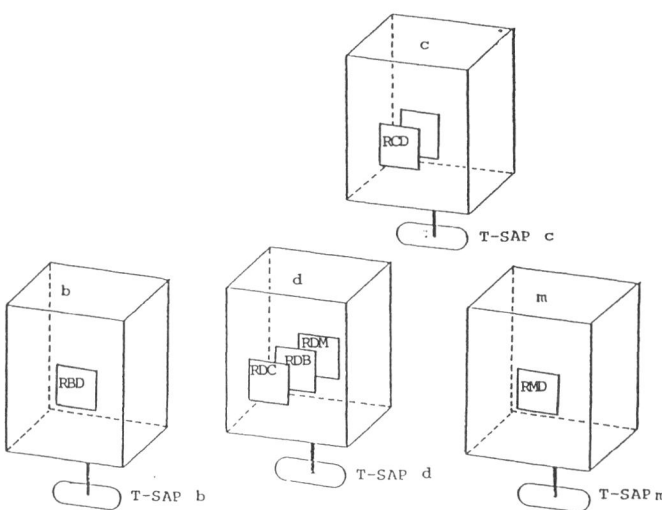

Fig. 2.1 Permanent TS-user instances c, d, b, m

1 (and 3) are local connection endpoint identifiers at TSAP c (and TSAP d), by
means of which TS user c (TS user d) can access the transport connection globally
identified by the pair (c.1, d.3).

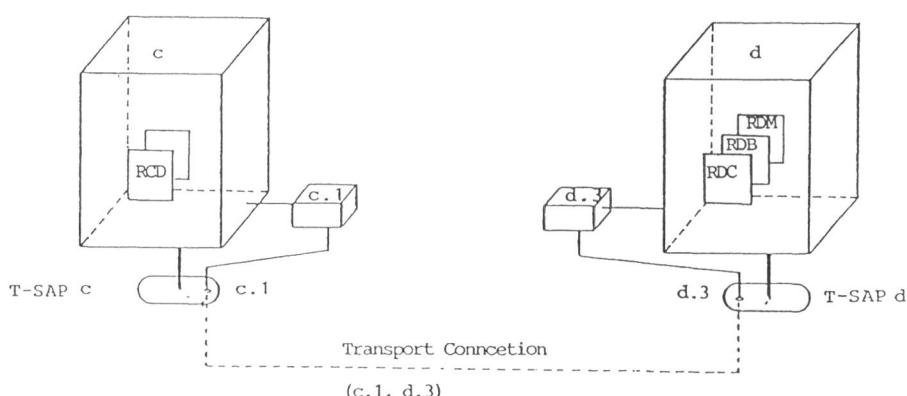

Fig. 2.2: A Transport connection between Transport Service Users c and d modelled
as interrelationship among permanent and temporary TS-User Instances

From this point of view, a Transport-Connection as shown in fig. 2.2 can be
interpreted as a correlation of the two collections of cooperation capabilities or
roles inherent in the corresponding TS-users.

The pairing of specific roles is performed by TS-users in their permanent basis roles following interaction between them in their temporary basis roles. This is visualised in fig. 2.3, which shows Transport Service user c and d at a point in time, where they have successfully passed phase 2 and are now in phase 3, which we call a session, i. e. it shows the corresponding TS-users playing beside their permanent and temporary basis roles a selected pair of temporary cooperation goal specific roles /3/.

In fig.2.3 playing of role RCD by c and playing of the corresponding role RDC by d is illustrated by the small boxes on top of boxes c.1 and d.3.

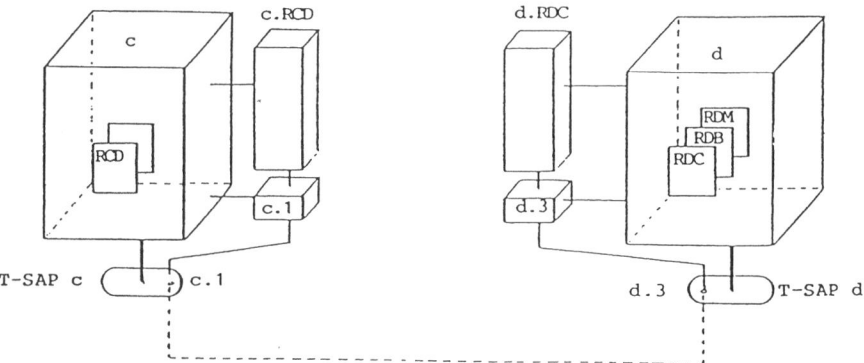

Fig. 2.3 A session (c.RCD, d.RDC) on top of a Transport Connection (c.1, d.3)

The effect of executing phase four and five should be evident to the reader.

Remark: To model n-party applications composed of two-party applications, it is necessary to add (to the permanent basis role) the capability of coordinating two party applications /1,2/.

3. A brief introduction to product nets

A global protocol specification defines cooperation among autonomous systems, by identifying all roles and their interrelationships. It is a requirement to specify a single role as well as the composition of all roles in a unique and formal fashion. This is a prerequisite for mathematical analysis and facilitates the formal transformation between different levels of abstraction. This requirement is fulfilled by Product nets since roles as well as their composition can formally be represented by them.

3.1. Definition of IE-nets as basic class of product nets

Let a net be a triple N = (S,T,F) such that
(a) S ∩ T = ∅   (b) F ⊆ (SxT) ∪ (TxS).
The elements of S, T, F are called places, transitions, arcs, respectively.
In the design and analysis of services and protocols we found it useful to consider
- as special arcs distinct from those in F - inhibitor arcs $(s,t)_\varepsilon I ⊆$ SxT and
erase arcs $(s,t)_\varepsilon E ⊆ S \times T$.
IE-nets (I for Inhibitor and E for Erase) are therefore
quintuples N = (S,T,F,I,E) such that, in addition to (a) and (b):
(c) I ∪ E ⊆ SxT   (d) F ∩ I = F ∩ E = E ∩ I = ∅.
Graphically, we represent places by circles or triangles, transitions by boxes,
arcs by arrows. Inhibitor arcs we draw as —//▸, erase arcs as —▸▸.

Let for all t $_\varepsilon$ T   °t : = {s_εS | $(s,t)_\varepsilon$F ∪ I ∪ E} and t° : = {s_εS | $(t,s)_\varepsilon$F}.
°t (t°) is called the set of input (output) places of t.

A marking of N is a mapping M from S into $\mathbb{N}_0$ (the positive integers including zero).

Under a marking M a transition t is enabled, if its F-input- places are positively
marked and its I-input-places are empty. Transition t fires by destroying one token
in each F-input-place and all tokens in each E-input- place creating one at each
output-place.

3.2. Extension of IE-nets to full product nets

If the 'contents' of a place under a marking is considered to be not merely a
constant but a collection of structured tokens, which may be distinguished by transitions, the modelling power of IE-nets increases considerably. IE-nets with these
additional features have been introduced under the name product nets in /5/.
Roughly speaking, a product net is a IE-net combined with arc labels and transition
inscriptions, which are as formally constructed as this is done in case of terms,
predicates, atoms and formulas in the first-order logic /4/. In first-order logic,
an interpretation of a formula is a well known concept. By introducing adequate
restrictions with respect to the syntax of arc labels and transition inscriptions,
the formal semantics of nets is derived from the interpretation of arc labels and
transition inscriptions.

A product net N is (informally) defined by the rules (1) to (4):
(1) N is a IE-net (S,T,F,I,E)

(2) For each place s∈S, the object-type D_s is a finite carthesian product of non empty sets $\mathcal{M}_i^{\Delta}$:

$$D_s \subseteq \mathcal{M}_1^{\Delta} \times \ldots \times \mathcal{M}_{n_{\Delta}}^{\Delta} \, , \; D_s \neq \emptyset.$$

With $n_s \geq 1$ the dimension of the object-type of place s is denoted.

(3) For all arcs $f \in F \cup I \cup E$, an arc label $K(f)$ is introduced (as a formal sum of n_s-tuples of terms):

$$K(f) := \sum_{j \in I_f} w_j(f) \langle t_1^j(f), \ldots, t_{n_{\Delta}}^j(f) \rangle, \; w_j(f) \in \mathbb{N}$$

I_f is a finite set (of indices). $t_i^j(f)$ are terms (built as usual in first-order logic /4/).

For each $f \in I \cup E$ $K(f)$ is restricted to a tuple of symbols for constants and/or variables:

$$K(f) := w_1(f) \langle t_1^{\Delta}(f), \ldots, t_{n_{\Delta}}^{\Delta}(f) \rangle, \; w_1(f) = 1$$

and $t_k^{\Delta}(f)$ is a symbol for a constant or a variable.

All symbols for variable occuring in terms at output arcs of transition t have to occur in terms at F-input arcs of t.

(4) For each transition $t \in T$ a formula Δ_t is introduced (built as usual in first-order logic /4/). All symbols for variables occuring in Δ_t have to occur in labels (in terms) of F-input arcs of t.

All variables occuring in labels of F-input arcs are called <u>bound variables.</u> All other variables are called <u>free variables.</u> It follows from the definiton above that all variables occuring in formulas and in labels (within terms) of output arcs are bound variables.

Example 3.1:

$D_{s1} = \mathbb{N}_0$

$D_{s2} = \mathbb{N}_0$

$D_{s3} = \mathbb{N}_0^2$

$D_{s4} = \mathbb{N}_0^2$

$K(s_1, t): = \langle X \rangle$

$K(s_2, t): = \langle Y \rangle$

$K(s_3, t): = \langle X, Z \rangle$

$K(t, s_3): = \langle Y, Y+1 \rangle$

$K(s_4, t): = \langle Y, W \rangle$

$\Delta_t : = X = Y$

X and Y are bound variables, Z and W are free variables.

(end of example)

A <u>marking</u> in a product net is a mapping M : $\bigcup\limits_{s \in S} (D_s \times \{s\}) \to \mathbb{N}_o$.

Let $s_i \varepsilon S$. Then $M(x,s_i) = m \varepsilon \mathbb{N}_o$ denotes the multiplicity of the object $x \varepsilon D_{si}$ of marking M. For $n > 0$ this fact is represented by the notation $n \,{}^{\langle}x^{\rangle} \varepsilon\, M(s_i)$.

<u>Example 3.2:</u>

A marking M of the product net
of example 3.1 is given as shown
in the picture:

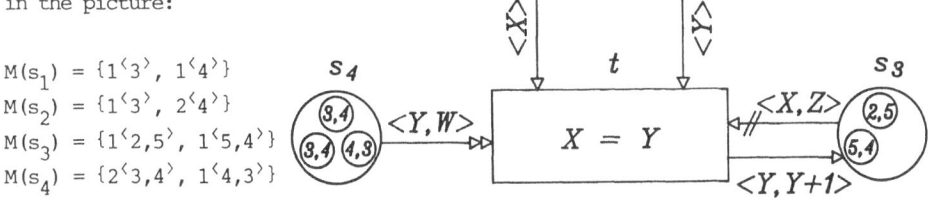

$M(s_1) = \{1^{\langle 3 \rangle}, 1^{\langle 4 \rangle}\}$
$M(s_2) = \{1^{\langle 3 \rangle}, 2^{\langle 4 \rangle}\}$
$M(s_3) = \{1^{\langle 2,5 \rangle}, 1^{\langle 5,4 \rangle}\}$
$M(s_4) = \{2^{\langle 3,4 \rangle}, 1^{\langle 4,3 \rangle}\}$

(end of example)

For transition $t \varepsilon T$ an <u>interpretation</u> δ_t consists of an assignment of "values" to each bound variable occuring in the formula $\mathbf{A}_t$ and/or in arc labels K(f) of arcs adjacent to t ('binding' of the variable). Starting with this assignment, values are computed for each term and for the formula $\mathbf{A}_t$ (for $\mathbf{A}_t$ this computation results in the value true or false).

<u>Example 3.3:</u>

An interpretation of the product net of example 3.1 is given.
The following assignment is an interpretation δ_t:

$\delta_t(1) = 1 \qquad \delta_t(X) = \delta_t(Y) = 3$
$\delta_t(Y+1) = 4 \qquad \delta_t(true) = true \qquad \delta_t(false) = false$
$\delta_t(X=Y) = \delta_t(\delta_t(X) = \delta_t(Y)) = \delta_t(3=3) = \delta_t(true) = true$
(end of example)

With regard to each interpretation δ_t of transition $t \varepsilon T$ of a product net with $\delta_t(\mathbf{A}_t) = true$ a marking $M^{\delta t}$ is derived, which is called the <u>characteristic (or threshold) marking</u> with respect to δ_t. Roughly spoken, this marking $M^{\delta t}$ consists of:

- objects in all places $s \varepsilon°t$ where $(s,t) \varepsilon F$, representing the
  interpretation δ_t at the level of markings.
- no objects for all other places of the net.
Example 3.4 will illustrate this relation.

Example 3.4:

The threshold marking $M^{\delta t}$ with respect to interpretation δ_t in example 3.3 is
given as:

$$M^{\delta t}(s_1) = M^{\delta t}(s_2) = \{1\langle 3\rangle\} \qquad M^{\delta t}(s_3) = M^{\delta t}(s_4) = 0$$

(There is an object $\langle 3 \rangle$ with multiplicity 1 in place s_1
and the same is true for place s_2).   (end of example)

Transition t is enabled with respect to a marking M and an interpretation δ_t
(with $\delta_t(\Delta_t) = $ true), iff
(a) For each place $s \varepsilon S : M^{\delta t}(s) \le M(s)$
(b) For each place $s \varepsilon°t$ with $f = (s,t) \varepsilon I$:
    No object with multiplicity > 0 is found in M(s) which is derived
    from interpretation δ_t (where arbitrary values may be considered
    for free variables).

Note: For $s \varepsilon°t$ and $(s,t) \varepsilon E$ M(s) has no effect with respect to the
      enabling condition.

Enabled transitions may fire. If M is the marking of a product net and if the
enabled transition $t \varepsilon T$ (with respect to M and δ_t) fires, a new marking M' is
obtained. $M^{\delta t}$ is subtracted from M for input places of t and for all output
places of t the interpreted arc weight $\delta_t(K(t,s))$ /5/ is added to M
respectively. For places $s \varepsilon°t$ with $(s,t) \varepsilon E$ all objects of M(s) are subtracted
which 'coincide' with δ_t for those components, which are defined by constants
or bound variables:
For each place $s \varepsilon°t \cup t°$ with no arc $(s,t) \varepsilon E$ leading from s to t:

- $\forall s \varepsilon°t \setminus t°: \quad M'(s) = M(s) - M^{\delta t}(s)$

- $\forall s \varepsilon t°: \qquad M'(s) = M(s) - M^{\delta t}(s) + \delta_t(K(t,s))$

For each place $s \varepsilon°t \cup t°$ with an arc $(s,t) \varepsilon E$ leading from s to t:

- Let $K(f) = K(s,t) = \langle t_1^1(f), \ldots, t_{n_0}^1(f) \rangle$ be the corresponding arc label.
  Let $E_s^{\delta t}$ be a marking with $E_s^{\delta t}(s) \le M(s)$ consisting of

-- all objects $(x_1,\ldots,x_n)$ of marking M found in place s with

$x_j := \delta_t(t_j^{\wedge}(f))$ for each component x_j which consists in a constant or a bound variable

-- no objects for all other places of the product net

Then $\forall s\epsilon°t \setminus t° : M'(s) = M(s) - E_s^{\delta t}(s)$

$\quad\forall s\epsilon°t \cap t° : M'(s) = M(s) - E_s^{\delta t}(s) + \delta_t(K(t,s))$

Else

- $M'(s) = M(s)$

Example 3.5:

In the example above transition t is enabled and may fire. M' is obtained as shown below:

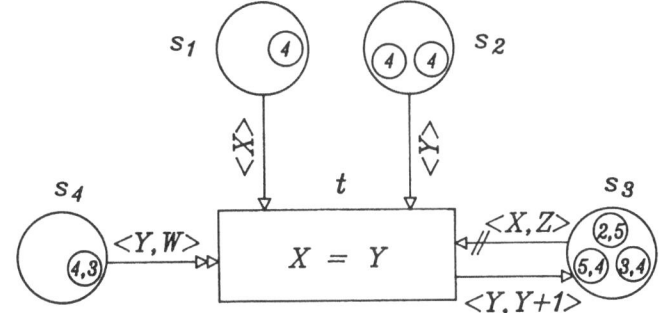

(end of example)

4. The mapping of modelling concepts onto nets

4.1 Net representations of the overall structure

Now let us consider the relations between the concepts pointed out in section 2 and those introduced in section 3. To illustrate these relations, a sketch of a refinement of the left part of figure 2.3 is given in fig. 4.1.)

Roles are represented by (sub-)product nets. Roles are activated by assigning tokens - identifying members of the community - to special places of the product net.

Performing a role is represented by the dynamics of a marked (sub-)product net.

Members in their permanent roles are represented by (sub-)product nets, which are permanently marked. Members in their temporary roles are represented by (sub-)-product nets, which are only temporarily marked.

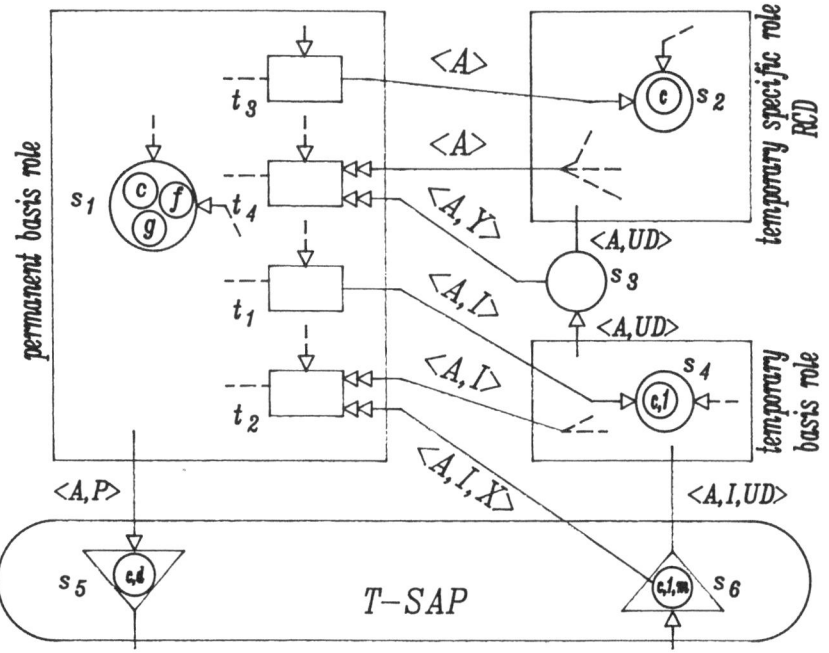

Legend to fig. 4.1

$t_3(t_1)$: activation of role RCD (of the temporary basis role)

$t_2(t_4)$: deactivation of the temporary basis role (of role RCD)

Places representing members of the community in their initial state
        of the permanent basis role   (s_1)
        of the temporary spec. role RCD   (s_2)
        of the temporary basis role   (s_4)

s_3:   coupling place between the respective temporary roles (e.g. for the transport of user data (UD) between instances of roles)

s_5, s_6: places belonging to TSAP's, which uniquely identify the members of the community by their respective address (e.g. token (c, d) in place s_5 may be interpreted as an actual T-CONNECT request from member c to member d).

Fig. 4.1 Illustration of the relations among modelling constructs and their formalization with Product Nets

Members playing identical roles, are distinguished by the marking, thus avoiding duplication of topology.

The set of temporary roles, a specific member is capable to play within an actual cooperation, is a set of unmarked (sub-) product nets. A net in this set gets its initial marking from the net representing the member in its permanent basis role. Playing their roles, members exchange messages.

Deactivation of roles is modelled by deleting the actual marking from the corresponding (sub-)product nets by the product nets representing corresponding permanent basis roles. This deletion is performed unconditionally, i. e. the deletion
of the marking depends on the actual state of the basis role, only. After finishing
the cooperation, all cooperation partners are in consistent final states;
examples can be found in /1,2/.

Based on the same concepts of permanent and temporary roles, the TS-provider is
modelled as (sub-) product net, too. Thereby, each T-SAP is regarded as a firm and
unique relationship between a permanent TS-user basis role and a permanent
TS-provider basis role.

This relationship has the function of a signalling path, which TS-user and TS-provider use to negotiate the establishment of connections. The existence of the latter
is restricted by the available resources (T-SAP endpoints), which are administered
by the provider. In the provider-net, resources are modelled by tokens, residing
in appertaining places.

The coupling of roles defining the behaviour of TS-users as well as of the TS-provider is modelled by places, connecting the (sub-)product nets.

A T-SAP is defined as a set of places modelling the data types (TS-primitives) by
means of which TS-user and TS-provider are communicating. A token in such a place
indicates that a TS-primitive has been initialized, but has not yet been terminated.

4.2 Superimposing a global specification with implementation-specifications

A global (and implementation independent) protocol specification identifies all
protocol functions (or roles) and establishes their logical relationships. It is,
therefore, characterized by a high degree of concurrency.

In contrast, an implementation specification - derived from the former - determines
how a single system performs the roles which were assigned to it.

The fundamental concept for the derivation of implementation specifications from
global protocol specifications is the superposition of the latter with an implementation structure (which may be different for different implementations).

If product nets are used for protocol specification, a partition of the transitions
of the net will provide for this purpose /8/.

From a net point of view, a partition cuts up the complete set of transitions of a product net specification into blocks of transitions, where each transition belongs to exactly one block. From our modelling point of view, each block is composed of the transitions of one or more sub-product nets, where each sub-product net represents a role which can be assigned to one member of the community.

To support the implementation process further, we introduce systems as collections of TS-user roles and the corresponding TS-provider parts. A system represents a common location for a subset of members of the community operating on the same common resources. Whereas a net is a rather mathematical model with a more or less implicite mapping onto elements of the real world, a net together with a partition establishes a formal relationship between the model and the real world:
- All transitions of the blocks of a partition which belong to a system
  will be mapped onto operations of the system.
- All places s of the net specification (together with D_s) which are accessed
  by transitions of a single block, only, will be mapped onto block-internal
  data structures.
- All places s of the net (together with D_s) which are accessed by transitions of
  different blocks belonging to the same system, are system internal interface places, which are mapped onto system internal data structures. System
  internal interface places represent interaction mechanisms within a system
  which are assumed to be secure.
- All places s of the net (together with D_s) which are accessed by transitions
  of blocks belonging to different systems are system external interface places
  which are mapped onto global data structures common between these systems.
  System external interface places represent channels between systems, which
  are not necessarily assumed to be secure.

The reader will easily find, that the 'implementation' of inhibitor arcs $(s,t) \varepsilon I$, where s is a system-internal place, is performed in a straightforward manner (e.g. by an IF-statement). This would not be true for inhibitor arcs $(s,t) \varepsilon I$, where s is a system-external interface place, i.e. it depends on the system cuts chosen whether or not an inhibitor arc is an implementable construct.

4.3 Selected problems in the field of distributed system's design

Bearing in mind the cooperation model of section 2, we select subsequently two problems pertinent to the design of cooperating systems, to illustrate how aspects of the real world will influence the modelling discipline.

4.3.1 The problem of global conflicts
_____

Let A, B and C be names of three TS-users belonging to different systems. A will
send a message to B or to C. In doing so A has to address the receiver. This is an
A-internal action. The TS-provider gets its information from the calling user
(user A in this case) as a pair of T-SAP adresses (calling user adress, called user
adress). If transition t_1 (t_2) models the send operation within TS-user A to
TS-user B (C) and transition t_3 (t_4) models the receive operation within TS-user B
(C), then fig. 4.2 is a sketch for a correct design of this problem:

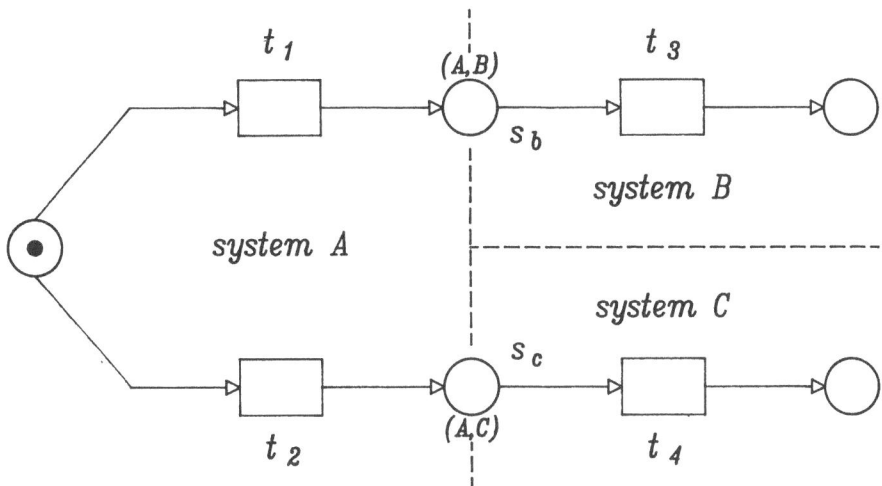

Fig. 4.2 Exchange of messages via external interface places representing
          a passive TS-provider

The semantics of fig. 4.2 is given as follows: The firing of t_1 (t_2) indicates that
B (C) is the receiver of the message. The places s_b and s_c rather model pairs of
T-SAPs than a single T-SAP.

Let us now look to fig. 4.3:

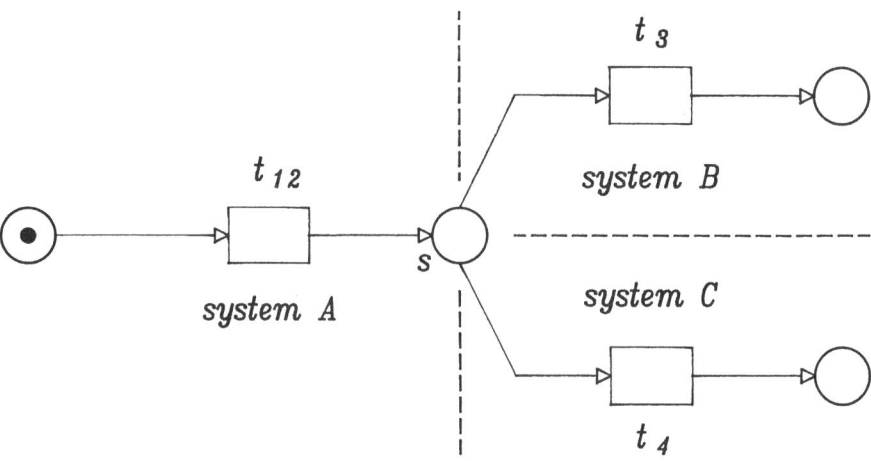

Fig.: 4.3 Illustration of a global conflict at an external interface place

At first sight, fig. 4.3 seems to be an adequate model of this problem too, but a token in place s has no addressing information: so in refinement of place s towards a TS-provider this provider would have to decide whether B or C is the receiver of the message, in contradiction to our assumption that a real provider is a passive medium.

With the concepts outlined above we are able to decide formally whether an implementation specification can be realized or not:

If place s of a net is an external interface place, (with respect to a given partition) then all transitions at output-arcs of s have to belong to one system of the partition only, otherwise there exists a global conflict. Implementation specifications, which contain global conflicts, are incomplete, since they do not define the negotiation mechanisms between systems necessary to resolve these conflicts.

4.3.2 The problem of modelling channels

The discussion above leads to system-external interface places , which represent one (or more) pairs of T-SAP's where the sending T-SAP's may be different and the receiving T-SAP's have to be identical. If we want to represent by an inter- face place one T-SAP instead of a pair of T-SAPs, then we have to refine inter- face places as shown in fig. 4.4:

111

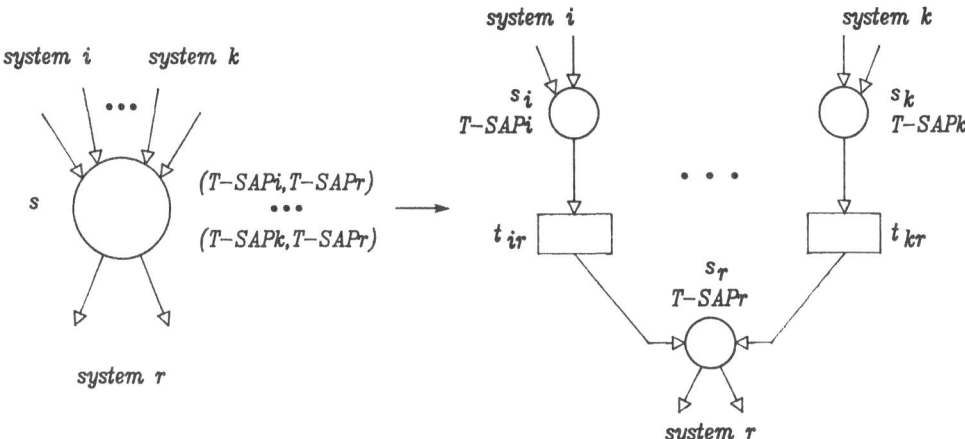

Fig. 4.4 Refinement of external interface places

$t_{ir}, \ldots, t_{kr}$ represent 'distributed transitions', belonging to the sending system as well as to the receiving system. The transitions represent the service of the transport of user data from sender to receiver without any failure.

Speaking in terms of an implementation, places $s_i, \ldots, s_k$ may be interpreted as send buffers and s_r as receive buffer. Send and receive buffers are locally controlled resources. A channel couples send and receive buffer without any feedback. At this level of abstraction a channel is modelled by the structure given in fig. 4.5:

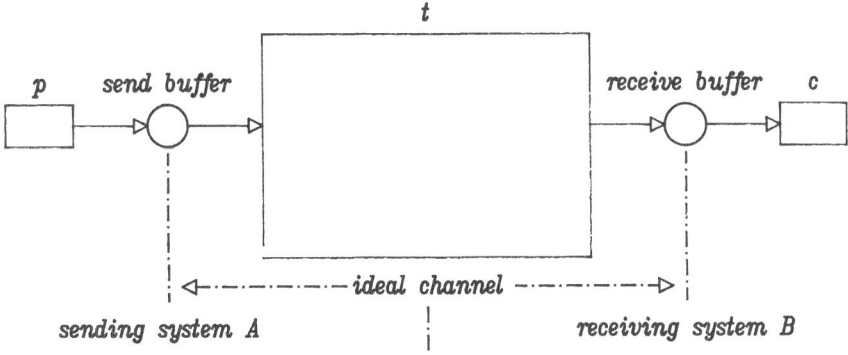

Fig. 4.5 Introduction of the notion of an "ideal channel"

112

Transition p produces messages (in system A) and transition c consumes messages
(in system B). The notion 'ideal channel' in fig. 4.5 emphasizes the reliability of
the data transfer (error free and without losses or duplicates). Usually, the
access of a producer of messages to a channel is controlled by 'ready'-signals.
However, this type of control does not prevent the receiver from becoming
congested. Fig. 4.6 models this structure:

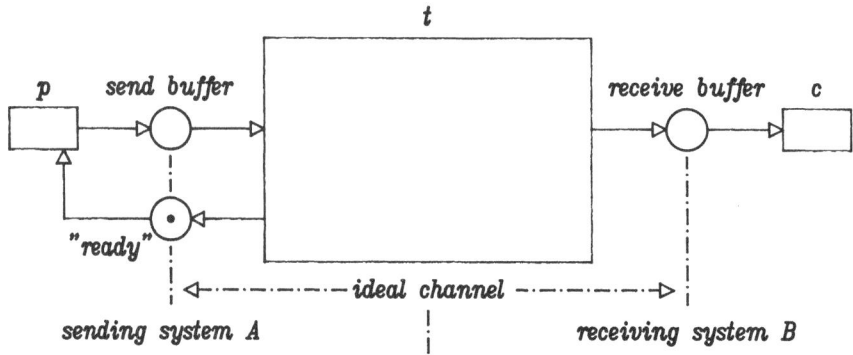

Fig. 4.6 Refinement of a channel by introduction of flow control on the
send buffer

As the firing of t does not consume time, the model ignores the aspect of reality
where a send operation has finished but the associated receive operation has not
yet started. To model this aspect we refine transition t as given in fig. 4.7:

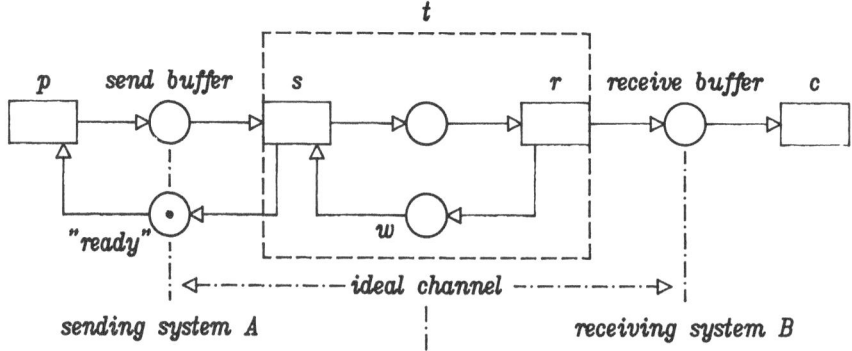

Fig. 4.7 Further refinement of a channel by separation of send and receive
operations

The number of tokens in place w is assumed to be ≥ 1.

Transition s(r) models the send-(receive-) operation. Of course, this model does not prevent the receiver from becoming congested, too. Now the reader might argue that the introduction of a place k at the receiver side with arcs (c, k) and (k, r) and a token in k would be a solution of the problem.

But we assumed the channel to be feedback free (this assumption holds true for all 'technical' channels), so this structure can not be realized. If the receiver wishes to control the rate of information, a second channel of the same type has to be introduced (in reverse direction). Now flow control can be established and as a consequence, congestion of the receive buffer can be avoided.

From the discussion above the reader should conclude that the 'realization' of a place with bounded capacity is quite different depending on the fact whether such a place is a system external interface place or not. In the first case two channels have to be introduced and a flow control mechanism has to be established upon these channels (including production and consumption-operations of the partners, not only send- and receive-operations!). In the latter case any programming language can be used without any difficulty to implement this structure.

5. Conclusion

It was our intention with this paper to point to the trivial, but nevertheless widely ignored, understanding that - for an engineer - a model obtains its 'natural' semantics from its relation to the real world, of which it is an abstraction. In this sense, a formal description technique is usually not associated with a natural semantics, even if a formal semantics is defined for it.

Therefore, using formal description techniques within a constructive approach makes only sense, if the formally described objects are declared in their natural semantics within the framework of a model.

To be regarded as well suited, a formal technique has to allow to express all aspects of the natural semantics in a 'natural' way.

In this sense, we found PETRI nets /6,7/, highly appropriate, - especially in their tailored form of product nets - for formally specifying cooperation.

References

/1/ B. Baumgarten, H.J. Burkhardt, P. Ochsenschläger, R. Prinoth: The Signing
    of a Contract - a Tree-Structured Application Modelled with Petri Net
    Building Blocks, in: Advances in Petri Nets 1985, G. Rozenberg (ed.),
    Springer LNCS 222, 1986

/2/ B. Baumgarten, P. Ochsenschläger, R. Prinoth: Synchronization in Tree-
    Structured Transactions - a Case Study, to appear in: Protocol Specifica-
    tion, Testing, and Verification, VI, North-Holland

/3/ H.J. Burkhardt, H. Eckert, R. Prinoth: Modelling of OSI-Communication
    Services and Protocols using Predicate/Transition Nets, in: Protocol
    Specification, Testing, and Verification, IV, Y. Yemini et al.(ed.),
    North-Holland, 1985

/4/ Chang, C.-L.; Lee, R.: Symbolic Logic and Mechanical Theorem Proving
    Academic Press, 1973

/5/ H. Eckert, R. Prinoth: Produktnetze - Definition eines PROSIT-Beschrei-
    bungsmittels, Arbeitspapiere der GMD Nr. 92, 1984

/6/ C.A. Petri: Concepts of Net Theory, in: Mathematical Foundations of
    Computer Science - Proceedings, High Tatras, 1973

/7/ C.A. Petri: Communication Disciplines, in: Computing System Design -
    Proceedings, Newcastle upon Tyne, 1977

/8/ R. Prinoth: Specification and Verification of Communication Protocols,
    in: Proceedings of the Workshop 'Introduction of High Level Protocol
    Standards for Open Systems Interconnection', BNI and AFNOR, Paris, 1983

The Communication Disciplines of CHAOS

F. De Cindio, G. De Michelis, C. Simone
Dipartimento di Scienze dell'Informazione
Università di Milano

Abstract

Although Petri's Communication Disciplines have little influenced the scientific community till now, they offer a powerful theoretical framework for dealing with the pragmatics of human communication.

In particular theoretically founded Communication Disciplines can be effectively embodied in Office Computer-Based Tools improving the flexibility of the communication protocols and their adaptability to changes in the office structure.

The paper discusses this claim by presenting CHAOS (Commitment Handling Active Office System), an 'intelligent' system supporting the coordination of activities inside the office, and its Communication Disciplines.

1. Introduction

Carl Adam Petri wrote his papers on Communication Disciplines in 1977 [Pet77a], [Pet77b] and [Pet79]. Up to now they have remained widely unknown, or at least little studied and considered, although in the same period Petri Nets became very popular in the Computer Science community.

It is our opinion that there is no specific reason for this negligence, and that, on the contrary, Communication Disciplines offer an original and effective framework for dealing with the problems of system design in the area of computer support of cooperative work. In this paper we aim to show how Communication Disciplines have contributed to the development of a software package whose purpose is to coordinate office work.

Human communication pragmatics can be the framework within which co-operation between human beings can be fully understood in all its main features. Within this framework, many efforts have been made recently to analyse organizational systems (mainly offices) and, as a result, new computer-based tools have been proposed (see, for example, the Proceedings of the CSCW (Computer-Supported Cooperative Work) Conference recently held in Austin, USA [CSCW86]).

Two are the main points of view from which human communication pragmatics has been characterized within offices.

1) By focusing on the mutual commitments office members make coordinating their work.

From one hand, when the attention is focused on the way mutual commitments are made, the Speech Acts Theory developed by Austin [Aus62] and Searle [Sea69], [Sea79] can be the basis for commitments analysis. In his well known work, Searle gives a taxonomy of speech acts (Assertives, Directives, Commissives, Declarations, Expressives) from the point of view of their illocutionary point, i.e., in terms of the (commitment-) relationships they create between the two involed interlocutors. For example, "John, can you give me your pencil ?" cannot be characterized by its propositional content. Its illocutionary point, in that a directive, characterizes the relationship between the speaker and the hearer it creates. Flores [FL81] and Winograd [WF86], [Win86] apply Searle's Speech Acts Theory to office work and design a package for office conversations handling: the Coordinator [AT84].

From the other one, when the attention is focused on the way coordination integrates roles and activities, the basis is the analysis of task assignment. Holt [Hol79], [Hol86] and Cashman [CH80], [SC84] characterize the coordination as it is performed by task distribution and integration.

In this tradition, our main endeavour is to analyse organizational systems in terms of the rules their members follow by mutually inter-acting in the linguistic domain. The characterization of offices as linguistic games (see Wittgenstein [Wit53]) has been the main result of our research [DDS85], [DDS86a]. The game rules of an office define the possible speech acts each member, depending on her/his role, can do at any moment within the conversations s/he is involved in. In turn the speech acts performed by the office members induce (in a perturbation-compensation mode) a change in the office rules. This basic circularity in office life allows one to consider the office as a particular type of closed, autonomous system [MV80], i.e., a living social system, organizationally closed, which maintains its identity by changing its structure (its rules) in order to compensate for the perturbations (the speech acts of its members) its structure is able to distinguish [DeM86].

2) By focusing on the communication disciplines the office members follow in their inter-action.

The above mentioned pioneeristic work by Carl Adam Petri moving the observer of organizational systems from outside to inside the system opened a new perspective in the analysis of human communication. As Petri claims in [Pet77b] communication disciplines are disciplines in both its senses: "that of schema of the same science" and that of "restraint of behavior". While in the second sense any organization disciplines communication, it is necessary a research effort in order to develop the theoretical framework of the communication disciplines and to transform the verbal description of any theoretically founded piece of a communication discipline into a mathematical theory. In this direction Richter and Voss [RV86] use net models to integrate information and resources management within the office; Holt [Hol86] bases on the Role/Center model his Coordination Technology, an office system supporting the coordination of tasks within the organizations; through the definition of GAMERU [DDS87] and the design of the CHAOS system [DDS86] we are developing a system-theoretical approach to organization behaviour.

The two points of view give different, although strictly related, images of the organizational systems and of the inter-actions between their members: in short, the first one emphasizes communication as a sense-making activity, while the second one emphasizes communication as a socially disciplined activity.

It is our opinion that they offer complementary, mutually influencing insights on organizational behaviour. In fact, the rules of organizational games are necessarily implemented by using a well-defined set of rules within each communication disciplines, and therefore the rules determine how the organizational game rules are actually followed in any organizational system.

Changes in the organizational rules induce changes in the rules within the communication disciplines since the former become ineffective in supporting the organization behaviour; e.g., the growth of an organizational system can induce more formal communication protocols, as the rather informal previous protocols become more and more ineffective. In turn, changes in the rules within the communication disciplines induce changes in the organization rules since office members experiment how new comunication patterns can be effectively used; e.g., the introduction of an efficient and well designed PABX can induce new communicative behaviours of the office members, and through them, the creation of new organizational rules.

As this last example shows, the rules within the communication disciplines of an organizational system can be (partially) embodied in a set of tools: e.g., some rules of the addressing discipline can be embodied in the Telephon Index, in the Yellow Pages, and in a personal address book together with the Public Telephone System; some rules of the synchronization discipline can be (partially) embodied in the telephon system with automatic secretaries.

It is important to emphasize that the embodiment of the rules of the communication disciplines in a set of tools is effective as well as those communication disciplines are based on well defined theories. The embodiment in a tool of a purely heuristic rule introduces rigidities in the organization from two points of view: on the one hand, the communicative behaviors of the organization members are restrained to the most frequent observed protocols; on the other one, the rules of the communication disciplines embodied in the tools can be modified only after the heuristic identification of the emerging communication protocols. On the contrary, as far as the communication disciplines have a theoratical foundation the disciplines themselves can be embodied in the tools avoiding the two above mentioned drawbacks.

The information techonologies offer to communication disciplines a wide range of possibilities, largely un-exploited till now, of being embodied. The recent efforts in the area of computer support of cooperative work (see [DJM86], [MGL86], [Win86] in the already mentioned [CSCW86]) show some interesting steps in this direction. Furthermore they make clear that the focus in the pragmatics of human communication is fundamental to the design of effective communication disciplines rules, and to their embodiment in computer systems.

Since 1980 we have been working in the area of organizational systems analysis and design, along the above proposed lines.

First, we developed the GAMERU language for the analysis and design of the organizational processes [DDS87]. GAMERU is based on a class of Petri Nets, namely Superposed Automata Nets [DDPS82], and supports the representation of the Game Rules of organizational systems, i.e., of the rules that define the domain of the possible speech acts te members of the organization can do within the conversations they play with their colleagues.

Secondly, we proposed [DDS85], [DeM86] a characterization of organizational systems as closed autonomous systems, i.e., as systems modifying their structure (their rules) to compensate for the commitments their members make within them. The CHAOS (Commitments Handling Active Office System) project currently under development at the Dipartimento di Scienze dell'Informazione of the Universita' degli Studi di Milano by a research group composed by the authors together with Raffaela Vassallo and Annamaria Zanaboni, is aimed at the development of a tool that supports office coordination [DDS86].

CHAOS is designed as a model of the above mentioned basic circularity of organizational systems. Its rules have been specified by means of GAMERU models. CHAOS design has been for the authors the occasion to investigate practically the relations between office game rules and the rules within communication disciplines.

The paper is organized as follows: after a short presentation of the CHAOS project, the communication disciplines it embodies are presented. A conclusive section discusses the missing disciplines and the problems their absence leaves open.

2. CHAOS: Commitment Handling Active Office Systems

CHAOS is the name of a family of office support systems, presently under development, aimed at supporting the network of conversations occuring inside an office, more in general inside an organization, and, in particular, aimed at supporting the network of commitments mutually undertaken by the office members through these conversations.

In this section first a short presentation of the approach is given (2.1) for introducing the basic terminology. Then the main modeling (2.2) and architectural (2.3) characteristics of the implemented package are briefly summarized. A more detailed presentation of the package can be found in [DDS86].

2.1 Conversing in the Offices

While performing their work people spend the most of their time communicating, more precisely developing conversations with people both inside and outside their office in order to make commitments for an effective coordination of the activities. Coordinating means opening new conversations to deal with the breakdowns affecting both the activities the office members are involved in and the conversations that are going on in the office. From this point of view, the office is a network of conversations for future possibilities and/or committed activities.

A conversation between A and B is a sequence of related utterances. The utterances within a conversation cannot be characterized in semantic terms (what is the meaning of a request? is it true or false?), but can be classified from the pragmatic point of view in some basic categories of Speech Acts [Aus62], [Sea69] on the basis of their illocutionary point [Sea79]: namely, directives (e.g. Requests, Acceptance or Rejecting of a Promise), commissives (e.g. Promises, Counter-offers, Acceptance or Rejecting of a Request), declaratives (e.g. Withdrawing of a Commitment, Declaration of new fileds of activitiy, Delegation of responsibility).

CHAOS conversations occur between an Actor, i.e., the person who opens the conversation, and a Partner, her/his interlocutor. Other members can be involved in some subconversation of the main one if, due to the declared responsibilities, they are asked to approve the commitment between the Actor and the Partner. There are two main types of Conversation occurring in any office.

The first is the Conversation for Action, characterized by the (possibly unsuccessful) definition of a commitment for doing an action. There are, for instance, conversations opened by a Cooperation request, where the actor asks the partner for some activity. The partner has, in general, various reply possibilites: to accept the request, to renege it, or to make a counter-offer. The conversation proceeds through a negotiation between the two interlocutors until one of them closes it in a negative way or they find an agreement. In this second case the conversation is still open until the negotiated action is accomplished or the commitment   is revoked. Conversations for Action can embed subconversations, e.g., when the person responsible for an area of activity is asked to give his/her approval about a commitment concerning that area.

The second is the Conversation for Possibilities, where the two interlocutors discuss new possibilities for the office, in terms of its structure, responsibilities distribution, subjects of interest and the like. These conversations are effective, i.e., actually modify the relationships between the members of the office and their domain of possibilities, when they end with a declaration which reflects an agreement about its content among the members of the office involved in the conversation. This is, for instance, the case of the delegation of a responsibility to some office member by someone having the authority for performing the delegation: it is effective only if the candidate accepts taking the new responsibility upon her/himeself and if the other members involved through subconversations give their approval.

Let us note that both Conversations for Action and for Possibilities are positively closed if the interlocutors find a mutual agreement. The accent we put on the need of agreement is aimed to improve organization liveliness and transparency. In fact, declarations performed on the basis of the necessary authority but without consensus give rise to a gap between the declared responsibility structure and the responsibility roles actually played. In an analogous way, task attribution without consensus will probably give rise to quantitative (e.g. delays) and qualitative problems.

Both Conversations for Actions and for Possibilities can either concern areas of already structured acitivities or not. In this second case the patterns of conversations model the protocols (or better the rules of the linguistic game) followed by human beings when involved in conversations in whatever social systemwithout any regard to their role.

For instance Fig. 2 in the next section models a conversations opened by a "generic" cooperation request. Here "generic" means that the Actor has not associated the conversation to any field of responsibility and therefore no approvals must be obtained for taking the commitment.

On the contrary the first case concerns conversations associated to structured areas of activity. Here the patterns of conversations characterize the roles played by the different interlocutors inside the organization. That is, they depend on the particular role distribution determined both by the effective declarations closing conversations for possibilites and by the commitment taken within conversation for action and then fulfilled or currently under fulfillment.

For instance, if the chief of a software research laboratory, by a declaration which closes a Conversation for Possibilities, attributes to a system designer, say X, the responsibility upon A.I. activities and the designer accepts, then a rule is created such that any commitment in the A.I. field taken by any laboratory member is subject to X's approval, so that X's role acquires a new characterization in terms of responsibility in the A.I. field.

If a software designer, say Y, is making, inside a Conversation for Action, the commitment of designing in Common Lisp a new shell ("A.I." being the area of responsibility and "Common Lisp" the area of expertise characterizing that commitment), since the commitment falls into the "A.I." area of responsibility, Y must ask the approval of X as "A.I." responsible. If X gives the ok and Y then fulfill the taken commitment, then a rule is created such that Y gives, sooner or later, an answer to any information request concerning Common Lisp, so that her/his role acquires a new caracterization in term of expertise in this subject.

CHAOS embeds a characterization of office roles, based on a distinction between responsibility and expertise. Fields of responsibility, such as areas of activity, customers, people and groups of people, must be explicitly declared and then other declarations can attribute to a person a responsibility role in the field. Expertise in a field is acquired in the actual fulfillment of a commitment which requires this expertise in the associated field (see next section).

On the basis of this distinction, CHAOS characterizes three roles of responsibility:
-    The role of supervisor is characterized by the fact that the supervisor of a field is the only one having the authority (authorized) to issue effective declarations concerning the field, such as: declaring a new subfield; re-structuring some subfields.
-    The role of (control) manager implies the responsibility of a close control over the commitments concerning the field. This means that if the manager of a field is declared, then, before a commitment in the field is made inside a Conversation for Action, her/his approval must be obtained.
-    The role of operative is characterized by the fact of being a privileged executor of tasks concerning the field. This implies that if one looks for cooperation in a field without explicitly indicating the partner, CHAOS identifies as partner one of the operatives on the field, if s/he exists (see section 3.5).

Furthermore, CHAOS identifies the role of expert of a field, which belongs to those who have made and fulfilled commitments in the field. If one looks for cooperation in a field without explicitly indicating the partner and an operative on the field has not been declared, then CHAOS addresses the request to the most expert in the field.

2.2 Conversations Models

In order to support people conversing and making commitments in the offices, CHAOS supports various kinds of Conversations for Action and of Conversation for Possibilites (see Fig. 1).

CONVERSATIONS FOR ACTION
- two types of conversations opened by a Request:
  - Cooperation Request;
  - Information Request;
- two types of conversation opened by a (conditional) Promise:
  - Cooperation Offer;
  - Offer to oneself.

CONVERSATIONS FOR POSSIBILITES
- conversations for delegating responsibility by declaring a supervisor, a manager, an operative in a field of responsibility;
- conversations for revoking responsibility;
- conversations for declaring a new member;
- conversations for creating a new field of responsibility
- conversations for re-structuring fields of responsibility

Fig. 1

Each type of conversation is modeled by a Gameru model [DDS87] which represents the combination of elementary speech acts it consist of (see Fig. 2a and Fig. 2b). The transitions enabled under the current marking give the set of speech acts open to each interlocutor at a particular moment in whatever conversation of each type.

The choice of representing conversations by means of Gameru models fits the need for having models which represent the set of the possible behaviours of the various members involved in conversations and subconversations without forcing constraints on their behaviour [DDS85].

In addition, CHAOS structures the content of each speech act inside a Conversation for Action in terms of:
- the illocutionary point, which expresses the type of the speech act in accordance with Searle's Speech Act Taxonomy (Assertives, Directives, Commissives, Declarations, Expressives) in terms of the relationship it creates between the two involed interlocutors;
- the propositional content, in turn given by: the *action* to be performed, augmented by some information characterizing it: e.g., an action of selling a product can be detailed by specifying the related 'contract' and 'customer'; the field(s) of *responsibility* and the field(s) of *expertise* to which the action belongs;
- the satisfaction conditions, in particular:
  a) the *expiration time* indicating the time-out for waiting for an answer;

CONVERSATION OPENED BY A COOPERATION REQUEST

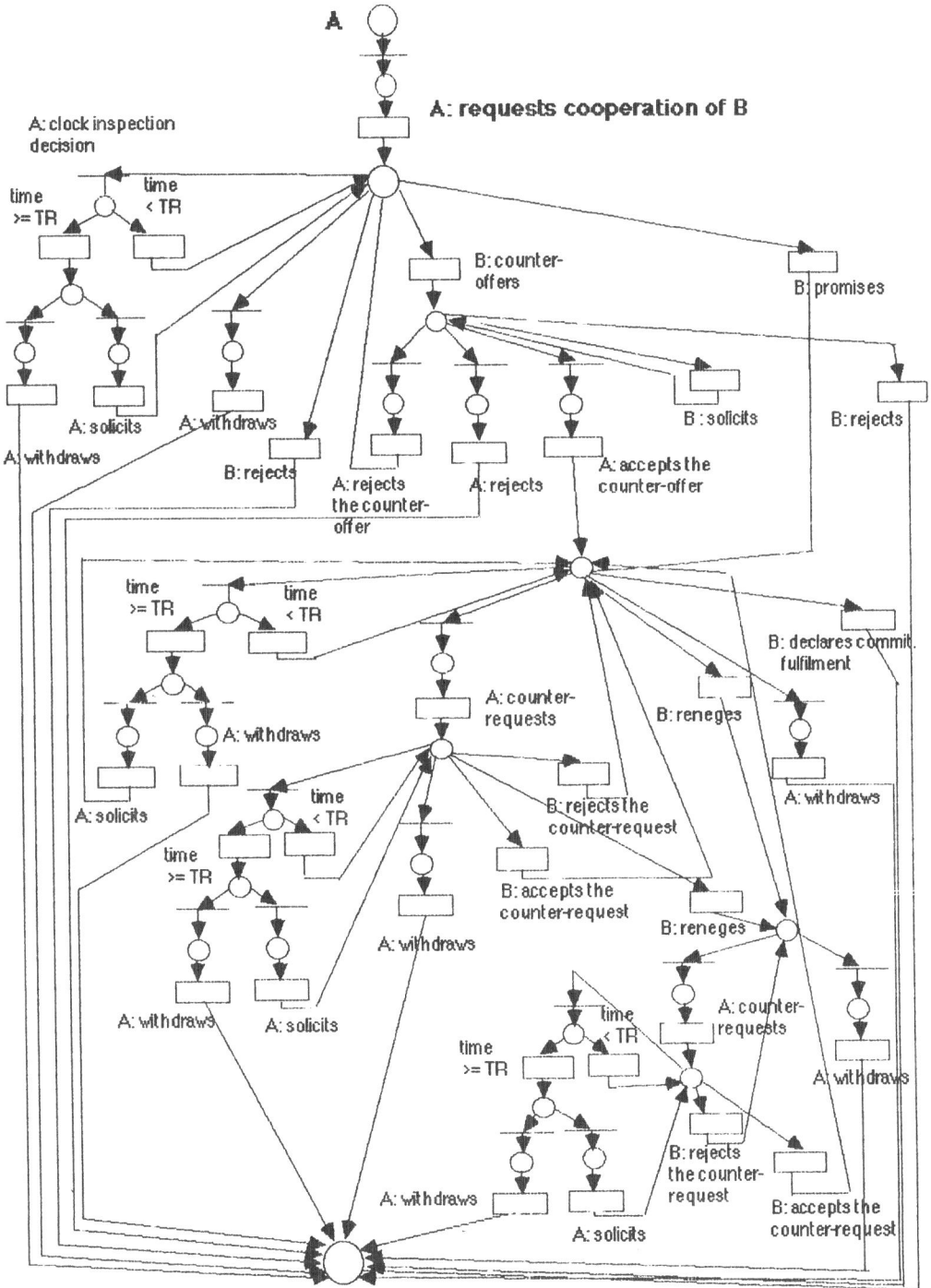

Fig. 2a: The Net modeling the Actor's behaviour

CONVERSATION OPENED BY A COOPERATION REQUEST

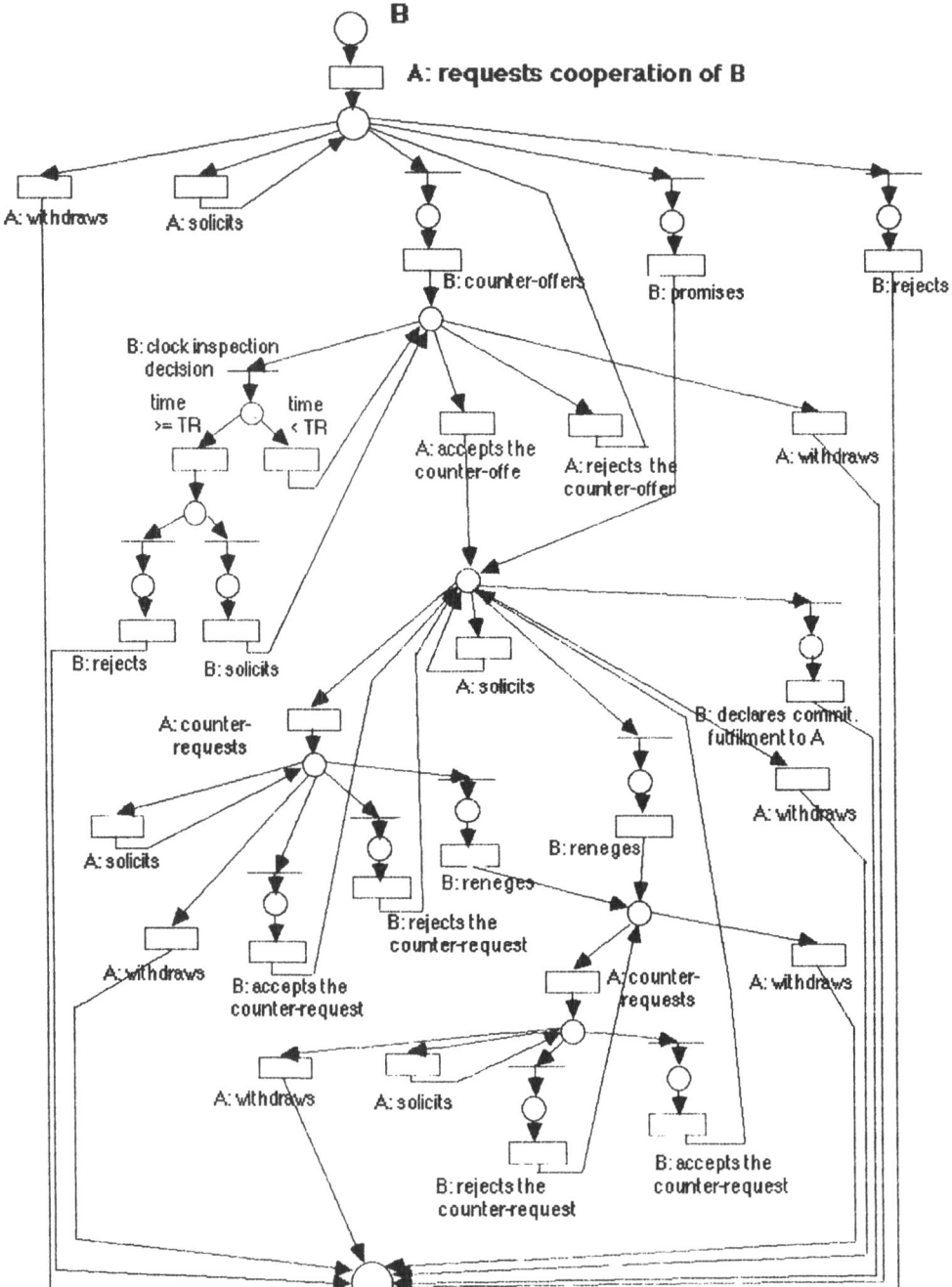

Fig. 2b : The Net modeling the Partner's behaviour

LEGENDA: each unlabelled transition represents the decision which precedes the interaction represented by the subsequent transition

b)   the *environment conditions* declaring if the action is *part of* another action; if the action has one or more *precondition,* i.e. if it can start only after some other actions are finished,

c)   the *completion times* indicating:
  - the beginning time, when the action cannot start before a given date;
  - the time-out for the final fulfillment of the commitment ;
  - the overall time required for accomplishing the action;

-   some comments.

The structure of speech acts contained inside a Conversation for Possibilites is analogous, except for the propositional content which depends on the particular conversation.

If we consider a Petri Net as an object, each particular conversation of a given type is represented by an instance of that object. The set of possible speech acts open to each partner at a particular moment in a conversation of a given type, is given by the status (marking) of the corresponding conversation instance (see Fig. 3).

==

CONVERSATION FOR ACTION

- ACTOR
- PARTNER
- TYPE OF CONVERSATION
- STATUS OF THE CONVERSATION
- HISTORY (the sequence of the performed speech acts)

==

SPEECH ACT (of a Conversation for Action)

- ILLOCUTIONARY POINT
- PROPOSITIONAL CONTENT
  - ☐ ACTION
  - ☐ FIELD(S) OF RESPONSIBILITY
  - ☐ FIELD(S) OF EXPERTISE
- SATISFACTION CONDITIONS
  - ☐ EXPIRATION TIME
  - ☐ ENVIRONMENT CONDITIONS
  - ☐ COMPLETION TIMES
- COMMENTS

==

Fig. 3

2.3 CHAOS Architecture

The CHAOS family is currently under development. The first prototype, called CHAOS-1, runs on the VAX 750 under Unix® Operating System and is written in FranzLisp plus Pearl [DFW82]. We presently are developing a new release which overcomes some of the CHAOS-1 limits: in fact, it deals with the responsibility structure by handling the Conversations for Possibilities; with the commitments structure by handling the mutual relationships between conversations; and with a more flexible user interface by allowing a disciplined use of a semi- structured messages in natural language. Till now, all the topics concerning communication/distribution of the application are neglected and simulated by means of shared data.

The focus is here on the overall CHAOS architecture which consists of three main modules:

- the User Interface Module;
- the Conversations Handler Module;
- the Knowledge Builder.

They are briefly sketched here below.

The User Interface module

The User Interface allows the user to:

a) perform speech acts, consistent with the set of speech acts available to her/him in the current state of the selected conversation. Once the user has selected one of her/his possibilities among the ones displayed on the screen, the User Interface guides her/him to express in a structured way the speech actcontent, following the pattern shown in Fig.3. Since the 'action' is described in a semi-natural language, the User Interface invokes the Knowledge Builder for its interpretation.

b) select a subset of all her/his conversations which satisfy certain properties: for instance, the conversations in which the user is waiting for an answer, the conversations having a certain partner, the converstions in which the user is waited for a commitment completion, an other ones.

c) execute some operations on a specific conversation such as: to read from her/his own mailbox (buffer) the messages carrying a speech act already arrived, but not yet consumed; to visualize or verify some properties on it; to continue the conversation, possibly after the buffer contents have been visualized. All these operation requests are passed to the Conversations Handler which retrieves the suitable data, and returns them back to the User Interface for displaying.

The Conversation Handler

The Conversations Handler handles conversations by sending, receiving, storing and retrieving the speech acts they consist of.

Every time a user opens a conversation, the Conversations Handler creates an instance of the object corresponding to the conversation type (a CONVERSATION INSTANCE) and stores it into the user database. If the partner is partially specified, the Conversation Handler calls the Knowledge Builder to identify the appropriate destination.

Every time a user performs a speech act inside a conversation, the Conversations Handler updates the current status of the corresponding CONVERSATION INSTANCE and calls the Knowledge Builder in order to update the Knowledge Base with the new information contained in the performed speech act.

Furthermore, the Conversation Handler is activated by the User Interface every time a user asks to continue a conversation, and every time the User Interface needs information about the conversations (status, contents, properties) or about the messages not yet consumed by the user.

The Knowledge Builder

CHAOS does not require *ad hoc* input by its users, but derives knowledge automatically by the conversations it handles. The Knowledge Builder "observes" each speech act in a conversation and uses the information contained in the speech act to update the Knowledge Base. This latter contains:

● The Responsibility Structure of the organization, i.e., the set of declared responsibility fields together with the relationships connecting them expressed in terms of the relationship "is subfield of";

● The Expertise Distribution inside the organization consisting of a set of facts which correlate organization members to fields of expertise by keeping track of the degree of expertise the various members have in the different fields and by associating to each area the set of members having expertise in it.

● A Dictionary containing the definition of the objects and actions referred to during conversations, in terms of their attributes; a Thesaurus containing the relationships between them [Sco86]: specifically in the case of the actions, the two above mentioned relationships of being 'precondition' or 'part of' another action.

● The Committed Action Network which keeps track of the relationships among different actions committed inside interrelated conversations together with the completion times.

This Knowledge Base, updated by the Knowledge Builder by "observing" the conversations occurring in the office, is used for supporting and disciplining communication inside the organization, as it will be shown in the next section.

3. How CHAOS learns from conversations and disciplines communication

The communication disciplines embodied in CHAOS are based on a theory, presently under development |DeM86|, which considers organizations as closed autonomous systems in the linguistic domain. In this frame the events relevant inside an organization are the agreements reached within conversations by the organization members, both at the commitments and at the commitment satisfaction level. The rules within communication disciplines are derived from the conversations in a formal and unambiguous way, fully described in the specifications of the new CHAOS release. Hints about this are presented in the following subsections in an intuitive way.

First of all, when conversations for possibilities are considered, the declaratives which close their successful termination modify the domain of possibilities open to the office members, possibly changing role distribution, introducing new areas of activities, and the like. Mainly involved here are the disciplines of authorization and delegation.

Secondly, when conversations for action are considered, the specific issue are the commitments, how they are interrelated both from the viewpoint of the authority structure and of the use of resources. Mainly involved here are the disciplines of authorization and synchronization.

Thirdly, the professional language characterizing the office work is shown in the action part of any speech act occurring in both kinds of conversations. In this framework, the linguistic games binding the office members can be analyzed in order to understand how names are given to objects and to activities manipulating them. Then, the discipline of naming is involved here.

As a nice consequence of how all the previous communication aspects are disciplined, both sources and destinations inside communications can be understood and properly connected even in presence of partially specified information. These aspects refer to the disciplines of identification and addressing.

Finally, the way in which CHAOS supports the organizational changes defines a discipline of reoganization, in accordance with the organizational analysis approach underlying the proposed technology.

In the following subsections all the quotations are taken from Petri's [Pet77b].

3.1 Delegation

"Delegation: this means the delegation of tasks from one agency to another... If somebody delegates something, this has, of course, formally comprehensible consequences for synchronization, addressing and other disciplines."

Applying this perspective to offices, CHAOS disciplines delegation on the basis of the roles characterization presented in Section 2.1. CHAOS supports delegation of responsibilities and delegation of tasks in the way sketched in Fig. 4.

Any delegation attributes to the delegated member a well defined role which s/he then plays inside the subsequent Conversations for Action and Conversations for Possibilities, as described more in detail in the following Sections.

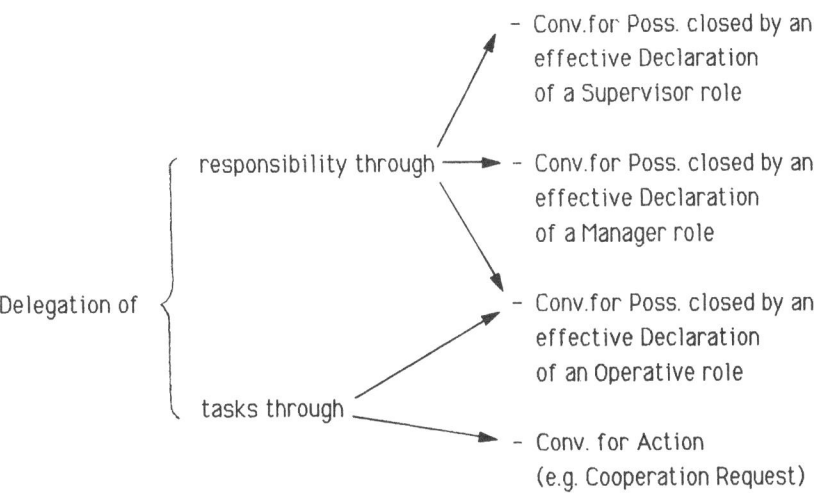

Fig. 4

Responsibility Delegation

Only the Supervisors have the authority to delegate whatever responsibility on the field of their competence. More precisely, the supervisor A of field F has the authority:

- to delegate supervision responsibility, i.e., to declare a new supervisor, on subfields of F, but not on the whole F. The only one who has the authority to perform such declaration is the supervisor of the field which has F as its subfield;

- to delegate responsibility of lower degree, i.e. to attribute to someone the role of manager or operative, both on the whole field F and on any of its subfields.

For delegating a role of responsibility to an office member B the competent supervisor A (called "the Delegating") opens a Conversation for Possibilities with B (called "the Delegated"). CHAOS supports the conversation by opening the subconversations for obtaining by other supervisors and managers their approvals which are necessary for making the specific delegation effective.

For instance, if A is the supervisor of the area of activity "SHELLS" with subfields "ART" and "KEE", and wants to delegate supervision responsibility to B who was before just control manager of the field 'Kee", then A must obtain the agreement of B and the approval of the supervisor of area "A.I.", which has "SHELLS" as its subfield.

If the necessary approvals are given and B accepts, either immediately or after some counter-offer cycle, to take upon her/himself the responsibility, then A closes the conversation by an effective declaration, i.e., by a declaration which is accepted by the organization, or better, by its involved part, and which therefore actually changes the domain of possibilities of its members.

Task Delegation

The "pure" delegation of tasks can be performed by any member who is engaged in a commitment, and wishes to obtain the cooperation of somebody else for dividing the committed action in more subactions. S/he can open as many Cooperation Request Conversations as s/he needs.

CHAOS supports the conversation by opening the subconversations for obtaining, by the people responsibles for the involved fields, their approvals, which are necessary (see section 3.2) in order for the particular commitment to be taken. The task delegation is effective if the two interlocutors find an agreement about the subaction and its temporal satisfaction conditions, and the necessary approvals are given on it.

Let us note that the delegation of operative responsibility on a field also concerns delegation of tasks. In fact (see section 3.5), if one looks to find cooperation in a field and does not indicate the partner, CHAOS adresses the request to the declared Operative responsible, if s/he exists . In other words, the Operative Responsible is the preferential partner in Cooperation Request Conversations.

3.2 Authorization

"Authorization: this discipline is concerned not only with assigning and schematically representing access rights but also with scheduling obligations which are consistently connected with access rights and authorizations for issuing directives, and with the rules for an adequate basis of supervision."...

CHAOS disciplines authorization by defining how the office members who have a delegated responsibility perform their role inside Conversations for Action and Conversations for Possibilities.

We have seen in the previous section that only the Supervisors have the authority to open Conversations for Possibilities finalized at responsibility delegation, and we have also mentioned that managers can be asked for approving commitments and declarations which fall in their field of competence. We can now be more precise.

Supervisor S of field F:

a) authorizes declarations concerning the field which are not performed by her/himself, as in the situation described in the previous section concerning delegation of supervision responsibility on the field 'KEE";

b) exercises a form of weak control inside Conversation for Action concerning field F . S/he is made aware by the system of the conversations which discuss commitments concerning the field and has access right into these conversations. If s/he does not exercise this right within a fixed amount of time, then "by default" her/his authorization to make the commitment is assumed.

Manager M of the field F:

a) authorizes the making of commitments concerning field F, i.e., having F indicated as field of responsibility inside the speech act (see Fig. 3). S/he can execise this control in different ways, from the strongest one, in which M makes directly any commitment concerning F, to the weakest one, in which M is asked for the approval when all negotiation about the commitment is successfully finished. Let us note that in this last case if the two interlocutors do not reach an agreement, the manager remains completely in the dark about the conversation concerning field F.

b) is asked for her/his opinion about declarations concerning field F. For instance, if supervisor S of F wants to open a new subfield of F, s/he asks the opinion of M to obtain her/his consensus since afterward M exercises control also over the new subfield.

Note that, beyond the different formulation, the essence of the two situations is the same, and is therefore correct to speak of authorization in both cases. In fact, M can always reject the proposal: in one case, by denyning approval; in the other, by giving a negative opinion. As said in section 2.1, goal of CHAOS is to enhance organization liveliness by supporting negotiation within conversations for action and for possibilities until an agreement is found, and by allowing all the people involved in it because of their role of responsibility to deny authorization. Nevertheless the possibility of dealing with (persistent) authorization refusals is preserved. In the above mentioned situation, if M continues to reject any proposal of adding subfields to F, then supervisor S can revoke the control responsibility of M.

3.3 Synchronization

"This discipline is concerned with getting proper timing restraints for different activities. In some cases, the use of clocks may serve for synchronizing all activities. But what should we generally understand by the term 'synchronization' ? My answer is that the precise definition of the term should be based upon the partial ordering in terms of causality as opposed to an ordering in terms of time."

Here we have to distinguish between two different orders of questions. The first one concerns synchronization among the different speech acts constituting a conversation. This synchronization is naturally disciplined in CHAOS, where the models of conversations as combination of elementary speech acts are expressed in terms of Gameru models, i.e., Petri net, models. Therefore, only a partial ordering among speech acts is assumed.

Furthermore, also the relationships among the actions committed in different but related conversations are expressed in terms of a casual relation. We have already mentioned that for indicating, for instance, that action A, say coding, cannot start before another action, say specification, is finished, the user must indicate as environment condition: B is-pre-of A. Therefore, only a partial ordering among actions is assumed.

Nevertheless, inside organizations, as Petri himself observes, "In some cases, the use of clocks may serve for synchronizing all activities". CHAOS assumes a time scale common to the whole organization in order to give a way for indicating time constraints, namely, the *expiration time* associated to any speech act and the *completion times* of the overall commitment (see Fig. 3).

This allow CHAOS to support a further kind of synchronization among interrelated activities which must be accomplished in order to fulfill a commitment. In fact, on the basis of precondition relations among actions and of the completion times of each one of them, the system is able to derive a PERT of the related activities and to support the handling of the consequences of the possible delay in the fulfillment of some of them.

3.4 Naming.

Naming "is conceived as the act of giving names, and will be subject to a discipline because not everybody may arbitrarily name entities ... A typical question would be: how can we understand the incompatible naming of files in different computer systems as a consequence of only one naming discipline? .... exactly how much freedom exists when names are given? ..."

Inside an office the questions can be specified as follows:
1)   how new names and relations among them and with the old ones are created   and possibly shared?
2)   when and how names used by any two ineterlocutors inside a conversation have to be compatible with the names shared between (any group of) office members?
3)   to what extent ambiguities arising in conversations should   be avoided and office members helped in solving them?

Let us explain naming discipline of CHAOS by trying to answer these questions.

1) Since professional languages are involved [Nyg84] , a set of basic names can be defined to build an <u>initial dictionary</u>. For example, let's take the software production process as a possible linguistic domain. In this domain concepts such as project, module, requirements, programming language, coding, testing, and the like can be defined, using suitable attributes (frames) to capture their overall meaning. Furthermore, these concepts can be linked together to construct <u>an initial thesaurus</u>, using some relationships as: part of, precondition of, and the like [Sco86].

<u>New</u> objects with their <u>names and relations</u> with other objects are introduced in the dictionary and in the thesaurus, and then become knowledge shared among the office members, at three moments in office life:

-    when a commitment concerning an action is made, all the names and relations defining its propositional content (in our domain, they generally define projects or subprojects) enter the shared knowledge, following the rules implied by the authority structure;

-    when a report is done an analogous process takes place: the propositional content of the fulfilled commitment becomes a common experience. Notice that in general the content of a commitment and of the related report can differ due to possible negotiations after the commitment definition.

-    finally, upon declarations of new objects, such as new fields of responsibility, areas of activities and the like.

In such a way as the professional languages evolve as professions do, the dictionary and the thesaurus are updated accordingly, by means of functions as: creation, enrichment, specialization, and the like. For example, prototyping, software production techniques derived from the use of AI tools modify the relationships between the production steps also depending on the kind of project under consideration.   Old methodologies have to coexist with new ones.

2)   The (possible) conflict between a shared lexicon and the linguistic conventions used by two interlocutors is handled in CHAOS on the basis of the following principles:

•   at any speech act occurring inside a conversation, CHAOS keeps track of any (new) relationship between objects or activities created by the ineterlocutors. This is considered as a local convention, valid as far as these two ineterlocutors are concerned. For example, A and B may agree that 'design of CHAOS' means the detailed specification of data and fuctions before the implementation, while inside the organization 'design of something' means the overall production process involving this 'something'.

•   this freedom can be limited only when names have to be used in different contexts: e.g., inside a diagram describing a project activities scheduling which is handled by another office member, let's say the project responsible. In this case, an inconsistency is signalled. The two interlocutors either accept the lexicon shared inside the organization: in the previous example A and B do not use improperly the word 'design'. Or define suitable synonyms (aliasing): in the example, the local name 'design' is considered as synonim of the shared name 'detailed architecture'.

3)   The acceptability of any support to human communication depends on a reasonably free use of language, with ambiguities arising from anaphores of different kinds. CHAOS aims at solving these situations for two kinds of reasons: from the system point of view, the necessity of 'understanding' in order to learn from what is going on in view of future inquiry (selection and updating of conversations); from the user point of view, the possibility of reducing the amount of communication necessary to solve ambiguities between any two interlocutors, by providing upon request the intended meaning of ambiguous sentences.

This is done by a process which incrementally uses some specialized views of what CHAOS learns during the conversations: first, by considering the knowledge collected during the conversation containing the ambiguous sentence; secondly, by considering the knowledge collected during past conversations between the two members; thirdly, by doing the same on conversations between the actor or the partner with any other office member in similar circumstances. In the last two cases, a user model is needed, in order to use the formalized   knowledge the interlocutors of the conversation have mutually constructed in the past (communicated knowledge). This knowledge consists of shared partial views of the dictionaries and of information (beliefs and suppositions) deduced by means of a set of inference rules based on the structure of the conversations and on the type of speech acts they have done.

The final result of the process, possibly its failure, is (generally) shown to the user for an acknowledgement or for a more precise specification, which becomes part of the system and can be provided at any time.

3.5 Identification and Addressing

Identification "involves demonstrating the identity of the source and destination of phenomena .....covers the question of pattern recognition as well as problems of proving the competences of agencies with regard to certain actions."

Addressing is the "description of routes or system paths through a net of channels and agencies".

These two disciplines are obviously strictly related to one another: if the source or destination is not identified, any addressing is impossible; once the source or destination is identified, the message has indeed to be properly addressed.

CHAOS disciplines   identification and addressing on the basis of the following principles.

●   Since any speech act occurs inside a specific (sub-)conversation and is univocally positioned inside it (i.e., it corresponds to a unique transition in the net which models the (sub-)conversation), the identification of source and destination   follows automatically, since they are known after the opening speech act, and cannot change during the conversation.

• CHAOS allows any actor to partially define the destination of the message in the speech act opening the conversation without affecting the previous point, since it is able to retrieve the complete information from the knowledge base (the responsibility structure, the committed actions network, the expertise distribution, the dictionary and thesaurus). For example, the user can indicate as destination 'an expert of the area Common Lisp', 'the person responsible for the project CHAOS', 'who will implement procedure Search of the module User Interface', and the like. CHAOS is not only able to identify the correct partner but also delivers to him/her the appropriate messages.

• CHAOS supports the user in the identification of the office members whose approval is necessary about a commitment under discussion. In fact, when the two interlocuotrs reach an agreement about it, the user who actually takes the commitment can simply express his/her willingness to prosecute the conversation and the system, on the basis of the model of the authority structure governing the commitments making and fulfillment shows which subconversation has to be opened and with whom (see the examples in 3.1).

It is worth noting that CHAOS disciplines addressing at the 'people' level by helping the office members to follow the office game rules. In regard to addressing at the 'message' level, CHAOS rests on the rules embodied in the underlying mail/communication system. It is not difficult to imagine an integration toward a system which is able to reach the partner also in the case of his/her dynamic location, once the system is informed about the new one (possibly using the flexibility provided by the information contained in the Knowledge Base: e.g., 'I'm going to the boss', 'I'm moving to implement the User Interface Module', 'I'm looking for authorization to start project CHAOS').

3.6 Reorganization

"Suppose you have a system of pipes and tubes in a chemical factory, and that you wish to attach a new tube to the system. A specification that a certain subset of valves must be closed in order to cut off flow from the part of the system that is being altered is inadequate; it must also be specificed that while the new tube is being fitted the valves must remain under control in order that nobody can open them, that is, that no other independent activity be allowed to interfere disastrously."

The discipline of reorganization is the one that is at the deepest level influenced by the theoretical approach underlying CHAOS.

As it has been sketched above, we consider any organizational system as a linguistic game, i.e., as a closed social system whose structure defines the space of possibility (in terms of speech acts) of the members within it, compensating the perturbations (the speech acts) its members actually perform according to its structure, by means of a modification of its structure itself, reorganizing itself. There is therefore a basic circularity in the organizational systems between structure and behavior, and this circularity disciplines the reorganization of each organizational system.

CHAOS embodies a model of this basic circularity, appearing to its users as a self-reorganizing system. In fact, CHAOS reorganizes itself by compensating for the perturbations that are the mutual commitments of its members w.r.t. the actions to perform and/or the new possibilities to experiment in terms of responsibility delegation, and modifies its rules so that they embody the derived expertise and authority distribution.

The reorganization discipline of CHAOS emulates in a partial way the basic circularity between organizational rules and conversations which characterizes the organizations as social systems, by reducing it to the two-level hierarchy of facts and inference rules in its Knowledge Base. This is the reason why CHAOS shows differences w.r.t. Petri's guidelines, which suggest to consider reoganization as an intervention by an external agent.

4. The missing disciplines

Copying, cancellation and composition are disciplines which consider objects flowing in the office as "documents" and define the rules for handling them: they are naturally related to authorization and delegation, which state who, when and how is allowed to manipulate any kind of office documents.

Up to now CHAOS has not dealt with these problems, since its attention is concentrated on the commitments made by its members, and not on the actions those commitments refer to. Two are the major reasons for this choice.

From one point of view, the disciplines of document handling have up till now been defined on an empirical basis, and lack a sound theoretical foundation. Their integration in CHAOS is made difficult by this situation, because the architecture of CHAOS is based on a sound and coherent theoretical framework.

From another point of view, the extension of CHAOS to document handling is logically subsequent to the full development of its commitments handling kernel. In fact, the disciplines of copying, cancellation and composition contribute to a richer characterization of office roles, but do not change the linguistic essence of the basic circularity of the office, through which roles are attributed and changed.

In regard to valuation discipline, which Petri claims "must treat the exchangeability of resources and its modalities", we agree with Petri that it "is perhaps the most important one since we know least about it, compared with others". We are now investigating if it has not as its primary concern the implementation issues in a communication system. Any scheduling algorithm in a distributed system, in fact, defines implicitly a valuation of the scheduled processes. From this point of view the current implementation of CHAOS embodies some rules of the valuation discipline w.r.t. the expertises , the degrees of responsibility, etc.

In [Pet77b] C. A. Petri claims that "<u>modelling</u> is done according to a scheme which normally has come into existence through a historical process running independent of conscious methodical technology". This is the very situation in which CHAOS has been developed: it reflects part of its environment since it is a model of the basic circularity of the office, but it is not based on any general schema for deriving models from systems. It is therefore not possible to claim that CHAOS embodies in any sense *the rules of* the discipline of modelling.

5. Acknowledgements

Many persons have contributed to project CHAOS. Among them, we would like in particular mention Raffaela Vassallo and Annamaria Zanaboni, who have participated in the development of CHAOS from its very beginning. Giacomo Ferrari is helping us in respect to the linguistic aspects of CHAOS with discussions, suggestions and criticisms. Alfonso Gerevini and Alessandro Bottarelli are deeply involved in the project, on which they are preparing their Masters theses. Klaus Voss provided useful suggestions, in particular for what concerns the distinction between communication disciplines and their rules.

This research has been supported by Ministero Della Pubblica Istruzione on a national contract.

6. References

[AT84] Action Technologies Inc., *Coordinator - User Manual and Report*, San Francisco, 1984

[Aus62] J. Austin, *How to do Things with Words*, Oxford University Press, London, 1962

[CH80] P.M. Cashman, A.W. Holt, A communication-oriented approach to structuring the software maintenance environment, *ACM SIGSOFT, Software Engineering Notes*, 5:1, January, 1980

[CSCW86] CSCW'86, *Proceedings of the Conference on Computer-Supported Cooperative*

[DDPS82] F. De Cindio, G. De Michelis, L. Pomello, C. Simone, Superposed Automata Nets, in: *Application and Theory of Petri Nets*, IFB 52, Springer–Verlag, Berlin, 1982

[DDS85] F. De Cindio, G. De Michelis, C. Simone, Formal Computer systems in social organizations, in: *Proc. Working Conference on Development and Use of Computer-based Systems and Tools*, Aarhus, 1985

[DDS86] F. De Cindio, G. De Michelis, C. Simone, R. Vassallo, A. Zanaboni, CHAOS as a Coordination Technology, in [CSCW86]

[DDS87] F. De Cindio, G. De Michelis, C. Simone, GAMERU: a language for analysis and design of human communication pragmatics within organizational systems, in: *Advances in Petri Nets 1986*, LNCS, Springer–Verlag, Berlin, 1987 (to appear)

[DeM86] G. De Michelis, Sistemi autopoietici del terzo ordine: il caso degli uffici, *Metamorfosi* 2, 1986

[DFW82] M. Deering, J. Faletti, R. Wilensky, *Using PEARL AI Package*, Computer Science Div. EECS Dept, University of California, 1982

[DJM86] R. Dunham, B. M. Johnson, G. McGonagill, M. Olson, G. M. Weaver, Using a Computer Based Tool to Support Collaboration: A Field Experiment, in [CSCW86]

[FL81] C.F. Flores, J.J. Ludlow, Doing and Speaking in: the Office, in F. Fish,   R. Sprague (eds.), *DSS: Issues and Challenges*, Pergamon, New York, 1981

[Hol79] A. W. Holt, Net Models of Organizational Systems, in Theory and Practise, in C. A. Petri (ed.), *Ansätze zur Organizationstheorie Rechnergestutzer Informationssysteme*, Oldenburg, Munchen, 1979

[Hol86] A. W. Holt, Coordination Technology and Petri Nets, in *Advances in Petri Nets 1985*, LNCS 222, Springer–Verlag, Berlin, 1986

[MGL86] T. W. Malone, K. R. Grant, K-Y. Lai, R. Rao, D. Rosenblitt, Semi-structured messages are surprisingly useful for Computer-supported Coordination, in [CSCW86]

[MV80] H. Maturana, F. Varela, *Autopoiesis and Cognition*, Reidel,   Dordrecht, 1980

[Nyg84] K. Nygaard, Profession Oriented Languages, in *Proc. Medinfo Europe 84*, Bressels, 1984

[Pet77a] C. A. Petri, Modelling as a Communication Discipline, in: H. Beilner, E. Gelenbe (eds.) *Measuring, modelling and evaluating computer systems*, North Holland, New York, 1977

[Pet77b] C. A. Petri, Communication Disciplines, in: B. Shaw (ed.), *Proc. of the Joint IBM-Univ. of Newcastle upon Tyne Seminar*, Univ. of Newcastle upon Tyne Computing Lab., 1977

[Pet79] C. A. Petri, Kommunicationsdisziplinen, in C. A. Petri (ed.), *Ansätze zur Organizationstheorie Rechnergestutzer Informationssysteme*, Oldenburg, Munchen, 1979

[RV86] G. Richter, K. Voss, Toward a comprehensive Office Model integrating Information and Resources, in *Advances in Petri Nets 1985*, LNCS 222, Springer Verlag, Berlin, 1986

[Sco86] D.S.Scott, Capturing concepts with Data Structures, [preliminary version, paper to be revised], Dept. of Computer Science, CMU, nov. 86

[SC84] S. Sluzier, P.M. Cashman, XCP : An experimental tool for supporting office procedures, *IEEE 1984 Proceedings of the First International Conference on Office Automation*, Silver Spring, MD:IEEE Computer Society, 1984

[Sea69] J.R. Searle, *Speech Acts : An essay in the philosophy of language*, Cambridge University Press, Cambridge, 1969

[Sea79] J.R. Searle, A Taxonomy of Illocutionary Acts, in: J.R. Searle (ed.) *Expression and Meaning: Studies in the Theory of Speech Acts*, Cambridge University Press, Cambridge, 1979

[Win86] T. Winograd, A Language Perspective on the Design of Cooperative Work, in: [CSCW86]

[WF86] T. Winograd, C. F. Flores, *Understanding Computers and Cognition*, Ablex, Norwood, 1986

[Wit53] L. Wittgenstein, *Philosophische Untersuchungen*, Blackwell, Oxford, 1953

ON THE STRUCTURE OF DEPENDENCE GRAPHS

A. Ehrenfeucht

Department of Computer Science

University of Colorado at Boulder

Boulder, Colorado, USA

and

G. Rozenberg

Department of Computer Science		Department of Computer Science
University of Leiden	and	University of Colorado at Boulder
Leiden, The Netherlands		Boulder, Colorado, USA

ABSTRACT

Dependence graphs (which are special kinds of acyclic directed node labeled graphs) are very fundamental objects in the *theory of traces* approach to *concurrent systems* and in the *theory of graph grammars*.

In this paper we characterize dependence graphs and naked dependence graphs - where a naked dependence graph is the unlabeled graph obtained from a dependence graph by removing its node labels.

INTRODUCTION

There are essentially two developments (that have started independently) that have led to the notion of a *dependence graph.*

The first one is within the *theory of concurrent systems.* The *theory of traces* tries to reconcile the sequential character of observations of a concurrent system with its concurrent nature. Basically (the behaviour of) a concurrent system is given by a (regular) language K over the alphabet of events Σ (K represents all sequential observations of the behaviour) and a binary symmetric and reflexive relation $D \subseteq \Sigma \times \Sigma$ (representing the dependence between events of the system). Hence if a string $w \in K$ is of the form $w_1 abw_2$ where $(a,b) \notin D$ then the fact that b follows a in w represents the sequential nature of the observation only - a and b are really *independent* in the system; consequently $w_1 baw_2$ is also in K. In this way commuting (any number of times) pairs of adjacent independent letters in strings from K yields the sets of equivalent strings - each such set, called a *trace,* represents different sequential observations of the same concurrent behaviour. All strings in one trace of K can be represented by *one* directed node labeled (by Σ) acyclic graph - such a graph is a ($\Gamma-$) *dependence graph* (where $\Gamma = (\Sigma, D)$ - it is called a *dependence alphabet).* The theory of traces deals with properties of (regular) *trace languages* - i.e., (regular) sets of traces. It is quite obvious (to us - see also [AR1]) that, in many considerations, it is more convenient to deal with (regular) sets of (dependence) graphs than to deal with (regular) sets of sets of strings. The theory of traces as sketched above was initiated in [M] - it has however also other independent origins (see [CF], [FR] and [K]).

The second development is within the *theory of graph grammars* (see [ENR]). The theory of NLC *grammars* (see, e.g., [JR1]) constitutes an attempt to provide an elegant mathematical framework for dealing with graph grammars (generating languages of undirected node labeled graphs). An attempt to formalize NLC grammars that are analogous to "regular" string grammars led to the so called BNLC *grammars* (see [RW]).

The variant of NLC grammars to generate languages of *directed* graphs, the so called DNLC *grammars* was introduced in [JR2]. A "directed" analogue of a subclass of BNLC grammars, the so called RDNLC *grammars,* corresponds quite closely to right-linear string grammars - in [AR2] a characterization of regular languages of dependence graphs by a subclass of RDNLC grammars is given. In this way dependence graphs are objects as basic for the theory of graph grammars as strings are for the theory of string grammars.

This paper investigates the structure of dependence graphs and in particular it investigates *naked dependence graphs* - i.e., directed unlabeled graphs obtained from dependence graphs by removing node labels.

The paper is organized as follows.

In Preliminaries we introduce basic notation and terminology. In particular the formal notion of a dependence graph is recalled.

In Section 1 we provide a characterization of the so called *stable labelings* of (directed) graphs - the property of stability is a property satisfied by dependence graphs.

In Section 2 we assume that a dependence alphabet Γ is given and we characterize the so called Γ-labelings of (directed) graphs - labelings of dependence graphs over Γ are Γ-labelings.

In Section 3 we use the above mentioned two results to characterize naked dependence graphs.

O. PRELIMINARIES

In this section we will introduce basic notation and terminology to be used in our paper.

For a set X, $\# X$ denotes its cardinality and $Id(X) = \{(x, x) : x \in X\}$; $\emptyset$ denotes the empty set. For sets X, Y, $X-Y$ denotes their difference. If X is a set, $\mathbf{P}$ a partition of X and $x \in X$, then $[x]_{\mathbf{P}}$ is the class of $\mathbf{P}$ containing x.

For a function $\phi : X \longrightarrow Y$, $\mathbf{R}_\phi = \{y \in Y : \phi(x) = y$ for some $x \in X\}$; for a $Z \subseteq X$, $\phi \mid_Z$ denotes the restriction of ϕ to Z .

In this paper we consider finite alphabets only. An injective function between alphabets is referred to as a *coding*.

A finite nonempty directed graph is referred simply as a *graph*. A graph g is specified in the form (V , E) where V is the set of nodes of g and E is the set of edges of g ; we also use V_g to denote V and E_g to denote E .

For a graph $g = (V , E)$ and a nonempty $U \subseteq V$ the *subgraph of g induced by U* (i.e. the graph $(U, \{(u,v) \in E : u, v \in U\})$) is denoted by $g(U)$; a *subgraph* of g is always the subgraph of g induced by a subset of V_g .

For a graph $g = (V , E)$ the *symmetric closure of g* , denoted $sym(g)$, is the graph (V , E') where for $u, v \in V$, $(u,v) \in E'$ iff either $(u,v) \in E$ or $(v,u) \in E$. We say that g is *symmetric, a s graph* for short, if $g = sym(g)$. The *symmetric and reflexive closure of g* , denoted $symr(g)$, is the graph (V , E') where for $u, v \in V$, $(u,v) \in E'$ iff either $(u,v) \in E$ or $(v,u) \in E$ or $u = v$. We say that g is a *symmetric and reflexive graph, a sr graph* for short, if $g = symr(g)$.

Let g be a s graph and let $U \subseteq V_g$. We say that U is a *clan of g* if for all u_1 , $u_2 \in U$ and each $v \in V_g - U$, $(v , u_1) \in E_g$ iff $(v , u_2) \in E_g$. (The notion of a clan is from [ER]).

A s graph g is *discrete* if $E_g = \emptyset$, *complete* if $E_g = V_g \times V_g$ and *weakly complete* if $V_g \times V_g - Id (V_g) \subseteq E_g$. A (weakly) complete subgraph of a s graph g is called a *(weak) clique* of g .

Let $g_1 = (V_1 , E_2)$, $g_2 = (V_2 , E_2)$ be graphs. A function $\phi : V_1 \longrightarrow V_2$ is a *homomor- phism* of g_1 into g_2 iff , for all $u, v \in V_1$, $(u,v) \in E_1$ iff $(\phi(u), \phi(v)) \in E_2$; ϕ is an *isomor-*

phism iff ϕ is also onto and injective. We use $HOM \, (g_1 \, , \, g_2)$ to denote the set of all homomorphisms of g_1 into g_2 and $ISOM \, (g_1 \, , \, g_2 \,)$ to denote the set of all isomorphisms of g_1 onto g_2 .

A *labeling of a graph* g *is a (total) function on* V_g . A *node labeled graph* is a system $(V \, , \, E \, , \, R_\phi \, , \, \phi)$ where $(V \, , \, E)$ is a graph and ϕ is a labeling of $(V \, , \, E)$; $(V \, , \, E)$ is denoted by $und(h)$.

Given an alphabet Σ , a graph $\Gamma = (\Sigma \, , \, D)$ is a *dependence alphabet* if Γ is a sr graph. For a word $w \in \Sigma^+$, $w = a_1 \cdots a_n$, $n \geq 1$, $a_i \in \Sigma$ for $i \in \{1,...,n\}$, the *canonical* $\Gamma-$*dependence graph of* w is the node labeled graph $g = (V \, , \, E \, , \, R_\phi \, , \, \phi)$ where $V = \{1,...,n\}$, $\phi(i) = a_i$ for $i \in \{1, \, ..., \, n\}$ and for all $i, \, j \in V, \, (i,j) \in E$ iff $i < j$ and $(a_i,a_j) \in D$. A node labeled graph isomorphic to g is a $\Gamma-$*dependence graph of* w . A node labeled graph is a $\Gamma-$*dependence graph* if it is a $\Gamma-$*dependence graph* of w for a $w \in \Sigma^+$.

A node labeled graph is a *dependence graph* if it is a Γ-dependence graph for a dependence alphabet Γ. A *naked* $(\Gamma-)$ *dependence graph* is a graph g such that, for a $(\Gamma-)$ dependence graph h , $g = und(h)$.

Example 0.1. Let $\Gamma=(\{a,b,c,d\} \, , \, D)$ be the following dependence alphabet :

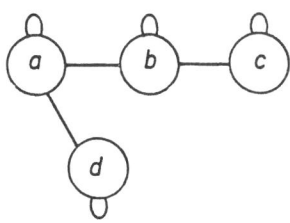

Figure 0.1

(since Γ is a s graph we represent it in the obvious way using undirected edges) .

146

Let $w = a\ c\ b\ d\ a\ c\ b$. Then the canonical Γ-dependence graph of w is as follows :

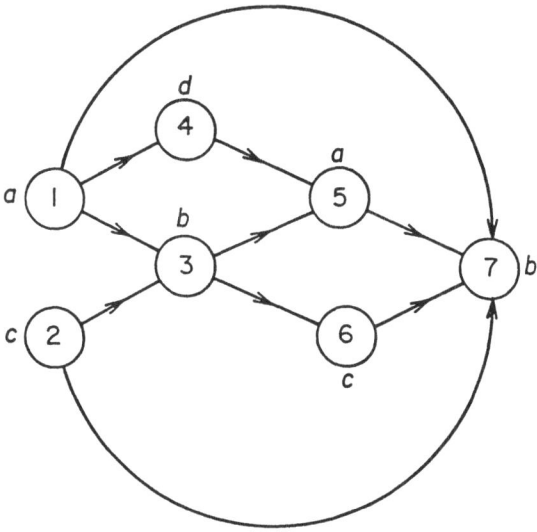

Figure 0.2

(in a graphical representation of a graph we will indicate the identity of a node inside the node and the label of a node is indicated outside, next to the node). Thus the following graph :

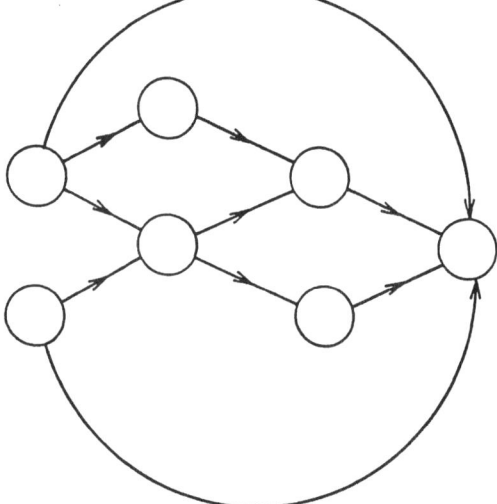

Figure 0.3

is a naked $(\Gamma-)$ dependence graph. $\square$

The following basic property of Γ-dependence graph is from [AR1].

Proposition 0.1. Let $\Gamma = (\Sigma, D)$ be a dependence alphabet. A node labeled graph $g = (V, E, R_\phi, \phi)$ is a Γ-dependence graph iff (V, E) is acyclic and for all $u, v \in V$, $(u,v) \in E$ iff $(\phi(u), \phi(v)) \in D)$. $\square$

In particular, since a dependence alphabet is a sr graph, this result implies the following property.

Corollary 0.1. Let $g = (V, E, R_\phi, \phi)$ be a dependence graph. If $u, v \in V$ are such that $u \neq v$ and $\phi(u) = \phi(v)$, then $(u, v) \in sym((V, E))$. $\square$

1. STABLE LABELINGS

In this section we consider labelings of graphs satisfying a specific "consistency" condition. This condition is satisfied by dependence graphs and investigating labelings of graphs satisfying this condition constitutes the first step in our effort to characterize (naked) dependence graphs.

Definition. Let g be a graph and let $h = sym(g)$. A labeling ϕ of g is *stable* (*on* g) iff for all pairs (v_1, v_2), (v_1', v_2') of nodes from V_g such that $v_1 \neq v_2$ and $v_1' \neq v_2'$, if $\phi(v_i) = \phi(v_i')$ for $i \in \{1,2\}$, then $(v_1, v_2) \in E_h$ iff $(v_1', v_2') \in E_h$. $\square$

In order to relate various labelings of the same graph we need the notion of a refinement.

Definition. Let g be a graph and let ϕ, ψ be labelings of g. We say that ψ is a *refinement of* ϕ if there exists an onto coding $\lambda: R_\psi \to R_\phi$ such that, for each $v \in V_g$, $\phi(v) = \lambda(\psi(v))$. $\square$

The following lemma expressing some basic properties of stable labelings follows directly from the above two definitions.

Lemma 1.1. Let g be a graph and ϕ , ψ be labelings of g .

(1) ϕ is stable on g iff ϕ is stable on $sym(g)$.

(2) If ϕ is stable on g and ψ is a refinement of ϕ , then ψ is stable on g . $\square$

In the next two sections, unless explicitly stated otherwise, we will consider only symmetric graphs. Since we will be interested in properties of stable labelings, according to Lemma 1.1(1) this will not lead to a loss of generality.

Definition. Let g be a s graph. The *similarity relation of* g , denoted sim_g is the binary relation on V_g defined by : for all v_1 , $v_2 \in V_g$, $(v_1 , v_2) \in sim_g$ iff $\{v_1 , v_2\}$ is a clan of g . $\square$

Theorem 1.1. For each s graph g , sim_g is an equivalence relation.

Proof.

(1) *Reflexivity.*

Obviously for each $v \in N_g$, $(v,v) \in sim_g$.

(2) *Symmetry.*

Obviously, for all v_1 , $v_2 \in N_g$, $(v_1 , v_2) \in sim_g$ implies $(v_2 , v_1) \in sim_g$.

(3) *Transitivity.*

Consider v_1 , v_2 , $v_3 \in N_g$ and assume that (v_1 , v_2) , $(v_2 , v_3) \in sim_g$.

If $\# \{v_1 , v_2 , v_3\} \leq 2$, then from (1) and (2) it follows that $(v_1 , v_3) \in sim_g$.

Hence assume that $\#\{v_1 , v_2 , v_3\} = 3$. Consider $v \notin \{v_1 , v_3\}$.

(i) Assume that $v = v_2$.

Since $(v_1 , v_2) \in sim_g$, $(v_3 , v_1) \in E_g$ iff $(v_3 , v_2) \in E_g$.

Since $(v_2 , v_3) \in sim_g$, $(v_1 , v_3) \in E_g$ iff $(v_1 , v_2) \in E_g$.

Hence $(v_2 , v_1) \in E_g$ iff $(v_2 , v_3) \in E_g$ and consequently $(v, v_1) \in E_q$ iff $(v, v_3) \in E_g$.

(ii) Assume that $v \neq v_2$.

Since $(v_1 , v_2) \in sim_g$, $(v, v_1) \in E_g$ iff $(v, v_2) \in E_g$.

Since $(v_2 , v_3) \in sim_g$, $(v, v_2) \in E_g$ iff $v, v_3) \in E_g$.

Hence $(v, v_1) \in E_g$ iff $(v, v_3) \in E_g$.

From (i) and (ii) the transitivity of sim_g follows also when all of v_1 , v_2 , v_3 are different

nodes. □

For a s graph g and a node $v \in V_g$ we write $[v]$ to denote the equivalence class of sim_g

containing v and $\mathbf{P}_g$ denotes the set of equivalence classes of sim_g. Clearly each element of

$\mathbf{P}_g$ is a clan in g.

Definition. Let g be a s graph. The *canonical labeling of* g, denoted cal_g, is defined by

: for all $v \in V_g$, $cal_g(v) = [v]$. □

Example 1.1. Let g be a s graph with $V_g = \{1, \ldots, 11\}$ and E_g as follows :

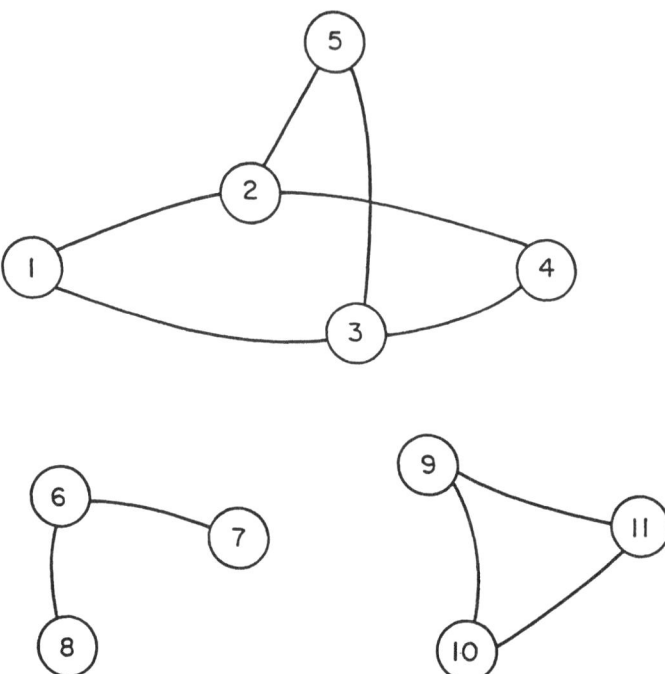

Figure 1.1

The equivalence classes of sim_g are :

$a = \{1,5,4\}$, $b = \{2,3\}$, $c = \{6\}$, $d = \{7,8\}$, $e = \{9,10,11\}$.

The subgraphs of g induced by these equivalence classes are as follows :

g(a) :

Figure 1.2

g(b) :

Figure 1.3

g(c) :

Figure 1.4

g(d) :

Figure 1.5

g(e) :

Figure 1.6

The canonical labeling of g is as follows :

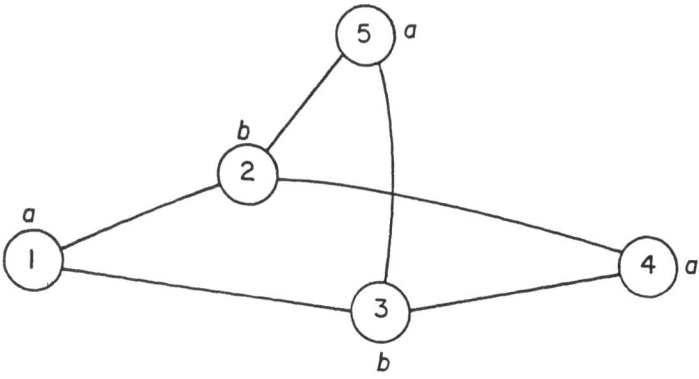

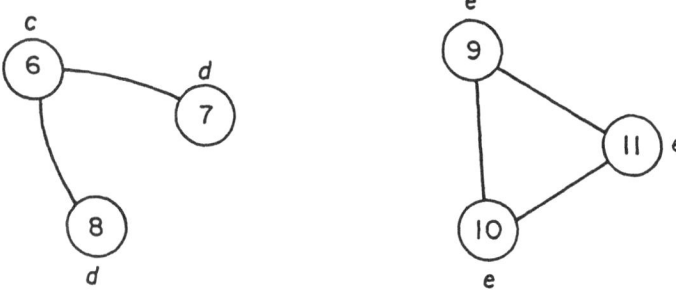

Figure 1.7

As we can see each of the subgraphs $g(a), \ldots, g(e)$ from the above example is either discrete or a weak clique. As a matter of fact this is a general phenomenon as shown by the following result.

Lemma 1.2. Let g be a s graph and let $v \in V_g$. Either $g([v])$ is discrete or $g([v])$ is a weak clique of g.

Proof.

Let g be a s graph. Consider $g([v])$ for an arbitrary $v \in V_G$.

If $\# [v] \leq 2$ then the conclusion of the lemma obviously holds.

If $\# [v] \geq 3$ then consider arbitrary three different nodes v_1 , v_2 , v_3 from $[v]$.

(i) Assume that $(v_1, v_2) \in E_g$.

Since $(v_2 , v_3) \in sim_g$, $(v_1 , v_3) \in E_g$. Hence, because $(v_1 , v_2) \in sim_g$,

$(v_2 , v_3) \in E_g$. Consequently, $g(\{v_1 , v_2 , v_3\})$ is a clique.

(ii) Assume that $(v_1 , v_2) \notin E_g$.

By the reasoning as above we conclude that $G([v])$ is discrete.

From (i) and (ii) it follows that the result holds also if $\#[v] \geq 3$. $\square$

Since, for a s graph g and a $v \in V_g$, $[v]$ is a clan, the following result on clans depicts

the structure of (connections between the elements of) $\mathbf{P}_g$.

Lemma 1.3. Let g be a s graph and V_1 , $V_2 \subseteq V$ be two   disjoint   clans of g .

Then for all v_1 , $u_1 \in V_1$, v_2 , $u_2 \in V_2$, $(v_1 , u_2) \in E_g$ iff $(u_1 , v_2) \in E_g$.

Proof.

Let g be a s graph and V_1 , $V_2 \subseteq V$ be two   disjoint   clans of g . If either

$\# V_1 = 1$ or $\# V_2 = 1$ then the lemma obviously holds.

So assume that both $\# V_1 \geq 2$ and $\# V_2 \geq 2$ and let

v_1 , $u_1 \in V_1$, v_2 , $u_2 \in V_2$. Again, if either $v_1 = u_1$ or $v_2 = u_2$ the conclusion of the lemma

obviously holds. So assume that $v_1 \neq u_1$ and $v_2 \neq u_2$.

Since $(v_2 , u_2) \in sim_g$, $(u_1 , v_2) \in E_g$ iff $(u_1 , u_2) \in E_g$.

Since $(v_1 , u_1) \in sim_g$, $(u_2 , u_1) \in E_g$ iff $(u_2 , v_1) \in E_g$.

Hence $(v_1 , u_2) \in E_g$ iff $(v_2 , u_1) \in E_g$.

Thus the lemma holds. $\square$

Our interest in the canonical labelings is motivated by the following result.

Theorem 1.2. For each s graph g , cal_g is stable on g .

Proof.

Let g be a s graph and consider v_1 , $v_2 \in V_g$ such that $v_1 \neq v_2$.

We consider separately two cases.

(1) Assume that $cal_g(v_1) \neq cal_g(v_2)$.

Consider v_1' , v_2' such that $cal_g(v_1) = cal_g(v_1')$ and $cal_g(v_2) = cal_g(v_2')$; hence $[v_1] = [v_1'] \neq [v_2] = [v_2']$.

Since $(v_2 , v_2') \in sim_g$, $(v_1' , v_2') \in E_g$ iff $(v_1' , v_2) \in E_g$.

Since $(v_1 , v_1') \in sim_g$, $(v_2 , v_1') \in E_g$ iff $(v_2 , v_1) \in E_g$.

Consequently $(v_1' , v_2') \in E_g$ iff $(v_1 , v_2) \in E_g$.

(2) Assume that $cal_g(v_1) = cal_g(v_2)$.

Consider v_1' , v_2' such that $cal_g(v_1') = cal_g(v_1)$ and $cal_g(v_2') = cal_g(v_2)$; hence $[v_1] = [v_1'] = [v_2] = [v_2']$.

By Lemma 1.2 it immediately follows that $(v_1' , v_2') \in E_g$ iff $(v_1 , v_2) \in E_g$.

The theorem follow from (1) and (2) . $\square$

Hence, for each s graph g , cal_g is a stable labeling. Moreover, we can explain now why cal_g is referred to as the canonical labeling of g .

Lemma 1.4. Let g be a s graph. Every stable labeling ϕ of g is a refinement of cal_g .

Proof.

Let g be a s graph and ϕ a stable labeling of g .

Clearly, it suffices to show that, for all $v_1 , v_2 \in V_g$, $\phi(v_1) = \phi(v_2)$ implies $cal_g(v_1) = cal_g(v_2)$.

To this aim assume to the contrary that there exist v_1 , $v_2 \in V_g$ such that $\phi(v_1) = \phi(v_2)$ but $cal_g(v_1) \neq cal_g(v_2)$. Since $cal_g(v_1) \neq cal_g(v_2)$, $[v_1] \neq [v_2]$ and so we can assume that there exists a $v \in V_g^v - \{v_1 , v_2\}$ such that $(v , v_1) \in V_g$ and $(v , v_2) \notin V_g$. This however contradicts the stability of ϕ (because $\phi(v_1) = \phi(v_2)$) .

Consequently, for all v_1 , $v_2 \in V_g$, $\phi(v_1) = \phi(v_2)$ implies $cal_g(v_1) = cal_g(v_2)$.

This allows us to define an onto coding $\lambda : \mathbf{R}_\phi \rightarrow \mathbf{P}_g$ by : for all $a \in \mathbf{R}_\phi$, $\lambda(a) = cal_g(v)$ where $v \in V_g$ is such that $\phi(v) = a$. $\square$

We are ready now to prove the main results of this section.

Theorem 1.3. Consider an arbitrary (not necessarily symmetric) graph h and let $g = sym(h)$. A labeling ϕ of h is stable iff ϕ is a refinement of cal_g .

Proof.

Follows directly from Lemma 1.1, Theorem 1.2 and Lemma 1.4 . $\square$

Example 1.2. Let h be the following graph :

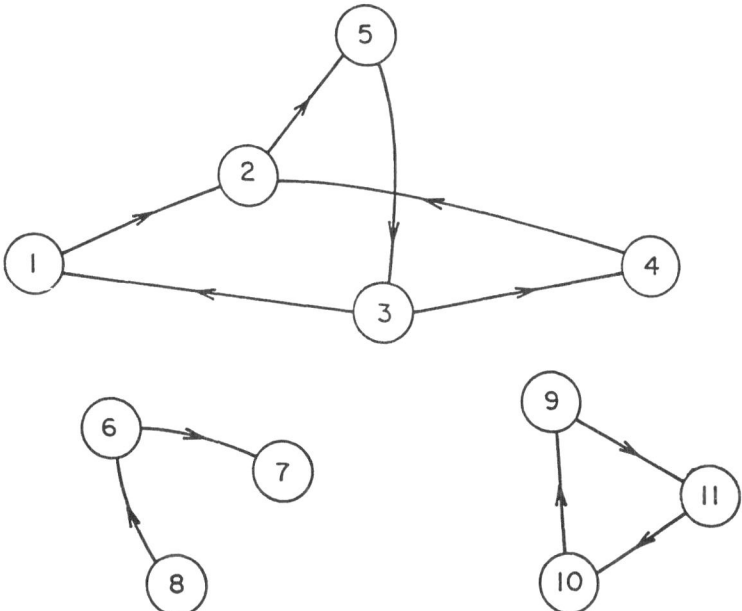

Figure 1.8

Clearly $g = sym(h)$ is the graph from Example 1.1 .

(i) Consider the following labeling of ϕ of h :

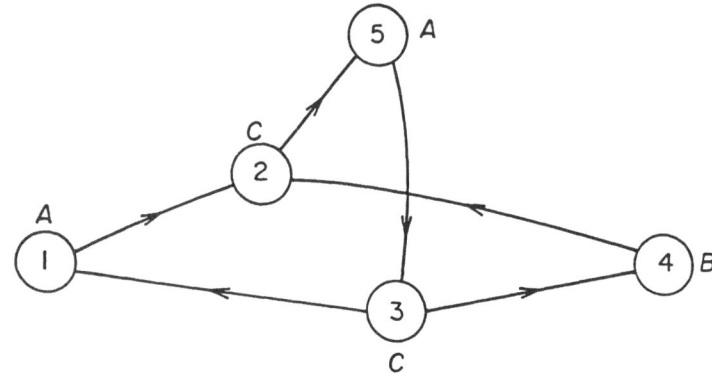

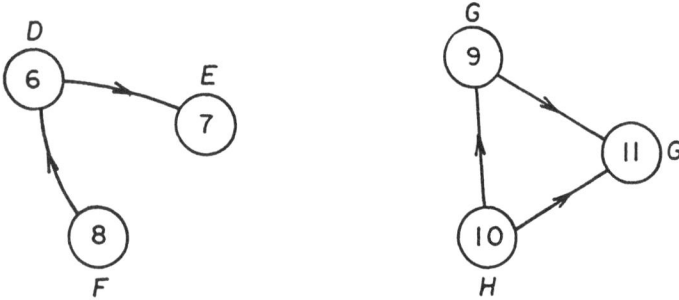

Figure 1.9

It is easily seen that ϕ is a refinement of cal_g : consider the coding λ such that

$\lambda(A) = \lambda(B) = a$, $\lambda(C) = b$, $\lambda(D) = c$, $\lambda(E) = d$, $\lambda(F) = d$, $\lambda(G) = \lambda(H) = e$.

Consequently ϕ is stable (which can be also easily seen) .

(ii) Consider the following labeling ψ of h :

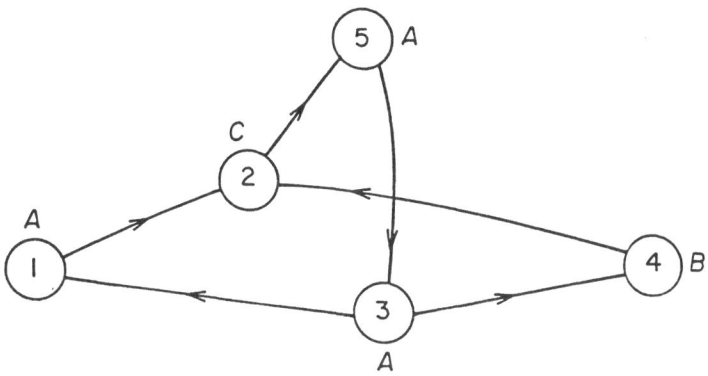

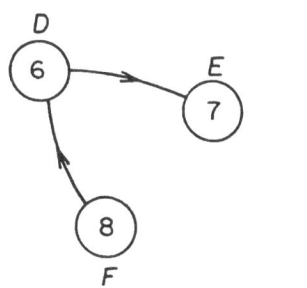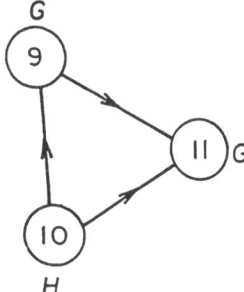

Figure 1.10

It is easily seen that ψ is not a refinement of cal_g : $[3] \neq [5]$ while $\psi(3) = \psi(5)$. Hence ψ is not stable; indeed $(3,4) \in E_g$ and $(5,4) \notin E_g$ while $\psi(3) = \psi(5)$. $\square$

2. Γ–LABELINGS

In this section we will consider the problem of labeling a graph "consistently with" a given dependence relation.

Throughout this section we fix a labeling alphabet Σ and a symmetric and reflexive relation $D \subseteq \Sigma \times \Sigma$; the dependence alphabet (Σ , D) is denoted by Γ .

The labeling of a graph g "consistently with" Γ is formally defined as follows.

Definition. Let g be a graph and $h = sym(g)$. A labeling ϕ of g is a $\Gamma-labeling$ of g

iff $\quad \mathbf{R}_\phi \subseteq \Sigma$ and, for all $v_1 , v_2 \in V_g$ such that $v_1 \neq v_2 ,$ $(v_1 , v_2) \in E_h$ iff

$(\phi(v_1) , \phi(v_2)) \in D$. $\square$

Clearly for a graph g each Γ-labeling of g is stable. It is easily seen that (by Proposition 0.1) the labeling of each Γ-dependence graph $(V , E , \mathbf{R}_\phi , \phi)$ is a Γ-labeling of (V , E) .

The following obvious result will allow us to consider sr graph only.

Lemma 2.1. Let g be an arbitrary (not necessarily symmetric) graph and let $h = symr(g)$. A labeling ϕ of g is a Γ-labeling of g iff ϕ is a Γ-labeling of h. $\square$

In the rest of this section, unless explicitly stated otherwise, we will consider sr graphs only. Since we will be interested in properties of stable labelings, according to Lemma 2.1 this will not lead to a loss of generality.

We can reformulate now in more algebraic terms the notion of a Γ-labeling of a sr graph.

Lemma 2.2. Let g be a sr graph. A labeling ϕ of g is a Γ-labeling iff $\phi \in HOM(g , \Gamma)$.

Proof.

Follows directly from the definition of a Γ-labeling and from the definition of a homomorphism between graphs. $\square$

We consider now clans of sr graphs generated by homomorphisms of these graphs.

Lemma 2.3. Let g_1 , g_2 be sr graphs, $\phi \in HOM(g_1 , g_2)$ and $v \in V_{g_2}$.

(1) $g_1(\phi^{-1} (v))$ is a clique of g_1 .

(2) $\phi^{-1}(v)$ is a clan of g_1 .

Proof.

(1) This follows directly from the fact that g_2 is reflexive.

(2) Consider a $z \in V_1 - \phi^{-1}(v)$ and a $u \in \phi^{-1}(v)$. Since ϕ is a homomorphism, $(z,u) \in V_{g_1}$ iff $(\phi(z), v) \in V_{g_2}$. Consequently $\phi^{-1}(v)$ is a clan. $\square$

For sr graphs g_1, g_2 and a $\phi \in HOM(g_1, g_2)$ we use $\mathbf{P}(\phi)$ to denote the set $\{\phi^{-1}(v) \neq \emptyset : v \in V_{g_2}\}$. We call a clan *complete* if the subgraph induced by it is a clique.

Lemma 2.4. Let g_1, g_2 be sr graphs and $\phi \in HOM(g_1, g_2)$. Then $\mathbf{P}(\phi)$ is a partition of V_{g_1} into complete clans.

Proof.

Follows directly from Lemma 2.3. $\square$

Hence homomorphisms of sr graphs generate partitions of these graphs into complete graphs. The situation holds also the other way around: partitions of sr graphs into complete clans induce homomorphisms of them.

Lemma 2.5. Let g be a sr graph and $\mathbf{P}$ a partition of V_g into complete clans. Let ϕ be the function, $\phi : V_g \rightarrow \mathbf{P}$, defined by : for all $v \in V_g$, $\phi(v) = [v]_{\mathbf{P}}$. Then $\phi \in HOM(g,h)$ where h is the graph $(\mathbf{P},F)$ such that, for all $X,Y \in \mathbf{P}$, $(X,Y) \in F$ iff there exist $x \in X$, $y \in Y$ such that $(x,y) \in E_g$.

Proof.

Follows directly from Lemma 1.3. $\square$

Thus Lemma 2.4 and Lemma 2.5 together allow us to identify homomorphisms of sr graphs with partitions of sr graphs into complete clans.

For a s graph g we will use $\hat{P}_g$ to denote the following partition of V_g :

for a $v \in V_g$ such that $[v]_{P_g}$ is complete, $[v]_{\hat{P}_g} = [v]_{P_g}$,

for a $v \in V_g$ such that $[v]_{P_g}$ is discrete, $[v]_{\hat{P}_g} = \{v\}$.

Then $\widehat{cal}_g$ is defined by : for all $v \in V_g$, $\widehat{cal}_g (v) = [v]_{\hat{P}_g}$ and $\hat{g} = (\hat{P}_g , \hat{E}_g)$ where for

$X , Y \in \hat{P}_g$, $(X , Y) \in \hat{E}_g$ iff there exist $x \in X , y \in Y$ such that $(x , y) \in E_g$.

Example 2.1. Let g be the s graph from Example 1.1.

Then $\hat{P}_g = \{a_1 , a_4 , a_5 , b_2 , b_3 , c , d_7 , d_8 , e\}$, where c, e are as in Example 1.1 and

$a_1 = \{1\}, a_4 = \{4\}, a_5 = \{5\}, b_2 = \{2\}, b_3 = \{3\}$, and $\hat{g}$ is as follows :

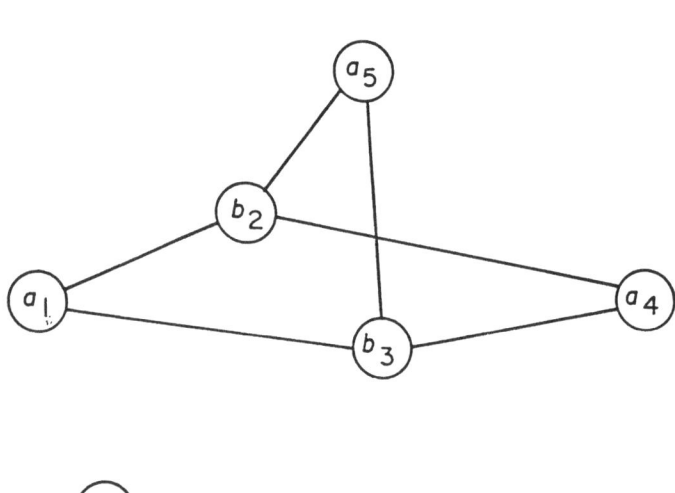

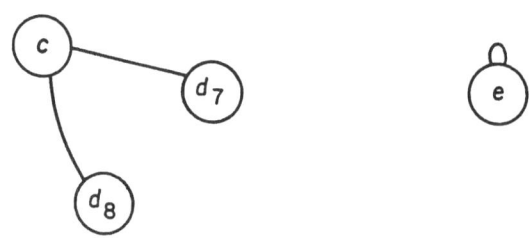

Figure 2.1

Example 2.2. Let g be the following sr graph :

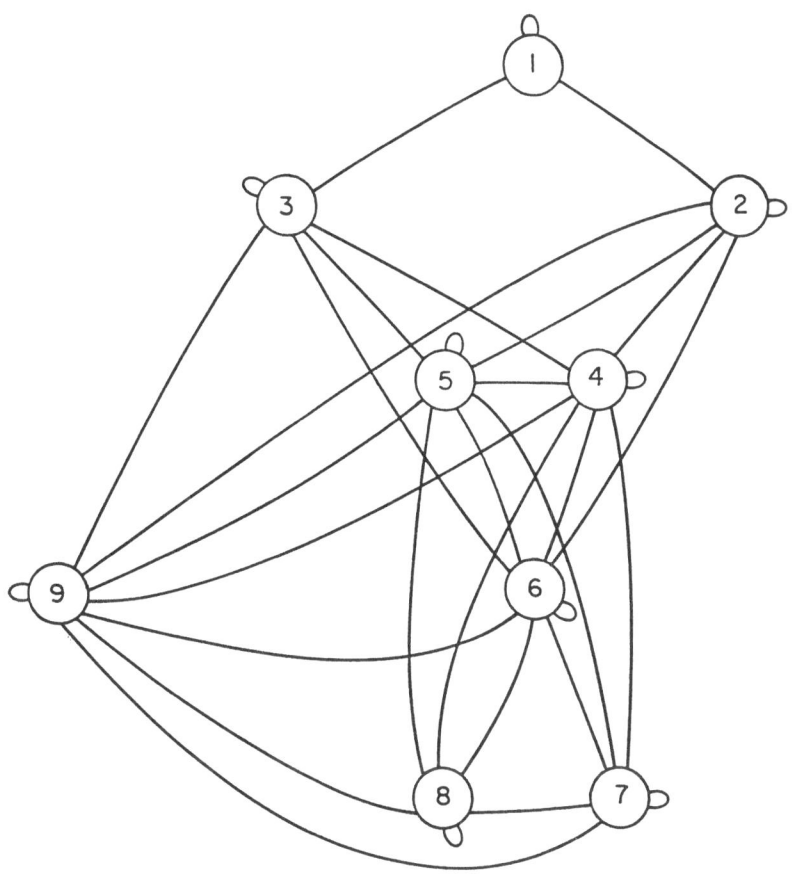

Figure 2.2

Then $\hat{\mathbf{P}}_g = \{a\,,b_1,b_2,c\,,d\,,e\}$, where $a = \{1\}$, $b_1 = \{2\}$, $b_2 = \{3\}$,

$c = \{4,5,6\}$, $d = \{7,8\}$ and $e = \{9\}$, and $\hat{g}$ is the following graph :

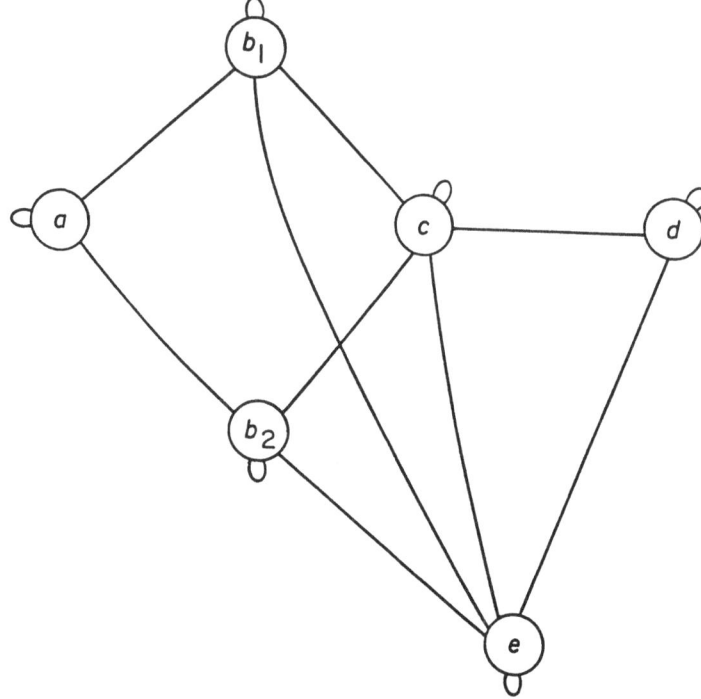

Figure 2.3

The following properties of $\hat{\mathbf{P}}_g$ for a sr graph g illustrated by the above two examples
follow directly from the definition of $\hat{\mathbf{P}}_g$.

Lemma 2.6. Let g be a sr graph.

(1) Each element of $\hat{\mathbf{P}}_g$ is a complete clan of G .

(2) $(X,X) \in \hat{E}_g$ for each $X \in \hat{\mathbf{P}}_g$.   □

The partition $\hat{\mathbf{P}}_g$ has a very useful maximality property.

Definition. Let g be a sr graph. A partition $\mathbf{P}$ of V_g into complete clans is
maximal if for every partition $\mathbf{P}'$ of V_g into complete clans we have: $\mathbf{P}' \subseteq \mathbf{P}$.

□

Lemma 2.7. Let g be a sr graph. Then $\hat{P}_g$ is a maximal partition of V_g into complete clans.

Proof.

Let g be a sr graph and consider $\hat{P}_g$. By Lemma 2.6, each $X \in \hat{P}_g$ is a complete clan. Let $\mathbf{P}$ be a partition of V_g into complete clans.

Assume that $\mathbf{P} \subseteq \hat{P}_g$ does not hold. Hence there exist $X, Y \in \hat{P}_g$ and $Z \in \mathbf{P}$ such that $Z \cap X \neq \emptyset$ and $Z \cap Y \neq \emptyset$. Let $x \in Z \cap X$ and $y \in Z \cap Y$. Since x, y are in different classes of $\hat{P}_g$, $\{x, y\}$ is not a clan. Consequently there exist a $t \in V_g$ such that either $(t, x) \in E_g$ and $(t, y) \notin E_g$ or $(t, x) \notin E_g$ and $(t, y) \in E_g$. Since Z is a clan, $t \notin V_g - Z$ and since Z is a complete clan $t \notin Z$; a contradiction.

Consequently $\mathbf{P} \subseteq \hat{P}_g$ and so $\hat{P}_g$ is maximal.

Hence the lemma holds. $\square$

The following technical lemma will be needed to prove the main result of this section.

Lemma 2.8. Let $g_1 = (V_1, E_1)$, $g_2 = (V_2, E_2)$ be sr graphs and $\phi \in HOM(g_1, g_2)$. There exists a $U \subseteq V_1$ such that
$$\phi|_U \in ISOM(g_1(U), g_2(R_o)).$$

Proof.

Consider $\mathbf{P}(\phi)$ and let $U \subseteq V_1$ be such that it contains exactly one element from each class of $\mathbf{P}(\phi)$. The result follows now from Lemma 1.3 and Lemma 2.4. $\square$

We are ready now to prove the main result of this section.

Theorem 2.2. Let h be an arbitrary (not necessarily sr) graph and let $g = symr(h)$. There exists a Γ-labeling of h iff $\hat{g}$ is isomorphic with a subgraph of Γ.

Proof.

(i) Let ϕ be a Γ-labeling of h . Thus, by Lemma 2.1, ϕ is a Γ-labeling of g .

By Lemma 2.7, $\mathbf{P}(\phi) \subseteq \hat{\mathbf{P}}_g$ and consequently there exists a $\psi \in HOM(\Gamma(R_\phi), \hat{g})$ which is onto. By Lemma 2.8, there exists a subset $U \subseteq R_\phi$ such that $\psi|_U \in ISOM(\Gamma(U), \hat{g})$.

Consequently $\hat{g}$ is isomorphic with a subgraph of Γ.

(ii) Assume that $\hat{g}$ is isomorphic with a subgraph of Γ and let ψ be an isomorphism mapping $\hat{g}$ onto a subgraph of Γ.

Then obviously the composition ρ of $\hat{cal_g}$ with ψ is a Γ-labeling of g and hence (by Lemma 2.1) ρ is a Γ-labeling of h .

The theorem follows from (i) and (ii). $\square$

Example 2.3. Let h be the following graph :

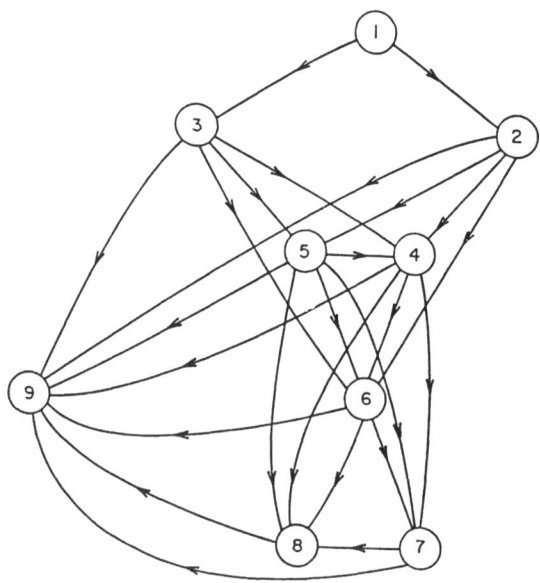

Figure 2.4

165

and let $\Gamma = (\Sigma, D)$ be the following dependence alphabet :

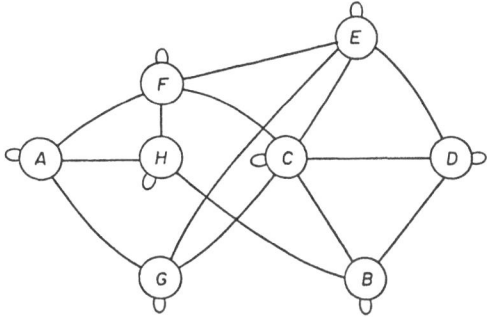

Figure 2.5

Obviously *symr* (h) is the graph g from Example 2.2.

Now if we consider $\Gamma(\Sigma-\{B,H\})$, then we easily notice that $\hat{g}$ is isomorphic to $\Gamma(\Sigma-\{B,H\})$. Consequently there exists a Γ-labeling of h and the obvious isomorphism between $\hat{g}$ and $\Gamma(\Sigma-\{B,H\})$ yields the following Γ-labeling of h :

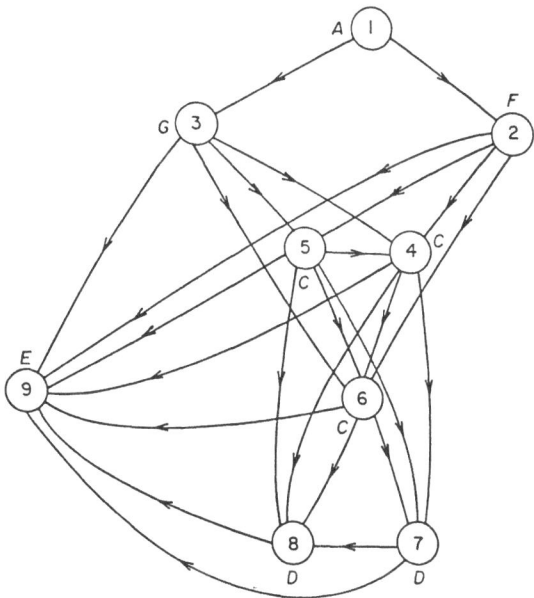

Figure 2.6

If Γ' is the dependence alphabet resulting from Γ by removing any (symmetric) edge to E from a node different from E , then the maximal degree of any node in Γ' becomes 4 and consequently $\hat{g}$ is not isomorphic with a subgraph of Γ'. Thus there is no Γ-labeling of h .

□

3. DEPENDENCE GRAPHS

We turn now to dependence graphs. In particular we will use our main results, Theorem 1.3 and Theorem 2.2, to characterize dependence graphs and naked Γ-dependence graphs for a given Γ .

Theorem 3.1. Let $h = (V, E, R_\phi, \phi)$ be a node labeled graph and let $g = und(h)$.

Then h is a dependence graph iff g is acyclic and ϕ is a refinement of $\widehat{cal}_g$.

Proof.

Let $h = (V, E, R_\phi, \phi)$ be a node labeled graph and let $g = und(h)$.

(i) Assume that h is a dependence graph.

By Proposition 0.1 , h is acyclic. By Proposition 0.1 , ϕ is stable on g and hence, by

Theorem 1.3 , ϕ is a refinement of cal_g . Thus, by Corollary 0.1 , ϕ is a refinement of $\widehat{cal}_g$.

(ii) Assume that g is acyclic and ϕ is a refinement of $\widehat{cal}_g$, and let $f = symr(g)$. Define the pair $\Gamma = (R_\phi, D_g)$ by : for all $a, b \in R_\phi$, $(a, b) \in D_g$ iff there exist $u, v \in V_g$ such that $\phi(u) = a$, $\phi(v) = b$ and $([u], [v]) \in E_{\hat{f}}$.

From Lemma 2.6 it follows that D_g is symmetric and reflexive — consequently Γ is a dependence alphabet.

Since h is acyclic, the definition of D_g, Lemma 1.2, Lemma 1.3 and Proposition 0.1 imply that g is a Γ-dependence graph. $\square$

Theorem 3.2. Let $\Gamma = (\Sigma, D)$ be a dependence alphabet, let g be a graph and let $f = symr(g)$. Then, g is a naked Γ-dependence graph iff g is acyclic and $\hat{f}$ is isomorphic with a subgraph of Γ.

Proof.

Let Γ, g, f be as in the assumption of the theorem.

(i) Assume that g is a naked dependence graph ; let $h = (V, E, R_\phi, \phi)$ be a node labeled graph such that h is a Γ-dependence graph and $g = und(h)$.

By Proposition 0.1, ϕ is stable on f. Hence, by Theorem 1.3, ϕ is a refinement of cal_f and

so, by Corollary 0.1, ϕ is a refinement of $\widehat{cal_f}$. By Lemma 1.3, Lemma 2.6 and Proposition 0.1 , $\hat{f}$ is isomorphic to $\Gamma(R_\phi)$.

Hence by Proposition 0.1, g is acyclic and $\hat{f}$ is isomorphic to a subgraph of Γ.

(ii) Assume that g is acyclic and that $\hat{f}$ is isomorphic to a subgraph of Γ. Let ψ be an isomorphism of $\hat{f}$ onto a subgraph of Γ. If we label now each $v \in V_g$ by $\psi([v]_{\hat{P_f}})$, then, because g is acyclic and ψ is an isomorphism (and by Lemma 2.6(1)), the so obtained graph satisfies Proposition 0.1 .

Hence g is a Γ-dependence graph. $\square$

4. DISCUSSION

In this paper we have succeeded in providing a characterization of *(naked) dependence graphs* — this was achieved in a broader framework of characterizing certain kinds of labelings of s and sr graphs.

There is something "philosophical" about our main result (Theorem 3.2). Let's consider a concurrent system $S = (K, D)$ as it is modeled in the theory of traces (sketched in the introduction to this paper). A dependence graph g in S represents a possible behaviour of (a *process* in) S and it is obtained by taking a sequential observation from K (a $w \in K$) and "breaking it" into a nonsequential object using the information about the system as given by D. Our result says that g when viewed "properly" (it is: seen as $\hat{g}$) is an image of a part of the system! ($\hat{g}$ is isomorphic to a subgraph of $\Gamma = (\Sigma, D)$ where Σ is the set of events of the system). Hence "the observational point of view" and the "image of the system point of view" — two popular approaches to define processes of concurrent systems "coincide" within the theory of traces !!!

The results on *naked dependence graphs* as presented in this paper and the (rather obvious) characterization of *dependence graphs* as given by Proposition 0.1 together give a rather complete picture of dependence graphs. An obvious continuation of the research presented here is to consider *languages* of (naked) dependence graphs which would express the set of all behaviours of the system; thus for a system $S = (K, D)$ as above its language of (naked) dependence graphs consists of *all* dependence graphs obtained by taking strings from K and breaking them by D.

A characterization of languages of (naked) dependence graphs in the case when K is regular is our next goal.

ACKNOWLEDGEMENTS

The authors gratefully acknowledge the support by the National Science Foundation under grant number MCS-8305245.

REFERENCES

[AR1]   Aalbersberg, IJ.J. and Rozenberg, G., Theory of traces, Dept. of Computer Science, University of Leiden, Techn. Report No. 86-16, 1986.

[AR2]   Aalbersberg, IJ.J. and Rozenberg, G., Traces, dependence graphs and DNLC grammars, *Discrete Applied Mathematics,* v. 11, pp. 299-306, 1985.

[BMS]   Bertoni, A., Mauri, G. and Sabadini, N., Equivalence and membership problems for regular trace languages, *Lecture Notes in Computer Science,* v. 140, pp. 61-71, 1982.

[CF]   Cartier, P. and Foata, D., Problemes combinatoires de commutation et rearrangements, *Lecture Notes in Mathematics,* v. 85, 1981.

[CP]   Cori, R. and Perrin, D., Automates et commutations partielles, *RAIRO, Informatique Theorique,* v. 19, pp. 21-32, 1985.

[ENR]   H. Ehrig, M. Nagl and G. Rozenberg (eds.), Graph grammars and their applications to computer science, *Lecture Notes in Computer Science,* v. 153, 1983.

[ER]   Ehrenfeucht, A. and Rozenberg, G., Structures and clans, Dept. of Computer Science, University of Colorado at Boulder, Technical Report No. CU-CS-328-86, Boulder, 1986.

[FR]   Fle, M.P. and Roucairol, G., On serializability of iterated transactions, *Proc. ACM SIGACT-SIGOPS Symp. on Principles of Distributed Computing,* pp. 194-200, 1982.

[JR1]   Janssens, D. and Rozenberg, G., On the structure of node-label controlled graph languages, *Information Sciences,* v. 20, pp. 191-216, 1980.

[JR2]   Janssens, D. and Rozenberg, G., A characterization of context-free string languages by directed node-label controlled graph grammars, *Acta Informatica,* v. 16, pp. 63-85, 1981.

[K]   Keller, R.M., A solvable program-schema equivalence problem, *Proc. 5th Annual Princeton Conference on Information Sciences and Systems,* pp. 301-306, Princeton, 1971.

[M]   Mazurkiewicz, A., Concurrent program schemes and their interpretations, Dept. of Computer Science, University of Aarhus, Technical Report No. PB-78, Aarhus, 1977.

[O]   Ochmanski, E., Regular trace languages, Ph.D. Thesis, Dept. of Mathematics, University of Warsaw, 1985.

[RW]   Rozenberg, G. and Welzl, E., Boundary NLC grammars. Basic definitions, normal forms and complexity, *Information and Control,* v. 69, pp. 136-167, 1986.

SOME REMARKS ON D-CONTINUITY

César Fernández   Agathe Merceron*
GMD-F1P, Postfach 1240 , 5205 St.Augustin 1
Federal Republic of Germany

1   Introduction

One afternoon in 1978 (or was it in 1977 ?), C.A. Petri called one of the authors of this paper to his room. He wanted to talk about some "axioms of concurrency". By that time, we knew already that a non-sequential process –viewed as a record of conditions-holdings and events-occurrences of a system– could be modelled using a special type of net called occurrence net, [7],[6]. Since at the behavior level, cycles are unrolled, one may associate to each occurrence net a partially ordered set, poset for short, which is an appropriate mathematical machinery to study non-sequential processes.

On the one hand, since a non-sequential process can be described by an occurrence net, hence by a poset, one can study the properties of the posets obtained from some concurrent system. In this way, the starting point is a concurrent system and one studies its semantics using posets.

On the other hand, one can propose a model of a non-sequential process, i.e. start with a partially ordered set $(X, \prec)$ (or even with a more general structure (X, co) where co is a relation of disorder) and impose to this structure a set of well motivated axioms, called "axioms of concurrency", so that this poset becomes suitable to represent a run of a concurrent system. That was the idea of C.A. Petri.

C.A. Petri thought that starting with a rope (a poset satisfying the concurrency axioms) one could reconstruct the occurrence net. Furthermore, he thought that ropes have a special property called by him: Generalised Dedekind Continuity (D-continuity for short).

D-continuity is a generalisation to posets of the completeness property of the reals.

If we think that the reals have no jumps (unlike the integers) and no gaps (unlike the rationals), then, the idea of "free of jumps" and "free of gaps" should be transported to posets.

The way to do this is not unique. C.A. Petri suggested in that afternoon several possibilities. Later, in 1979, C.A. Petri presented his Concurrency paper [8], in which the concurrency axioms and the generalised Dedekind continuity idea appear for the first time.

Since that paper, several authors have studied the D-continuity property ([9],[2],[5],[1]). In particular, some links between D-continuity and the axioms of concurrency are exhibited in [2] and [5].

In this paper we would like to discuss D-continuity, mainly using examples.

The structure of the paper is the following:

Section 2 contains basic definitions about posets.

In section 3 we discuss an example which has been permanently used by several authors but not yet been published .

Section 4 is devoted to D-continuity as defined in [8] and [9].

The paper finishes, in section 5, with some final words.

---

*Also L.R.I. Université Paris-Sud 91405 Orsay Cedex France

2   Basic definitions

Since we have to deal with partially ordered sets (posets), let us start with their definition.

Definition 2.1 *Poset*

> *A pair $(X, \prec)$ is called a partially ordered set (poset for short) iff*
>
>> 1. $\prec \subseteq X \times X$;
>>
>> 2. $\prec$ *is irreflexive, i.e.* $\forall x \in X\colon \neg(x \prec x)$;
>>
>> 3. $\prec$ *is transitive, i.e.* $\forall x, y, z \in X\colon x \prec y \wedge y \prec z \Rightarrow x \prec z$.

As usual we have: $\preceq = \prec \cup\, id_X,\ \succ = \prec^{-1}$ and $\succeq = \succ \cup\, id_X$.
Graphically we shall denote by dots ($\bullet$) the elements of X and $x \prec y$ will be denoted by
$x\, \bullet\!\!\sim\!\!\sim\!\!\sim\!\!\bullet\, y$. If $x \prec y$ and there is no $z \in X$ such that $x \prec z \prec y$ then we shall draw
$x\, \bullet\!\!\!-\!\!\!-\!\!\!\bullet\, y$.
In this paper we shall assume $X \neq \emptyset$.
We would like to interpret a poset $(X, \prec)$ as a process in which the elements of X represent basic
occurrences and $\prec$ represents the relation of sequentiality between basic occurrences. If $x \prec y$ we
would say that the occurrence x has happened "before" the occurrence y.
The relation $\prec$ will allow us to define two new relations which are going to turn out to be important
in what follows. We introduce them in the next definition.

Definition 2.2 *The relations li and co, lines and cuts*

> *Let $(X, \prec)$ be a poset.*
>
>> 1. $li = \prec \cup \succ \cup\, id_X$; $co = (X \times X \setminus li) \cup id_X$.
>>
>> 2. *Let $l \subseteq X$;*
>>   *l is a li-set iff $\forall x, y \in l\colon x\ li\ y$;*
>>   *l is a line iff it is a maximal li-set, i.e. it is a li-set and*
>>   *$\forall z \in X \setminus l\ \exists y \in l\colon z\ co\ y$;*
>>   *L denotes the set of lines of X.*
>>
>> 3. *Let $c \subseteq X$;*
>>   *c is a co-set iff $\forall x, y \in c\colon x\ co\ y$;*
>>   *c is a cut iff it is a maximal co-set, i.e. it is a co-set and*
>>   *$\forall z \in X \setminus c\ \exists y \in c\colon z\ li\ y$;*
>>   *C denotes the set of cuts of X.*

A li-set l is a subset of X in which the basic occurrences of l occur sequentially. A line may be
viewed as a sequential subprocess of the process $(X, \prec)$.
In the same way, a co-set c is a subset of X in which the basic occurrences of c occur concurrently.
The concept of a cut may be viewed as replacing the notion of time point.
In this paper we shall assume the axiom of choice. Using this, we may prove that each co-set may
be embedded in a cut and also that each li-set may be embedded in a line.
Some important notations are introduced now.

Definition 2.3

    1. $\prec \ = \ \preccurlyeq \ \setminus \ \preccurlyeq^2$;

    2. If $x \in X$ then:
$$^\bullet x = \{y \in X \mid y \prec x\}$$
$$x^\bullet = \{y \in X \mid x \prec y\}.$$

The last notation is extended to subsets of X as follows: Let $A \subseteq X$, then $^\bullet A = \bigcup_{a \in A} {}^\bullet a$ and $A^\bullet = \bigcup_{a \in A} a^\bullet$.

The D-continuity idea is a general one, i.e. it can be applied to dense posets as well as to combinatorial ones. We define formally what these terms mean.

Definition 2.4 *Dense poset*

    A poset $(X, \prec)$ is called dense iff $\prec \ = \emptyset$.

In other words, if $x \prec y$ then there exists always a $z \in X$ such that $x \prec z \prec y$.
The notion of dense posets contrasts with the notion of combinatorial posets. Formally:

Definition 2.5 *Combinatorial poset*

    A poset $(X, \prec)$ is called combinatorial iff $\prec \ = (\prec)^+$ where $(\prec)^+ = \bigcup_{n \in \{1,2,3,\ldots\}} (\prec)^n$.

If we exclude the trivial case in which $li = id_X$, then it is easy to see that if a poset is dense it cannot be combinatorial and vice-versa. Definition 2.5 means that in a combinatorial poset, if $x \prec y$ then there exists a finite li-set $\{x_1, x_2, \ldots, x_n\}$ such that:
$$x = x_1 \prec x_2 \prec \ldots \prec x_{n-1} \prec x_n = y.$$
It is time now to proceed to the next section.

3   An example

In this section we would like to introduce an example in order to clarify (hopefully!) the concepts introduced so far.
This example was given to us – a long time ago – by C.A. Petri. It has not been published anywhere (as far as we know), in spite of the fact that many authors –working on D-continuity– have used it.
We start with the set $\mathbf{R}^2 = \{x = (x_1, x_2) \mid x_1, x_2 \in \mathbf{R}\}$.
We define the following order relation: Let $x = (x_1, x_2)$, $y = (y_1, y_2)$ be elements of $\mathbf{R}^2$. Then:
$x \prec y$ iff $x_1 < y_1 \wedge x_2 < y_2$.
The pair $(\mathbf{R}^2, \prec)$ is clearly a poset.
$(\mathbf{R}^2, \prec)$ is dense and, hence, not combinatorial. A geometrical line:
$x_2 = ax_1 + b$ could be a line or a cut in the sense of definition 2.2. If a is a positive real number, then $x_2 = ax_1 + b$ is a line. If a is a negative real number or if $a = 0$ then $x_2 = ax_1 + b$ is a cut.
Geometrical lines of the type $x_1 = k$, $k \in \mathbf{R}$ are also cuts in the sense of definition 2.2. The set $\{x = (k, k) \mid k \in \mathbf{Z}\}$ is a li-set which is not a line since it is not a maximal li-set.
Similarly, the set $\{x = (k, -k) \mid k \in \mathbf{Z}\}$ is a co-set which is not a cut.
In fact, any monotonically increasing continuous function $x_2 = f(x_1)$ is a line and any monotonically decreasing continuous function is a cut.
As a matter of exercise, we would like –at this point– to recall some of the concurrency axioms introduced by C.A. Petri in [8], and see if they are fulfilled by $(\mathbf{R}^2, \prec)$.

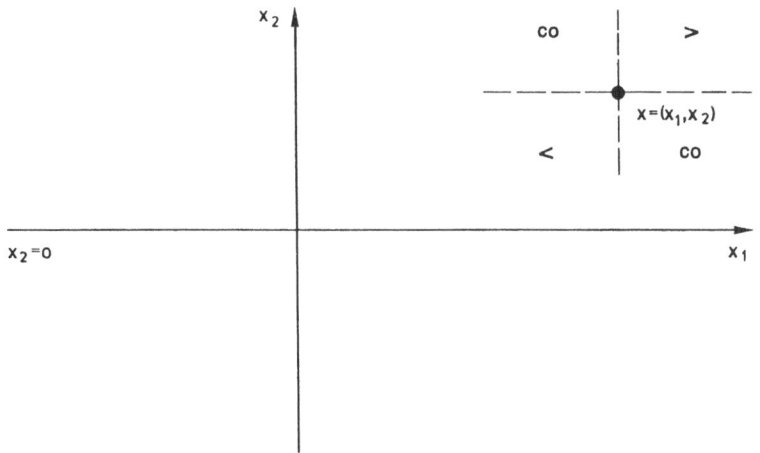

Figure 1: $(R^2, \prec)$

Definition 3.1 *Axioms of concurrency*

1. $(X, \prec)$ *is li-reduced iff* $Li(x) = Li(y) \Rightarrow x = y$ *where* $Li(x) = \{z \in X \mid z \; li \; x\}$.
2. $(X, \prec)$ *is co-reduced iff* $Co(x) = Co(y) \Rightarrow x = y$ *where* $Co(x) = \{z \in X \mid z \; co \; x\}$.
3. $(X, \prec)$ *is coherent iff* $co^* = li^*$ *where* $R^* = R^+ \cup id_X$ *for* $R \subseteq X \times X$.
4. $(X, \prec)$ *is K-dense iff* $\forall l \in L \; \forall c \in C: l \cap c \neq \emptyset$.

It is easy to see that $(\mathbf{R^2}, \prec)$ is li-reduced and co-reduced. $(\mathbf{R^2}, \prec)$ is also coherent (we omit the proof) but it is not K-dense since the line $x_2 = e^{x_1}$ and the cut $x_2 = 0$ do not meet each other (they are asymptotic).

We shall come back to this example in the sequel.

4 D-continuity

We have said in the introduction that D-continuity is a generalisation to posets of the completeness property of the real numbers. We would like to explain this generalisation starting with the old construction of cuts made by R. Dedekind.

In order to do that, let us consider a totally ordered set $(X, \prec)$, i.e. a poset in which for all $x, y \in X$ we must have $x \preceq y$ or $y \preceq x$. (Any two elements are comparable. In this case $co = id_X$.)

Definition 4.1 *Dedekind cuts in totally ordered sets*

Let $(X, \prec)$ be a totally ordered set and $A, \overline{A} \subseteq X$. The pair $(A, \overline{A})$ is a Dedekind cut iff:

1. $\emptyset \neq A \neq X$ and $\emptyset \neq \overline{A} \neq X$;
2. $A \cup \overline{A} = X$ and $A \cap \overline{A} = \emptyset$;
3. $\forall x \in A \; \forall y \in \overline{A}: x \prec y$.

The set of Dedekind cuts, or D-cuts for short, is denoted by D.

From 4.1, it follows that $\overline{A} = X \setminus A$. Sometimes we make an abuse of language and we say that A –instead of $(A, \overline{A})$– is a D-cut.

One equivalent (for totally ordered sets) form of 4.1.3 –which is going to be important while doing the generalisation of that definition to posets– is:

$3'$. $\forall x \in A \; \forall y \in \overline{A}$: $\neg(y \prec x)$.

Let $(A, \overline{A})$ be a D-cut of a totally ordered set $(X, \prec)$. A $(\overline{A}$, resp.$)$ could have one or none maximal (minimal) element; these elements are important to "complete" the set. Formally:

Definition 4.2 *Maximum and minimum*

> Let $(X, \prec)$ be a totally ordered set and A be a D-cut. Then:
>
> 1. $Max\ A = \{x \in A \mid \not\exists z \in A : x \prec z\}$.
> 2. $Min\ \overline{A} = \{x \in \overline{A} \mid \not\exists z \in \overline{A} : z \prec x\}$.
> 3. $M(A) = Max\ A \cup Min\ \overline{A}$.

In fact, $|Max\ A| = 0$ or 1 and $|Min\ \overline{A}| = 0$ or 1, i.e. $|M(A)| = 0$, 1 or 2.

In the case of posets, $Max\ A$ and $Min\ \overline{A}$ could have zero or several elements, as we shall see later.

Since we consider the case that $(X, \prec)$ is a totally ordered set, X has only one line, X itself, and every element of X forms a cut.

If $|M(A)| = 2$ we say that between A and $\overline{A}$ there is jump.

If $|M(A)| = 0$ we say that between A and $\overline{A}$ there is a gap (hole).

Consider as an example $(X, <)$, where X is a set of real numbers with the usual order. We take the following D-cut:

$$A = \{x \in X \mid x < 0 \ \vee \ x^2 < 2\}; \ \overline{A} = X \setminus A.$$

If $X = \mathbf{N}$ then $Max\ A = \{1\}$ and $Min\ \overline{A} = \{2\}$, hence $|M(A)| = 2$, there is a jump.

If $X = \mathbf{Q}$ then $Max\ A = \emptyset = Min\ \overline{A}$ and of course $|M(A)| = 0$, there is a gap.

We can extend the set $\mathbf{Q}$ filling the gaps; in this example "filling the gap" means to add the new element $\sqrt{2}$. In this way, we may complete the rationals and obtain the set $\mathbf{R}$. Indeed, if $X = \mathbf{R}$, we have $\forall A \in D: |M(A)| = 1$, which means that $\mathbf{R}$ is free of gaps and free of jumps or, in other words, $\mathbf{R}$ is complete (D-continuous), which leads to the following definition:

Definition 4.3 *D-continuity for totally ordered sets*

> Let $(X, \prec)$ be a totally ordered set.
> $(X, \prec)$ is said to be D-continuous iff $\forall A \in D: |M(A)| = 1$.

Now, we would like to generalise this idea to posets which are not necessarily dense and, of course, not totally ordered.

Let $(X, \prec)$ be a poset. A is a D-cut iff definition 4.1' is satisfied, where definition 4.1' is obtained by replacing 3 by 3' in definition 4.1.

$Max\ A$, $Min\ \overline{A}$ and $M(A)$ are defined exactly as in definition 4.2.

We are looking for a definition of D-continuity –for posets– which coincides with the definition 4.3 in the case of totally ordered sets. One simple possibility would be to adopt definition 4.3 for every line, i.e. $(X, \prec)$ is said to be D-continuous iff $\forall l \in L \ \forall A \in D: |M(A) \cap l| = 1$. It is easy to

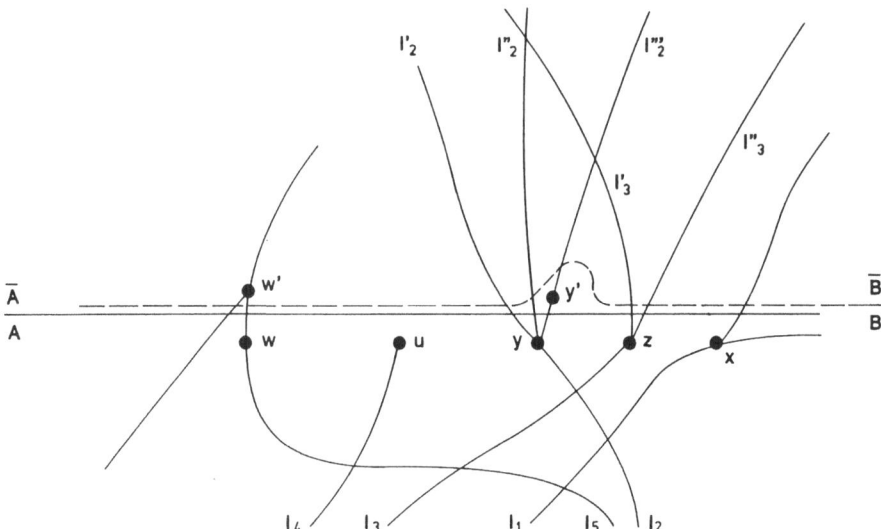

Figure 2: An illustration for the definition of D-continuity.

prove that this definition implies the density of the poset (see [2] for a formal proof of this fact). For this reason, the generalisation of D-continuity to posets cannot be so simple.

One way to solve this difficulty is to refine the sets $Max\ A$ and $Min\ \overline{A}$, i.e. to consider subsets of $Max\ A$ and $Min\ \overline{A}$ when defining D-continuity. But then, how can one choose these subsets? We present now such a choice (made by C.A. Petri in [8]) together with its motivation.

Let $(X, \prec)$ be the poset shown in Figure 2 and $(A, \overline{A})$ a D-cut of X. We could imagine a "river" separating A and $\overline{A}$. In Figure 2, several subprocesses (lines) are represented. Assume that a car is moving along a line, starting somewhere in A and going to $\overline{A}$. When the car arrives at the border of the river, it has three possibilities:

1. Either it goes directly to $\overline{A}$ or it continues its travel along the border of the river, looking for a nicer bridge spanning the river. That is the case of a car travelling along l_1 when it has reached x.

2. The car must cross the river. That is the case of cars travelling along l_2, l_3 or l_5. (We implicitly do not admit that a car returns or stops if it can go ahead).

3. The car arrives at a point where the road ends. That is the case of the car travelling along l_4 which ends in u.

In cases 2 and 3, the car either must cross the river from A to $\overline{A}$ or it cannot cross the river at all. Note also that y, z, w and u belong to $Max\ A$. Since we want to refine $Max\ A$, we must consider only those points which belong to $Max\ A$. For this reason we do not consider the possibility 1, since x does not belong to $Max\ A$.

When the car travelling along l_2 arrives to y, it has three possibilities to cross the river: l'_2, l''_2 and l'''_2. Similarly, the car travelling along l_3 has two possibilities to cross the river. In some sense, these two cars have chosen unfriendly roads because for a friend waiting on the other side of the river (in $\overline{A}$) it is not clear where he/she has to wait in order to shake hands.

Now, the car travelling along l_5 has a unique bridge in order to go from A to $\overline{A}$. The car traveling along l_4 cannot cross the river, since at u, it notices that it has no bridge going to $\overline{A}$. In some sense, l_5 and l_4 are friendly roads since a friend waiting in $\overline{A}$ knows exactly what to do: if the car travels along l_5, he/she has to stay in w' in order to shake hands, if the car travels along l_4, he/she may go home since the car will not be able to cross the river.

At this point, our idea is to consider the subset $Obmax\ A$ of $Max\ A$ formed by the points which lie on friendly roads. These points have either a unique way to cross the river or no way at all. In a sense, they are "objectively" maximal because their maximality does not depend on the choice of a D-cut in their "neighbourhood". In order to clarify this idea, let us consider a D-cut $(B, \overline{B})$ as indicated in Figure 2. For this D-cut we have: $y' \in Max\ B$ and $y \notin Max\ B$. However this D-cut still separates y from some of its immediate successors, i.e. y is not in $Max\ B$ but it is still –for instance– in $Max\ (l'_2 \cap B)$. In other words, the maximality of y depends on the "choice of a D-cut in its neighbourhood", therefore y is not an "objectively" maximal.

The same idea can be applied to the construction of $Obmin\ \overline{A}$ starting from $Min\ \overline{A}$.

We formalise these ideas in the following definition:

Definition 4.4 *Obmax, Obmin*

> Let $(X, \prec)$ be a poset and $A \in D$.
>   1. $Obmax\ A = \{x \in Max\ A \mid \forall B \in D\ \forall l \in L: x \in Max(l \cap B) \Rightarrow x \in Max\ B\}$.
>   2. $Obmin\ \overline{A} = \{x \in Min\ \overline{A} \mid \forall B \in D\ \forall l \in L: x \in Min(l \cap \overline{B}) \Rightarrow x \in Min\ \overline{B}\}$.
>   3. $c(A) = Obmax\ A \cup Obmin\ \overline{A}$.

$c(A)$ represents the points of $Max\ A \cup Min\ \overline{A}$ which are on one side of the river and from which either there is a unique way to go to the other side of the river or there is no way at all.

It is not difficult to prove (see for instance [2]) the following proposition:

Proposition 4.5 *Characterisation of Obmax and Obmin*

> Let $(X, \prec)$ be a poset, $A \in D$, $x \in Max\ A$ and $y \in Min\ \overline{A}$. Then:
>   1. $x \notin Obmax\ A \Leftrightarrow \exists z \in X,\ x \prec z,\ \exists l \in L: l \cap [x, z] = \{x\}$;
>   2. $y \notin Obmin\ \overline{A} \Leftrightarrow \exists z \in X,\ z \prec y,\ \exists l \in L: l \cap [z, y] = \{y\}$.
>   Where, as always, $[x, z] = \{u \in X \mid x \preceq u \preceq z\}$.

For combinatorial posets, we obtain:

Proposition 4.6 *Characterisation of Obmax and Obmin in combinatorial posets*

> Let $(X, \prec)$ be a combinatorial poset. Then:
>   1. $Obmax\ A = \{x \in Max\ A \mid |x^\bullet| \leq 1\}$.
>   2. $Obmin\ \overline{A} = \{x \in Min\ \overline{A} \mid |^\bullet x| \leq 1\}$.

We can now generalize the definition 4.3 in the following way:

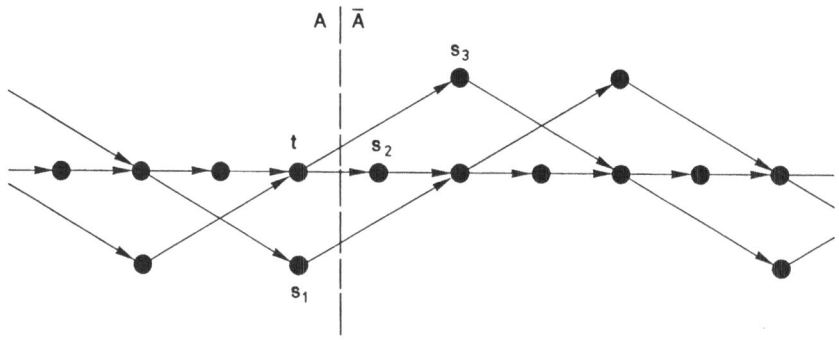

Figure 3: The four season rope of C.A. Petri

Definition 4.7 *D-continuity for posets*

Let $(X, \prec)$ be a poset. Then:

1. $(X, \prec)$ is jump-free iff $\forall A \in D \; \forall l \in L: \; |c(A) \cap l| \neq 2$.
2. $(X, \prec)$ is gap-free iff $\forall A \in D \; \forall l \in L: \; |c(A) \cap l| \neq 0$.
3. $(X, \prec)$ is D-continuous iff $\forall A \in D \; \forall l \in L: \; |c(A) \cap l| = 1$.

It follows immediately from the definition and from the fact that $|c(A) \cap l| \leq 2$, that a D-continuous poset is gap-free and jump-free.

If we consider the poset attached to the well known rope of C.A. Petri (see Figure 3) and the D-cut as indicated in the figure, we have:

$Max \; A = \{t, s_1\}$, $Min \; \overline{A} = \{s_2, s_3\}$, $M(A) = \{t, s_1, s_2, s_3\}$.

$Obmax \; A = \{s_1\}$ (t has two bridges in order to go from A to $\overline{A}$), $Obmin \; \overline{A} = \{s_2, s_3\}$, $c(A) = \{s_1, s_2, s_3\}$.

The rope in Figure 3 is D-continuous.

Consider the poset $(\mathbf{R}^2, \prec)$ introduced in section 3.

$A = \{x = (x_1, x_2) \in \mathbf{R}^2 \mid x_2 \leq 0\}$ is a D-cut, $Max \; A = Obmax \; A = \{x = (x_1, x_2) \mid x_2 = 0\}$, $Min \; \overline{A} = Obmin \; \overline{A} = \emptyset$ and $c(A) = \{x = (x_1, x_2) \mid x_2 = 0\}$.

If we take the line $l = \{x = (x_1, x_2) \in \mathbf{R}^2 \mid x_2 = e^{x_1}\}$ then $c(A) \cap l = \emptyset$ and we have a gap, hence $(\mathbf{R}^2, \prec)$ is not D-continuous.

From definition 4.7, it follows that if we want a poset to be D-continuous, we have to eliminate jumps and gaps.

A jump is pictorially and locally characterised in Figure 4. Every line which contains x contains also y and vice-versa. In this case $x, y \in c(A)$ ($x \in Obmax \; A$, $y \in Obmin \; \overline{A}$) and $|c(A) \cap l| = 2$.

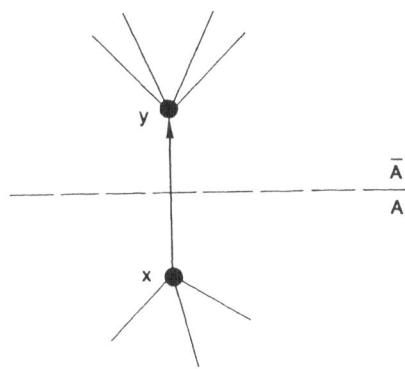

Figure 4: A jump

We could eliminate jumps postulating one of Petri's axioms, namely li-reducedness, which has been defined in section 3. The following proposition has been proved in [2]:

Proposition 4.8 *Li-reducedness implies jump-freeness*

Let $(X, \prec)$ be a poset.
If $(X, \prec)$ is li-reduced, then $(X, \prec)$ is jump-free.

The converse of the above proposition does not hold, which means that the absence of jumps is not equivalent to li-reducedness. The absence of jumps is equivalent to a weaker property, the non- single degree property, which turns out to be implied by li-reducedness.

Definition 4.9 *The non-single degree property*

Let $(X, \prec)$ be a poset.
$(X, \prec)$ has the non-single degree property iff
$\forall x, y \in X: x \prec\!\!\cdot\, y \Rightarrow \exists z \in X, x \neq z \neq y: (x \prec z \text{ co } y) \lor (x \text{ co } z \prec y).$

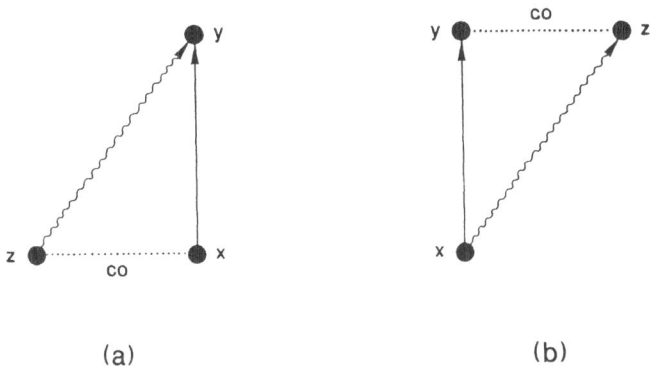

(a)   (b)

Figure 5: Illustrating the non-single degree property

In some sense, we have a jump between x and y when the bridge between x and y is a dangerous one. If we assume that the bridge is narrow and can contain only one car, then, in a jump situation, a car coming from x to y is going to collide with a car going from y to x at the same time. The non-single degree property eliminates the danger of collision of Figure 4 by adding an alternative bridge like in Figure 5.

The non-single degree property ensures that if $x \in Obmax\ A$ then $y \notin Obmin\ \overline{A}$ and vice-versa, i.e. there is no jump between x and y.

The following theorem has been proved in [1]:

Theorem 4.10 *Jump-freeness and the non-single degree property.*

Let $(X, \prec)$ be a poset.
$(X, \prec)$ is jump-free iff it satisfies the non-single degree property.

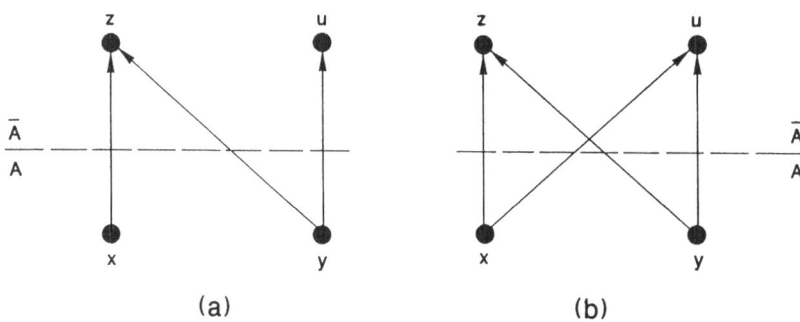

Figure 6: Local types of gaps

Let us now analyse the gaps of a poset.

There are different types of gaps.

In Figure 6 we have two cases of gaps: In (a), $c(A) = \{x, u\} \neq \emptyset$, but $l \cap c(A) = \emptyset$ when $l = \{y, z\}$. In (b), $c(A) = \emptyset$ and of course $l \cap c(A) = \emptyset$ for all lines l. We call these possibilities local cases because they concern only a local area of the line.

Figure 7 illustrates what we call global cases of gaps.

The possibility (a) is characterised by the existence of a line l such that $Min\ l = \emptyset$ and $l \cap A = \emptyset$.

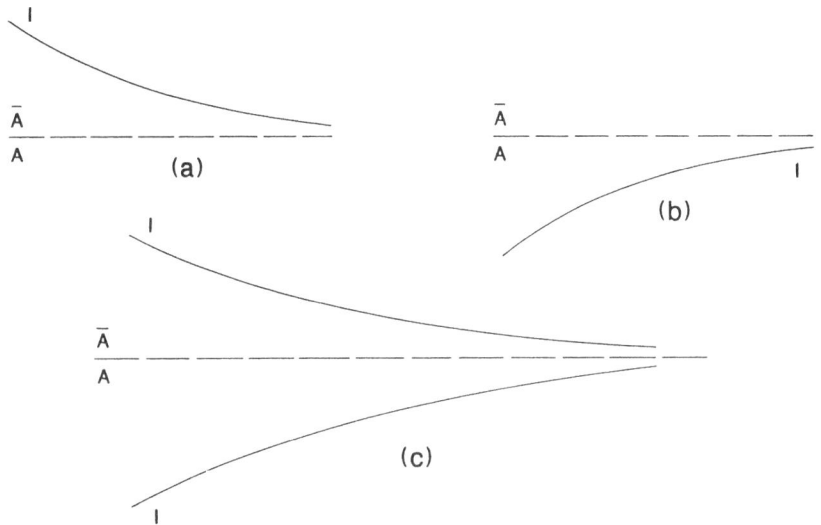

Figure 7: Global cases of gaps

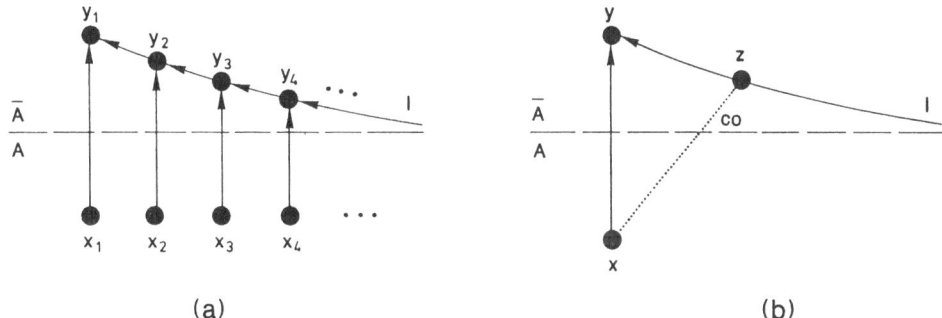

Figure 8: Two details of a global gap

Figure 8 gives two examples of the situation (a) of Figure 7.

If Figure 8 (a), $c(A) = Obmax\ A = \{x_1, x_2, x_3 \ldots\}$ and there exists a cut c such that $c \subseteq c(A)$ (in fact $c = c(A)$). Since $c(A) \cap l = \emptyset$, the poset considered there not only has a gap, but it is also not K-dense (see section 3 for the definition of K-density).

In Figure 8 (b), $c(A) = Obmax\ A = \{x\}$. In this situation, $c(A)$ does not contain a cut and the gap has not the same nature as the one in (a). In other words, the cases 8(a) and 8(b) are distinguished by the set $c(A)$ containing a cut or not respectively. A symmetrical subcase is illustrated in Figure 7(b) where $Max\ l = \emptyset$ and $l \cap \overline{A} = \emptyset$.

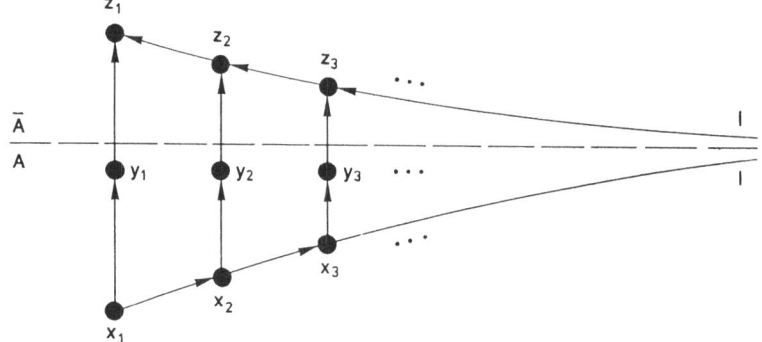

Figure 9: A gap where $l \cap \overline{A} \neq \emptyset \neq l \cap A$

If Figure 7(c), the possibility $l \cap \overline{A} \neq \emptyset \neq l \cap A$ and $c(A) \cap l = \emptyset$ is depicted.
One example of this situation is given in Figure 9.
Here, $c(A) = Obmax\ A = \{y_1, y_2, y_3, \ldots\}$ (which is a cut).
However $l = \{x_1, x_2, x_3 \ldots\} \cup \{z_1, z_2, z_3 \ldots\}$ is a line and $c(A) \cap l = \emptyset$.
In the three global cases, $c(A) \neq \emptyset$ but $|c(A) \cap l| = 0$ for some line l.
These types of gaps are eliminated postulating two properties, namely, K-density and strong
cut-boundedness.

Definition 4.11 *Strong cut-boundedness*

    A poset $(X, \prec)$ *is strongly cut-bounded iff* $\forall A \in D$: $A \subseteq\downarrow c(A) \wedge \overline{A} \subseteq\uparrow c(A)$.

$(\downarrow Y = \{x \in X \mid \exists y \in Y\colon x \preceq y\}$ and $\uparrow Y = \{x \in X \mid \exists y \in Y\colon y \preceq x\}$ for $Y \subseteq X$.)

In other words, definition 4.11 means that $\forall x \in X \; \exists y \in c(A)$: $x \preceq y$ (if $x \in A$) or $y \preceq x$ (if $x \in \overline{A}$).
The following proposition was proved in [4]:

Proposition 4.12 *Strong cut-boundedness and cuts*

    Let $(X, \prec)$ *be a strongly cut-bounded poset. Then:*

        1. $c(A) \neq \emptyset$;

        2. $\exists c \in C\colon c \subseteq c(A)$.

So, postulating the strong cut-boundedness property, the local case in Figure 6 (b) is eliminated
(since there we have $c(A) = \emptyset$). The situation in Figure 8(b) is also eliminated and its corresponding subcase of Figure 7(b) as well. The remaining cases are eliminated postulating K-density. We
then have:

Theorem 4.13 *Characterisation of gap-freeness*

    A poset $(X, \prec)$ *is gap-free iff it is K-dense and strongly cut-bounded.*

The proof of this theorem can be retrieved in [2].
Summing up, we have:

Theorem 4.14 *Characterisation of D-continuity*

> A poset $(X, \prec)$ is D-continuous iff it is K-dense, strongly cut-bounded and has the non-single degree property.

For combinatorial posets, we may replace the strong cut-boundedness property by two properties which are closer connected with the different cases of gaps studied above. We have:

Theorem 4.15 *Characterisation of D-continuity for combinatorial posets*

> A combinatorial poset $(X, \prec)$ is D-continuous iff:
>
> 1. $(X, \prec)$ has the non-single degree property;
> 2. $(X, \prec)$ is K-dense;
> 3. $\forall l \in L \; \forall A \in D: Max\ l = \emptyset \Rightarrow \overline{A} \cap l \neq \emptyset$ and $Min\ l = \emptyset \Rightarrow A \cap l \neq \emptyset$;
> 4. $\forall x \in X: |x^\bullet| > 1 \Rightarrow \exists y \in x^\bullet: |{}^\bullet y| = 1$ and $|{}^\bullet x| > 1 \Rightarrow \exists y \in {}^\bullet x: |y^\bullet| = 1$.

Remark

In theorem 4.15, (3) eliminates the global cases 7(a) and (b), (4) eliminates the local case 6(b) and (2) eliminates the global case 7(c) and the local case 6(a).

Proof: $\Rightarrow$

Assume $(X, \prec)$ is D-continuous.

Then, by theorem 4.14, (1) and (2) are satisfied.

Let us prove: D-continuity $\Rightarrow$ (3).

Let $A \in D$ and $l \in L$ such that $Max\ l = \emptyset$.

By D-continuity we have $c(A) \cap l = \{x\}$, $x \in X$.

If $x \in Obmin\ \overline{A}$ then $x \in \overline{A}$ and $l \cap \overline{A} \neq \emptyset$, so we are done.

If $x \notin Obmin\ \overline{A}$ then $x \in Obmax\ A$, i.e. $x \in Max\ A$.

But $Max\ l = \emptyset$ means $\exists y \in l: x \prec y$.

$x \in Max\ A \Rightarrow y \in \overline{A}$ and, again, $l \cap \overline{A} \neq \emptyset$.

The case $Min\ l = \emptyset$ is similar and, therefore, omitted.

We prove now D-continuity $\Rightarrow$ (4).

Let $x \in X$ such that $|x^\bullet| > 1$.

Assume $\forall y \in x^\bullet: |{}^\bullet y| > 1$.

Take a D-cut A such that $x \in Max\ A$, i.e. $y \in \overline{A}$, for example $A = \downarrow x$.

Clearly $x, y \notin c(A)$ by $|x^\bullet| > 1$, $|{}^\bullet y| > 1$ and prop. 4.5.

Therefore, any line l such that $x, y \in l$ does not meet $c(A)$ contradicting D-continuity.

The case $|{}^\bullet x| > 1$ is similar.

$\Leftarrow$

$(X, \prec)$ is free of jumps since it is of non-single degree.

We have to prove that $(X, \prec)$ is free of gaps.

By theorem 4.13, we know that $(X, \prec)$ is free of gaps iff it is K-dense and strongly cut-bounded.

$(X, \prec)$ is K-dense by hypothesis, i.e. it is enough to show that $(2) \wedge (3) \wedge (4) \Rightarrow (X, \prec)$ is strongly cut-bounded i.e. $\forall A \in D: A \subseteq \downarrow c(A) \wedge \overline{A} \subseteq \uparrow c(A)$.

We prove only $A \subseteq \downarrow c(A)$, the other case follows similarly.

Let $x \in X$. Take $l \in L$ such that $x \in l$.

<u>**Case 1**</u> $l \subseteq A$.

If $l \subseteq A$ then $l \cap \overline{A} = \emptyset$.
By (3) $Max\ l \neq \emptyset$. Let $z \in Max\ l$.

Claim $z \in c(A)$.
(a) $z \in Max\ A$.
If not, $\exists y \in A$: $z \prec y$. But $z \in Max\ l$ implies $y \succ u\ \forall u \in l$, i.e. $y \in l$, a contradiction with $z \in Max\ l$.
Therefore $z \in Max\ A$.
(b) $z \in c(A)$.
$z^\bullet = \emptyset$ follows by the same argument given in (a).
By proposition 4.6 we have $z \in Obmax\ A$, i.e. $z \in c(A)$.

But $x \preceq z$, then $x \in\ \downarrow c(A)$ and $A \subseteq \downarrow c(A)$.

Case 2 $l \nsubseteq c(A)$.
Since $x \in\ A \cap l$, it follows that $l \cap A \neq \emptyset \neq l \cap \overline{A}$.
Let $y \in \overline{A} \cap l$.
$x \prec y$ since $x, y \in l$, $x \in A$ and $y \in \overline{A}$.
$(X, \prec)$ combinatorial implies $\exists x_1, x_2, \ldots x_n \in X$ sucht that
$x = x_1 \prec x_2 \prec x_3 \prec \ldots \prec x_n = y$.
Let $x_i \in A$ and $x_{i+1} \in \overline{A}$.

Claim $x_i \in c(A)$ or $x_{i+1} \in c(A)$.
If $|x_i^\bullet| = 1$ then $x_i^\bullet = \{x_{i+1}\} \subseteq \overline{A}$ which means $x_i \in Max\ A$.
But then, by proposition 4.6, it follows that $x_i \in Obmax\ A \subseteq c(A)$ and we are done.
By analogy, if $|^\bullet x_{i+1}| = 1$ then $x_{i+1} \in Obmin\ \overline{A} \subseteq c(A)$ and again we are done.
We may assume $|x_i^\bullet| > 1$ and $|^\bullet x_{i+1}| > 1$.
By (4), $\exists x' \in x_i^\bullet$: $|^\bullet x'| = 1$.
Similarly, $\exists x'' \in\ ^\bullet x_{i+1}$: $|x''^\bullet| = 1$.
x_i, x_{i+1}, x' and x'' are shown in Figure 10.

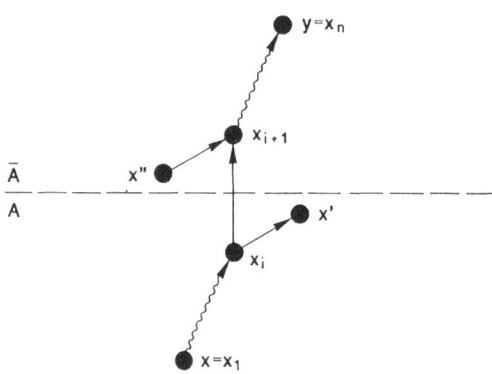

Figure 10: Illustrating the proof of theorem 4.15

$x' \prec x_{i+1}$ contradicts $x_i \prec x_{i+1}$.
$x_{i+1} \prec x'$ contradicts $x_i \prec x'$.
Hence $x'\ co\ x_{i+1}$.
Similarly: $x''\ co\ x_i$.
$x' \prec x''$ contradicts $x'\ co\ x_{i+1}$.

$x'' \prec x'$ contradicts the fact that $|x''^\bullet| = 1$ and x' co x_{i+1}.

Then: x' co x''.

Let $l \in L$ such that $x_i, x_{i+1} \in l$ and let $c \in C$ such that $x', x'' \in c$.

Clearly $c \cap l = \emptyset$, which is a contradiction with the fact that $(X, \prec)$ is K-dense.

Hence either $x_i \in c(A)$ or $x_{i+1} \in c(A)$.

But then, $\exists z \in c(A)$ $(z = x_i$ or $z = x_{i+1})$ such that $x \preceq z$, i.e. $x \in \downarrow c(A)$ and again $A \subseteq \downarrow c(A)$.

For the special case of posets attached to occurrence nets, the characterisation given in theorem 4.15 can be simplified.

We recall first some definitions:

Definition 4.16 *Net, occurrence net*

1. *A triple $N = (S, T, F)$ is called a net iff*

   (a) $S \cap T = \emptyset$;

   (b) $S \cup T \neq \emptyset$;

   (c) $F \subseteq (S \times T) \cup (T \times S)$; $dom(F) \cup cod(F) = S \cup T$.

2. *A net $N = (S, T, F)$ is an occurrence net iff*

   (a) $\forall s \in S$: $|s^\bullet| \leq 1$ and $|^\bullet s| \leq 1$;

   (b) F^* *is acyclic:* $\forall x, y \in S \cup T$: $(x, y) \in F^* \wedge (y, x) \in F^* \Rightarrow x = y$, where F^* is the transitive closure of F.

As always, $x^\bullet = \{y \in S \cup T \mid (x, y) \in F\}$ and $^\bullet x = \{y \in S \cup T \mid (y, x) \in F\}$.

It is well known that to each occurrence net one can attach a poset in the following way:

$X = S \cup T$ and $\prec = F^+$.

If $N = (S, T, F)$ is an occurrence net, it is easy to see that $(X, \prec)$, the attached poset, is combinatorial and satisfies the property (4) of theorem 4.15.

Hence the following theorem:

Theorem 4.17 *D-continuous occurrence nets*

Let $N = (S, T, F)$ be an occurrence net and $(X, \prec)$ its attached poset.
$(X, \prec)$ is D-continuous iff:

1. $(X, \prec)$ *is of non-single degree;*

2. $(X, \prec)$ *is K-dense;*

3. $\forall l \in L \, \forall A \in D$: $Max \, l = \emptyset \Rightarrow \overline{A} \cap l \neq \emptyset$ and $Min \, l = \emptyset \Rightarrow A \cap l \neq \emptyset$

5 Final words

In this paper we have tried to clarify the notion of D-continuity mainly looking at some examples and analysing in some details the notions of jumps and gaps.

The connection between the definition given in section 4 –which is the earliest introduced by C. A. Petri– and system properties (we recall that an occurrence net may be used to represent a process of a marked net, i.e. of a concurrent system), has been worked out in [3] where results linking marked nets and D-continuous occurrence nets are contained. Furthermore, current research seem to indicate that there are also nice connections between D-continuous occurrence nets and lattice theory.

Acknowledgement

We wish to thank Eike Best for reading this manuscript and for his useful comments.

References

[1] E. Best and C. Fernández
A Petri Net Theory of Processes and Systems.
Manuscript (1986)

[2] E. Best and A. Merceron
Concurrency Axioms and D-continuous Posets.
LNCS Vol.188 pp.32-47 (1985)

[3] E. Best and A. Merceron
Frozen Tokens and D-Continuity: A Study in Relating System Properties to Process properties.
LNCS Vol.188 pp.48-61 (1985)

[4] E. Best and A. Merceron
D-continuity and some Axioms of Concurrency for Nonsequential Process Models.
Submitted paper.

[5] C. Fernández and P.S. Thiagarajan
D-Continuous Causal Nets: A Model of Non-Sequential Processes. TCS Vol. 28 pp.171-196 (1984)

[6] H.J. Genrich, K. Lautenbach and P.S. Thiagarajan
Elements of General Net Theory.
LNCS Vol.84 pp.21-164 (1980)

[7] C.A. Petri
Non-Sequential Processes.
GMD-ISF Report 77.05 (1977)

[8] C.A. Petri
Concurrency.
LNCS Vol.84 pp.251-260 (1980)

[9] C.A. Petri
State Transition Structures in Physics and in Computation.
International Journal on Theoretical Physics, Vol. 21(12) pp.979-992 (1982)

[10] C.A. Petri
Concurrency and Continuity.
7th European Workshop on Application and Theory of Petri Nets, Oxford (1986)

Numerical Simulations with Place/Transactor-Nets

Hans Fuss

Institut für methodische Grundlagen (F1)
Gesellschaft für Mathematik und Datenverarbeitung m.b.H. (GMD)
D – 5205 Sankt Augustin, Federal Republic of Germany

Abstract: *Place/Transactor nets, or P/Ta-nets for short, are a special development of Petri nets. Their prominent feature is a change of second order: The token flow as the change of the contents of places (first order change of states) is well-known from other net models; in P/Ta-nets we have on top of that a variation of the* intensity *of every flow (a change of second order), depending on the contents of some selected Places. Such Places, so-called K-Places, are adjoined like an inscription to every flow arc. P/Ta-nets are described in detail, as well as their (historical) background, and their significance for numerical simulations. Some examples of programming with P/Ta-nets in the field of applications are given. Relations to other types of nets are sketched, and some general causal features in simulation, especially on the use of random numbers, are reflected.*

1 Introduction

1.1 Motivation

Theory and practice are too often two different kettle of fish – and that applies to net theory and simulation practice as well. This fact is not a bad thing *per se*, for both disciplines follow their own claims and demands, but it becomes disadvantageous for both parties if they stay separate for too long. One may have the impression that this situation has arisen now, if one skims 'our' and 'their' newsletters and the literature in both fields.

This paper is written primarily for those readers who are familiar with one side and interested in views and arguments of the other, in order to try to bridge the gap between these two sides a little more.

1.2 Notion and Notation

In net theory, a net is viewed upon as an object of mathematics, and research is done on this object. In simulation practice, the net model is seen as a mapping of a real system, the behaviour of one corresponding to the behaviour of the other, so that conclusions can be drawn from one to the other. This correspondence is sometimes seen so strongly, that an occasional reader might perhaps get confused by not being aware of *what* system is actually being spoken of *now*: the real, the simulated, or the simulation system (not to speak of the simulating system!)

We want to distinguish here between all these systems and their models and their simulation; of highest practical relevance to us is the system behaviour and its simulation.

Let $\mathcal{U}$ be the *universe*. In some cases, it might be convenient to be a little less universal and to restrict the discussion just to the *real world* $\mathcal{W}$, that part of the universe which is known to us. We may think of $\mathcal{U}$ and $\mathcal{W}$ as being the same, for actually it does not make any difference here. But we should note that even $\mathcal{W}$ is not quite the real world 'as it really is', but a mental image

(i. e. already a model) of it, for every person can have a different $\mathcal{U}$ or $\mathcal{W}$ in mind. Moreover, we know from modern physics that there are cases where only *different* views of the same reality *will* be known to different observers.

Important however is that in $\mathcal{W}$, there is one part of it, namely $\mathcal{R}$, the *real system*, which is of special interest to us; important to us because we want to make an image (a model) of it. It is equally important to focus our view on the complement of $\mathcal{R}$ in $\mathcal{W}$ (or $\mathcal{U}$), to the *environment* $\mathcal{E}$ of our real system. Communication between $\mathcal{R}$ and $\mathcal{E}$ will take place at certain points $\mathcal{I}$, the *interfaces*. An interface may be of the type of a mailbox, where the system or the environment deposit messages (or goods) for the other one, or of the type of an interpreter, who translates from one language (or one currency) into the other.

This *real system* $\mathcal{R}$ in question is mapped by virtue of the model builder via a mapping φ to a *model system*, or in short: to a **model** $\mathcal{M}$. This mapping process is called *modeling*.

The model itself, of course, is represented by some *real objects* (a model car, a set of differential equations on paper, a computer program, a net, or just thoughts in a head ...) Therefore, it is possible to have models of models.

Usually it is not sufficient to model only the objects of a system, for in order to make some predictions it is necessary to model the actions, i. e. the behaviour, in the system as well.

To let the model system $\mathcal{M}$ run automatically, i. e. to let the model do all or some of its possible actions according to its inherent rules, maybe with feeding external data through an interface to it, is called *simulating* the real system $\mathcal{R}$.

1.3   The Information Conservation Principle

Originally, practical simulation purposes were predominant when P/Ta-nets had been developed. The correspondence of variables and counters is obvious, and *tokens* should represent the simulated items and objects. It seemed to be good practice to trace the simulated items through their course in the simulation run instead of re-calculating their numbers at every new instance. In net theoretical terms, this led to the paradigm that information is an entity similar to mass or energy – which can neither be destroyed nor generated, but transformed – which follows certain natural laws, e. g. some conservation principles. Then it is clear that **a system cannot produce more information than it knows itself,** – it can only process and transform information[†]. Therefore instead of speaking of sources and sinks of information one can observe how information *enters* and *leaves* a (closed) system at certain points (which are the so-called sources and sinks); and this applies to the model system as well.

This paradigm will become important in connection with the transaction rule (q.v.), especially with enabling rule E 1.

2   History

2.0   From the beginning ...

In 'ancient' days of net theory, one did not much differentiate between different types of nets, just 'Petri nets' was the only class of nets one was aware of, and more than 90 percent of the nets consisted of what is known today as condition/event systems and place/transition systems. Nine tenths of the rest were channel/agency nets (Kanal-Instanz-Netze), the rest mixtures and net like constructs. The main research was done via the token game – the net itself was not very much an object of mathematical discussion.

[†] The corresponding inverse argument – which is important for simulations – is the following: if some original system information is omitted when building a model, then the simulation is likely to produce poorer results.

2.1 1-1−nets

Some useful applications however could be accomplished with what was already at hand at that time, e. g. in the simulation field it was possible to model a biological balance in a sea (Jensen); and in planning an input/output program system (today we would call it an operating system) for the IBM 7090/1410 computer system the specification description as a net model helped much in the design phase, and finally it helped to write a program system which ran without errors for years and years, 24 hours a day (Fuss / Schnurer).

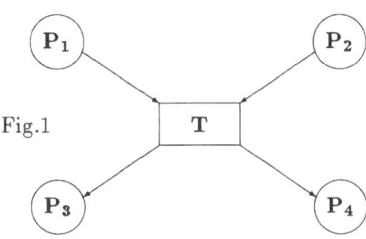

Fig.1

Nets of places and transitions ('Stellen', 'Transitionen') like this were drawn to represent real life systems, and they were of great benefit to describe the causal conditions between objects. Of special interest was: what happens to the tokens, if the transition is not a quadruped – like in this diagram – but a tripod.

The discussion was: are there 'sources' and 'sinks' of tokens – and hence: sources and sinks of information?

The notion 'live' made things clearer and led to the above conservation assumption.

2.2 Nets with flow weights

Pretty soon the question arose whether it was possible to apply net theoretical concepts to *numerical* simulation or construction, too. To a certain extent, this was accomplished by associating weights $k_1 \ldots k_n$ to the flow lines (i. e. to the arcs); this class of nets is now known under the name 'place-transition systems' or P/T-systems.

But solving one problem created several others: one consequence of 'bigger' flows (bigger than 1) was, that the P's had to have capacities bigger than 1. And the problem of imbalance of tokens gained a new aspect. In the above type of nets (here in a sloppy way named 1-1-nets), where all the k's are equal to 1 (one), one could think of the tokens as pegs in checkers and, in the case of an uneven number of input- and output-arcs at a transition, one could pack one token on top of the other and unpack them in due course.

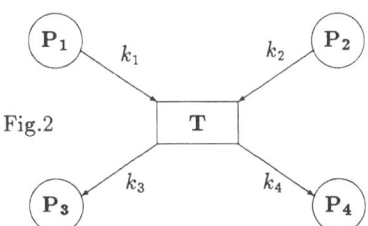

Fig.2

Here in this class of nets, this idea does not work that easily. What happens to the tokens, if $k_1 + k_2 \neq k_3 + k_4$? The problem of conflict became an additional aspect, too: e. g. capacities > 1 may allow situations of repetitive firing; or n tokens in a place, enough for firing both conflicting transitions, but not enough if one of them is multiple enabled. The notions 'live' and 'safe' were elaborated, and invariants calculated.

But a canonical transformation from one type of nets to the other was not available.

2.3 Parameterisation of the flow constants

For more powerful numerical simulations [K/Fu] the problem arose how to make the nets more flexible, e. g. how to parameterize the flow. This meant finding a suitable way of how to make a parameter λ out of the constant k.

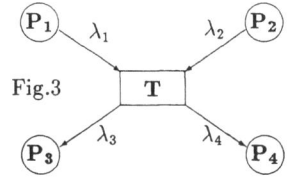

Fig.3

One simple solution often used is to calculate the k's by an arithmetical routine and assign the new values to k between the simulation cycles. But a 'cleaner' solution would be a construction purely in terms of nets, i.e. to vary the value of the parameter λ solely *by net elements* during the token game. The solution offered here took advantage of what is varying anyway in a net in the course of the token flow: the contents of a place.

The intellectual step to get this new type of nets was: *draw a circle around the above λ's, thus turning them into places, and take the contents of a place as the value of the corresponding λ.*

Due to the fact that a modification of the flow weights is done by the net itself, P/Ta-nets have been named *'self-modifying nets'*, too [Va]. Though this is a very colourful name, this term has its shortcomings: it could foster the misunderstanding that the *structure* of the net can be modified by the net itself, whereas it is only the token width of the arcs which is variable here.

3 Definition of a P/Ta-Net

3.0 Naming

The original name of this new type of nets was "Puffer-Transaktor-Netz" – as the first publications were in German [Fu1], [Fu2]. The word 'place' ("Stelle") seemed at that time to be a name that should better be reserved for places of capacity 1 (one), i.e. for conditions. Obviously the T-element was different from a transition, its name was inspired from econometrics.

Please not the difference in the spelling of 'Place' and 'place', which indicates whether the object is an element of a P/Ta-net (in the first case), or an element of a Petri net in general.

3.1 The Statics of a P/Ta-Net

As P/Ta-nets have been developed for practical simulation purposes we assume all 'nice' properties like: being connected, non-empty, countable elements, etc.

A Place-Transactor-Net (P/Ta-Net) is defined as follows:

P/Ta-Net = { $\mathcal{P}, \mathcal{K}, \mathcal{T}; F; M; R$ },   with

> $\mathcal{P}$ the set of Places, *(containing values)*
>
> $\mathcal{K}$ a subset of Places, $\mathcal{K} \subset \mathcal{P}$, *(i.e. K-Places may appear in two roles)*
>
> $\mathcal{T}$ the set of Transactors,
>
> F the flow relation, $F_1 \subset \{\mathcal{P} \times \mathcal{K} \times \mathcal{T}\}$, $F_2 \subset \{\mathcal{T} \times \mathcal{K} \times \mathcal{P}\}$, $F = F_1 \cup F_2$,
>
> M the marking, a mapping of $\mathcal{P}$ to $\mathcal{N}_0$, $\mathcal{N}_0 = \{0,1,2,3,...\}$,
>
> R an elaborate and token-preserving transaction rule.

Numerical values (related to the number of tokens) will be given by inscriptions in natural numbers.

We define for all $P_n \epsilon \mathcal{P}$:   Let...
$C_n = c(P_n)$ be the value of the function $c : P_n \longmapsto \mathcal{N}_1$ for the argument P_n, with $\mathcal{N}_1 = \{1,2,3...\}$, describing the <u>capacity</u> (maximal contents) of a Place P_n,

$i(P_n)$ be a mapping $P_n \longrightarrow \mathcal{N}_0$, $\mathcal{N}_0 = \{0,1,2...\}$, with $0 \leq i(P_n) \leq C_n$, describing the actual <u>contents</u> of P_n, i.e. the number of tokens <u>in</u> P_n,

$\triangle(P_n)$ be the <u>defect</u> (the space currently available in a Place) $\triangle(P_n) = C_n - i(P_n)$.

The Places in the ternary flow relation F, according to their position in F, are graphically represented by different locations at the arc (which is the graphical representation of F):

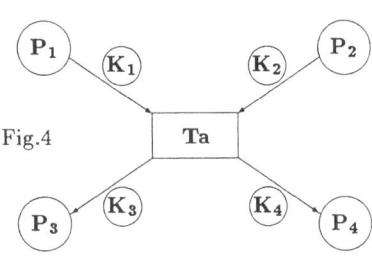

Fig.4

(1) A Place $P_0 \epsilon \mathcal{P}$ which appears in an end position of F_1 or F_2, correspondingly in the graphical representation appears in an end position of an arc. It describes a **variable**, the *flow of tokens* is between P-Places.

(2) A Place $P_0 \epsilon \mathcal{K}$, so defined by its appearance in the middle position of the flow relation F, is drawn in a middle position between a Transactor and a P-Place. It describes a **parameter** which has an *influence* upon the flow according to the transaction rule (q.v.).

A Place itself is always the same object, only its **pragmatic status** varies with its position in the flow relation in that Transactor, i.e. with its usage. (A similar situation applies to a noun in a sentence: it may be an object (P), or a subject (K), *the K's rule the P's*.)

In a different Transaction, a K-Place can be used as a P-Place and hence its contents can be modified, subject to some other K-Places.

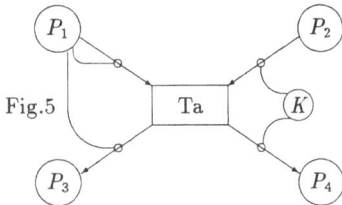

Fig.5

Regarding the K-type Place, there is another graphical representation, equivalent to the above one. This one has its advantages if we have multiple use of a K-Place (like in the right part of the diagram), or if self-relations are involved (like in the left part). In this diagram we have a synchronisation of a flow of k tokens from (P_2) to (P_4) with the flow of *all* tokens from (P_1) to (P_3).

Actually it does not make much sense to raise the question (far from answering it) whether *the same* tokens arrive in P_3 (or P_4) that leave P_1 (or P_2). Tokens represent integers in counters, they are not individuals. But sometimes it is of some help in a simulation model to think of tokens as travelling objects, especially if P_1 and P_3 represent *the same* objects before and after an action – e.g. business income before and after tax deduction.

3.2 The Dynamics of a P/Ta-Net – The dynamisation of the token flow

The token flow between P-Places is ruled by the contents of the associate K-Places at the 'time' of the transaction (i.e. its occurrence) in a *one-to-one* way: the number of tokens moved to or from a P-Place by a transaction is equal to the number of tokens which are the contents of the associate K-Place **at that time**. 'During' the transaction the contents of such Places remain unchanged – as far as this role is concerned.

According to the conservation principle for information stated above we define the transaction rule R as being token-preserving, i.e. input and output (in their sum) have to be balanced.

To enable a transaction, two different conditions have to be fulfilled: one (E2-E3) is the usual condition for concession in nets and concerns the (full) <u>contents</u> (and (empty) spaces) of the involved P-Places, we call it having concession <u>P-wise</u>; the other one (E1) concerns the involved K-Places and takes care of the <u>flow balance</u>, which requires that the sum of the contents in the K-Places on the input-side (F_1), (i.e. the input flow) has to be equal to the sum on the output-side (F_2), (i.e. to the output flow), we call it having concession <u>K-wise</u>.

Let the index '1' be used with Places where the flow relation F_1 applies[†], and let the index '2' be used where F_2 applies.

[†] i.e. in the notation proposed in [B/F]: (index 1) in $^\bullet Ta$ and (index 2) in $Ta^\bullet$

The Transaction Rule R is defined as follows:

The Transactor T_0 is *enabled* if the following 3 conditions hold simultaneously:

(E 1) $\sum_j i(K_{1,j}) = \sum_j i(K_{2,j})$ for all j around T_0 (see footnotes 2 and 3)   *(flow balance?)*[1]

(E 2) $0 < i(K_{1,j}) \leq i(P_{1,j})$ for every input[2] pair $(P,K) \subset F_1$ *(enough input tokens?)*

(E 3) $0 < i(K_{2,j}) \leq \Delta(P_{2,j})$ for every output[3] pair $(K,P) \subset F_2$ *(enough output space?)*

If the Transactor fires, the **result of the transaction** is (in the said one-to-one way):

(TR 1) $i(P_{1,j}) - i(K_{1,j}) \Longrightarrow i(P_{1,j})$ for every input pair $(P,K) \subset F_1$ <u>and</u>

(TR 2) $i(P_{2,j}) + i(K_{2,j}) \Longrightarrow i(P_{2,j})$ for every output pair $(K,P) \subset F_2$

Please note: due to the above condition E 1, the transaction is *token-preserving*.

As said before, we have $\mathcal{K} \subset \mathcal{P}$, hence K-Places can be used in the same way as a 'normal' P-Places too, i.e. their contents can be changed in the normal way – and this change implies the desired variability of the flow.

So we have two types of dynamics. The first one is well-known from all nets, the token flow produces a <u>change of states</u> (change of the contents of Places); the other one is like a derivative: a change of the contents of *some special Places*, namely K-Places, causes a *change of changes* (in a cyclic net: instead of their repetition), i.e. a <u>variability of the (intensity of the) flow</u>.

4 Application in Modelling, Examples

4.1 Some practical notations and restrictions

Inscriptions

A certain amount of inconsistency is tolerated: The name of a Transactor is usually written *into* the box. Similarly, the name of a Place is written into the circle as well. But sometimes, one wants to distinguish between the *name* of the variable and its current or initial *value*. In such cases we prefer to write the *contents into* the Place, and its *name at its side*.

A third kind of inscriptions which can be important in a few cases, is the capacity of a Place.

Text shorthand

As places are usually represented by circles $\bigcirc$ and transitions by boxes $\square$, we may in the running text sometimes write '(a)' instead of 'Place a', and '$[b]$' instead of 'Transactor b'.

Initial values

Unless otherwise specified, an initial value of zero (i.e. empty contents) is assumed for all Places.

Special feature of the AFMG program

P/Ta-nets have been developed at the same time [Fu1] and in conjunction with the Asynchronic Flow Model Generator [Fu2] program, i.e. theory and practice were strongly connected and influenced each other. It seemed to be very convenient (from the CPU time consumption point of view) to let the program work for as long as possible once it has started processing a Transactor. Anyway, *one* CPU can only simulate a concurrent operation of *several* Transactors by interleaving them. This yields *one* behaviour out of several possible ones. Therefore in principle it does not do any (more) harm to specialize the implemented version of 'being enabled' in the following way: when the processor finds a Transactor enabled, it checks the same Transactor after its firing again and – if it is still enabled – it fires the same one again and again.

This special feature will reflect on the examples given here, especially if they become more complex.

---

[1] with reference to figure 4: is $i(K_1) + i(K_2) = i(K_3) + i(K_4)$?
[2] input pair (P,K): for every pair $(P_{1,i}, K_{1,j})$, where $(P_{1,i}, K_{1,j}, T_0) \epsilon F_1$
[3] output pair (K,P): for every pair $(P_{2,i}, K_{2,j})$, where $(T_0, K_{2,i}, P_{2,j}) \epsilon F_2$.

4.2 Division of Flow

The following diagram is an example of modeling the definition and calculation of the value of two variables, Wn and Tx, from one, Wg.

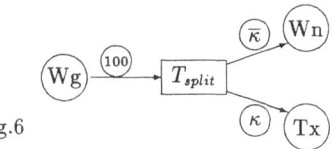

Fig.6

This example shows the splitting of the flow of entities into two flows according to the percentage ratio (=parameter) r = κ : (100-κ). An interpretation is: to separate the income tax (Tx) from the gross wages (Wg), leaving the net wages (Wn).

This seems to be like modeling with place-transition nets – if not κ were a parameter, the value of which can be altered by an Transactor T_{alter}. This is shown in the following chapter.

4.3 Varying a Flow Parameter

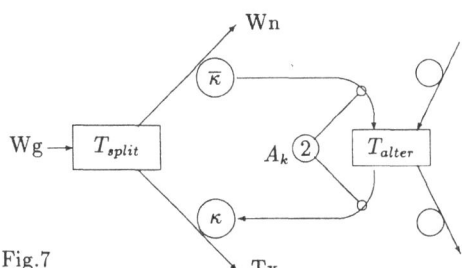

Fig.7

In this example, the value of the above κ is increased constantly by 2 at every firing of $[T_{alter}]$. Since κ and $\overline{\kappa}$ add up to 100 %, $\overline{\kappa}$ has to be decreased by 2 at the same time. So we take *one* Place A_k, the current contents of which is $i = 2$ in our example, to shift two tokens from $\overline{\kappa}$ to κ whenever the Transactor T_{alter} fires.
If desired, it is possible to change the value '2' of (A_k) by some other Transactor in turn.

P/K-Conflicts[†]
We want to remember that the possibility of changing parameters of the net (i. e. K-Places) during run time by the net itself is not for free: The two Transactors (here T_{split} and T_{alter}) which use the same Place (here κ, as well as $\overline{\kappa}$) in two different roles at the same time (one as a K-Place, the other as a P-Place) can be enabled concurrently. This creates a new type of conflicts, a P/T-conflict, at this Place, due to its double use. (In our example: the ratio $Wn : Tx$ depends upon the decision of this conflict, whether $[T_{split}]$ or $[T_{alter}]$ was first.)

4.4 Time Lag

The following example shows the situation which is quite familiar in economic models: the introduction of a time lag (of one time step); i.e. to push the present value of a variable into the background and use it for further computation in the next cycle.

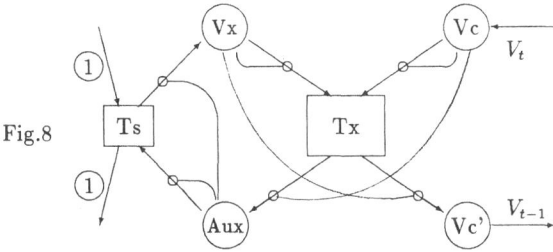

Fig.8

The Transactor [Tx] exchanges the current value V_t of the variable V, which is stored in (Vc), for the lagged one from (Vx) to (Vc'), and stores V_t into (Aux). From there it has to be transferred to (Vx) to be available for the next processing cycle, e.g. by the Transactor [Ts] which synchronises the process to the time step.

A time lag of two periods is handled by a repetition of the above.

[†] see also text in Chapter 3.2, first paragraph: '...as far as this role is concerned', etc.

4.5 Numerical Calculations (Multiplication, Division)

Numerical calculations shall be demonstrated in a typical task in simulations: to calculate a value y_{t+1} from its previous one y_t, according to $y(t+1) = \lambda \times y(t)$. In our example, we have as such a variable the price P of some goods – maybe manual labour – in mind which is increased from one year to the next by say four and a half percent. Because of the demanded precision, and because tokens come in integers, we have to calculate in *per thousand*, so the value of k=4.5% must be in our calculation $k = 45$, which is stored in Place (k).

We further assume that there is a certain reservoir of tokens in Place (p^+), from where the increase of the value of P_t to P_{t+1} finally will come, and that there are 'enough' tokens (=numbers) available in Place (KP) – this is a consideration which is new in numerics, normally it is expected implicitly that there are always enough numbers at hand!

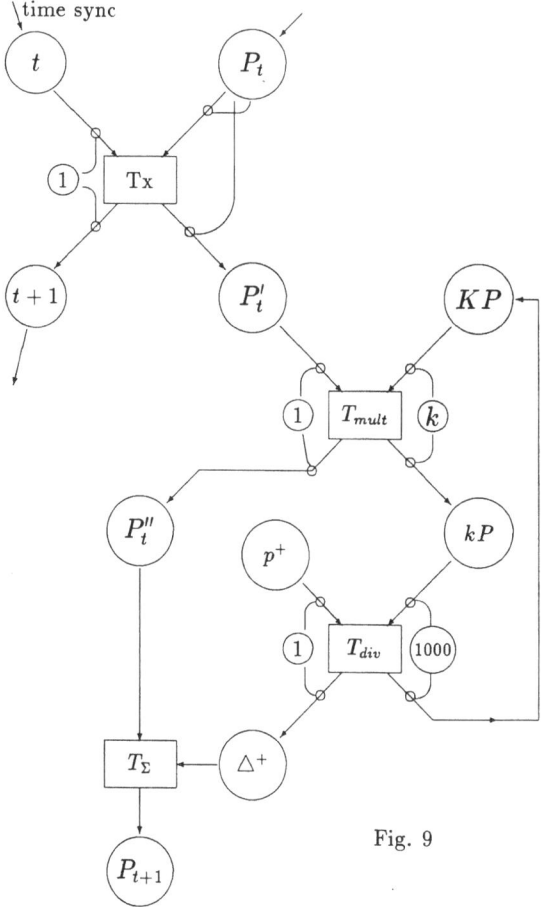

Fig. 9

The calculation will be performed as $(k \times P_t)/1000 + P_t \Rightarrow P_{t+1}$.

In the upper left hand corner we see a synchronisation line going through [Tx], it can be used to separate cycles, like the advance of the simulated time (e.g. the year number from t to t+1). Then [Tx] frees the value of (P_t) for further computation by producing a copy of this value to the Place P_t' (which by general assumption was empty initially – and will be empty soon again for the next cycle).

Transactor T_{mult} performs the multiplication by 45 in the following way: for every one token which T_{mult} shifts from (P_t') to (P_t'') (which yields another copy of P_t in P_t'') it puts k=45 tokens from the reservoir (KP) to Place (kP), which is the multiplication of (P_t) by 45.

The division by 1000 is done by Transactor T_{div}: for every 1000 tokens which it shifts from (kP) back to (KP), it puts 1 token from (p^+) to $(\triangle^+)$.

Finally T_Σ sums up.

Tokens at (kP), unprocessed by $[T_{div}]$ – corresponding to rounding errors – are kept in (kP) for the next cycle.

In due course, the tokens from P_{t+1} have to be shifted up to Place P_t in order to be available for the next processing cycle.

5 Relations between P/Ta-nets and P/T-nets

5.0 Mappings, Foldings, Substitutions

In the text so far, the emphasis was on the token game – which is quite natural in a paper on numerical simulations. Moreover, it was the same in the historical development, problems in *one* net were discussed first, general questions on nets were discussed later. So we do the same...

Among many other interesting questions about relations between different classes of nets, one question is easy to ask: Are there – canonical – transformations from one type to the other?

5.1 P/T-nets to P/Ta-nets

Trivial. To represent a P/T-system by a P/Ta-net, one has simply to encircle the inscriptions at the arcs, i.e. to replace these inscriptions by places with the same contents. Actually this is – as already mentioned in the section 'history' – where the idea of P/Ta-nets came from.

5.2 P/Ta-nets to P/T-nets

Hard. A mapping is possible, the solution is a combinatorial one. The mapping of (P/Ta-net) Places to (P/T-net) places of the same capacity is one-to-one. A Transactor however, due to the k-wise concession, will be replaced by a (possibly huge) number of transitions. The problem has already been discussed [Fu3], the main ideas shall be sketched here in an example.

Let us suppose there are two input Places P_1 and P_2 to a Transactor T_0, and two output Places P_3 and P_4 from T_0. The corresponding K-Places are $K_1 \ldots K_4$, their capacities $c_1 \ldots c_4$.
We assume at this stage that there are no further restrictions due to the capacities of the P-Places. Otherwise, we have to consider only the minimum $min(c_{P_1}, c_{K_1})$ or $min(c_{P_2}, c_{K_2})$ respectively. These conditions however are static ones and can be checked at net generation time.

At a certain time[†], there shall be the contents of $i_1 \ldots i_4$ in the K-Places.
(The contents of the P-Places are irrelevant for this discussion, for they matter only for the p-wise concession, that part will stay the same in P/T-nets.)

Every *feasible* combination (see enabling rule E 1) of the above $i_1 \ldots i_4$, not violating the conservation principle (in more detail: all combinations where $i_1 + i_2 = i_3 + i_4$) define one transition among $\{P_1, P_2; P_3, P_4\}$, obviously with the same flow direction.

We further specify our example: let the capacities of the K-Places be $c_1=3$, $c_2=5$, $c_3=2$, $c_4=3$.

Possible input flows are from P_1: 1, 2, or 3; from P_2: 1, 2, 3, 4, or 5. This allows the following input combinations: (1,1), (1,2), (1,3), (1,4), (1,5); (2,1), (2,2), ... (3,4), (5,5). But only those where the sum of the flow is at most 5, are feasible (for the output flow cannot be bigger).

Let us consider just one output case, the maximal one: (2,3). Corresponding input flows can be (1,4), (2,3) and (3,2). We now have broken down this part of the Transactor to three transitions with the inscriptions (1,4), (2,3) and (3,2) respectively. Which one finally will be enabled for firing will be decided by the contents of the K-Places. Therefore, places named 'content counters' have to be introduced to all places, those of the relevant K-Places are linked as side conditions to the transitions. These pointers make it possible to select *the* one transition which is meant to be active among several possible and possibly equivalent transitions. One content counter is

---

[†] This argument seems to assume the knowledge of a global state at a certain time for a distributed system – which does not exist. But it is not that bad, it only assumes a fixed state of the structural relations regardless of whether the system works or is at a standstill.

associated to every possible content of a K-Place, and just this value is inscribed to the arc between the associated place and the transition in the P/T-net. (Of course, the same content counter is linked to a number of transitions, as many as there are feasible combinations.) Initially, exactly one out of every group of content counters has to be marked, naturally the one that indicates the initial contents of this K-Place.

Obviously, a transition must change the content counters as well as the contents of places.

5.3 P/Ta-nets to other net models

Different. This is a challenging topic, interesting from the net theoretical point of view. From the simulation aspect however, where items are tranported through a real world, the benefit will probably be at a second consideration. The conservation principle of mass and energy is predominant at first sight.

6 General Considerations about Simulations with Nets

6.0 Comparison to other models

We compare net models here, which are still considered to be new to the simulation community, to other, 'normal' or 'traditionally known' types of simulation models. With the latter we think of models built up from (partial) differential or difference equations, feed-back control models, input/output models with tables, sets of linear equations, and the like.

6.1 Re-Programming?

Net models in general, and especially P/Ta-nets, have one special feature which makes them very suitable for simulation models, and this feature is of special value during the construction phase (building, testing, altering, fine tuning) of the model: With every aggregation or refinement of the model we stay within the category of nets, and we stay in the simple numerical methods which are connected to them (i. e. the transition rule); whereas with any alteration of just numerical values (of parameters) in a 'traditional' model, one could alter even the *structure* of the model so deeply that the computational solvability of the model would be affected: for example, a change of values in a matrix can turn it from regular to singular (or vice versa).

In other words: Every slight change of the model, even only in its numerical values (which could be derived empirically, i. e. by measurement) could cause an enormous amount of re-organisation and re-programming in the 'normal' models, whereas in net models (so in P/Ta-net models, too) *no* alteration would change the numerical methods involved – for there is only one numerical method that is used in nets: the application of the transition rule, here the transaction rule.

An altered net is already the altered program for any net compiler – i. e. no re-writing of any program code is necessary.

·6.2 Reasons of Reasons of Reasons

If we study the flow of tokens in a P/Ta-net (with the net itself being the object of our interest) we know that the (local) reason for the amount of flow is the current contents of the adjoined K-Place. Unless the value is a constant of the net, we can argue further that the reason for the current value of the K-Place is that the flow of tokens into or out of this K-Place happened when the K-Place appeared earlier in the role of a P-Place, where the flows have been controlled by other K-Places, and so on (see chapter 4.3 too). We can construct chains of causal dependencies of values (contents) of Places, alternating the role from K-type to P-type for the Places involved. (Finally there will be a constant at the end of the hierarchy, or a self-reference, or a communication with the environment, or the dependencies are cyclic.) But through all levels of refinement we do not get to any objects of higher decomposition of whatever kind, but we stay in the class of Places (K- and P-type Places) and Transactors and the flow relation F.

6.3 Time-Reversal Simulation

In planning and in straight forward simulations it sometimes so happens that one overshoots the mark. Then one would have to return to a check-point and try a new computation with a different (set of) parameter(s) – which is a rather rigid concept. In some cases it might be sufficient to go back just a little and try again. Sometimes it may be even more desirable to leave most things as they are and go backwards just with a very limited part of the model for one or more (time) steps, if only some local influences in the original system have changed.

In principle the method of backward simulation could be applied to all system representations which are neutral to the direction of time (i.e. time-symmetric or time-reversal), in particular to interaction system models [We], [MS&], and to the vast multitude of (Petri) net models (see bibliography).

In 'normal' simulation models it is virtually impossible to do a simulating step backwards in time. This will be no problem at all for P/Ta-net models (as with all net models): these models can simulate backwards and forwards just by reversing the flow (i.e. exchanging input and output). The application of the transaction rule in the reversed order is a transaction again; in the transaction rule, only the roles of 'contents' and 'defect' have to be reversed.

6.4 Conflicts

Nevertheless, when actually trying to do reversal simulation, the net processor most likely will not work – which usually happens with the processing part of the AFMG program as well. This is because of conflicts, especially because of backward conflicts.

We shall not discuss a conflict situation in general here, for many other papers, certainly in this volume too, discuss it at length. The *decision of conflicts* is of importance to all systems, hence to all simulation models, too. But the notion of conflict is unknown to most of the other simulation methods and languages – and this can arouse suspicions. The question arises: if it is possible on the one hand for some real systems to be modeled in a way where conflicts **are** mapped (e.g. in net models) – and if on the other hand it is possible to model the same real system into a mathematical model **without** conflicts – so what about the difference in results?

This difference is of importance only if the original system embodies parallelism or concurrency, distribution of control or decision, partial independence and things like that. This will typically happen in spatially or temporally distributed systems and in systems where human beings are part of the system. The technical difference between net simulation models and the others seems to be that in the one (net) case **all conceivable courses of events have been provided for**, though they may never occur; in the other case **all possible alternatives have been decided**.

It might happen that the results are numerically identical in both simulation methods (with consideration of conflicts and without); and this will be so *if the selected paths* through the system *are identical.* In a normal simulation run of a model, the mapping of just *one behaviour,* i.e. *one* sequence of states and actions out of the possibly large number of 'behaviours' of the real system $\mathcal{R}$, has been chosen.

But for planning or for forecasting purposes, it is not sufficient to consider just *one* future possibility, and this will be of special delicacy if $\mathcal{R}$ is a socio-economic system. This should give rise to give a second thought to the reliability of (traditional) simulation models. Surely not all bad simulation results can be surpassed by improving just the numerical accuracy of the parameters. We can take it for granted that the deficiency of many simulations does not lie in their capacity to handle the numerical precision properly. It is the theoretical background of the models itself that produces poor results. Here seems to be a good starting-point for considerations about amelioration in the art of modeling when prior attempts of improvement have failed.

Another pragmatic aspect which questions simulation of socio-economic systems in general – quite in contrast to automatic systems – is the following: we cannot read peoples' thoughts, so we cannot predict their decisions. Our choice of the alternatives can be different from theirs. And in addition to that, very often people, especially those who have to make important (economic, political etc.) decisions, will not disclose their future decisions anyway, because they feel that this will impair their own chances, intentions and interests.

A behaviour like that definitely shows some limitations for doing simulations.

6.5   The Use of Random Numbers[†]

The most common way of making decisions of conflicts in a simulation run, or of dealing with uncertainty, is to use statistics and probabilities. But to do so is only satisfactory where the *law of large numbers* applies. Sometimes, e. g. for planning purposes, statistics are of no help, because they are not yet available. And in the case of processing a possible sample of real data in a simulation, one would wish that this sample (benchmark) is not 'too far away' from a real input.

The problem is that in some cases – and one never knows in which ones (maybe just in this case) – a plain random number generator will produce a simulated result that is way out from anything one can expect in reality. So one will always be in doubt whether the result is a 'good', usable, and reliable one, or if the random number generator tells stories.

In a recent paper [Fu6], the usefulness of random numbers *under the condition of some fairness*, i. e. that the random numbers are *not too random* for simulation purposes, has been discussed for the two main applications: **decision making** and **sampling**.

The proposal was, not only to just avoid stray numbers and outliers but to guarantee that randomness is kept below some preset limits. This can be achieved by using *event counters* instead of plain random numbers, i. e. to make use of the *synchronic distance* in the first case, and to supply random input data in a limited *multiplicity* only in the second case. After m times this class is empty and a future choice of it would not yield a value, so that there has to be a re-draw.

7   Conclusion

The use of *Petri nets* for modelling the causal and flow structures of a system has the well-known structural advantages. Their disadvantage was their largeness which proved unwieldy as soon as numerical values of practical problems were involved. Large amounts of tokens can now be handled in P/Ta-nets by *one single* constructive element: a Place in a specific parameter function, the Place serving as a *K-type* Place. If a Place serves in a P-type function it represents a (normal) variable. In both cases, the value of the parameter or the variable is represented by the contents

---

[†] This chapter is not related to P/Ta-nets only – random numbers are used in simulations all over. So the problem of how to work with random input can arise in the case of simulations with all net models, especially in the case of numerical simulations with P/Ta-nets, too. Therefore a short view upon this subject is given here.

of the Place. The introduction of *K-type Places* and their parameter function adds a second degree of freedom, *a differential element*, to the net model, so that now a good deal of numerical calculations can be performed in the otherwise rather rigid net model. This can be noted as the major progress which makes it possible to use the notions of the theory of concurrency for the simulation of distributed systems.

A typical application of simulations with P/Ta-Nets is any system where *money* is involved, because in such simulations money usually comes in big numbers.

8   References

Most papers related to this one in general are listed in the following 2 bibliographies:

S.Drees/D.Gomm/H.Plünnecke/W.Reisig/R.Walter: *Bibliography of Net Theory.* 101 p. Arbeitspapiere der GMD No. 212; GMD (1986)

E.Pless/H.Plünnecke: *A Bibliography of Net Theory,* 2nd Ed., GMD Report ISF-80.05 (1980)

Of special interest are the following papers:

[B/F] E.Best/C.Fernandez: *Notations and Terminology on Petri Nets.* Arbeitspap. d.GMD No. 195 (1986)

[Fu1] H.Fuss: *P-Ta-Netze zur Simulation von asynchronen Flüssen.* in: LNCS 26, pp.326-335 (1975)

[Fu2] H.Fuss: *AFMG - Ein asynchroner Fluss-Modell-Generator.* Berichte d. GMD 100. GMD (1975)

[Fu3] H.Fuss: *Abbildungen zwischen Puffer-Transaktions-Netzen und einfacheren Netzen.* Techn. Report ISF. GMD (1981)

[Fu4] H.Fuss: *Reversal Simulation with Place-Transactor-Nets.* in: H.Wedde (Ed.): *Adequate Modeling of Systems.* Proc. Int. Working Conf. on Model Realism, pp.222-232, Springer (1983)

[Fu5] H.Fuss: *Improvement of Simulations by Place-Transactor-Nets.* in A.Jávor (Ed.) *Simulation in Research and Development,* pp.85-91 IMACS. N.Holland (1985)

[Fu6] H.Fuss: *Simulating 'Fair' Random Numbers.* in: Proc. 2.Eur. Sim. Congr., pp.252-257, SCS (1986)

[Jef] D.Jefferson: *Virtual Time.* in: ACM Transact. Progr. Lang. & Syst., Vol.7, No.3, pp.404-425 (1985)

[K/Fu] W.Krelle/D.Beckerhoff/H.Langer/H.Fuss: *Ein Prognosesystem für die wirtschaftllche Entwicklung der Bundesrepublik Deutschland.* 355 S. A.Hain, Meisenheim/Glahn (1969)

[MS&] A.Maggiolo-Schettini/H.Wedde/J.Winkowski: *Modeling a Solution for a Control Problem in Distributed Systems by Restrictions.* TCS 13, pp.61-83, N.Holl. (1981)

[P1] C.A.Petri: *General Net Theory.* in: B.Shaw (Ed.): *Computing System Design,* pp.131-170, Univ. Newcastle/Tyne (1976)

[P2] C.A.Petri: *Concurrency.* in: LNCS 84, pp.251-260; Springer 1980

[P3] C.A.Petri: *Concurrency Theory.* in: *Advanced Course on Petri Nets* (Sept.'86) Bad Honnef, pp.1-21 (Preprints as course material to the participants, GMD (1986); proceedings to appear at Springer 1987)

[Re] P.Reynolds (Ed.) *Distributed Simulation 1985,* Proceedings, 111p. SCS Simulation Series Vol.15, No.2, Simulation Councils Inc. (1985)

[Va] R.Valk: *Self-Modifying Nets, a Natural Extension of Petri Nets.* pp.464-476 in: LNCS 62, (1978)

[Vo] K.Voss: *Using Predicate/Transition-Nets to Model and Analyze Distributed Database Systems.* Transact. on Softw. Engin., Vol.Se-6, pp.539-544. IEEE 1980

[We] H.Wedde: *Lose Kopplung von System-Komponenten.* Berichte d. GMD 96; GMD (1975)

Net Models of Dynamically Evolving Data Structures

Hartmann J. Genrich
Institut für Methodische Grundlagen
Gesellschaft für Mathematik und Datenverarbeitung
Sankt Augustin, B.R.D.

Abstract: An asynchronous push-down stack designed by Petri [6] is revisited and used as a basis for constructing two registers that can be used as a queue and a double-ended queue, respectively. These registers show several appealing features. They are highly concurrent arrays of processor elements and storage cells. Their size is unbounded; they can be repeatedly extended at their front end without affecting their internal structure or operation. Their latency is independent of the size. Their internal processor elements perform reversable operations; the implementation would not depend on energy dissipation.

Introduction

In his doctoral thesis [6], Petri outlines a theory of information processing systems that could supersede the theory of sequential automata as the theoretical basis of computer science. To quote the abstract of the English translation, Petri's main train of thought is as follows:

"The theory of automata is shown not capable of representing the actual physical flow of information in the solution of a recursive problem. The argument proceeds as follows:

1. We assume the following postulates: a) there exists an upper bound on the speed of signals; b) there exists an upper bound on the density with which information can be stored.

2. Automata of fixed, finite size can recognize, at best, only iteratively defined classes of input sequences. (See Kleene [4] and Copi, Elgot, and Wright [1].)

3. Recursively defined classes of input sequences that cannot be defined iteratively can be recognized only by automata of unbounded size.

4. In order for an automaton to solve a (soluble) recursive problem, the possibility must be granted that it can be extended unboundedly in whatever way might be required.

5. Automata (as actual hardware) formulated in accordance with automata theory will, after a finite number of extensions, conflict with at least one of the postulates named above.

Suitable conceptual structures for an exact theory of communication are then discussed, and a theory of communication proposed.

All of the really useful results of automata theory may be expressed by means of these new concepts. Moreover, the results retain their usefulness and the new procedure has definite advantages over the old ones.

The proposed representation differs from each of the presently known theories concerning information on at least one of the following essential points:

1. The existence of a metric is assumed for neither space nor time nor for other physical magnitudes.

2. Time is introduced as a strictly local relation between states.

3. The objects of the theory are discrete, and they are combined and produced only by means of strictly finite techniques.

The following conclusions drawn from the results of this work may be cited as of some practical interest:

1. The tolerance requirements for the response characteristics of computer components can be substantially weakened if the computer is suitably structured.

2. It is possible to design computers structurally in such a way that they are asynchronous, all parts operating in parallel, and can be extended arbitrarily without interrupting their computation.

3. For complicated organizational processes of any given sort the theory yields a means of representation that with equal rigor and simplicity accomplishes more than the theory of synchronous automata."

To justify the above claims, Petri presents a design of a Turing machine whose two-way infinite tape is realized as a pair of unbounded push-down stacks. The stack design - being the critical part - shows several important features.

1. It does not conflict with the two postulates of *bounded signal speed* and *bounded information density*.

2. It does not conflict with a third postulate of *bounded fan-in/fan-out*. At every level of decomposition, the number of direct neighbours to which a component is connected is bounded.

3. The push-down stack is an arbitrarily extendible array of identical simple processor elements and storage cells.

4. The internal processor elements perform reversable operations; their realization in actual hardware would no require energy dissipation.

5. The latency of the stack is independent of its size.

The original presentation of the push-down stack in [6] is not easy to follow. Using a special net theoretical system model called information flow graphs [7], Petri himself revised the presentation of the design several times [9]. After that it was not too difficult to present the design in a systematic top-down fashion using interpreted bipolar synchronization schemes as the underlying net model [3].

The purpose of this note is to apply the ideas guiding Petri's design of a stack with unbounded size but bounded latency to the design of two other basic types of dynamically evolving data structures, namely queues and double-ended queues.

By unbounded size we denote here the possibility of repeated extension that does not affect the internal structure and operation. This 'on-the-fly' extendibility of constructions is a paradigm for the possibility of re-organizing parts of a large system while other parts may continue working without being disturbed.

There are many reasons why such an on-the-fly reorganization may be needed. In large technical systems, eg, repairing or replacing faulty or dated components or, adding components to increase capacities must be possible while the operation goes on. In socio-technical systems, re-organization of a department, partial automation of processes, extension of production facilities etc. should happen without affecting the whole.

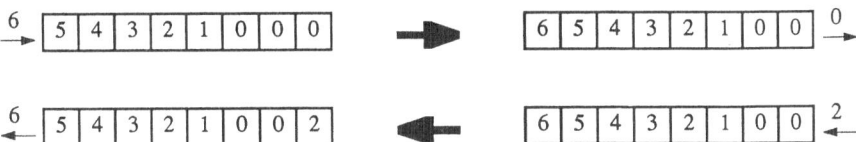

Figure 1: Two-way shift register

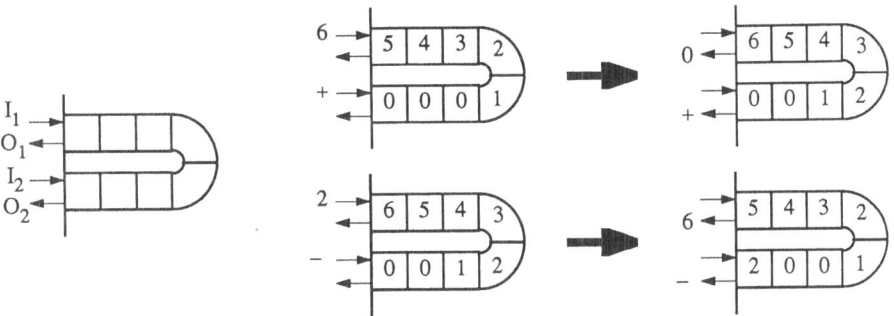

Figure 2: Bent shift register

For single hardware elements like VLSI chips, extendibility itself is not the issue. Rather, its guiding principles lead to designs of very regular arrays of processor and storage elements that allow fast interactions because of their inherent concurrency.

The organization of this note is as follows. In the first section, the top-down design of the stack, as presented in [3], is repeated in a slightly different form (using predicate/transition nets instead of bipolar schemes). In sections 2 and 3 an unbounded queue and an unbounded double-ended queue, respectively, are constructed using the stack as a starting point.

1 The Stack

Petri's stack is based on a two-way shift register. Such a register is a one-dimensional array of storage cells that can be operated from both its ends. Whenever an item is entered from the left end, all items shift one position to the right; the right-most item drops out. For a left shift, a new item is entered at the right end. All items move one position to the left; the left-most item drops out. This is shown in figure 1.

The first step in the construction is to coordinate the two ways of operating the register. This is achieved by bending the register so that the two ends form a common front end with two input ports I_1 and I_2 and two output ports O_1 and O_2. The bent shift register is displayed in figure 2.

The new data item is placed on the input port I_1 and a control signal is sent along I_2 to indicate the kind of shift that is desired: a '+' for a right shift and a '−' for a left shift. (When the register is used as a stack later, + will indicate a push and − a pop.) The item that drops out as the result of a shift appears on output port O_1. In addition, a copy of the request received at I_2 appears on output port O_2.

Since the bent shift register follows a *first-in/first-out* discipline, we denote for any integer $n \geq 0$ the register with $2n$ cells as LIFO[n]. It will appear in a PrT-net as a transition-like component as shown in figure 3. The variables x, x' denote the data items and r denotes the type of request.

By denoting the register with $2n$ double-cells as LIFO[n], we wish to suggest that LIFO[n] is best viewed as consisting of n double-cells ('slices' of the bent register). This will become clearer

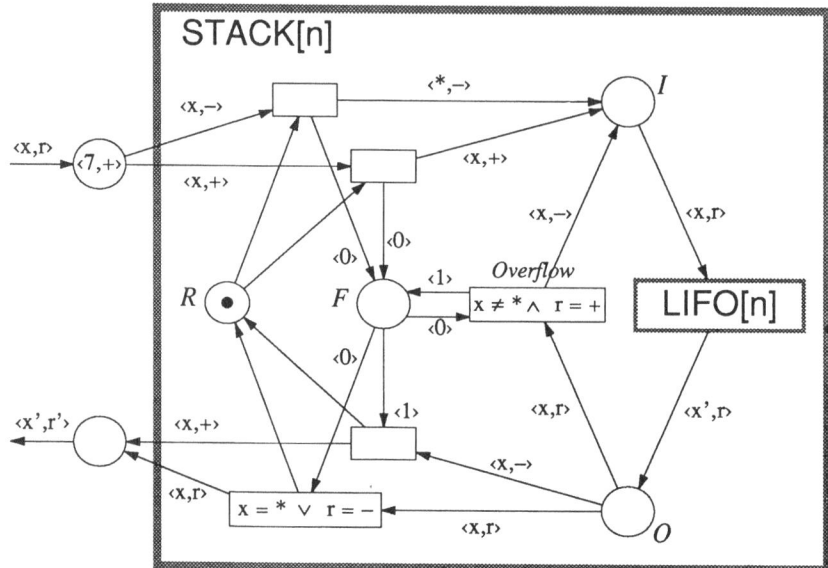

Figure 3: Using LIFO as a stack

when we start considering the internal structure of the LIFO register.

For now, we observe that LIFO[n] can be used as a finite *stack* (of length $2n$) if we reserve one special value, say '*', to indicate an empty cell. Thus the stack is empty iff each cell contains a *, and it is full iff each cell contains an item different from *.

A *push* consist of placing a data item on I_1 and a + on I_2. In the resulting right shift, a * will appear on O_1 if the stack is not full, and a + will appear on O_2 as an acknowledgement.

A *pop* consists of placing a * on I_1 and a − on I_2. In the resulting left shift, the left-most (*top*) item will appear on O_1 and a copy of the − on O_2.

If the stack is empty, a *pop* will return a *. This might not cause much harm. If the stack is full, however, a *push* will cause a data item to drop out of the stack and appear on O_1. Fortunately, such an overflow can be detected since normally, a * is expected to appear on O_2 as a result of a *push*. Once an overflow has been detected, it can be mended immediately by following up with a pop − but with a twist. The item that was pushed out is placed on I_1 (instead of a * in the normal case) and a − on I_2. As a result, the item that caused the overflow will return on O_1.

A device operating the LIFO register in this fashion to realize a stack is shown as a PrT-net in figure 3. It works briefly as follows. If a new request consisting of a data item x and control item r arrives and the device is ready to process it (token on place R), the overflow flag (place F) is reset to 0. If the request is a *pop* the data item is set to *. Then a shift operation is performed as described above. The result appearing on place O may be processed in two different ways: (a) If there was no overflow of the LIFO register, the result is transferred to the output ports of the stack. (b) If an overflow has occurred, the flag F is set to 1 and the data item is put back at the end of the LIFO register by a *pop* operation. As a result, the item that caused the overflow comes out; it is returned to the environment and the device is ready for processing the next request.

(As an exercise, you may add an isempty test operation.)

As promised, we shall now show how LIFO[n] is put together. The construction is displayed in figure 4 using a recursive definition given in pictorial form.

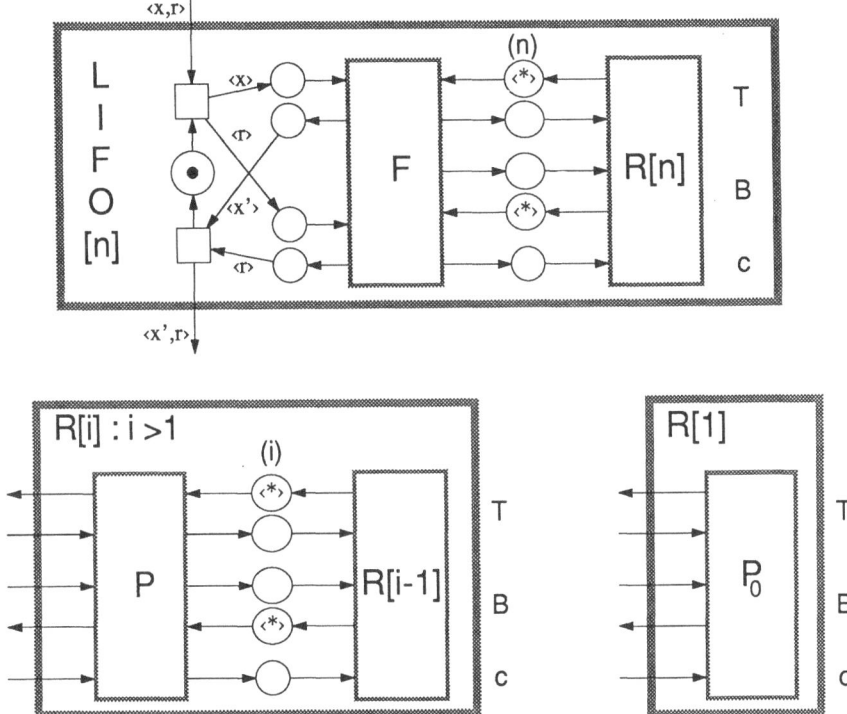

Figure 4: Recursive definition of LIFO

At its front end, LIFO[n] consists of a simple control mechanism that sequentializes the incoming requests and distributes their parts to the different ports of the first processing element labelled as F. F is connected to the body of the register, $R[n]$, by five lines; four lines for data transfer back and forth and one for the control information.

$R[n]$ is expanded recursively depending on the value of n. It consists of an one-dimensional array of $n-1$ processing elements named P and a final element P_0.

The basic idea is to form the i-th double-cell by pairing together cell $\#(n-i+1)$, denoted as T_i (T for top), with cell $\#(n+i)$, denoted as B_i (B for bottom). Thus the n-th double cell holds the left-most (top) item in T_n and the right-most (bottom) item in B_n. Each cell is represented by a pair of lines (places) indicating whether the data item will next take part in an operation of its left or its right neighbour element, respectively.

In addition, each double cell is capable of keeping a control signal ($+$ or $-$) in place c_i. The initial distribution of data and control items is that of an empty stack, all double-cells being ready to interact with their respective left neighbours.

The interactions between cells are represented by transitions labelled as P. At the front end, the interaction between the register and its environment is represented by the transition F. For the right-most double cell ($i = 1$), the interaction to the right is a special one denoted as P_0.

It turns out that F, P, and P_0 are simple transfer operations depending only on the value of the control signal coming from the left. In figure 5, we show how the items are shuffled as a result of applying F, P, and P_0. The basic operation is the conditional choice between two items depending on the value of the control signal. It is denoted as $[r : u, v]$ and defined as [IF r = $+$ THEN u ELSE v]. In the P element, there are two pairs of related conditional choices. Each pair represents the *conditional interchange* between two data lines - a generalisation of the *Quine*

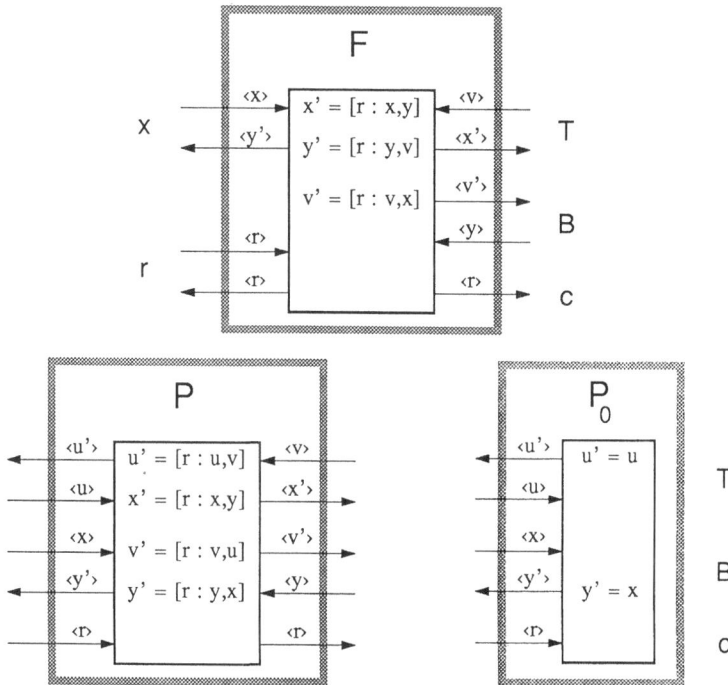

Figure 5: Definition of the interactions of LIFO

transfer of information flow graphs [7,8,10]. (The CE-net representation of the Quine transfer in [8] provided the *logo* of the present volume.)

At P, the control signal passes through without being affected; at P_0, it is erased. At F, a copy of the control signal is returned as an acknowledgement.

The key to understanding the operation of this array is the following idea. If a cell is ready for an interaction to the left it contains in T the right answer to a *pop*, and in B the right answer to a push. If it is ready to interact to the right, it is in a transient state; it contains an item that is temporarily 'parked' either in B, during a *push*, or in T, during a *pop*.

Each double-cell interacts alternatingly with its left and right neighbour; this is the only restriction to the concurrent occurrence of interactions. It is not difficult to verify that through F, P and P_0 we have indeed realized a two-way LIFO register that operates in a highly concurrent fashion. A bit of hand simulation with, say 4 double-cells, should convince the interested reader.

To facilitate simulation, another presentation of the interactions is given in figure 6. It replaces the PrT-net notation by a kind of wiring diagram that depicts quite clearly the symmetry of P and P_0 with respect to T and B. It also shows another interesting fact: the P element is *time-reversal invariant*. No information is gained or lost in an operation of the P element; its implementation does not depend on energy dissipation.

To conclude this section, we wish to point out some of the interesting properties of the LIFO register. Firstly, the register is an highly concurrent array of simple processor elements and storage cells. The internal structure is independent of the size of the register. Each element has at most two neighbours to which it is connected through at most five pairs of ports.

Secondly, the latency of the register is bounded, ie the maximal amount of time the environment has to wait between two requests, is independent of the size or the previous history. A new request

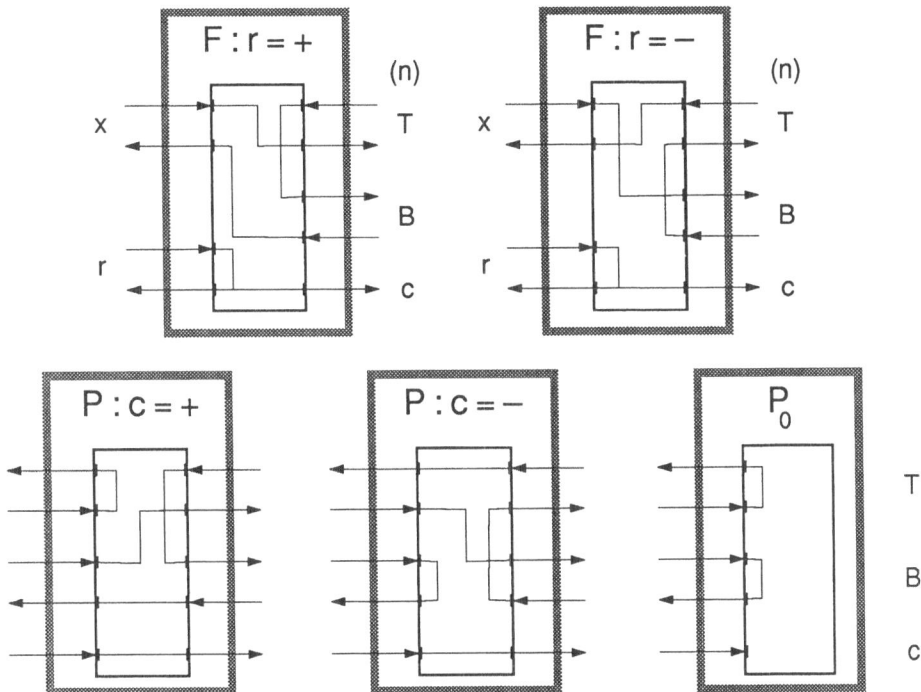

Figure 6: Wiring diagrams of the interactions of LIFO

can be processed at the front end as soon as its right neighbour has processed the previous request. In general, a new request will be processed concurrently with a large number of previous requests still trickling through the register.

Finally, the register is unbounded; it can be extended *on the fly* at the front end while the body of the register can continue processing earlier requests concurrently. Whenever an overflow is detected and mended in the manner described above, an empty double-cell may be inserted as indicated in figure 7. Note that the * in the top cell is misplaced (parked). The new double-cell will first interact with is right neighbour. The *pop* (−) causes the parked * to move into its proper, last position but one and the top element to move into the new top position.

2   The Queue

The next storage device we wish to have a closer look at is a buffer that connects a *sender* of data items (on the left end) with a *receiver* of items (on the right end). The idea is that items may be sent and received at different rates but in the same order. A straightforward way of realizing such a buffer is shown in figure 8. It consists of an array of data cells that are arranged in a pipeline manner.

Whenever the sender wants to put an item into the buffer, it takes the ready signal from the left output port O_S and places the item on the input port I_S. When the receiver wants to get an item from the buffer, it places a control signal on the right input port I_R and waits for the item to appear on the output port O_R. If the buffer is full, there will no ready signal appear on port O_S. If the buffer is empty, no item will appear on port O_R. There is no test for overflow or underflow.

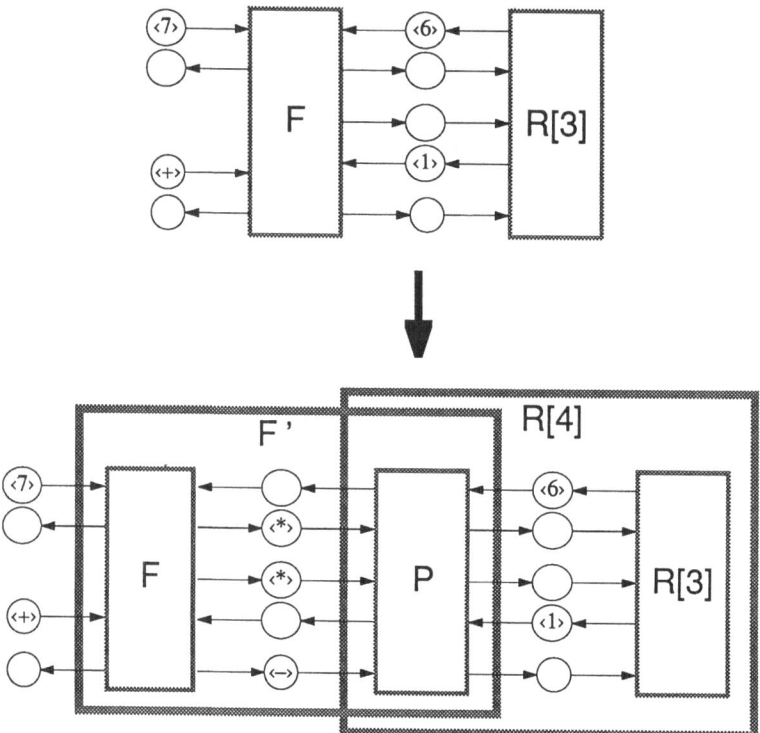

Figure 7: On-the-fly extension of LIFO

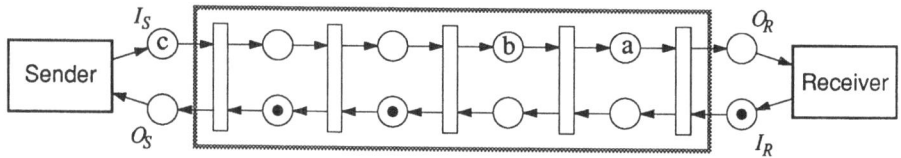

Figure 8: Buffer between sender and receiver

It is obvious that this buffer is highly concurrent, all cells interacting alternatingly with their two neighbours only. It is extendible without conflicting with any of the postulates we have assumed.

However, there is one characteristic magnitude that is not independent of the size of the buffer: The time a single item needs to travel from the sender to the receiver increases with the number of cells. This distance between sender and receiver becomes larger and larger in a similar way as in a Turing machine design with moving tape head, the distance between the head and the control unit may become unboundedly large (or the distance between control unit and the user if the control moves along with the head).

What we are after is a buffer of unbounded size but bounded distance between the two ends, ie bounded latency. To this end we design a register that is quite similar to the bent two-way shift register but observes a *first-in/first-out* (FIFO) rather than *last-in/first-out* (LIFO) discipline. Hence it can serve as a queue in the same way as the LIFO register can serve as a stack.

We start again with an one-dimensional array of cells that can be operated from both its ends; it is bent so that the two ends form a common front with two input ports I_1 and I_2 and two output ports O_1 and O_2. This is shown in figure 9.

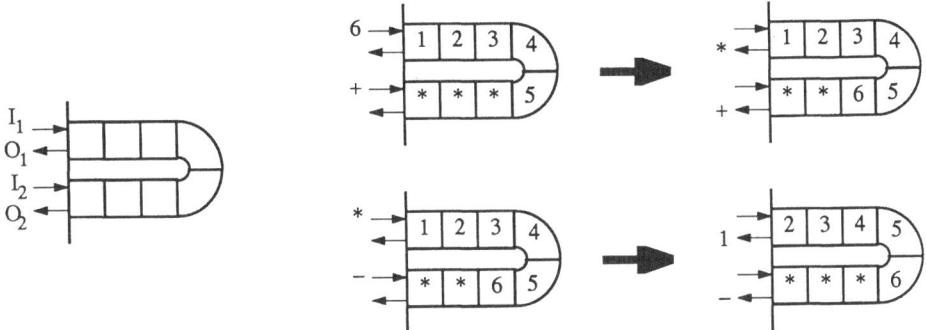

Figure 9: A FIFO register

Assuming that initially the register is empty (all cells contain $*$), each new data item is *put* into the queue by placing it on port I_1 and placing a $+$ signal on port I_2. Because of the FIFO discipline, the new data item has to be inserted behind the item that was put before. As a result, a $*$ is returned on port O_1 and a copy of the $+$ signal on port O_2.

To *get* a data item from the FIFO register, a $*$ is placed on port I_1 and a $-$ on port I_2. As a result, the left-most (top) item will appear on port O_1 and a copy of the $-$ on O_2.

If the register is empty, a *get* will return a $*$. If the register is full, a *put* will return the new item itself. Both cases are easy to detect and might not cause much harm. There is no need for mending an overflow as there was for the stack. (There is, however, a need for arbitration between the sender and receiver since both may access the register concurrently but arbitration is not our concern in this paper.)

For any integer $n \geq 1$, we denote such a register with $2n$ cells as FIFO$[n]$. Again we wish to indicate that the following construction is easy to understand, once FIFO$[n]$ is viewed as consisting of n double-cells (slices of the bent register).

The construction of FIFO$[n]$ is displayed in figure 10 using again a recursive definition given in pictorial form. Once more we form the i-th double-cell by pairing together cell $\#(n - i + 1)$, denoted as T_i, with cell $\#(n + i)$, denoted as B_i. Place c_i is to hold the control signal ($+$ or $-$).

In addition to the layout of the double-cells of LIFO, however, the double-cells of FIFO are capable of holding two bits in places t_i and b_i, respectively. They are used to control the interactions between neighbouring double-cells that no longer depend on the value of the control signal only but in addition, on the status of the two cells of the respective right double-cell. The current values of t_i and b_i indicate whether the respective data cells are empty (0) or full (1). (If the number of connections between processor elements and storage cells is a critical issue - as it is in VLSI design - the t/b part of the register may be replaced by a test for equality with $*$.)

Initially, when all data cells contain $*$, all t_i and b_i contain 0. During operation, the t/b part behaves as a LIFO of bits and serves as a unary counter. The number of 1's it holds is the current length of the FIFO. The first 0 after the initial sequence of 1's indicates the position where a new data item has to be inserted.

The key to understanding the interactions between double-cells as defined in figure 11 (this time we present them immediately as wiring diagrams) is that a new item moves along the empty part of the bottom half of the register until it touches either a full B cell or an empty T cell. It is 'parked' in the position that became vacant by returning the empty symbol from the end of the FIFO. There are three different cases for performing a *put*. First, if the upper right cell is empty (and hence the cell below is empty as well), the item has to get there its final position. Second,

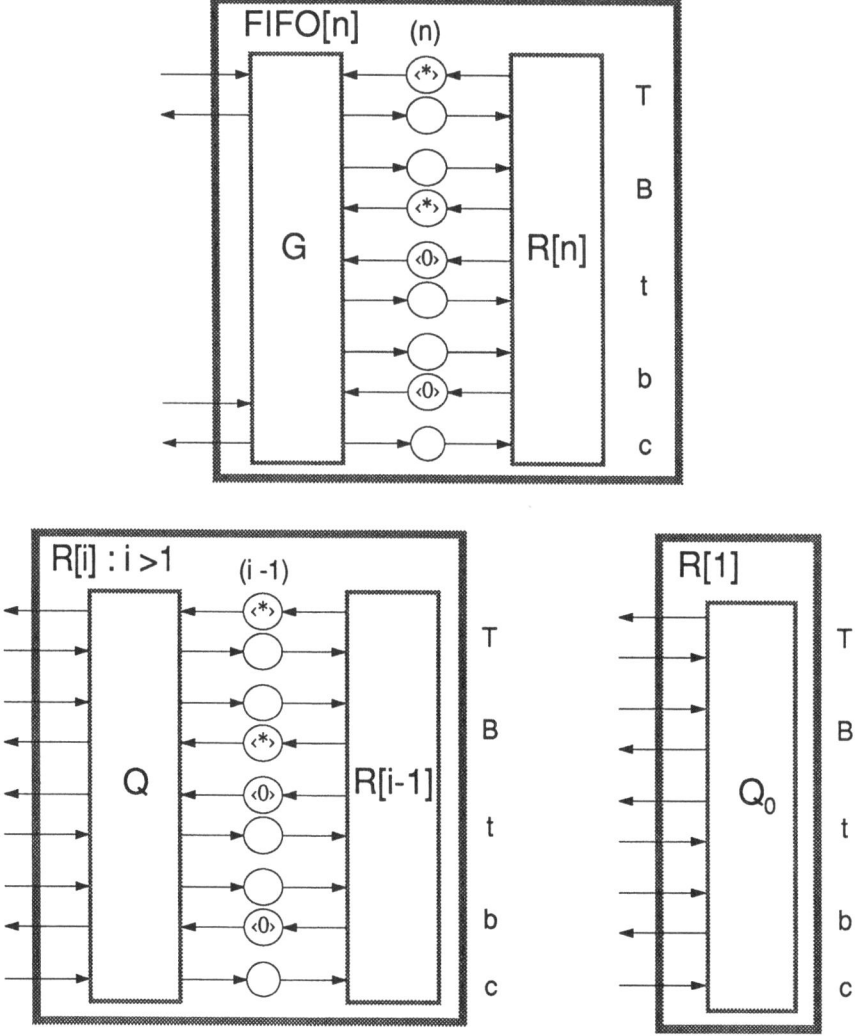

Figure 10: Recursive definition of FIFO

if the right upper cell is full but the cell below is still empty, the data item continues travelling along the empty part of the lower half. Finally, if the right lower cell is full (and the cell above as well), the item is already in its final position. For a *get*, the register behaves in the same way as the LIFO register of the previous section.

The interactions between double-cells are represented by transitions labelled by Q. At the front end, the interaction of the register with its environment is represented by the transition G. For the right-most double-cell, the interaction to the right is a special one denoted as Q_0. G,Q and Q_0 are simple transfer operations depending on the value of the control signal on the left and the status of the two cells on the right represented by the values of t and b.

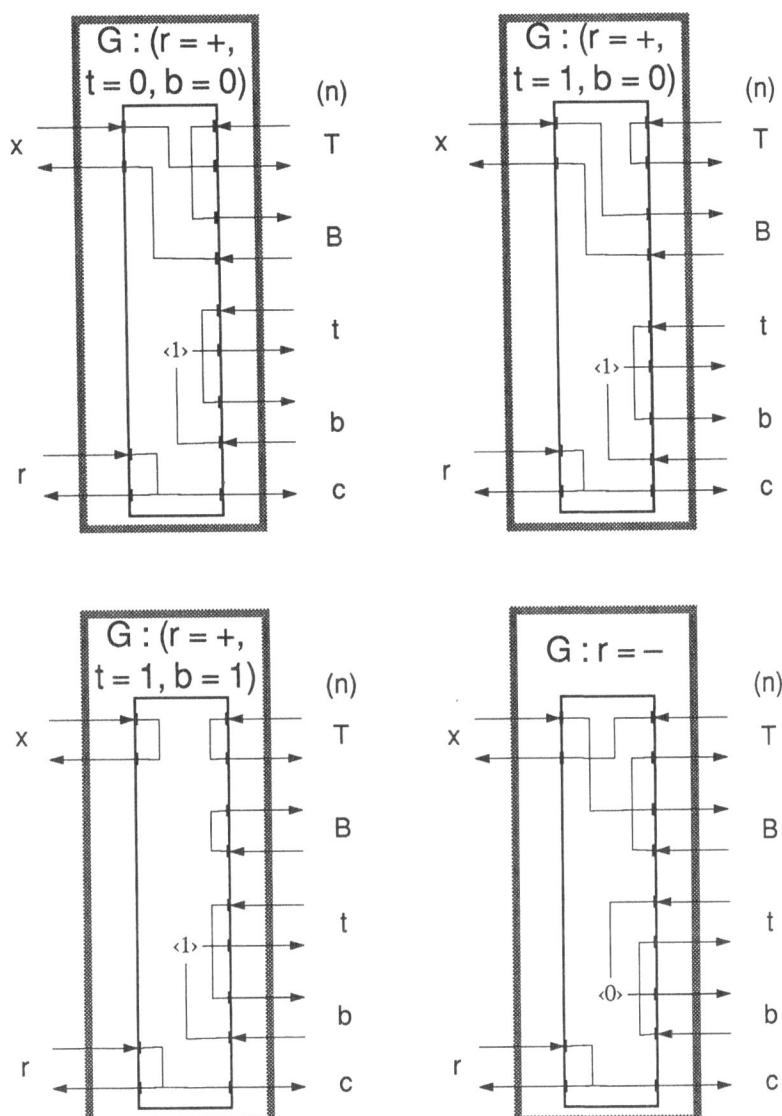

Figure 11: The processor elements of FIFO (page a)

212

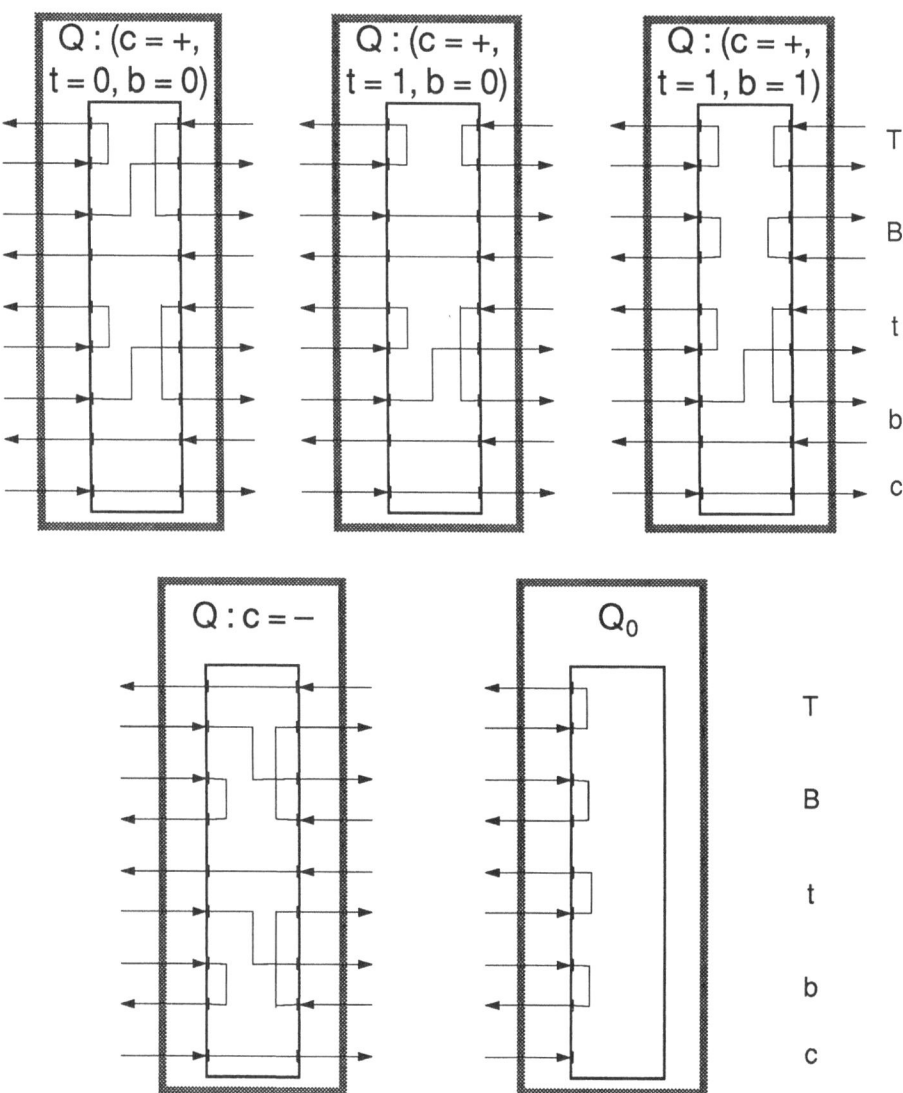

Figure 11: The processor elements of FIFO (page b)

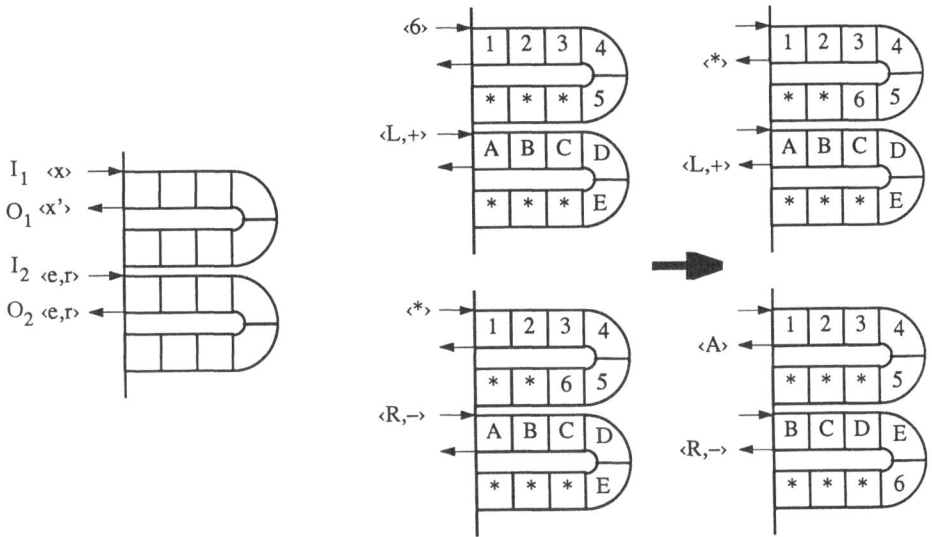

Figure 12: A double-ended FIFO register

It should not be too difficult to verify that through G, Q and Q_0 we have indeed realized a register that functions like a FIFO when used in the described manner. A bit of hand simulation will do.

To conclude this section, we wish to point out that the FIFO register has the same interesting properties as the LIFO register that were listed at the end of the previous section. The on-the-fly extension works in the same way as for the LIFO register (see figure 7). An empty double-cell is inserted that is initialized with $*$ in the data cells T and B, a $-$ in c and 0 in t and b.

3 The Double-Ended Queue

The last extendible register we wish to design is a combination of the two previous ones that can be used as a *double-ended queue*. Such a register can be operated independently at two ends by putting $(r = +)$ and *getting* $(r = -)$ data items. Operated from one end only, it functions as the LIFO of section 1. If on one end, only *puts* occur and on the other end, only *gets* occur, it functions as the FIFO of section 2.

Figure 12 shows such a double-ended FIFO register that consists of two LIFO registers for data items that are glued together in such a way that each slice consists of 4 cells. A control signal now consists of two bits indicating not only the kind of request $(put = +$ or $get = -)$ but also the end from where the request comes $(left = L$ or $right = R)$.

The definition given in figure 13 shows for each 4-cell two pairs of data cells, (TL_i, BL_i) and (TR_i, BR_i), and the control line c_i. In addition, two unary LIFOs (one for each data LIFO) are used to represent the status of the data cells $(full = 1$ or $empty = 0)$, quite in the same way as in the FIFO of the last section. The recursive expansion of the register body has been omitted this time since it follows the same pattern that we know from LIFO (figure 4 and FIFO (figure 10).

The key idea to understanding the interactions between neighbouring 4-cells as defined in figure 14 is that the two LIFOs are kept equally long (as indicated by the examples in figure 12). The numbers of items stored in each LIFO never differ by more than 1. Hence there are cases when an item has to be transferred from one LIFO to the other. In this way, the two LIFOs together form a *double-ended* FIFO.

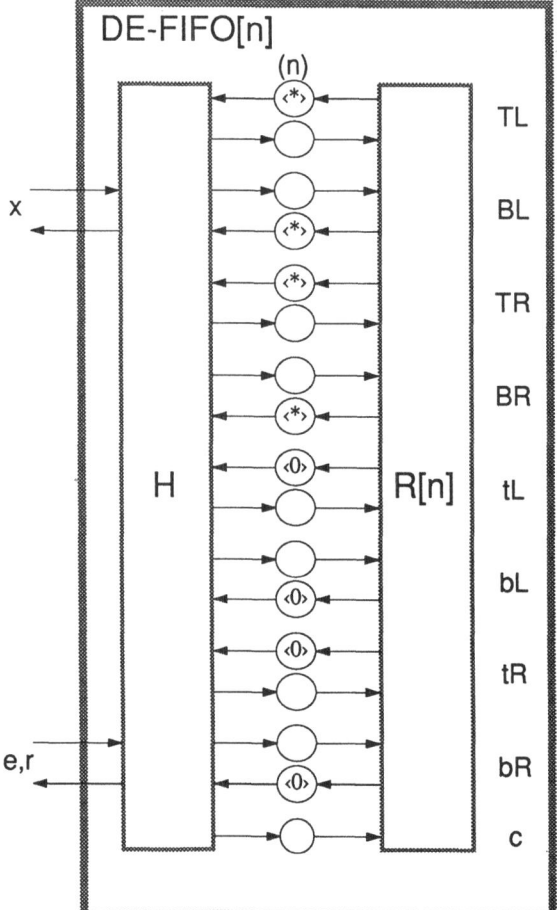

Figure 13: The lay-out of the double-ended FIFO

We show only the data part of the interactions between neighbouring 4-cells for requests coming from the left end ($c = \langle L, + \rangle$ or $c = \langle L, - \rangle$). From this the treatment of the status cells and the operations of the front end and right end can be derived easily. The operations are symmetrical in L and R.

There are three different operations each for a *put* and a *get* from the left. The situations in which these operations are applied are listed below the wiring diagrams. Only seven of the sixteen conceivable situations are possible if the register functions properly starting from the state in which all cells are empty. In five situations, the L-part behaves just as the LIFO in section 1 and the R-part remains unchanged. In two situations, an item is transferred from the L-part to the R-part, in a *put*, or in the reverse direction, in a *get*.

An overflow has to be mended in the same way as for the LIFO register in section 1.

Also the double-ended FIFO is an highly concurrent array of processing elements that have to perform conditional transfer operations. It is on-the-fly extendible like the LIFO and the FIFO.

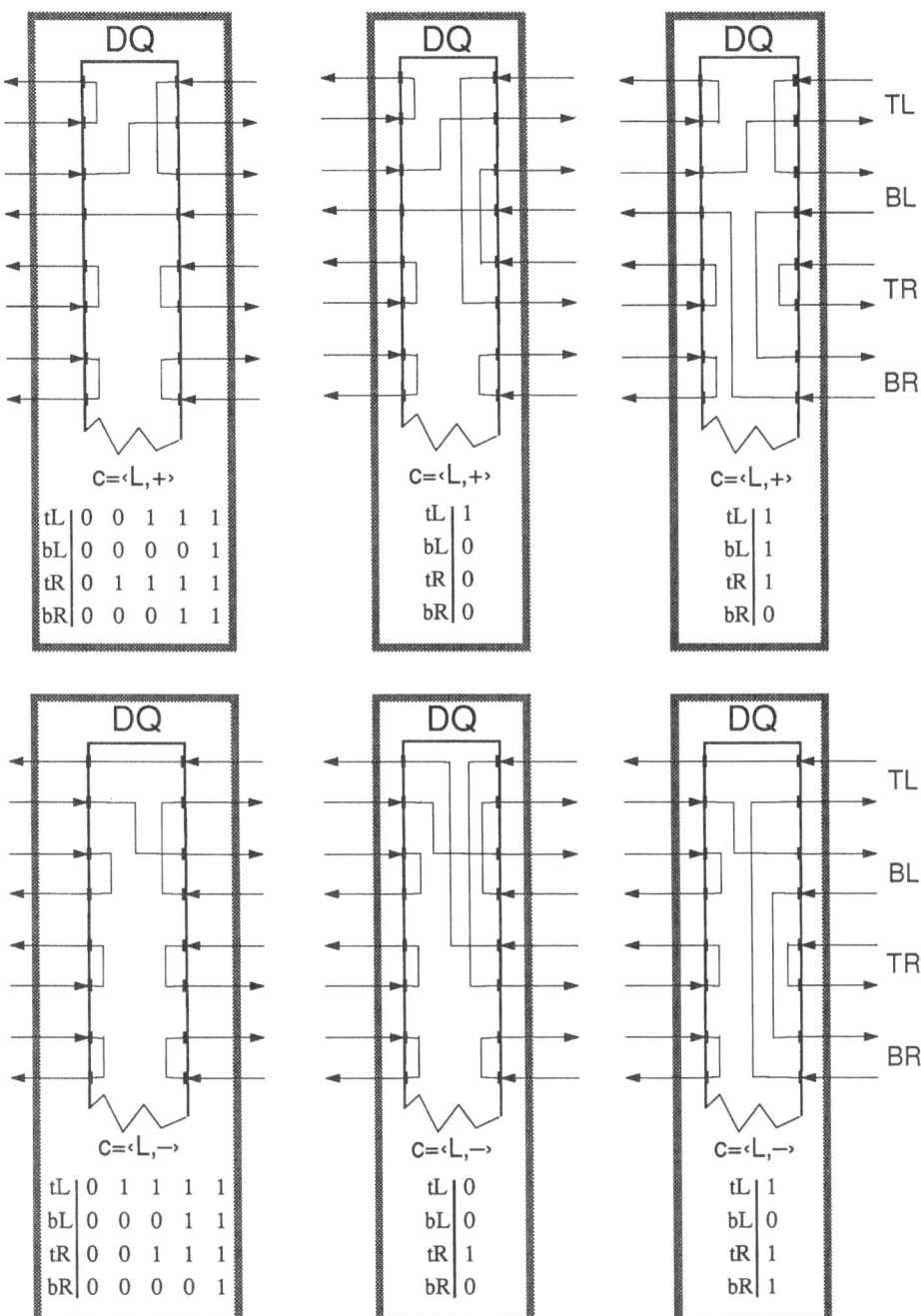

Figure 14: The internal processor elements of the double-ended FIFO

There is only one difference to note: since also a 4-cell can hold at most one control signal at a time, it cannot be initialized with two misplaced *, both in the upper half and the lower half. Hence at least one of the top elements of the two LIFOs must be popped out and stored in the new 4-cell before that is inserted.

Conclusion

In this note, we have reminded the reader of the roots of net theory. Twenty-five years ago, Petri stressed the fact that devices performing computations are multi-component systems and that the design of such systems should be based on an intelligent and skillful treatment of concurrency. Since then, new architectural principles have emerged. To mention but two that are closely related to the subject of this paper: *Systolic Arrays* [5] and *Conservative Logic* [2]. It seems that Petri's design of an asynchronous, arbitrarily extendible stack and moreover, the principles that supported this design and have guided the development of net theory, did not lose in originality and actuality.

References

[1] Copi, I.M.; Elgot, C.C.; Wright, J.B.: *Realization of Events by Logical Nets.* Journal of the ACM 5, 181-196 (1958)

[2] Fredkin, E.; Toffoli, T.: *Conservative Logic.* Int. Journal of Theoretical Physics, Vol. 21, Nos. 3/4, pp. 219-253 (1982)

[3] Genrich, H.J.; Thiagarajan, P.S.: *Well-Formed Flow Charts for Concurrent Programming.* Formal Description of Programming Concepts - II (D. Bjørner, Ed.), North Holland Publ. Comp., pp. 357-380 (1983)

[4] Kleene, S.C.: *Representation of Events in Nerve Nets and Finite Automata.* Automata Studies (C.E.Shannon and J. McCarthy, Ed.), Princeton University Press, pp. 3-41 (1956)

[5] Kung, H.T.: *Why Systolic Architectures.* IEEE Computer, Jan. 82, pp. 37-46 (1982)

[6] Petri, C.A.: *Kommunikation mit Automaten.* Schriften Nr. 2 des IIM, Rhein.-Westf. Institut für Instrumentelle Mathematik, Bonn (1962)
also: Technical Report RADC-TR-65-377, Vol. 1, Suppl. 1, Rome Air Development Center, Griffiss Airforce Base, New York (1966) [English transl.]

[7] Petri, C.A.: *Grundsätzliches zur Beschreibung diskreter Prozesse.* 3. Kolloquium über Automatentheorie, Birkhäuser Verlag Basel, pp. 121-140 (1967)
also: Technical Report ISF-82.04, Gesellschaft für Mathematik und Datenverarbeitung, Sankt Augustin (1982) [English transl.]

[8] Petri, C.A.: *"Forgotten" Topics of Net Theory.* In: Proceedings of the Advanced Course on Petri Nets (Brauer, Reisig, Rozenberg, Eds.), Lecture Notes in Computer Science, Springer-Verlag (to appear)

[9] Petri, C.A.: *Private Communication.*

[10] Smith, E.: *Reversible Logic and the Petri Gate.* Newsletters of the GI SIG Petri Nets and Related System Models, No. 22, pp. 7-12 (1985)

On Condition/Event Representations of
Place/Transition Systems

Ursula Goltz

GMD-F1P

Postfach 1240

D-5205 St. Augustin 1

1. Introduction

One of the fundamental principles in Net Theory is that each representation of a system as a marked net should be traced back to a common
basic interpretation - nets of *conditions* and *events* ([Pe]). However,
this immediately implies a number of questions.

- What is the basic idea for this condition/event representation of
  higher level nets (place/transition systems or nets with individual
  tokens)?

- Which equivalence notion is suitable for comparing a marked net
  with its condition/event representation?

- Is there a unique or canonical condition/event representation which
  may be derived systematically for any given system?

- How do basic notions like contact and conflict translate between
  condition/event systems and higher level nets?

- What is the relationship between the well-established process notion
  for condition/event systems and process notions for higher level
  nets? Are they consistent with respect to a certain condition/event
  representation?

Some of these questions have been adressed in [LSB],[Be],[Ge],[Go1],
[Du],[SmRe]. In this paper, we consider these questions directly for
place/transition systems without first representing them as predicate/
transition systems by associating individualities with tokens as suggested in [Ge] and [Du]. Hopefully, it will turn out that both
approaches are consistent with each other. This point will be discussed
in section 5.

There are two controversial ideas how to represent a place/transition
system by conditions (carrying at most one token) and events. Both

start with the idea that the number of tokens on places must be spread over an appropriate number of conditions associated with each place. The first idea, called *pipelining*, achieves this by introducing "silent" events which organise this distribution but leaves the transitions of the P/T-system otherwise unchanged (this idea may, for example, be found in [LSB]).

The second idea came up somewhat later. Here again, with each place, a number of conditions is associated but the distribution of tokens over these conditions is achieved by associating with each transition several events, one for each possible choice of token distribution.

Example 1

Let   N = y 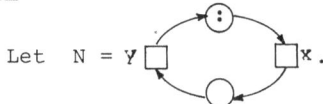 x .

According to the second approach mentioned above, this may be represented as

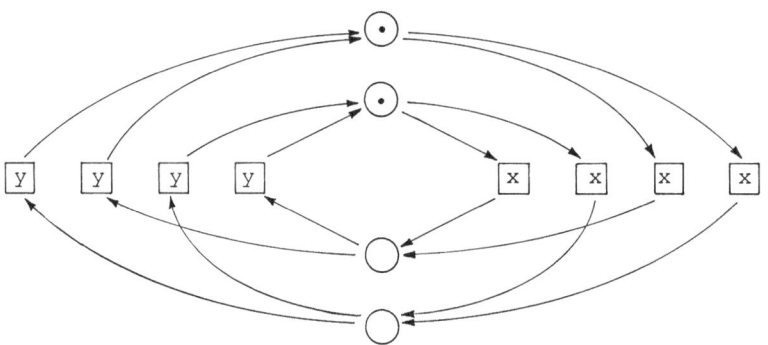

where the association of events with transitions is indicated by event labels.

This construction was the basis of the process notion for place/transition systems in [GoRe] and has been discussed in [Be] and [Go1]. In [Be], it is compared to the pipelining idea by discussing an example. A full formal treatment of any of the two approaches has not yet been given.

In this paper, the second construction is formalised and investigated. The notion of (step-)bisimulation equivalence [Go1, Vo, Go2] is used to relate P/T-systems to their representatives.

The construction we have chosen certainly does not give a simplest or minimal C/E-representation for a given P/T-system, and it can be argued whether it is the most appropriate one. However, if we require any C/E-representation to be bisimulation equivalent to the P/T-system, then all possible C/E-representations will be equivalent to each other and the equivalence class generated by this "canonical" representation may be considered. The claim in this paper is that "a most appropriate" representation (if it exists) will be included in this equivalence class. Some discussion on this point, as well as some hints on the correspondence between the respective sets of processes, will conclude the paper. A deeper investigation of these issues must be left for further research.

2. Basic notions

A notation for nets which differs slightly from the usual one will simplify our constructions considerably. We will write transitions in the form (preset, postset). (Note that this exludes non-simple nets with respect to transitions, however this degree of extensionality should be imposed anyway.) Furthermore, we will use event labels to relate the events in our C/E-representation to their associated transitions. Hence events will be written as (preset, label, postset). This is formalised by the following definitions.

2.1 Definition $N = (S,T;K,M_o)$ is called a place/transition system
    (P/T-system) iff
        S is a set (of places),
        $T \subseteq \mathbb{N}^S \setminus \{O\} \times \mathbb{N}^S \setminus \{O\}$, where O denotes the empty mapping:
        $\forall s \in S : O(s) = O$ (transitions),
        $K : S \rightarrow \mathbb{N} \setminus \{O\}$ (place capacities),
        $M_o \in \mathbb{N}^S$ (initial marking).

We allow transitions with arc-weights, hence the pre- and postsets of transitions are mappings from S to $\mathbb{N}$. We extend the usual notation $\cdot t$, $t\cdot$ for the pre- and postset of a transition t to represent arc weights: $\cdot t = pr_1(t)$, $t\cdot = pr_2(t)$, hence $\cdot t, t\cdot \in \mathbb{N}^S$ (pr_i means projection to the ith component).

We do allow isolated places in nets (they do not cause any problems for our purposes) but we do not allow isolated transitions. Just for sake of simplicity, we even require $\cdot t \neq O \wedge t\cdot \neq O$. We will only consider finite nets (S finite and hence also T finite). To obtain finite C/E-

representations, we consider only systems with finite capacities for
all places.

The C/E-representations we will obtain will not be condition/event
systems in the strict sence. They will violate the requirement that any
event is enabled in some case, they may be non-simple, they may contain
self-loops, and only forward reachable cases will be considered.
All this arises naturally since P/T-systems not only abstract from
individualities of tokens but, beyond that, are much less restricted
with respect to the properties listed above. We may have all these
features in one-safe P/T-systems. As one-safe P/T-systems should coincide with their C/E-representations, we consider a less restricted
class of systems on the level of conditions and events, which has been
called elementary net systems in [RoTh].

2.2 Definition $\Sigma = (B,E;C_0)$ is called a (T-labelled) elementary
    net system (EN-system) iff

      B is a set (of conditions),

      $E \subseteq P(B) \setminus \{\emptyset\} \times T \times P(B) \setminus \{\emptyset\}$ where T is a set (labelled events),

      $C_0 \subseteq B$ (initial case).

$\cdot e$ and $e\cdot$ denote the pre- and postset of e, as usual. $l(e) := pr_2(e)$
denotes the label of e. Note that there are some slight deviations
between this definition and [RoTh]. For example, we do allow isolated
conditions but we exclude non-simple nets with respect to events.

The behaviour of systems is described by defining the notion of a step.
If a number of transitions (or events) is enabled at some marking without interfering with each other, then they may occur concurrently in
one step. For P/T-systems we allow that transitions occur concurrently with themselves, hence steps are multisets over T.

2.3 Definition   Let $N = (S,T;K,M_0)$ be a P/T-system, let $M,M' \in \mathbb{N}^S$.
    $A \in \mathbb{N}^T$ is enabled at M iff
       (i) $\forall s \in S : M(s) \geq \sum_{t \in T} \cdot t(s) \cdot A(t)$,
       (ii) $\forall s \in S : M(s) + \sum_{t \in T} t \cdot (s) \cdot A(t) \leq K(s)$.
    A is a step from M to M' (M[A>M']) iff
       (i) A is enabled at M,
       (ii) $\forall s \in S : M'(s) = M(s) - \sum_{t \in T} \cdot t(s) \cdot A(t) + \sum_{t \in T} t \cdot (s) \cdot A(t)$.

The definition of step for elementary net systems may be derived from this definition by assuming capacity 1 for all places. However, we state it here using the set notation of EN-systems.

2.4 Definition   Let $\Sigma = (B,E;C_o)$ be an EN-system, let $C,C' \in P(B)$.

$\quad$ $G \subseteq E$ is <u>enabled at</u> C iff

$\quad\quad$ (i) $\quad \forall e \in G : \ ^{\bullet}e \subseteq C$,

$\quad\quad\quad\quad \forall e_1,e_2 \in G : \ ^{\bullet}e_1 \cap \ ^{\bullet}e_2 = \emptyset$,

$\quad\quad$ (ii) $\forall e \in G : C \cap e^{\bullet} = \emptyset$,

$\quad\quad\quad\quad \forall e_1,e_2 \in G : e_1^{\bullet} \cap e_2^{\bullet} = \emptyset$.

$\quad$ G is a <u>step from</u> C <u>to</u> C' ($C[G{>}C'$) iff

$\quad\quad$ (i) $\quad G$ is enabled at C,

$\quad\quad$ (ii) $C' = (C \setminus {}^{\bullet}G) \cup G^{\bullet}$ where $\ ^{\bullet}G = \bigcup_{e \in G} {}^{\bullet}e, \ G^{\bullet} = \bigcup_{e \in G} e^{\bullet}$.

3. A C/E-representation of P/T-systems

We have explained the idea how to represent a P/T-system as an EN-system already in the introduction. With the notations of section 2, we are now ready to give a concise formalisation.

3.1 Definition   Let $N = (S,T;K,M_o)$ be a P/T-system.

$\quad$ $\Sigma = (B,E;C_o)$ is called a (<u>canonical</u>) <u>C/E-representation of</u> N iff

$\quad\quad$ $B = \bigcup_{s \in S} \{(s,i) \mid 1 \le i \le K(s)\}$,

$\quad\quad$ $E = \bigcup_{t \in T} \{e \mid l(e) = t, \ ^{\bullet}e, e^{\bullet} \subseteq B$ with

$\quad\quad\quad\quad\quad \forall s \in S : \ ^{\bullet}t(s) = |^{\bullet}e \cap (\{s\} \times \mathbb{N})|$,

$\quad\quad\quad\quad\quad \forall s \in S : t^{\bullet}(s) = |e^{\bullet} \cap (\{s\} \times \mathbb{N})|\}$,

$\quad\quad$ $C_o \subseteq B$ with $\forall s \in S : M_o(s) = |C_o \cap (\{s\} \times \mathbb{N})|$.

The condition for E needs some explanation. For any event e associated with a transition t, it ensures that the correct number of tokens as specified by ${}^{\bullet}$t is collected from conditions. Each possible choice for this is represented by some event in E. Symmetrically this holds for the postset, and any combination of choice for the pre- and postset is represented by some event. This may be formalised as follows.

3.2 Lemma   Let Σ be a C/E-representation of N. For each $e \in E$, there are families of sets $(I^s_{\cdot e}, s \in S)$, $(I^s_{e\cdot}, s \in S)$ such that

$I^s_{\cdot e}, I^s_{e\cdot} \subseteq \{1, \ldots, K(s)\}$ and

(i) $\cdot e \cap (\{s\} \times \mathbb{N}) = \{s\} \times I^s_{\cdot e}$,

$e\cdot \cap (\{s\} \times \mathbb{N}) = \{s\} \times I^s_{e\cdot}$,

(ii) $l(e) = t \Rightarrow |I^s_{\cdot e}| = \cdot t(s)$, $|I^s_{e\cdot}| = t\cdot(s)$,

(iii) e is determined uniquely by these sets,

$(\forall s \in S : I^s_{\cdot e_1} = I^s_{\cdot e_2} \wedge I^s_{e_1\cdot} = I^s_{e_2\cdot}) \Leftrightarrow e_1 = e_2$.

Proof

Straightforward with $I^s_{\cdot e} = pr_2(\cdot e \cap (\{s\} \times \mathbb{N}))$,

$I^s_{e\cdot} = pr_2(e\cdot \cap (\{s\} \times \mathbb{N}))$.

□

Intuitively, $I^s_{\cdot e}$ corresponds to the choice to collect tokens from conditions belonging to s for the event e, $I^s_{e\cdot}$ corresponds to the postset distribution. By our particular notation for conditions, we may reduce this to choosing subsets of integers from $\{1, \ldots, K(s)\}$ with the only requirement that the cardinalities correspond to $\cdot t(s)$, $t\cdot(s)$ if $l(e) = t$.

This yields a method how to construct a C/E-representation for a given P/T-system.

-   The set of conditions is determined by S and K (see definition 3.1).

-   For each transition t, we make all possible choices for subsets $I^s_{\cdot e}$, $I^s_{e\cdot}$ from $\{1, \ldots, K(s)\}$ for all $s \in S$, where the cardinalities of these subsets are given by $\cdot t$, $t\cdot$. Each combination of two such families $(I^s_{\cdot e}, s \in S)$, $(I^s_{e\cdot}, s \in S)$ generates an event e.

This determines a unique net structure. For the initial case, definition 3.1 allows to choose how to distribute the tokens of M_o over conditions (if some place capacities are larger than necessary for M_o). For a unique representation, it could be required

$$C_o = \bigcup_{\substack{s \in S \\ M_o(s) > 0}} \{(s,i) \mid 1 \leq i \leq M_o(s)\}.$$

However, we will show in the next section that any C/E-representation is equivalent to the P/T-system, independently of the choice for C_o.

We conclude this section by an example showing that this construction treats self-loops correctly.

Example 2

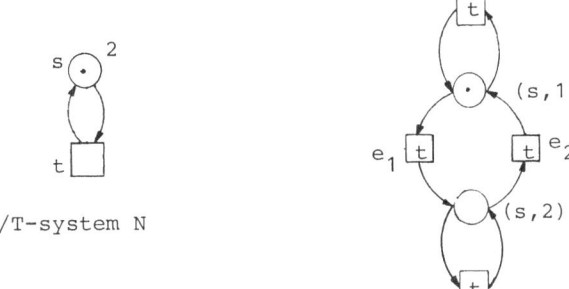

P/T-system N

C/E-representation Σ

At any reachable marking of N, t is enabled. In Σ this is
reflected by alternating enablings of e_1 and e_2.

However, if capacity 1 is assumed for s, then N coincides
with its C/E-representation (up to isomorphism) and both
systems are dead.

4. Equivalence

To compare the behaviour of P/T-systems and their C/E-representatives,
we adopt the concept of transition systems with the well established
equivalence notion called bisimulation equivalence ([Pa]).

A (labelled) transition system consists of a set of states Q with one
distinguished initial state q_0 and a family of transition relations $\xrightarrow{A}$.
Often A is chosen to be element from a set of atomic actions or corresponds to sequences of atomic actions. Then concurrency may only be
modelled by reducing it to non-deterministic interleaving. However, we
may as well choose A to be a set of actions (or even a multiset) taking
place concurrently.

4.1 Definition   Let T be a set (of <u>atomic transitions</u>).
Then $(Q, q_0, (\xrightarrow{A} \subseteq Q \times A \times Q, A \in \mathbb{N}^T))$, $q_0 \in Q$, is called a (<u>distribut-</u>
<u>ed) transition system.</u>

Distributed transition systems have already been used by several authors
(see e.g. [DeMo], [RoTh]).

Two transition systems are bisimulation equivalent if their states may
be related to each other such that the same steps are possible in both
systems and the resulting states are again related.

4.2 Definition   Two transition systems $(P, p_o, \rightarrow_1)$ and $(Q, q_o, \rightarrow_2)$ are
(bisimulation) equivalent iff there exists a bisimulation relation
$R \subseteq P \times Q$ with $(p_o, q_o) \in R$ and, if $(p,q) \in R$, then

-      $p \xrightarrow{A} p' \Rightarrow \exists q'$ with $q \xrightarrow{A} q'$ and $(p',q') \in R$

-      and symmetrically
       $q \xrightarrow{A} q' \Rightarrow \exists p'$ with $p \xrightarrow{A} p'$ and $(p',q') \in R$.

By associating transition systems with marked nets, we obtain an equivalence notion for net systems. The advantage of this procedure, instead of comparing the net systems directly, is that we have a uniform notion of equivalence for any kind of marked nets. We just have
to state, for any class of marked nets we want to consider, how to
derive a transition system.

The most obvious way is to take T to be the set of T-elements (transitions or events) in the net, and $\xrightarrow{A}$ to represent steps.

4.2 Definition   Let $N = (S,T;M_o)$ be a P/T-system.
Then $Q_N = (\mathbb{N}^S, M_o, (\xrightarrow{A}, A \in \mathbb{N}^T))$ is its associated transition system where

   $M \xrightarrow{A} M' :\Leftrightarrow M[A>M'.$

This yields exactly the full reachability graph of the P/T-system
where all markings with all possible steps between them are represented.

Another advantage of comparing marked nets not directly but by associating transition systems is that this gives a possibility for *abstraction*. By defining the transition relations appropriately, we may forget
details and consider the behaviour of the system from a more abstract
point of view. One possibility for this is to identify different T-
elements using labels and to consider labellings of steps. We will explain later why this is what we need for the comparison we are aiming
at.

4.3 Definition   Let $\Sigma = (B,E;C_o)$ be a T-labelled C/E-system.
      (i) $G \subseteq E$ is labelled by $A \in \mathbb{N}^T$ iff
            $\forall t \in T : A(t) = |G \cap 1^{-1}(t)|.$

      (ii) $Q_\Sigma = (P(B), C_o, (\xrightarrow{A}, A \in \mathbb{N}^T))$ is the associated transition
            system of Σ where $C \xrightarrow{A} C' :\Leftrightarrow \exists G \subseteq E$ labelled by A
                                                          and $C[G>C'.$

With these definitions, we may now relate arbitrary marked nets (of
the types considered in section 2).

4.4 Definition   Two marked nets are (bisimulation) equivalent iff their associated transition systems are bisimulation equivalent.

Applied to P/T-systems and their C/E-representatives, these definitions may be interpretated as follows. If a P/T-system N is bisimulation equivalent to its C/E-representation Σ then this means

- that any step A in N may be simulated by some step G in Σ (for each occurrence of a transition t in A there is exactly one occurrence of some event labelled by t in G),

- that, conversely, any step in the C/E-representation corresponds to some step in the P/T-system, namely exactly that given by the labelling.

The following theorem states that P/T-systems are related with their C/E-representations in this strong sense.

4.5 Theorem   Let N be a P/T-system, let Σ be a C/E-representation of N. Then N and Σ are bisimulation equivalent.

Proof

We have to establish a bisimulation $R \subseteq \mathbb{N}^S \times P(B)$.
We define

$$R = \{ (M,C) \mid \forall s \in S : M(s) = |C \cap (\{s\} \times \mathbb{N})| \}.$$

Obviously, we have $(M_o, C_o) \in R$ for all choices of C_o as allowed by definition 3.1.

Now let $(M,C) \in R$.
M and C determine a family of sets $(I^S_{M,C}, s \in S)$ such that

$$\forall s \in S : C \cap (\{s\} \times \mathbb{N}) = \{s\} \times I^S_{M,C}$$

and we have $\forall s \in S : |I^S_{M,C}| = M(s)$.
(with $I^S_{M,C} = pr_2 (C \cap (\{s\} \times \mathbb{N}))$).

Hence $I^S_{M,C}$ describes which conditions carry the tokens corresponding to M(s).

Let $M \xrightarrow{A} M'$. Then we have to show that there exists a step G labelled by A with $C[G>C'$ and $(M',C') \in R$.

We have $\forall s \in S : |I^S_{M,C}| = M(s) \geq \sum_{t \in T} {}^\cdot t(s) \cdot A(t)$, hence it is possible to choose, for each transition t, A(t) subsets $I^S_{\cdot e} \subseteq I^S_{M,C}$ with $|I^S_{\cdot e}| = {}^\cdot t(s)$ such that all chosen subsets are disjoint. This may be done for all $s \in S$.

Furthermore, $\forall s \in S : \sum\limits_{t \in T} t^{\cdot}(s) \cdot A(t) \le K(s) - M(s) = K(s) - |I^S_{M,C}|$.

Hence it is possible to choose, for each transition t, A(t) subsets I^S_e. of $\{1,\ldots,K(s)\} \setminus I^S_{M,C}$ with $|I^S_e\cdot| = t^{\cdot}(s)$ such that all chosen sub-sets are disjoint. Again, this may be done for all $s \in S$.

Using lemma 3.2, these chosen subsets determine a step $G \subseteq E$, G is enabled by C, hence C[G>C'. It remains to show that $(M',C') \in R$.

We have $|C' \cap (\{s\} \times \mathbf{N})| = |((C \setminus \bigcup\limits_{e \in G} {}^{\cdot}e) \cup \bigcup\limits_{e \in G} e^{\cdot}) \cap (\{s\} \times \mathbf{N})|$

$= |C \cap (\{s\} \times \mathbf{N})| - \sum\limits_{e \in G} |{}^{\cdot}e \cap (\{s\} \times \mathbf{N})| + \sum\limits_{e \in G} |e^{\cdot} \cap (\{s\} \times \mathbf{N})|$

$= M(s) - \sum\limits_{t \in T} {}^{\cdot}t(s) \cdot A(t) + \sum\limits_{t \in T} t^{\cdot}(s) \cdot A(t)$

$= M'(s).$

For the symmetric requirement, let $C \xrightarrow{A} C'$. We will show that M[A>M' and $(M',C') \in R$.

G is enabled by C
$\Rightarrow$ for all $s \in S$ (using lemma 3.2)

$\bigcup\limits_{e \in G} I^S_{{}^{\cdot}e} \subseteq I^S_{M,C}$ and $\bigcup\limits_{e \in G} I^S_{e^{\cdot}} \subseteq \{1,\ldots,K(s)\} \setminus I^S_{M,C}$.

Since there exist A(t) events in G with l(e) = t and $(e_1,e_2 \in G, e_1 \ne e_2 \Rightarrow I^S_{{}^{\cdot}e_1} \cap I^S_{{}^{\cdot}e_2} = \emptyset, I^S_{e_1^{\cdot}} \cap I^S_{e_2^{\cdot}} = \emptyset)$ and $|I^S_{{}^{\cdot}e}| = {}^{\cdot}t(s)$, $|I^S_{e^{\cdot}}| = t^{\cdot}(s)$, this implies $\sum\limits_{t \in T} A(t) \cdot {}^{\cdot}t(s) \le |I^S_{M,C}| = M(s)$,

$\sum\limits_{t \in T} A(t) \cdot t^{\cdot}(s) \le K(s) - |I^S_{M,C}| = K(s) - M(s)$.

Hence A is enabled by M and M[A>M' with

$M'(s) = M(s) - \sum\limits_{t \in T} {}^{\cdot}t(s) \cdot A(t) - \sum\limits_{t \in T} t^{\cdot}(s) \cdot A(t)$

$= |((C \setminus \bigcup\limits_{e \in G} {}^{\cdot}e) \cup \bigcup\limits_{e \in G} e^{\cdot}) \cap (\{s\} \times \mathbf{N})|$

$= |C' \cap (\{s\} \times \mathbf{N})|,$

$\Rightarrow (M',C') \in R.$

$\square$

5. Discussion

In the previous section we have established an equivalence between a P/T-system and the suggested C/E-representation. However, there are very fundamental issues which are hidden by the abstraction we have made for this.

Consider again example 1 from the introduction. The P/T-system we have
started with is obviously contact-free, moreover it can be considered
as conflict-free (there is exactly one maximal process of this net (up
to isomorphism) using the process notion of [GoRe]). However, the given
C/E-representation is full of conflicts and contacts. Let us try to
interpret these conflicts and contacts in even simpler (perhaps the
smallest possible) examples.

Example 3

Let N =

The (unique) canonical C/E-representation of N is

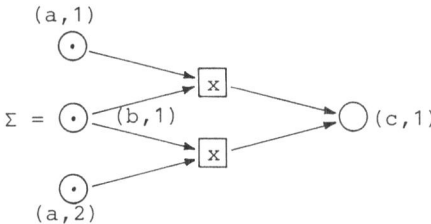

The reason for this at first sight a bit strange example is that in Σ
we have only (forward) conflict without any contact phenomena.

Even though there is no established notion of conflict for P/T-systems
N should clearly be considered as conflict-free. There is exactly one
maximal process with the process notion of [GoRe]. However in Σ, there
is a conflict in the initial case between two events corresponding to x.

But if we distinguish the two tokens on a in the initial marking of N
as suggested in [Ge] then there really is a choice which of these two
tokens should be used for x to occur. This can be seen as a "hidden"
conflict in the P/T-system, hidden by the abstraction from token individualities, and it is not so surprising that this shows up in the
"most detailed" C/E-representation.

Example 4

Let   N =

Its C/E-representation is

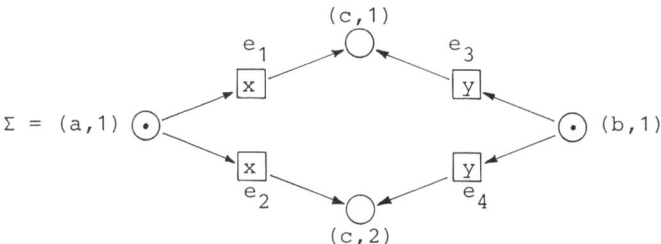

Again, N is conflict- and also contact-free. However, in the initial
case of Σ, a lot of conflicts and contacts arise, for example between the
two possible occurrences of x, but also between certain occurrences
of x and y by their common postsets. How may these conflicts be explained? Let us consider the idea of distinguishing tokens in P/T-sys-
tems more closely. Certainly this can only be done in a local way. The
token distinction is made with respect to places. But, to be consequent, we will not only distinguish tokens on places but we will also
distinguish the possibilities or "the room" for tokens on places. If a
place has capacity n, we will allow it to carry tokens of n sorts. And
if already a token of some sort is present then no more token of this
sort may be added. The sort of the token put by a transition to a place
has to be decided and this is again a hidden conflict. In our example,
we would allow two sorts of tokens on place c, one of each sort. Now
we have two ways for both x and y to occur depending on which sort of
token they choose to put on c. However, if both try to choose the same
sort, this again causes a conflict.

Now let us consider a case where one of x and y has occurred, for example the case {(c,1),(b,1)}. This means, the occurrence of x has produced a token of sort 1. Now we have contact for e_3. This corresponds
to the fact that y is now only allowed to produce a token of sort 2.

So the contact- and conflict-freeness of N is caused by the abstraction
from token individualities we make in a P/T-system. The claim to be
made here is the following. The abstraction we make when moving from
the C/E to the P/T level is exactly hiding this kind of forward and
backward conflicts.

To give a more formal basis for this, we could associate with a P/T-
system a strict predicate/transition system in the sense of [Ge] where
all tokens on places have a unique individuality.

<u>Conjecture</u>   The following diagram commutes with appropriate trans-
    formations:

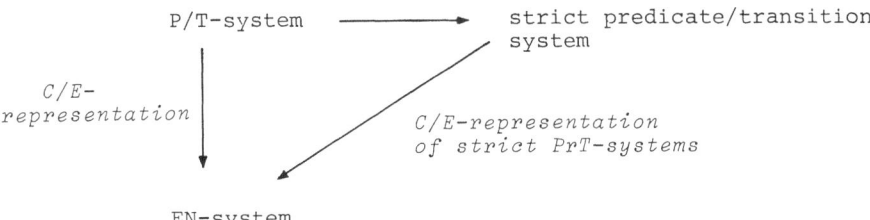

What we have done by now is to try to explain why the "most detailed"
C/E-representation we have given might be useful. However, when looking
at examples, we find that we often immediately see that there are other
ways of representing a P/T-system as a EN-system which are closer to
the P/T-system with respect to conflict and contact.

<u>Example 5</u>

    For the P/T-system of example 1, consider

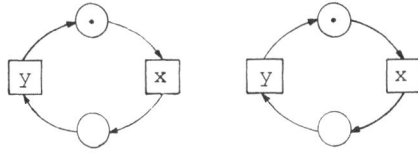

    For example 2, consider

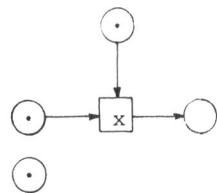

    For example 3, consider

All these EN-systems are contact- and conflict-free. Moreover, they
are bisimulation equivalent to the respective canonical C/E-representa-
tions given before using the bisimulation of section 4 for comparing
labelled EN-systems.

Now the following questions arise.

- Is it possible, for any contact-free P/T-system, to find an EN-system which is bisimulation equivalent (and hence also equivalent to the canonical C/E-representation) and is contact-free?
- The same question for conflict, where a suitable notion of conflict in P/T-systems still needs to be established. A candidate for this has been proposed in [Go2] and will be discussed in [Go3].

Trying to answer these questions must be left for further research.

We conclude this discussion with some remarks on the relationship with respect to processes. There is a well-established notion of process on the condition/event level (see, for example, [RoTh]) and for P/T-systems ([GoRe]). However, it is not true that the sets of processes of a P/T-system and its C/E-representation simply coincide (by forgetting token individualities); this problem has been discussed in [Be]. The reason is that contacts may arise in the C/E-representation as discussed above, and complements for conditions need to be added before processes of the EN-system may be considered. When comparing the sets of processes, we need to forget about these complements and the causal dependencies caused by them. To make this precise must again be left for further research.

References

[Be]   E. Best: "In Quest of a Morphism", Petri Net Newsletter 18, October 1984

[DeMo] P. Degano, U. Montanari: "A Model for Distributed Systems Based on Graph Rewriting", report 111, consiglio nazionale delle ricerche, progetto finalizzato informatica, Cnet, Nov. 1983

[Du]   R. Durchholz: "A Note on Multiplicity", Petri Net Newsletter 21, June 1985

[Ge]   H.J. Genrich: "Projections of C/E systems", in: Advances in Petri Nets 1985, LNCS 222, Springer-Verlag, 1986

[Go1]   U. Goltz: "Considering Nets as Distributed Transition Systems", Petri Net Newsletter 21, June 1985

[Go2]   U. Goltz: "Building Structured Petri Nets", Arbeitspapiere der GMD 223, October 1986

[Go3]   U. Goltz:"How Many Transitions may be in Conflict?", Petri Net
         Newsletter 25, Dec. 1986

[GoRe]   U. Goltz, W. Reisig: "The Non-sequential Behaviour of Petri
         Nets", Information and Control, Vol 57, Nos. 2-3, May/June 1983

[LSB]   P.E. Lauer, M.W. Shields, E. Best: "Formal Theory of the Basic
         COSY Notation", Computing Laboratory, University of Newcastle
         upon Tyne, Technical report No. 143, November 1979

[Pa]   D. Park: "Concurrency and Automata on Infinite Sequences",
         Proc. 5th GI Conf. on Theoretical Computer Science, LNCS 104,
         Springer-Verlag, 1981

[Pe]   C.A. Petri: "Interpretations of Net Theory", internal report
         GMD-ISF 75-07, revised version, Bonn 1976

[RoTh]   G. Rozenberg, P.S. Thiagarajan: "Petri Nets: Basic Notions,
         Structure, Behaviour", in: LNCS 224, Springer-Verlag, 1986

[SmRe]   E. Smith, W. Reisig: "The Semantics of a Net is a Net - An
         Exercise in General Net Theory", in this volume

[Vo]   K. Voss: "System Specification with Labelled Nets and the
         Notion of Interface Equivalence", Arbeitspapiere der GMD 211,
         June 1986

FINITE CONJUNCTIVE NONDETERMINISM
(Extended Abstract)

M. Hennessy, University of Sussex
G. Plotkin, University of Edinburgh

§1. Introduction

In process description languages such as CCS, [M], processes correspond
to terms over a given signature. Each operator of the signature, or
combinator, corresponds to a method for constructing new processes from
existing ones. In this paper we suggest that the choice of combinators
should be governed by the logical properties they induce. Specifically
they should be chosen so that the logical properties of the constructed
process are easily inferred from those of its constituents. Of course
we must ensure that these operators also have an acceptable operational
and denotational semantics. In other words, we would like to develop
an approach to the semantics of processes which reconciles the more
usual denotational and operational semantics with logic. In such a
framework we would expect the logic to determine the denotational semantics. More specifically, if $[\![p]\!]$ represents the denotation of p then
we would expect

$$[\![p]\!] = [\![q]\!] \quad \text{iff (for every } \phi, \ p \models \phi \iff q \models \phi) \qquad (*)$$

where ϕ ranges over formulae from some suitable assertion or property
language.

This just means that satisfying a property depends on the intrinsic
meaning of a process rather than on some particular syntactic forms it
might take. If the semantic domain is ordered (*) can be strengthened
in two different ways:

$$[\![p]\!] \leq [\![q]\!] \quad \text{iff (for every } \phi, p \models \phi \text{ implies } q \models \phi) \qquad (1)$$

This is perfectly reasonable if we view the semantic domain in an infor-
mation-theoretic manner. When increasing the information in a process
specifications which were satisfied remain satisfied. In our particular
case we will view $[\![p]\!]$ as encapsulating all the potential capabilities
of the process p. So for any formula ϕ, p will satisfy it by virtue of
having certain capabilities because under this interpretation any process with at least the same capabilities as p also satisfies ϕ. This
notion of satisfaction underlines the positive or live aspects of processes and we will call the resulting logic a liveness logic.

The dual of (1), which also implies (*) is:

$$[\![p]\!] \leq [\![q]\!] \quad \text{iff (for every } \phi, \ q \models \phi \text{ implies } p \models \phi) \tag{2}$$

Here the negative or safety aspects of processes are emphasised: a process satisfies a formula ϕ by virtue of <u>not</u> having certain capabilities because every process with fewer capabilities also satisfies ϕ. This logic will be called a <u>safety logic</u>.

We will pursue these ideas for a very simple process language. It contains a zero 0, prefixing by an action and one binary operator v, which operationally can be viewed as the joining of capabilities. We refer to it as <u>conjunctive nondeterminism.</u> The assertion language used is a simple variant of the parameterised modal language first studied in [HM]. We define two satisfaction relations $\models_L$ and $\models_S$ which give a liveness and safety interpretation respectively to the property language. In the former the operator on processes v acts like disjunction satisfying:

$$p \models_L \phi \quad \text{or} \quad q \models_L \phi \quad \text{implies } p \vee q \models_L \phi \tag{A}$$

while in the latter it acts like conjunction:

$$p \models_S \phi \quad \text{and} \quad q \models_S \phi \quad \text{implies } p \vee q \models_S \phi. \tag{B}$$

The converse of (B) is also true whereas the converse of (A) is false in general. It is these properties which make v a better behaved process combinator than, for example, the CCS + . This is reflected in the simple uniform embeddings of our process language into the assertion language. In both cases we also show that (*) above is satisfied and formulate axiomatisations of various logical systems although for the moment we have no completeness theorems.

The process language used is trivial but our main aim has been to develop a firm foundation for this logic-based approach to processes. Accordingly most of the paper is concerned with justifying the various choices we make when developing the semantics of the languages. The domain for the process language is algebraically characterised by using the notion of <u>indistinguishability by tests</u>. We work with algebras which are lower semilattices with operators (for the actions). Tests are certain semilattice morphisms to a test space. The domain is the initial algebra where indistinguishability is identity. It is also characterised as the solution to a recursive domain equation in the category of join semilattices much as in [HP] and following the ideas of Scott and Strachey. The semantics of the modal logic is then justified in terms of suitable Heyting algebras of liveness and safety properties, derived from the test morphisms. The choice of modalities is shown to arise in a standard way from the transition relation via a notion of

Kan corestriction, generalising the standard view of quantifiers as adjoints.

In future papers we hope to extend this approach to more complicated process and specification languages.

§2. Conjunctive Nondeterministic Processes

In this section we introduce a language for conjunctive nondeterministic processes and investigate its semantics. Let Act be an uninterpreted set of actions which processes can perform. The set of processes is defined by:

$$p ::= 0 \mid ap \mid p \vee p' \ .$$

Although an operational semantics can be defined in the usual way by parameterised next state relations, we prefer an alternative approach which uses a parameterised predicate, can-a, and a parameterised relation after-a, for each a ε Act. These are, respectively, the least predicates and relations which satisfy:

    i) bp can-a   if   a = b

   ii) p can-a implies p ∨ q   can-a

                       q ∨ p   can-a

and

    i) 0 after-a = 0

   ii) bp after-a = $\begin{cases} 0 \text{ if } a \neq b \\ p \text{ if } a = b \end{cases}$

   iii) (p ∨ q) after-a = p after-a ∨ q after-a

In analogy with bisimulation equivalence [M] we can define a behavioural equivalence ~ on processes as the largest relation which satisfies

           p ~ q   if for every a ε Act,

       i)   p can-a   iff   q can-a

 and ii)   p after-a   ~   q after-a

Theorem 1.1 ~ is the least congruence generated by the equations

$x \vee (y \vee z)$	$= (x \vee y) \vee z$	- associativity
$x \vee y$	$= y \vee x$	- commulativity
$x \vee x$	$= x$	- absorption
$x \vee 0$	$= x$	- zero
$a(x \vee y)$	$= ax \vee ay.$	- action distributivity

                                                  □

This theorem automatically gives a fully-abstract term model, relative to ~. This model has a very simple representation which we call DP.

This consists of the set of finite, non-empty, prefix-closed sets of strings from Act*; 0 is interpreted as $\{\varepsilon\}, a_{\underset{\sim}{DP}}(S) = \{as \mid s \varepsilon S\} \cup \{\varepsilon\}$ and $v_{\underset{\sim}{DP}}$ is set-theoretic union.

<u>Theorem 1.2</u>   a) $\underset{\sim}{DP}$ is isomorphic to the initial model
generated by the equations in Theorem 1.1.

    b) $\underset{\sim}{DP}[\![p]\!] = \underset{\sim}{DP}[\![q]\!]$ iff $p \sim q$.   □

$\underset{\sim}{DP}$ can also be viewed as a lower semilattice :the partial order is defined by $x \le y$ if $x \lor y = y$. This order also has a natural operational counterpart, either as an asymmetric version of our rendition of bisimulation equivalence, $\sim$, or as a testing preorder along the lines of [DH]. With each string w from Act* we associate a test can_w, a natural generalisation of can-a, and it follows that $\underset{\sim}{DP}[\![p]\!] \le DP[\![q]\!]$ if and only if for every w, p can-w implies q can-w.   In the sequel we will make use of the fact that $\underset{\sim}{DP}$, with this order, is a distributive lower semi-lattice:it satisfies

$$d \le e_1 \lor e_2 \text{ implies } d = d_1 \lor d_2 \text{ where } d_i \le e_i, \ i = 1,2.$$

For processes in our language p,q, we will write $p \le q$ to mean $\underset{\sim}{DP}[\![p]\!] \le \underset{\sim}{DP}[\![q]\!]$; note that the related equivalence is simply $\sim$.

§3. The Specification Language

Following the work of [HM], [S], [W1], a natural language for specifying properties of our processes would be a modal language, with parameterised modalities $\langle a \rangle$, $\boxed{a}$. However, these are difficult to interpret within the safety and liveness paradigms of the introduction. Instead we use the more neutral modal operator 0_a, which is associated with the relation after_a on processes, and the constant A, for each $a \varepsilon$ Act, which is associated with can_a. The specification language is given by:

$$\phi ::= \text{true} \mid \text{false} \mid \phi \lor \phi' \mid \phi \land \phi' \mid \phi \to \phi' \mid 0_a \phi \mid A.$$

A satisfaction relation between processes and formulae is defined by structural induction on formulae. It is designed to interpret formulae in a "liveness" manner.

$p \models_L \text{true}$ for all p , $p \models_L \text{false}$ for no p

$p \models_L \phi \lor \psi$ if $p \models_L \phi$ or $p \models_L \psi$

$p \models_L \phi \land \psi$ if $p \models_L \phi$ and $p \models_L \psi$

$p \models_L \phi \to \psi$ if for every $q \ge p$, $q \models_L \phi$ implies $q \models_L \psi$

$$p \models_L 0_a \phi \qquad \text{if } p \text{ after-a} \models_L \phi \; .$$

$$p \models_L A \qquad \text{if } p \quad \text{can-a}$$

Let $L(p) = \{\phi \mid p \models_L \phi\}$ and we write $\models_L \phi$ if for every p, $p \models_L \phi$.

According to this definition the operator $\vee$ on processes actually acts like a logical disjunction:

$$p \models_L \phi \quad \text{or} \quad q \models_L \phi \qquad \text{implies} \qquad p \vee q \models_L \phi \; ,$$

although the converse is not true. The semantics of the previous section also integrates nicely with the logic:

$$p \leq q \qquad \qquad \text{iff} \qquad \qquad L(p) \subseteq L(q) .$$

There is also a natural representation of processes in the logic. Let

$$\phi_L(0) \quad = \quad \text{true}$$

$$\phi_L(ap) \quad = \quad 0_a \phi_L(p) \wedge A$$

$$\phi_L(p \vee q) \quad = \quad \phi_L(p) \wedge \phi_L(q) \; .$$

Then

$$p \models_L \psi \quad \text{iff} \quad \models_L \phi(p) \to_L \psi$$

and

$$p \leq q \quad \text{iff} \quad \models_L \phi_L(q) \to \phi_L(p) .$$

A safety interpretation can also be given for the formulae: $\models_S$ is defined analogously to $\models_L$ except that:

$$p \models_S \phi \to \psi \qquad \text{if for every } q \leq p \qquad q \models_S \phi \text{ implies } q \models_S \psi$$

$$p \models_S A \quad \text{if} \quad p \text{ can-b} \quad \text{implies} \quad b = a$$

and the disjunction rule is replaced by

$$p \models_S \phi \vee \psi \quad \text{if } p_1 \models_S \phi \text{ and } p_2 \models_S \phi \text{ for some } p_1, p_2 \qquad \text{such that}$$

$$p \sim p_1 \vee p_2 .$$

In this interpretation the operator $\vee$ on processes acts like logical conjunction:

$$p \models_S \phi \quad \text{and } q \models_S \phi \qquad \text{iff} \qquad p \vee q \models_S \phi \; .$$

We also get connections between the semantics of processes and this satisfaction relation, analogous to the liveness case. However, the representation of processes is changed slightly:

$$\phi_S(0) \quad = \quad \text{false}$$

$$\phi_S(ap) \quad = \quad 0_a \phi_S(p) \wedge A$$

$$\phi_S(p \lor q) \quad = \quad \phi_S(p) \quad \lor \quad \phi_S(q) \cdot$$

Then it follows that

$$p \le q \qquad \text{iff} \qquad S(p) \supseteq S(q)$$

$$p \models_S \psi \qquad \text{iff} \qquad \models_{\overline{S}} \phi_S(p) \to \psi$$

$$p \le q \qquad \text{iff} \qquad \models_{\overline{S}} \phi_S(p) \to \phi_S(q)$$

where $S(p) = \{\phi \mid p \models_{\overline{S}} \phi\}$ and $\models_{\overline{S}} \psi$ means $p \models_{\overline{S}} \psi$ for every process p.

Many of these results are not easy to derive until we have investigated the semantics of the modal logic in §4.

We have not been very successful in axiomatising the various logical systems which arise naturally in this framework. They are

1. the process logic, with statements of the
form $p \le q$.

2. the specification logics, axiomatising
$\{\phi \mid \models_{\overline{L}} \phi\}$ and $\{\phi \mid \models_{\overline{S}} \phi\}$.

3. the satisfaction logic, axiomatising
$\{< p,\phi> \mid p \models_{\overline{L}} \phi\}$ and $\{<p, \phi> \mid p \models_{\overline{S}} \phi\}$.

The first is covered by theorem 1.1. The two specification logics have much in common with those investigated in [PS] but are much stronger. For example, in both we have the axioms

$$0_a(\phi \to \psi) \quad \leftrightarrow \quad (0_a\phi \to 0_a \psi)$$

and
$$0_a(\neg \phi) \quad \leftrightarrow \quad \neg 0_a\phi$$

where $\neg\phi$ represents $\phi \to$ false.

There are also axioms which are specific to the individual logics. For example

$$\neg(\phi \land \psi) \quad \to \quad \neg\phi \lor \neg\psi$$

is valid in the liveness case whereas

$$(\phi \to \psi) \lor (\psi \to \phi)$$

is valid in the safety case.

The satisfaction relations may be axiomatised as Gentzen systems with formulae of type p Liv ϕ and p Safe ϕ. A natural collection of rules suggest themselves but, as yet, we have no completeness results. Examples of the rules include

$$\frac{\Gamma \vdash p \text{ Liv } \phi}{\Gamma \vdash ap \text{ Liv } 0_a\phi} \qquad\qquad \frac{\Gamma, p \text{ Liv } \phi \vdash p \text{ Liv } \psi}{\Gamma \vdash p \text{ Liv } \phi \to \psi} \quad .$$

§4. Algebras for conjunctive nondeterminism

We consider algebras $\underset{\sim}{A} = (A, 0, \lor)$ with a distinguished element 0 and a binary operation $\lor$ to model conjunctive nondeterminism. For the pro-

cess language we consider such algebras with operators a: $\underset{\sim}{A} \to \underset{\sim}{A}$ for a
in Act. Intuitively an x in A is a set of capabilities, 0 the empty
set and ∨ is set union.

To single out the lower semilattices, and then the action-distributive
ones in an algebraic fashion we consider <u>tests</u> defined to be homomor-
phisms $\underset{\sim}{A} \overset{t}{\to} F(\underset{\sim}{1})$ of algebras. Here $\underset{\sim}{1} = \{*\}$ and $F(X)$ (for any set X) is
the lower semilattice of finite subsets of X ordered by subset. Intuit-
ively x passes t iff t(x) = {*}. That t is a homomorphism is then
that 0 does not pass and x ∨ y passes iff x or y does.

<u>Definition</u>: For any x,y in an algebra $\underset{\sim}{A}$, x and y are <u>test indistinguish-</u>
<u>able</u>, written $x \approx_T y$, iff for every test $\underset{\sim}{A} \overset{t}{\to} F(\underset{\sim}{1})$, t(x) = t(y). Then $\underset{\sim}{A}$
is <u>characterised by tests</u> if $\approx_T$ is equality.

<u>Theorem 4.1</u> $\approx_T$ is the least congruence on algebras generated by the
semilattice laws. □

<u>Corollary 4.1</u> $\underset{\sim}{A}$ is characterised by tests iff it is a lower semilatt-
ice.□

These results show how the semilattice notions arise naturally from an
algebraic notion of testing.

<u>Definition</u>: Let $\underset{\sim}{A}$ be an algebra with operators. A test, t, in $\underset{\sim}{A}$ is
<u>uniform</u> iff for any w ε Act* either t(wx) = {*} (for any x in $\underset{\sim}{A}$) or else
there is a test t_w such that $t(wx) = t_w(x)$ (for any x in $\underset{\sim}{A}$).

<u>Proposition 4.1</u> The uniform tests are the largest class U of tests
such that for any test, t, in U and for any a in Act, either t(a0) = {*}
or else there is a t_a in U such that $t(ax) = t_a(x)$ (for any x in $\underset{\sim}{A}$).□

Now defining the relation of <u>uniform test indistinguishability</u>, $\approx_{UT}$
and <u>characterisation by uniform tests</u> in the obvious way, we get:

<u>Theorem 4.2</u> $\approx_{UT}$ is the least congruence on algebras with operators gen-
erated by the semilattice and action distributivity laws. □

So we have $\underset{\sim}{DP} \cong \underset{\sim}{FA}/_{\approx_{UT}}$, where $\underset{\sim}{FA}$ is the free algebra with operators. To
relate the algebraic and operational views of testing, regard $\underset{\sim}{FA}$ as the
language of processes. Then for any w in Act$^+$, there is a uniform test
t_w where $t_w(x) = \{*\}$ iff x can-w. Now order the uniform tests point-
wise by:

$$t \le t' \quad \text{iff} \quad t(x) \le t(x'), \text{ all x in } \underset{\sim}{FA};$$

we obtain a complete lattice and:

<u>Theorem 4.3</u> If t is a uniform test on $\underset{\sim}{FA}$, then $t = \bigvee \{t_w | t(w0) = \{*\}\}$□

<u>The category, SL, of lower semilattices</u>. We now regard $\underset{\sim}{DP}$ as the solu-
tion to a domain equation, as traditional in semantics, following Scott
and Strachey, but in a different category $\underline{SL}$, appropriate to conjunctive

nondeterminism (cf[HP]). It is hoped that this idea can be extended to more complex languages.

First we survey some functors. The construction, F, of finite subsets is the left adjoint to the forgetful functor from $\underline{SL}$ to $\underline{Sets}$. It will also be useful to consider a slight generalisation $Fd:\underline{Pos} \rightarrow \underline{SL}$ left adjoint to the forgetful from $\underline{SL}$ to the category of posets. $Fd(\underset{\sim}{P})$ is the collection of finitely generated ideals in $\underset{\sim}{P}$ (where $I \subset \underset{\sim}{P}$ is such an ideal iff there are $x_1,\ldots,x_m$ ($m \geq 0$) in $\underset{\sim}{P}$ such that $I = \{x | x \leq x_i,$ for some i with $1 \leq i \leq m\}$

In $\underline{SL}$ sums and products coincide. We write $\underset{\sim}{A} \oplus \underset{\sim}{B}$ for the sum; it is just the cartesian product with coordinatewise operations. Also important is the $\underline{tensor\ product}$ $\underset{\sim}{A} \otimes \underset{\sim}{B}$. Say that a binary function $f:\underset{\sim}{A} \times \underset{\sim}{B} \rightarrow \underset{\sim}{C}$ is $\underline{bilinear}$ iff for all x, x', y, y' in $\underset{\sim}{A}$

$$f(0,y) \quad = \quad f(x,0) \quad = \quad 0$$
$$f(x \vee x',y) \quad = \quad f(x,y) \quad \vee \quad f(x',y)$$
$$f(x,y \vee y') \quad = \quad f(x,y) \quad \vee \quad f(x,y').$$

Then the tensor product is the range of the universal bilinear map $A \times B \xrightarrow{\otimes} \underset{\sim}{A} \otimes \underset{\sim}{B}$. It is constructed as $F(\underset{\sim}{A} \times \underset{\sim}{B})/\approx$ for a suitable congruence, $\approx$. Let $\underline{SL}^{-}$ be the category of semilattices without a zero, meaning structures $(\underset{\sim}{A}, \vee)$ which are associative, commutative and absorptive. Then the forgetful functor has a left adjoint $(\cdot)_0:\underline{SL}^{-} \rightarrow \underline{SL}$ which "adds a new 0 to $\underset{\sim}{A}$" (we have $\underset{\sim}{A}_0 = \underset{\sim}{A} \cup \{0\}$ with a suitable definition of the operations).

$\underline{Theorem\ 4.4}$ $\underset{\sim}{DP}$ is the initial solution to the recursive domain equation:

$$\underset{\sim}{A} \cong F(Act) \otimes \underset{\sim}{A}_0.$$ □

This is in the sense of [SP]: define a functor F on $\underline{SL}$ by :

$$F(\underset{\sim}{A}) \quad = \quad F(Act) \otimes \underset{\sim}{A}_0$$

then $\underset{\sim}{DP}$ yields an algebra $\alpha: F(\underset{\sim}{DP}) \rightarrow \underset{\sim}{DP}$ by $\alpha(a_1 \cdot x_1 \vee \ldots \vee a_m \cdot x_m) = a_1 x_1 \vee \ldots a_m x_m$ (where $m \geq 0$ and $a \cdot x = \{a\} \otimes x$). If $\beta: F(\underset{\sim}{A}) \rightarrow \underset{\sim}{A}$ is any other such then there is a unique $\theta: \underset{\sim}{DP} \rightarrow \underset{\sim}{A}$ with $\theta \circ \alpha = \beta \circ F(\theta)$.

One can also view $\underset{\sim}{DP}$ as the conjunctively nondeterministic version of the partial order Act^+ (with the subsequence ordering). The latter is the initial solution to the equation: $\underset{\sim}{P} \cong Act \times \underset{\sim}{P}_\perp$ in $\underline{Pos}$ (where $(\cdot)_\perp$ adds a new least element). Now $\times$ in $\underline{Pos}$ corresponds to $\otimes$ in $\underline{SL}$ via the natural isomorphism: $Fd(\underset{\sim}{P} \times \underset{\sim}{Q}) \cong Fd(\underset{\sim}{P}) \otimes Fd(\underset{\sim}{Q})$ and similarly $(\cdot)_\perp$ corresponds to $(\cdot)_0$. Then we have that $\underset{\sim}{DP} \cong Fd(Act^+)$, essentially recovering the original definition of $\underset{\sim}{DP}$.

5. Semantics of modal logics

We systematically justify our semantics for the modal logics in both
the liveness and safety cases. For liveness we assign an element
$[\![q]\!]_L \in L$, a complete Heyting algebra (cHa) to each formula q in such a
way that:

$$[\![q]\!]_L \;=\; \{\underset{\sim}{DP}[\![p]\!] \mid p \vDash_L q\} \; .$$

The cHa is a lattice of liveness properties. The definition of $[\![q]\!]_L$ is
as usual in the semantics of intuitionistic logic for the propositional
connectives. The modality is given as part of a general theory of modalities over cHa's using a notion of Kan construction, generalising
the usual treatment of quantifiers as adjoints [PS]; this also explains
why just one modality arises in this case. Safety is treated in just
the same way, via a cHa S.

Two Heyting Algebras . For liveness we use

$$L \;=\; \text{the upper closed subsets of } \underset{\sim}{DP}$$

ordered by subset. (For any poset P, a subset $X \subset P$ is upper closed
iff for any x in X if $x \leq y$ then y is in X; the collection $U(P)$ of
such sets is always a cHa.) For safety we use

$$S \;=\; \text{the ideals of } \underset{\sim}{DP}$$

ordered by subset. (For any lower semilatttice $\underset{\sim}{A}$ an ideal is a subset,
X, such that (i) 0 is in X (ii) $x \vee y$ is in X iff x and y are . The coll-
ection $I(\underset{\sim}{A})$ of such sets is a cHa iff $\underset{\sim}{A}$ is distributive.)
The elements are assigned in the usual way for the logical operations.
For example, $[\![q \rightarrow q']\!]_L = [\![q]\!] \Rightarrow_L [\![q']\!]_L$, where the implication operation
over L is

$$X \Rightarrow_L Y = \{x \in \underset{\sim}{DP} \mid \forall x' \geq x. \; x' \notin X \text{ or } x' \in Y\}.$$

and similarly for safety, where

$$X \Rightarrow_S Y = \{x \in \underset{\sim}{DP} \mid \forall x' \leq x. \; x' \notin X \text{ or } x' \in Y\}.$$

For the rest of the logic,

$$[\![A]\!]_L = \{(ax) \vee y \quad x,y \in \underset{\sim}{DP}\}$$

$$[\![0_a q]\!]_L = \{x \mid x/a \in [\![q]\!]_L\}$$

$$[\![A]\!]_S = \{ax \mid x \in DP\} \cup \{0\} \; .$$

$$[\![0_a q]\!]_S = \{x \mid x/a \in [\![q]\!]_S\}$$

(Here $x/a = \{w \mid aw \in x\} \cup \{\varepsilon\}$).
One can now express the relation between a process and its logic representation by the formulae:

$$[\![\phi_L(p)]\!]_L \quad = \{x \mid x \geq DP[\![p]\!]\}$$

$$[\![\phi_S(p)]\!]_S \quad = \{x \mid x \leq DP[\![p]\!]\} \; .$$

<u>Liveness and Safety Properties</u>   To each test $A \xrightarrow{t} F(\underline{1})$ (for any
lower semilattice A), we can associate a liveness test set, $t^{-1}(\{*\})$
(the points that pass the test, here considered a desirable state of
affairs) and a safety test set $t^{-1}(\phi)$ (the points that do not pass the
test, that now being thought desirable). To specify a process we imagine asking that it be in each of a collection of such sets. So say a
<u>liveness property</u> is an intersection of such liveness test sets and
similarly for <u>safety</u>.

<u>Proposition 5.1</u>   1. The liveness properties are the upper closed
                              sets.

                           2. The safety properties are the ideals.

This justifies our choice of L and S. For example for any action a
there is the test $\underset{\sim}{DP} \xrightarrow{a?} F(\underline{1})$ where $a?(x) = \{*\}$ iff $aw \varepsilon x$ for some a.
Then $[\![A]\!]_L$ is the associated liveness test set and $[\![A]\!]_S$ is the comple-
ment of the associated safety test set.

<u>The Modalities</u>   Given monotonic maps $Q \xrightarrow{f} P \xleftarrow{g} R$ of partial orders, a
map $R \xrightarrow{\Diamond} Q$ is the <u>left Kan corestriction</u> (of g along f) iff $\Diamond y$ is the
least x such that $g(y) \leq f(x)$. Also $R \xrightarrow{\Box} Q$ is the <u>right Kan corestrict-</u>
<u>ion</u> (of g along f) iff $\Box y$ is the greatest x such that $g(y) \geq f(x)$.
The idea is that $Q = R = $ the cHa of interest and f,g represents the relation corresponding to the modality. For example in the case of a relation $T \subset W \times W$ over a set of worlds we take $Q = P(W)$, the powerset of W,
$P = P(T)$ and $f = \pi_0^{-1}$, $g = \pi_1^{-1}$ and find:

$$\Diamond y = \{x \ \varepsilon \ X \mid \exists \ x'.x \ T \ x' \wedge x' \ \varepsilon \ y\}$$
$$\Box y = \{x \ \varepsilon \ X \mid \forall \ x'.x \ T \ x' \supset x' \ \varepsilon \ y\}$$

a slight reformulation of the usual Kripke semantics of modal logic.
Note that if T is a function then $\Diamond y = \Box y = T^{-1}(y)$.
In the case at hand we consider $T_a \subset \underset{\sim}{DP}^2$ (each a ε Act) where: $x \ T_a \ y$
iff $y = x/a$, a relation representing the transition function. Then in
the case of liveness $Q = L$, $P = U(T_a)$ (taking T_a as a subalgebra of
$\underset{\sim}{DP}^2$), $f = \pi_0^{-1}$, $g = \pi_1^{-1}$. Here T_a is a function and we find $\Diamond y = \Box y = $
$T_a^{-1}(y)$. But now we see that

$$[\![0_a q]\!]_L = T_a^{-1}([\![q]\!]_L)$$

justifying the choice of modality for the liveness logic and explaining
why only one modality arose.
For safety we work with $S \xrightarrow{\pi_i} F(T_a)$ (i=0,1) and find now $\Diamond y = \Box y = T_a^{-1}(y)$
and $[\![0_a q]\!]_S = T_a^{-1}([\![q]\!]_S)$ is expected and completing our programme of
justifying the semantics of our modal language.

§6. Related Work

Various authors have studied the relationship between process languages
and languages for modal assertions. (See [Pn], [GS], [W1], [W2], [S]).
For example, in [GS] a simple language for nondeterministic programs is
embedded into a language with the usual logical connectives ∧, ∨, ¬
and modalities. The resulting language has a non-logical connective +,
which corresponds to the nondeterministic choice in the process language. The semantics used for the process language is observational
equivalence [M], and each formula is interpreted as a set of equivalence
classes of processes. A sound and complete proof system is given for
the augmented assertion language. The embedding of processes into the
assertion language is faithful in that $p \approx q$ (i.e. are semantically equivalent) if and only if $\vDash p \equiv q$, where $\equiv$ is logical equivalence. However,
it is not uniform; in particular the embedding of p + q depends on the
syntactic structure of p and q. The problem is that the operator '+',
taken from CCS is not well-behaved logically. We overcome this problem
by replacing '+" with conjunctive nondeterminism '∨', which, although
it is not as expressive, is well-behaved. In particular our embedding
is uniform.

A somewhat similar programme is carried out in [W1], but the semantic
framework is at first sight quite different. Here assertions are taken
as basic and processes are interpreted as sets of assertions: intuitively the set of assertions they satisfy. The basis for a logic for statements of the form p Sat φ is presented. This is similar to our specification logics but the process language is much richer; it contains a
synchronous parallel operator ×. On the other hand, the assertion language is weaker:it does not contain negation or implication. Winskel
also presents a sound and complete proof system for a language which
mixes processes and assertions. This proof system is quite similar to
that in [S] except that Stirling maintains a distinction between processes and assertions. However there is implicit mixing in that the
satisfaction relation being axiomatised is parameterised by assertion
formulae; $p \vDash_A B$ could also be viewed as $p \times A \vDash B$.

In many ways this work is more advanced than that which we have reported
here. However our main aim is to re-examine the mathematical framework on which such work is based. This includes re-examining the process
language and its semantics so as to integrate it more naturally into
the logical assertion language. In the present paper we have introduced
what is essentially a logical operator, ∨, on processes and provided
this language with a natural operational and denotational semantics,
while at the same time retaining a logical significance for ∨. We hope

to continue this programme, by introducing another logical operator ∧; operationally this is similar to "internal nondeterminism" while logically is will act as a normal logical connective. To proceed with this program a new semantics for processes is required and the foundations have been laid in this paper.

Acknowledgements: The first author would like to acknowledge the support of SERC and the second author that of the British Petroleum Venture Research Unit.

REFERENCES

[DH]   DeNicola, R., and Hennessy, M. Testing equivalences for processes, TCS 34 (1984) pp 83-133

[M]   Milner, R., A Calculus for Communicating Systems, LNCS Vol.92 (1980)

[HM]   Hennessy, M., and Milner, R., Algebraic laws for nondeterminism and concurrency, JACM 32 (1985), pp 137-162

[W1]   Winskel, G., A complete proof system for SCCS with modal assertions, technical report No.78, Computer Laboratory, University of Cambridge (1985)

[S]   Stirling, C., Modal logics for Communicating Systems. Research report, Department of Computer Science, Edinburgh University, 1984, to appear in TCS

[GS]   Graf, S., and Sifakis, J., A logic for the specification of controllable processes of CCS, Acta Informatica, Vol.23(1986), pp 507-527.

[PS]   Plotkin, G., Stirling, C., Intuitionistic modal logic, to appear (1986)

[W2]   Winskel, G., On the Composition and Decomposition of Assertions, technical report No.59, Computer Laboratory, University of Cambridge (1985)

[HP]   Hennessy, M., and Plotkin, G., A Fully-Abstract Semantics for a Simple Programming Language, Mathematical Foundations of Computer Science, LNCS Vol. 74 (1979) pp. 108-120.

[GS2]   Graf, S., and Sifakis, J., A Logic for the Description of Nondeterministic Programs and Their Properties, Information and Control, Vol.68, Nos 1-3(1986), pp.254-270

[Pn]   Pneuli, A., Linear and Branching Structures in the Semantics and Logics of Reactive Systems, Proceedings of ICALP '85, LNCS Vol.194 (1985) pp 15-33

[SP]   Smyth, M., and Plotkin, G., The Categorical Solution of Recursive Domain Equations, Siam Journal of Computing 11, 4,(1982), pp. 761,783.

PETRI NET LANGUAGES AND ONE-SIDED

DYCK-REDUCTIONS ON CONTEXT-FREE SETS

M. Jantzen and H. Petersen

FB Informatik, Universität Hamburg

Introduction

In [2,6,8,9] cancellation grammars (or grammars related to them) are
defined and their relation to well-known families of languages are
studied. Savitch showed in [9] that the class of EOL languages can be
obtained from the context-free sets (CF) by iteratively and completely
cancelling one matching pair $x\bar{x}$ of parenthesis x and $\bar{x}$. This type of
reduction is here called a $Dyck_1$-reduction on a set L which can be taken
from any family of languages - not only the context-free sets - and thus
need not be definable by certain restricted classes of grammars as in
[2,9]. In this short note we will show that we get all (free) terminal
Petri net languages and all transition sequences from the context-free
sets by $Dyck_1$-reductions and, moreover, each non-erasing homomorphic
image thereof, the corresponding families denoted by L and P as in [7].

Moreover, we get by applying the $Dyck_1$-reduction to the context-free
sets not only the Petri net languages but also their intersection with
the context-free languages, their substitution into context-free languages, in fact their closure with respect to nested iterated substitution
and the algebraic extension of (or equivalently the least super AFL con-
taining) the family L ∧ CF = {L ∩ K | L ∈ L , K ∈ CF} . This family not
only contains the context-free sets (by definition) but also many other
languages which are neither context-free nor Petri net languages.

Definition

The $Dyck_1$-reduction could either be defined by introducing the special
rule $x\bar{x} \longrightarrow \lambda$ and adding it to the rules of the underlying grammars
generating a family K , as done in [2], or by restricting the class of
allowed context-free grammars and their derivation relation, see [9].

Here we consider it a special reduction similar to the one in [1] but not using all the terminology used when dealing with Church-Rosser Thue systems.

Definition 1

Let $X := \{x, \bar{x}\}$ be the alphabet of special parenthesis, where x denotes the opening, while $\bar{x}$ denotes its matching closing parenthesis. Let Y be any alphabet which does not contain x and $\bar{x}$, i.e., $Y \cap X = \emptyset$ and $Z := Y \cup X$. Cancelling a matching pair $x\bar{x}$ of parentheses within a string $w = ux\bar{x}v \in Z^*$ yields the string uv and this one-step reduction denoted by Δ, can be used possibly several times.

This will be expressed by using the transitive and reflexive closure Δ^* of this operation, which in fact specifies a binary relation on $2^{Z^*} \times 2^{Z^*}$ and is defined recursively by the following equations.

For any $L \subset Z^*$ let

$$\Delta^*(L) := \bigcup_{w \in L} \Delta^*(w) \text{ , where}$$

$$\Delta^*(w) := \bigcup_{i \geq 0} \Delta^i(w) \quad \text{with } \Delta^0(w) := \{w\} \text{ ,}$$

$$\Delta(w) := \{uv \mid ux\bar{x}v = w \text{ and } u,v \in Z^*\}$$

and recursively $\Delta^{i+1}(w) := \Delta(\Delta^i(w))$.

The set D_1 of all well-formed parenthesis then becomes
$$D_1 = \{w \in X^* \mid \Delta^*(w) = \lambda\} \text{ .}$$

The reduction Δ specifies a confluent semi-Thue system which is a certain Church-Rosser Thue system in the sense of Book, [1], and this means that the strings, which cannot be reduced further by cancelling $x\bar{x}$, the so-called irreducible strings modulo Δ, are the unique descendants of their ancestors and therefore the mapping $\mu: Z^* \longrightarrow Z^*$ with $\mu(w) := v$ where $v \in \Delta^*(w)$ and $\Delta(v) = \emptyset$ is well defined.

As an introductory example take $w = xa\bar{x}x\bar{x}bx\bar{x}$, then $\Delta^*(w) = \{xa\bar{x}x\bar{x}bx\bar{x}, \quad xa\bar{x}bx\bar{x}, \quad xa\bar{x}x\bar{x}b, \quad xa\bar{x}b\}$ and $\mu(w) = xa\bar{x}b$. If one thinks of x as a synchronization signal and $\bar{x}$ being the corresponding wait or request symbol, then one might wish to consider only those strings in $\mu(w)$ that do not contain any dangling parentheses. Thus, we consider the mapping $\tau: 2^{Z^*} \longrightarrow 2^{Y^*}$ given by $\tau(L) := \mu(L) \cap Y^* = \Delta^*(L) \cap Y^*$. Thus, if $L = \{w\}$ is a singleton, then so is $\tau(L)$ unless it is the empty set. Hence, τ can be considered a partial function on Z^* and we will shortly write $\tau(w) = v$ omitting the curly brackets. The mappings μ

(and likewise τ) are canonically generalized to languages L and families of languages $\underline{K}$ by $\mu(L):= \{v \mid v=\mu(w), w \in L\}$ and $\mu(\underline{K}):=\{\mu(L) \mid L \in \underline{K}\}$.

Let $\underline{FIN}$ ($\underline{REG},\underline{CF},\underline{EOL},\underline{REC},\underline{RE}$, resp.) denote the families of finite (regular, context-free, EOL, recursive, recursively enumerable, resp.) sets. In what follows we will write a context-free grammar as $G= (V_N,V_T,P,S)$, where V_N is the set of nonterminals with $S \in V_N$, V_T is the set of terminal symbols with $V_N \cap V_T = \emptyset$, $V:= V_N \cup V_T$, $P \subseteq V_N \times V^*$. The one-step derivation relation $\Rightarrow$ for G is defined as usual by $w \Rightarrow w'$, if $w = uAv$, $A \to \bar{v} \in P$ and $w'= u\bar{v}v$. The transitive, reflexive closure $\overset{*}{\Rightarrow}$ is used to determine the language L(G) generated by G as $L(G):= \{w \in V_T^* \mid S \overset{*}{\Rightarrow} w \text{ in } G\}$.

For any $R \in \underline{REG}$ it is well-known, [1], that $\Delta^*(R) \in \underline{REG}$. Since $\bar{x}^*x^*$ is the set of all strings irreducible modulo the cancellation Δ we also have $\mu(R)= \Delta^*(R) \cap \bar{x}^*x^* (Y \cdot \bar{x}^*x^*)^* \in \underline{REG}$ and obviously also $\tau(R) \in \underline{REG}$ for any regular set R . Let us use the notation from [4,7] to denote the families of Petri net languages we are dealing with, namely $\underline{L}^f$ and $\underline{L}$ for the family of free terminal and λ-free labelled terminal Petri net languages, $\underline{P}^f$ and $\underline{P}$ for the free and λ-free labelled transition sequences. As known from [7] we have $\underline{L}^f \subset \underline{L}$ and $\underline{L} \supseteq \underline{P}$ but $\underline{P}^f \not\subset \underline{L}^f$.

The Family $\tau(\underline{CF})$ and Petri Net Languages

In [2] it is shown that $\tau(\underline{CF})$ contains NP-complete sets and more examples of languages in $\tau(\underline{CF}) \smallsetminus \underline{CF}$ are contained in [6], like the following $L_{prod}:= \{a^n b^m c^{n \cdot m} \mid n,m \in \mathbb{N}\} \in \tau(\underline{CF})$ and $L_{copy}:= \{ww \mid w \in \{a,b\}^*\} \in \tau(\underline{CF})$, which are neither context-free nor Petri net languages.

The following result is from [6] and is the basis for the results we want to include in this note, even though we know of many other interesting results about the family $\tau(\underline{CF})$ that we will not mention here.

Theorem 1

Any Petri net language $L \in \underline{L}^f$ (or $L \in \underline{L}$, $L \in \underline{P}^f$, $L \in \underline{P}$, resp.) is an element of the family of $\tau(\underline{CF})$.

Proof

Let $\Sigma:= (S_\Sigma, T_\Sigma, F_\Sigma, W_\Sigma)$ be a P/T-net, $m_0 \in \mathbb{N}^S$ an initial marking, $m_f \in \mathbb{N}^S$ the final marking so that the free terminal Petri net language

$L_\Sigma \in \underline{L}^f$ is given by $L_\Sigma := \{w \in T_\Sigma^* \mid m_o[w > m_f\}$. Now define the subset

$T_o \subset T_\Sigma$ by $T_o := \{t \in T_\Sigma \mid m_o[t>\}$. Since we want to represent any

marking $m \in \mathbb{N}^S$ by a number of opening parenthesis x , we will define

the following encoding of $m = (n_1, n_2, \ldots, n_k) \in \mathbb{N}^S$ by $\mathrm{cod}(m) := \prod\limits_{i=1}^{k} p_i^{n_i}$,

where p_i is the i-th prime.

It is easy to check that $\mathrm{cod}(a+b) = \mathrm{cod}(a) \cdot \mathrm{cod}(b)$ and $\mathrm{cod}(a-b) = \dfrac{\mathrm{cod}(a)}{\mathrm{cod}(b)}$ for $a, b \in \mathbb{N}^S$, where the sum (difference) of vectors is taken component-wise. Obviously, this encoding allows for a unique decoding of the marking m from $\mathrm{cod}(m)$. As in [4] we will use for each transition $t \in T$ the vectors $t^- := W(\cdot, t)$ and $t^+ := W(t, \cdot)$ to denote the smallest marking enabling t , respectively the marking obtained then by its occurrence. Then, by the above, $m[t>$ iff $\mathrm{cod}(t^-)$ divides $\mathrm{cod}(m)$ is easy to verify.

With this notation we can define the context-free grammar G_Σ for which we show $\tau(L(G_\Sigma)) = L_\Sigma$. If L is taken from $\underline{p}^f$, then the necessary changes concern the rules in 6) and will be given later.

$G_\Sigma := (V_N, V_T, P, S)$ where

$V_N := \{S, H, T_0, H_t, F_t \mid t \in T_\Sigma\}$,

$V_T := \{x, \bar{x}\} \cup T_\Sigma$, and

P contains precisely the following rules:

1) $S \to T_0 H$ and $S \to t$, if $m_o[t > n_f$ for some $t \in T$; and for each $t \in T_\Sigma$ the following lists of rules are defined:

2) $H \to H_t H$

3) $H_t \to \bar{x}^{\mathrm{cod}(t^-)} H_t x^{\mathrm{cod}(t^+)}$

4) $H_t \to t$

5) $T_0 \to t x^{\mathrm{cod}(m)}$, whenever $m_o[t > m$

6) $H \to \bar{x}^{\mathrm{cod}(m)} t$, whenever $m[t > m_f$

In order to have a notation useful for proving the above claims, let us write

$u \underset{\mu}{\Rightarrow} w$ for strings $u, w \in V^*$ iff there exists $v \in V^*$ such that $u \Rightarrow v$ in G and $w = \mu(v)$. As usual $\underset{\mu}{\overset{*}{\Rightarrow}}$ denotes the reflexive transitive closure of $\underset{\mu}{\Rightarrow}$.

Then consider a leftmost derivation with respect to $\underset{\mu}{\Rightarrow}$, which produces a typical sentential form like $t_1 t_2 \ldots t_k x^{\mathrm{cod}(m)} H$ where $t_i \in T$ and prove that then $m_0[t_1t_2...t_k>m$ holds in Σ by induction on k. If $k = 1$, then this is true by definition of the rules for T_0. If the assertion is true for a fixed k, then we will show it for $k + 1$, which means $t_1...t_k x^{cod(m)} H \overset{*}{\underset{\mu}{\Rightarrow}} t_1...t_k t\ x^{cod(m')} H$ for $m[t>m'$. In detail we find by using leftmost derivations $t_1t_2...t_k\ x^{cod(m)} H \overset{}{\underset{\mu}{\Rightarrow}}$

$t_1...t_k\ x^{cod(m)} H_t H \Rightarrow t_1...t_k\ x^{cod(m)}\ \bar{x}^{cod(t^-)} H_t\ x^{cod(t^+)} H$. Since H_t

will necessarily terminate with terminal symbol $t \in T_\Sigma$, we have to

use the rule $H_t \rightarrow \bar{x}^{cod(t^-)} H_t\ x^{cod(t^+)}$ as often as necessary to get

$cod(m) = r \cdot cod(t^-)$ but then we have $r = \dfrac{cod(m)}{cod(t^-)}$ and consequently

$t_1...t_k\ x^{cod(m)} H \overset{*}{\underset{}{\Rightarrow}} t_1...t_k\ x^{cod(m)}\ \bar{x}^{r \cdot cod(t^-)}{}_t\ x^{cod(t^+)r} H \overset{*}{\underset{\mu}{\Rightarrow}}$

$t_1...t_k t\ x^{cod(m')} H$ with $cod(t^+) \cdot r = \dfrac{cod(m) \cdot cod(t^+)}{cod(t^-)} = cod(m-t^-+t^+) = $

$cod(m')$ for $m[t>m'$.

If in the last step we replace H by $\bar{x}^{cod(m)} t$, then this can only be terminated correctly if the marking reached so far in the net has exactly the same coding $cod(m)$ and will be transformed by t into the final marking m_f by $m[t>m_f$. Thus indeed, if $S \overset{*}{\underset{\mu}{\Rightarrow}} w$ is a leftmost, or any other, derivation (and we can always choose a leftmost one), then $w \in (V_T \setminus X)^*$ iff $w \in \tau(L(G_\Sigma))$ and $w \in L_\Sigma$ as shown above. On the other hand, for any transition sequence $t_1t_2...t_k \in L_\Sigma$ with $m_0[t_1t_2...t_k>m_f$ we can derive from S in G_Σ the following sentential form

$T_0 H_{t_1} H_{t_2}...H_{t_{k-1}}\ \bar{x}^{cod(m)} H$, so that $t_1...t_k \in \tau(L(G_\Sigma))$ is immediately

deduced by the previous considerations. This proves that $L_\Sigma \in \underline{L}^f$ implies $L_\Sigma \in \tau(\underline{CF})$ or $\underline{L}^f \subset \tau(\underline{CF})$. If we replace the terminal symbol $t \in T_\Sigma \subset V_T$ in the rules from the sets 4), 5) and 6) by the image $h(t)$ with respect to some nonerasing homomorphism, this gives the inclusion $\underline{L} \subset \tau(\underline{CF})$. In order to characterize languages from the families $\underline{P}^f$ and $\underline{P}$, where no terminal marking must be reached, it is enough to replace the rule 6) $H \rightarrow \bar{x}^{cod(m)} t$ by the rules $H \rightarrow F_t$, and

$F_t \rightarrow \bar{x}^{cod(t^-)} F_t$, $F_t \rightarrow t$ for each $t \in T$.

If $cod(m)$ is divisible by $cod(t^-)$, then this means that $m[t>$ is true and t may occur. The marking reached then would be $\dfrac{cod(m) \cdot cod(t^+)}{cod(t^-)}$

but is not simulated by the corresponding derivation in G_Σ since $x^{cod(t^+)}$ is not generated at all. This then shows $\underline{P}^f \subset \tau(\underline{CF})$ and

$\underline{P} \subset \tau(\underline{CF})$.

What is important in this construction is the fact that no sentential form within G ever begins or ends with parentheses x or $\bar{x}$. The same will be valid for the construction used to prove the next result which concerns the intersections of Petri net languages and context-free sets.

Lemma 2

If $K \in \underline{CF}$ and $L \in \underline{L}$ (or $\underline{L}^f$, $\underline{P}^f$, $\underline{P}$, resp.) are arbitrary, then $L \cap K \in \tau(\underline{CF})$.

Proof

Let $G = (V_N, V_T, P, S)$ be such that $K = L(G)$ and $\Sigma = (S_\Sigma, T_\Sigma, F, W)$, $m_o, m_f \in \mathbb{N}^S$ with $L_\Sigma = \{w \in T_\Sigma^* \mid m_o[w > m_f\} \in \underline{L}^f$, then let $Y := V_T \cap T_\Sigma$ and $L_\Sigma \cap K \subset Y^*$.

First construct G' from G such that essentially $L(G') = \{t_1' t_2 \ldots t_{k-1} t_k'' \mid t_1 t_2 \ldots t_k \in L(G)\}$, where t', t'' are new symbols for each $t \in T_\Sigma$. This can be done effectively and gives again a context-free grammar. If $T \in K \cap L_\Sigma$, then we will add the appropriate rules to generate the terminal "strings" t directly to the grammar G which generates $K \cap L_\Sigma = \tau(\bar{G})$.

Now, for each $t \in Y \subset V_T'$ substitute in G' the new nonterminal H_t within each rule. The primed symbol t' is replaced by the new nonterminal T_o whereas t'' is replaced by E_t in each and every rule of G' . Now $\bar{G}$ is constructed from g' by adding all the rules listed in 3), 4), 5) of the preceding construction adding the rules $E_t \to \bar{x}^{cod(m)} t$ whenever $m[t > m_f$. This guarantees that only those strings of $K = L(G)$ are taken which describe valid transition sequences within L_Σ .

This proves the lemma for the free terminal languages. If $L'_\Sigma = h(L_\Sigma) \in \underline{L}$ is used instead of L_Σ , we first have to construct the context-free grammar G'' for $h^{-1}(K)$ and then apply the construction to G'' instead of G' by using now the free terminal language L_Σ . The proofs for the language classes $\underline{P}^f$ or $\underline{P}$ are similar.

Since no sentential form is bordered by a parenthesis, we can substitute any language from the class $\underline{CF} \wedge \underline{L} = \{K \cap L \mid K \in \underline{CF} , L \in \underline{L}\}$ into any context-free set or into any language from the same class. It follows that the nested iterated substitution, compare [3], which in this case is equal to the algebraic closure, compare [5], of the family $\underline{CF} \wedge \underline{L}$, denoted by $(\underline{CF} \wedge \underline{L})^\nabla$, is also contained in $\tau(\underline{CF})$. Without formulating the detailed proof, which is not too complicated, we state this as

Theorem 3

$$(\underline{CF} \wedge \underline{L})^\nabla \subset \tau(\underline{CF})$$

The algebraic closure $\underline{K}^\nabla$ of a family $\underline{K}$ is obtained by constructing extended context-free grammars with rules $A \to K$ where $K \in \underline{K}$ are whole sets of strings from the family $\underline{K}$. The derivation relation is generalized accordingly, for details see [5], and the result is obvious from the two preceding constructions.

It is clear that the above result gives a large extension of the class of Petri net languages, and with these and the results in [2] we are just at the beginning to understand the new family $\tau(\underline{CF})$ better. What we know already is that the emptiness problem for this class is undecidable as is the co-emptiness problem. Right now, we only have partial results for the word problem: If the context-free grammar G for the language $K = L(G)$ has no rule $A \to w$ where A is a nonterminal which can be rewritten by a string $w \in \{x, \bar{x}\}^*$, then the word problem for $\tau(K)$ is shown to be decidable in [6]. There, also a number of examples of languages from the class $\tau(\underline{CF})$ with exponential or hyperexponential growth of the word lengths are presented.

Acknowledgements

We thank K.-J. Lange and M. Kudlek for stimulating discussions and developments of results not presented in this short work. Last not least our thanks go to W.J. Savitch whose idea might be used for another dozen of birthday books still to come.

References

[1]   R. Book, M. Jantzen, C. Wrathall: Monadic Thue Systems, Theoret. Comput. Sci., 19(1982) 231-251.

[2]   V. Geffert: Grammars with context dependency restricted to synchronization, Proc. MFCS 86, Lecture Notes in Comput. Sci. 233, Springer-Verlag (1986) 370-378.

[3]   S.A. Greibach: Full AFL's and nested iterated substitution, Information and Control, 16(1970) 7-35.

[4]   M. Jantzen: Language theory of Petri nets, Proc. Advanced Course on Petri Nets, Bad Honnef (1986), to appear 1987.

[5]   J. van Leeuwen: A generalization of Parikh's theorem in formal
        language theory, Proc. ICALP 74, Lecture Notes in Comput. Sci.
        14, Springer-Verlag (1974) 17-26.

[6]   H. Petersen: Klammer-Löschungs-Grammatiken und Dyckreduktionen
        auf kontextfreien Sprachen, Studienarbeit am FB Informatik,
        Univ. Hamburg (1986).

[7]   J.L. Peterson: Petri Net Theory and the Modeling of Systems,
        Prentice Hall (1981).

[8]   B. Rovan: A framework for studying grammars, Proc. MFCS 81,
        Lecture Notes in Comput. Sci. 118, Springer-Verlag (1981)
        473-482.

[9]   W.J. Savitch: Parantheses grammars and Lindenmayer systems, in:
        G. Rozenberg, A. Salomaa (eds), The Book of L, Springer-Verlag
        (1986) 403-411.

FROM NETS TO LOGIC AND BACK
IN THE SPECIFICATION OF PROCESSES

V.E. Kotov, L.A. Cherkasova

Computing Center

Siberian Division of the USSR Academy of Sciences

630090,Novosibirsk, USSR

Among formal models proposed to specify concurrent systems and
processes, three main groups emerged. They can be referred to as net
models, algebraic calculi and process logics. Each group supports specific abstraction methodology and possesses different descriptive and
analytical power. These two abilities are, to some extent, contradictory and usually are exploited separately in theoretical studies.
However, their combination is highly desirable in practical tools for
the verification and synthesis of systems. This stimulates the compa-
rative studies of different classes of models with cross-interpreta-
tion of their basic notions and primitives. Some efforts to combine
their advantages in the framework of unified theories are now in progress. The models based on Petri nets seem to be the most close to
the adequate description of "pure" concurrency. As a result, many net-
-based models were proposed for the formal specification and validation of concurrent systems: control structures in parallel programs,
circuitry design, net protocols, etc. However, these models, providing a good insight into structural properties of designed concurrent
systems, do not contain sufficient support for the validation of behavioural properties and equivalencies. This forces to have recources
to some "external" formalisms. For example, having a net description
of a system, we use firing sequences, traces, net languages, etc. to
define and analyze the net properties. It would be more convenient if
both descriptive and analytical parts of a modelling theory were based on the same set of basic notions and constructors.

The Petri's paper [1] and some subsequent works introduce and
study the net representation of concurrent processes (occurrence nets)
allowing to specify both systems and the processes, they generate,
in similar terms. It gives the possibility to establish some relations between system and process structures in more direct way. In
paper [2] we tried to extend this approach generalizing the notion
of the process in such a way, that it includes alternative actions.
That means that a generalized concurrent process can contain mutually

exclusive (conflicting) actions equally with sequential and concurrent ones. In this way, the additional possibility to specify non-determinism was included in the net representation of processes (generalized occurrence nets). As a result, the connection between structural properties of the net-processes and boundness, fairness and free-choice property of Petri nets was established.

The basic set of the relations between actions of a generalized process (succession, concurrency and alternative) can be also defined in terms of another specification model, namely by means of a process logic. The group of logic models is oriented at the solution of analytical problems. Therefore, logical relations are used in dynamic or temporal logics [3,4,5,6] exclusively for the specification of properties of systems and processes, which are introduced with the help of some totally different formalisms. In this paper we would like to combine the descriptive power of nets with the analytical power of logic. The hypothesis consists in the following: a rather simple modification of classical propositional logic augmented by temporal (or causal) operations allows both to specify (generalized) concurrent processes and, simultaneously, to describe their properties. The logical formula specifying a process is considered to be the most detailed, total description of the process properties. Having such a logical description of a process, we can directly derive its properties using axioms and inference rules of the same logic. If this logic is developed in such a way that it can describe both systems and their properties, then we would have the possibility to combine the construction and verification of systems in the framework of the same formalism.

We start with the introduction of a logic of finite (generalized) processes. Then the relation between the net and the logic representations of processes is considered. The set of equivalencies and inference rules for the deduction of process partial properties are formulated, then the rules for the total properties deduction are proposed. In the last section of the paper, the extension of the logic for infinite processes is made, and so, a logic for the system specification and verification is outlined.

1. FINITE PROCESS LOGIC FPL (DEFINITIONS)

Let $\mathcal{O}\mathcal{l} = \{a,b,c,\ldots\}$ be a finite alphabet of action symbols (the action basis of a process) and $\overline{\mathcal{O}\mathcal{l}} = \{\bar{a},\bar{b},\bar{c},\ldots\}$ be the dual to alphabet of the "negated" symbols denoting "non-actions", i.e. the symbols which point to the fact that the correspondent actions do not

occur in a process. The special symbol δ denotes a"wrong" or "empty"
action (deadlock, mistake, abort, etc.).

The basic logical connectives are: & (conjunction), $\vee$ (disjunc-
tion), $\rceil$ ("not occur"), $\triangledown$ (exclusive or, alternative) and / (succession).

A formula of FPL in a basis $\mathcal{O}$ is defined as follows:

1) a, $\bar{a}$ and δ, where $a\epsilon\mathcal{O}$, $\bar{a}\epsilon\bar{\mathcal{O}}$, are elementary formulae;

2) if A and B are formulae, then A&B, A$\vee$B, $\rceil$A, A$\triangledown$B, A/B are
formulae.

A formula in a basis $\mathcal{O}$ specifies a process in the basis $\mathcal{O}$;
an elementary formula specifies an elementary process consisting of
one action (or non-action). Viewed intuitively, the semantics of &
and $\vee$ is defined in the standard way: (a & b) means that both the action a and the action b occur; (a $\vee$ b) means that the action a
or the action b occurs, or both actions occur. The operation $\rceil$ is
a modified negation: $\rceil$A means that the process A does not occur,
i.e. no actions of A occur. The operation $\triangledown$ is a modification of
the exclusive or; the formula (A$\triangledown$B) defines a process in which if
the subprocess A occurs then the subprocess B does not occur, and
vice versa. The operation / is the only temporal operation which
orders actions and processes; the formula (A / B) defines the process
in which all actions of the subprocess B can occur only if the subprocess A· is completed. For the more formal definition of the semantics of the introduced formulae, we will decsribe a (generalized) pro-
cess as a set of strings in the joint alphabet $\hat{\mathcal{O}} = \mathcal{O} \cup \bar{\mathcal{O}}$. Let n
be the number of symbols in $\hat{\mathcal{O}}$. Let us denote by $\hat{\mathcal{O}}^{n}$ the set of
strings which satisfy the following condition:

(*) $\forall \sigma \epsilon \hat{\mathcal{O}}^{n}$, $\forall a \epsilon \mathcal{O}$: (a$\epsilon\sigma$ <u>or</u> $\bar{a}\epsilon\sigma$),

i.e. all strings of $\hat{\mathcal{O}}^{n}$ are of the same length and any symbol
of $\mathcal{O}$ enters any string in its direct or negated form.

With each formula A constructed in the basis $\mathcal{O}$, a set $C_{\mathcal{O}}(A)$
of execution sequences is associated in the following way.

1) Let a, $\bar{a}$ and δ are elementary processes. Then

$$C_{\mathcal{O}}(a) = (\hat{\mathcal{O}} \smallsetminus \bar{a})^{n}, \quad C_{\mathcal{O}}(\bar{a}) = (\hat{\mathcal{O}} \smallsetminus a)^{n}, \quad C_{\mathcal{O}}(\delta) = \emptyset, \text{ where } \emptyset \text{ denotes}$$

the empty set, i.e. the formula a in the basis $\mathcal{O}$ defines the
process in which the action a always occurs; similarly, the formula
$\bar{a}$ defines the process in which the action a never occurs; and formula δ defines the empty process. For example, if $\mathcal{O} = \{a,b\}$ then

the formula a defines the process $C_{\mathfrak{A}}(a) = \{ab, ba, a\bar{b}, \bar{b}a\}$, the formula b defines the process $C_{\mathfrak{A}}(b) = \{ab, ba, b\bar{a}, \bar{a}b\}$.

2) If a formula has a form A&B then $C_{\mathfrak{A}}(A\&B) = C_{\mathfrak{A}}(A) \cap C_{\mathfrak{A}}(B)$. For $\mathfrak{A} = \{a,b\}$ the formula a&b defines the process $C_{\mathfrak{A}}(a\&b) = C_{\mathfrak{A}}(a) \cap C_{\mathfrak{A}}(b) = \{ab, ba\}$, i.e. the process in which the actions a and b both occur and in an arbitrary order. So, the conjunction is treated as the concurrency of actions or (sub)processes. The formula a&b specifies the process shown in Figure 1 in the form of an occurrence net.

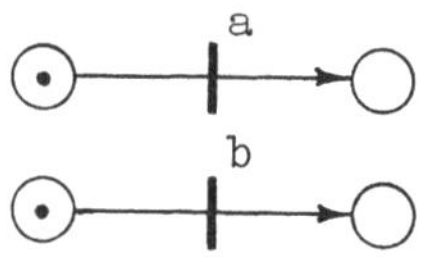

Figure 1

3) If a formula has a form A∨B then $C_{\mathfrak{A}}(A \vee B) = C_{\mathfrak{A}}(A) \cup C_{\mathfrak{A}}(B)$, i.e. it defines the process which contains the union of execution sequences of the process A and of the process B.

4) To describe the semantics of the succession operation /, we need some auxiliary definitions.

Let "$\cdot$" denote the concatenation of two strings. We extend this operation for sets of strings in the standard way: $A \cdot B \overset{def}{=} \{a \cdot b \mid a \in A, b \in B\}$.

Let $\|$ denote the shuffle operation for two strings:

$$a \| \lambda = \lambda \| a \overset{def}{=} \{a\},$$
$$a \cdot x \| b \cdot y \overset{def}{=} \{a\} \cdot (x \| b \cdot y) \cup \{b\} \cdot (a \cdot x \| y),$$

where a and b are symbols, λ is an empty string, x and y are strings.

The shuffle operation is extended for sets of strings:

$$A \| B \overset{def}{=} \{a \| b \mid a \in A, b \in B\}.$$

Let $\mathfrak{A}(A)$ denote the set of symbols from $\mathfrak{A}$ which are present in the formula A, $\bar{\mathfrak{A}}(A)$ denote the set of symbols from $\bar{\mathfrak{A}}$ which are present in A and let $\hat{\mathfrak{A}}(A) = \mathfrak{A}(A) \cup \bar{\mathfrak{A}}(A)$. (Note, that $\mathfrak{A}(\bar{a}) = \{a\}$).

Now, if a formula has a form A/B (and $\mathfrak{A}(A) \cap \mathfrak{A}(B) = \emptyset$) then

$$C_{\mathfrak{A}}(A/B) \overset{def}{=} (C_{\mathfrak{A}(A)}(A) \cdot C_{\mathfrak{A}(B)}(B)) \| (\hat{\mathfrak{A}} \setminus (\hat{\mathfrak{A}}(A) \cup \hat{\mathfrak{A}}(B))),$$

else $C_{\mathfrak{A}}(A/B) = \emptyset$.

For example, if $\mathfrak{A} = \{a,b,c\}$ then the formula ((a&b)/c) specifies the process {abc, bac} shown in the form of a net in the Figure 2.

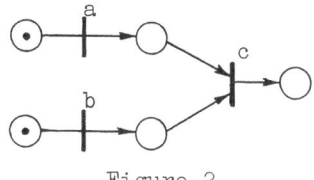

Figure 2

5) If a formula has a form $\rceil A$ then

$$c_{\mathfrak{A}}\,(\rceil A)\overset{def}{=}\,(\,\hat{\mathfrak{A}}\,\smallsetminus\,\mathfrak{A}(A))^{n},$$

i.e. the formula defines the process in which no action of A occurs.
The operation " $\rceil$ " can be defined also as follows:

$$\rceil a = \bar{a}$$
$$\rceil\bar{a} = \bar{a}$$
$$\rceil(A \circ B) = \rceil A\&\rceil B, \text{ where } \circ \in \{\&, \vee, \diagup\}.$$

6) The semantics of the operation "$\triangledown$" can be defined by means of the previously introduced operations:

$$A \triangledown B = (A\&\rceil B)\vee(\rceil A\&B).$$

For example, the formula $((a\triangledown b)\triangledown c)$ in the basis $\mathfrak{A} = \{a,b,c\}$ defines the (generalized) process $\{a\bar{b}\bar{c},\ b\bar{a}\bar{c},\ c\bar{a}\bar{b}\}$ shown in the form of a net in Figure 3.

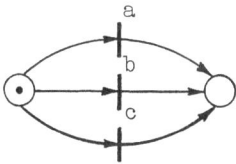

Figure 3

REMARK. The order of non-actions in a process is insignificant and, for example, the strings $a\bar{b}\bar{c}$ and $a\bar{c}\bar{b}$ differ only lexicographically. To simplify manipulations with strings, we assume that all strings are presented in the canonical form in which all negated symbols are written down at the end of a string in the alphabetical order.

Note also that, by the definition of $\rceil$ and $\triangledown$, the following equality takes place: $\rceil(A\triangledown B) = (\rceil A\&\rceil B)$.

2. DEDUCTION OF PARTIAL AND TOTAL PROPERTIES

The formula of the introduced logic specifies both processes and their properties. This situation is similar to the following two treatments of the predicate P(x) = true iff x is an even integer: P defines the set of all even integers; P defines the property "to be even". The formula specifying a process can be considered as a complete description of all its properties.

Two main classes of properties of generalized processes can be distinguished: the total and partial properties. The first ones are valid for any actual realization of the process; the second ones are valid at a subset of possible realizations. The second class of properties emerges because of including alternative actions in generalized processes. Intuitively, the total properties correspond to the notion of validity at a model, the partial properties correspond to the notion of satisfiability.

Let us consider the process defined by the formula (a∨b)&(c∨d) and interpreted by the net form as shown in Figure 4. The formula (a∨b) describes the total property of the process, namely the fact that its actions a and b are always alternative. The property described by the formula (a&c) is partial, as there exist realizations of the process in which both a and c occur (in any order) and there exist realizations in which neither a nor c occurs. The formula, which describes, for example, the property (a∨b) of the process specified by the formula (a∨b)&(c∨d) in the basis {a,b,c,d}, can be considered as a formula specifying the subprocess in the basis {a,b}.

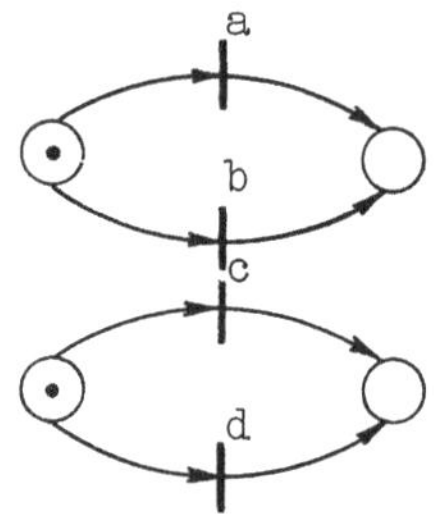

Figure 4

Two formulae $\mathcal{P}_1$ and $\mathcal{P}_2$ in a basis α are equivalent $(\mathcal{P}_1 = \mathcal{P}_2)$ if $C_\alpha(\mathcal{P}_1) = C_\alpha(\mathcal{P}_2)$.

The following equivalencies characterize the properties of the introduced operations:

1. Associativity

1.1. A&(B&C) = (A&B)&C

1.2. Av(BvC) = (AvB)vC
1.3. Av(BvC) = (AvB)vC
1.4. A/(B/C) = (A/B)/C.

2. Commutativity
2.1. A&B = B&A
2.2. AvB = BvA
2.3. AvB = BvA

3. Distributivity
3.1. (A&B)/C = (A/C)&(B/C)
3.2. A/(B&C) = (A/B)&(A/C)
3.3. (AvB)/C = (A/C)v(B/C), if $\mathcal{O}l(A) = \mathcal{O}l(B)$
3.4. A/(BvC) = (A/B)v(A/C), if $\mathcal{O}l(B) = \mathcal{O}l(C)$
3.5. Av(B&C) = (AvB)&(AvC)

4. Axioms for v and ⫪
4.1. AvB = A&⫪B v ⫪A&B
4.2. ⫪(A&B) = ⫪A& ⫪B
4.3. ⫪(AvB) = ⫪A& ⫪B
4.4. ⫪(A/B) = ⫪A& ⫪B
4.5. ⫪a = $\bar{a}$
4.6. ⫪$\bar{a}$ = $\bar{a}$

5. Other properties
5.1. $\bar{a}$/A = $\bar{a}$&A
5.2. A/$\bar{a}$ = $\bar{a}$&A
5.3. A&(A/B) = A/B
5.4. B&(A/B) = A/B
5.5. A/B/C = (A/B)&(B/C)
5.6. A&(AvB) = A.

6. Auxiliary
6.1. A&A = A
6.2. AvA = A
6.3. a&$\bar{a}$ = δ
6.4. A/B/A = δ
6.5. δ&A = δ
6.6. δ/A = δ
6.7. A/δ = δ
6.8. δvA = A
6.9. δvA = A
6.10. A = A&a v A&$\bar{a}$, if a∈ $\mathcal{O}l \smallsetminus \mathcal{O}l(A)$.

Two strings 6 and $6'$ will be called permutated ($6 \approx 6'$) if $6'$ can be obtained from 6 with the help of the permutation of some symbols.

A formula ϕ' defines a subprocess of the process defined by a formula φ if

1) $c_\alpha(\varphi)\!\restriction_{\hat{\alpha}(\varphi')} \geq c_{\alpha(\varphi')}(\varphi')$

2) $\forall 6 \in C_\alpha(\varphi)\!\restriction_{\hat{\alpha}(\varphi')},\ \forall 6' \in C_{\alpha(\varphi')}(\varphi') :\ 6 \approx 6' \Rightarrow 6 \in C_{\alpha(\varphi')}(\varphi')$,

where $c_\alpha(\varphi)\!\restriction_{\hat{\alpha}(\varphi')}$ is projection of strings of $c_\alpha(\varphi)$ on the set of symbols $\hat{\alpha}(\varphi')$.

Due to the second condition, all components in φ' are bound by the same relations as they are bound in φ. This definition of a sub-process is equivalent to the notion of the causal component [7]. A net $N'=(P',T',F')$, where P' is its set of places, T' is its set of tran-sitions, F' is its incidence relation, is a causal component of a net $N = (P,T,F)$ if $P' \subseteq P$, $T' \subseteq T$ and $\forall x,y \in P' \cup T' : F'^+(x,y) = F^+(x,y)$, where F^+ is the transitive closure of F. For example, the formulae $\varphi'_1 = (a\&d)$ and $\varphi'_2 = (b/d)$ define the subprocesses of the process defined by the formula $\varphi = (a \vee b)\&((c \vee b)/d)$. The correspondent net re-presentation of the process is shown in Figure 5.

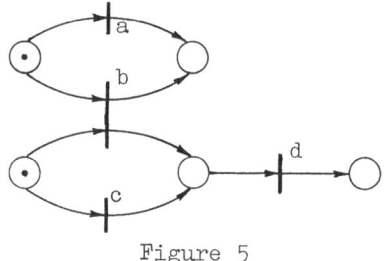

Figure 5

A property φ' is satisfiable (partial) for a process φ (we will write $\varphi \Vdash \varphi'$) iff φ' is a subprocess of φ. For example, $(a \vee b)\&$ $\&((b \vee c)/d) \Vdash (a\&d)$.

In the following inference rules for the deduction of partial pro-perties of processes we suppose that for formulae A and B the con-dition $\alpha(A) \cap \alpha(B) = \emptyset$ is valid:

I. $A \vee B \Vdash A$

$\widetilde{\text{I}}$. $A \vee B \Vdash A' \vee B'$, where $A \Vdash A'$ and $B \Vdash B'$

II. $A/B \Vdash A$ and $A/B \Vdash B$

$\widetilde{\text{II}}$. $A/B \Vdash A'/B'$, where $A \Vdash A'$ and $B \Vdash B'$

III. $A\&B \Vdash A$

$\widetilde{\text{III}}$. $A\&B \Vdash A'\&B'$, where $A \Vdash A'$ and $B \Vdash B'$

Due to the context restriction $\mathcal{O}(A) \cap \mathcal{O}(B) = \emptyset$ in all six rules, the conclusion that the process in the right part of rule is the sub-process of the process in the left part of the rule is quite evident.

To introduce one more inference rule, we describe the algorithm of transformation of an arbitrary formula to its canonical form.

At the first step of the algorithm, a formula $\mathcal{P}$ is transformed using the equivalency axioms to the form $\mathcal{P} = \mathcal{P}_1 \vee \mathcal{P}_2 \vee \ldots \vee \mathcal{P}_k$, where $\mathcal{P}_i$ contains only operations & and /.

At the second step, the axioms 3.1, 3.2, 5.1-5.5 are applied to delete some "redundant information". For example, we transform the formula (a&b)&(a/b) to the formula (a/b). The set of conjuncts $\{\mathcal{P}_i\}_{i=1}^k$ of the formula $\mathcal{P}$ presented in the canonical form can be treated as the set of all possible (maximal) parallel subprocesses of the (generalized) process $\mathcal{P}$.

The following inference rule is valid:

IV. $\mathcal{P} \Vdash \mathcal{P}_i$, where $\mathcal{P}_i$ is a conjunct of the $\mathcal{P}$ and $\mathcal{P}$ is presented in the canonical form.

Note, that inference rules III and $\widetilde{\text{III}}$ with conjuncts in their left parts are applied without any context restrictions.

EXAMPLE. $\mathcal{P}= (a \vee b)\&((b \vee c)/d)=(a\&\bar{b} \vee b\&\bar{a})\&((b\&\bar{c} \vee c\&\bar{b})/d)=$
$= (a\&\bar{b} \vee b\&\bar{a})\&((b\&\bar{c})/d \vee (c\&\bar{b})/d)= a\&\bar{b}\&((b\&\bar{c})/d) \vee$
$\vee b\&\bar{a}\&((c\&\bar{b})/d) \vee a\&\bar{b}\&((c\&\bar{b})/d) \vee b\&\bar{a}\&((b\&\bar{c})/d)=$
$= \delta \vee \delta \vee a\&(c/d)\&\bar{b} \vee (b/d)\&\bar{a}\&\bar{c} = a\&(c/d)\&\bar{b} \vee(b/d)\&\bar{a}\&\bar{c}.$

Now, $\mathcal{P}$ is in the canonical form.

$$\mathcal{P} \Vdash a\&(c/d)\&\bar{b} \Vdash a\&(c/d) \Vdash a\&c, \text{ where } (c/d) \Vdash c.$$

So, in the process $\mathcal{P}$ actions a and c can occur concurrently.
A property $\mathcal{P}'$ is valid (total) for a process $\mathcal{P}$ (we will write $\mathcal{P} \vDash \mathcal{P}'$) iff $C_{\mathcal{O}}(\mathcal{P})\upharpoonright_{\hat{\mathcal{O}}(\mathcal{P}')}= C_{\mathcal{O}(\mathcal{P}')}(\mathcal{P}')$, where $C_{\mathcal{O}}(\mathcal{P})\upharpoonright_{\hat{\mathcal{O}}(\mathcal{P}')}$ is projection of strings from $C_{\mathcal{O}}(\mathcal{P})$ on the set of symbols $\hat{\mathcal{O}}(\mathcal{P}')$.

In the following first four inference rules for the deduction of total properties we suppose that for formulae A and B the condition $\mathcal{O}(A) \cap \mathcal{O}(B) = \emptyset$ is valid:

I. $(A \vee B) \vdash A' \vee \daleth A'$, where $A \vdash A'$

$\widetilde{\text{I}}$. $(A \vee B) \vdash A' \vee B'$, where $A \vdash A'$, $B \vdash B'$

II. $(A/B) \vdash A$ and $(A/B) \vdash B$

$\widetilde{\text{II}}$. $(A/B) \vdash A'/B'$, where $A \vdash A'$ and $B \vdash B'$

III. $(A\&B) \vdash A$, if 1) $\mathcal{O}(A) \cap \mathcal{O}(B) = \emptyset$,

or 2) (A&B) does not contain symbol $\vee$, and is a canonical conjunct.

$\widetilde{\text{III}}$. $(A\&B) \vdash A'\&B'$, where $A \vdash A'$, $B \vdash B'$, and

        if 1) $\mathcal{O}(A) \cap \mathcal{O}(B) = \emptyset$.

        or 2) $A\&B$ is a canonical conjunct.

IV. Let $\mathcal{P} = \overset{k}{\underset{i=1}{V}} \mathcal{P}_i$ be the canonical form. If for any i ($1 \le i \le k$):

$\mathcal{P}_i \vdash A$ is valid, then $\mathcal{P} \vdash A$.

Examples 1) $\mathcal{P} = (a/(b\&c)/d)ve$

        $a/(b\&c)/d \vdash a/(b\&c) \vdash b\&c$

        $\mathcal{P} \vdash (b\&c) \vee \daleth(b\&c) = b\&c \vee \bar{b}\&\bar{c}$.

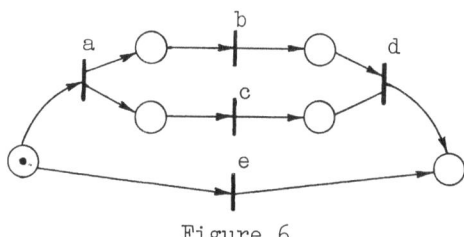

Figure 6

The formula $\mathcal{P} = (a/(b\&c)/d)ve$ specifies the process shown in Figure 6 in the form of generalized net-process. In this net, there are only two possibilities for transitions b and c: 1) b and c can fire; 2) neither b, nor c can fire; and exactly this property is deduced from the process formula $\mathcal{P}$.

        2) $\mathcal{P} = (avb)\&(cvd)$

        $avb \vdash a \vee \daleth a = a \vee \bar{a}$

        $cvd \vdash c \vee \daleth c = c \vee \bar{c}$

        $\mathcal{P} \vdash (a \vee \bar{a})\&(c \vee \bar{c}) = a\&c \vee \bar{a}\&c \vee a\&\bar{c} \vee$

                                        $\vee \bar{a}\&\bar{c}$.

The formula $\mathcal{P} = (avb)\&(cvd)$ specifies the process the net representation of which is shown in Figure 4. In this net (and process) actions a and c can or cannot occur in any combination. This property is deducted from the process formula $\mathcal{P}$ too.

The following assertion can be used in the equivalent formula transformations: if $\mathcal{P} \vdash A$, then $\mathcal{P}\&A = \mathcal{P}$. The validity of this assertion immediately follows from the definitions of $\vdash$ and $=$.

3. INFINITE PROCESS LOGIC IPL

By the notion of event we mean here a system component which generates actions (event occurrences) in the process describing the system functioning. An elementary event generates a unique action and can be identified with this action.

Let $\mathcal{O} = \{a,b,c,\ldots\}$ be a finite alphabet of event symbols, and $\overline{\mathcal{O}} = \{\bar{a},\bar{b},\bar{c},\ldots\}$ be the dual alphabet of the "negated" symbols for the events which generate no actions in a process. The special symbol δ denotes an empty event (deadlock, mistake).

The set of basic logical connectives (&, v, v, $\top$, /) is extended with new generalized operations &, $\underset{k}{\&}$, $\underset{\infty}{\&}$, /, $\underset{k}{/}$, $\underset{\infty}{/}$, $\mathcal{I}$, (k≥1), with the following common properties:

1) each operation singles out a set of "non-elementary" events and for each event it forms a corresponding set of actions, which are identified by the event symbol supplied by indeces.

2) each operation restricts in a special way the order of various occurrences of the same event, namely: $\underset{k}{\&} a$ specifies a process in which the actions $\{a_1,\ldots, a_k\}$ occur concurrently, and the actions $\{a_{k+1}, a_{k+2},\ldots\}$ do not occur; $\underset{\infty}{\&} a$ defines a process in which all actions $\{a_1, a_2,\ldots\}$ occur concurrently; $\underset{k}{/} a$ specifies a process in which actions $\{a_1, a_2,\ldots, a_k\}$ occur sequentially, and the actions $\{a_{k+1}, a_{k+2},\ldots\}$ do not occur. The formula $\underset{\infty}{/} a$ defines a process in which all actions $\{a_1, a_2,\ldots\}$ occur sequentially; a formula $\mathcal{I} a$ defines a process in which for each $i \geqslant 1$, the actions $\{a_1,\ldots, a_i\}$ occur sequentially and the actions $\{a_{i+1}, a_{i+2},\ldots\}$ are not realized.

A formula of IPL in the basis $\mathcal{O}$ is defined as follows:

1) a, $\bar{a}$ and δ, where $a \in \mathcal{O}$, $\bar{a} \in \overline{\mathcal{O}}$, are formulae,

2) if A and B are formulae, then A&B, AvB, $\top$A, AvB, A/B, $\underset{k}{\&} A$, $\underset{\infty}{\&} A$, $\underset{k}{/} A$, $\underset{\infty}{/} A$, $\mathcal{I} A$ are formulae.

To specify the process defined by a formula A, we will single out in the alphabet $\mathcal{O}$ a subset $\mathcal{O}_1$ of "non-elementary" events, i.e. those event symbols which enter subformulae $\underset{k}{\&}\varphi$, $\underset{\infty}{\&}\varphi$, $\underset{k}{/}\varphi$, $\underset{\infty}{/}\varphi$, $\mathcal{I}\varphi$. So, the initial alphabet of events $\mathcal{O} = \mathcal{O}_1 \cup \mathcal{O}_2$, where $\mathcal{O}_2 = \mathcal{O} \setminus \mathcal{O}_1$, is associated with the infinite action basis $\mathcal{O}_\infty = (\underset{a \in \mathcal{O}_1}{\cup} \{a_1, a_2, a_3,\ldots\}) \cup \mathcal{O}_2$, in which the symbols of non-elementary events $a \in \mathcal{O}_1$ are substituted by infinite sets of indexed actions $\{a_1, a_2, a_3,\ldots\}$.

Each formula A in the event alphabet $\mathcal{O}$ defines a process (a set of execution strings $C_{\mathcal{O}_\infty}(A)$) in the action basis $\mathcal{O}_\infty$ according to the following rules.

Let $\hat{\mathcal{O}}_\infty = \mathcal{O}_\infty \cup \overline{\mathcal{O}}_\infty$. Let $\hat{\mathcal{O}}_\infty^*$ denotes a set of infinite strings satisfying the requirement (*), i.e.

$$\forall \sigma \in \hat{\mathcal{O}}_\infty^* , \ \forall a \in \mathcal{O}_\infty : (a \in \sigma \quad \underline{or} \quad \bar{a} \in \sigma).$$

1) If a and $\bar{a}$ are elementary events, i.e. $a \in \mathcal{O}_2$, $\bar{a} \in \overline{\mathcal{O}}_2$, or a is δ, then their semantics is defined just as in FPL, i.e. $C_{\mathcal{O}_\infty}(a) = (\hat{\mathcal{O}}_\infty \setminus \bar{a})*$, $C_{\mathcal{O}_\infty}(\bar{a}) = (\hat{\mathcal{O}}_\infty \setminus a)*$, $C_{\mathcal{O}_\infty}(\delta) = \emptyset$.

If a and $\bar{a}$ are non-elementary events, i.e. $a \in \mathcal{O}_1$, $\bar{a} \in \bar{\mathcal{O}}_1$, then

$$C_{\mathcal{O}_\infty}(a) = (\hat{\mathcal{O}}_\infty \smallsetminus \{\bar{a}_1, \bar{a}_2, \bar{a}_3, \ldots\})*,$$

$$C_{\mathcal{O}_\infty}(\bar{a}) = (\hat{\mathcal{O}}_\infty \smallsetminus \{\bar{a}_1, \bar{a}_2, \bar{a}_3, \ldots\})*,$$

i.e. the formula a in the alphabet $\mathcal{O}$ describes a process in which all corresponding actions $\{a_1, a_2, a_3, \ldots\}$ occur. In a similar way, $\bar{a}$ defines a process in which event a does not occur.

2) The semantics of the operations $\&, v, \daleth, \triangledown$ is defined just as in **FPL**.

The semantics of the formula A/B can be defined by means of operations of FPL, if $\mathcal{O}(A) \cap \mathcal{O}_1 = \emptyset$, i.e. a formula A contains only elementary events. If $\mathcal{O}(A) \cap \mathcal{O}_1 \neq \emptyset$, then the formula A/B describes the empty process, i.e. $C_{\mathcal{O}_\infty}(A/B) = \emptyset$, since the formula A contains non-elementary events and, therefore, defines an infinite (or the empty) process.

3) To describe the semantics of new operations, we need some auxiliary operation of indexing $(A)_n$:

$$(A \circ B)_n = (A)_n \circ (B)_n, \text{ where } \circ \in \{\&, v, \triangledown, /\},$$

$$(\daleth A)_n = \daleth (A)_n,$$

$$(a)_n = a_n, \text{ if } a \in \mathcal{O},$$

$$(\bar{a})_n = \bar{a}_n, \text{ if } \bar{a} \in \mathcal{O}.$$

The semantics of new operations is defined in the following way:

$$\underset{k}{\&} A \overset{\text{def}}{=} ((A)_1 \& (A)_2 \& \ldots \& (A)_k) \& \daleth ((A)_{k+1} \& (A)_{k+2} \& \ldots), \quad k \geqslant 1.$$

$$\underset{\infty}{\&} A \overset{\text{def}}{=} (A)_1 \& (A)_2 \& \ldots \& (A)_k \& (A)_{k+1} \& \ldots .$$

Thus, the operation $\underset{\infty}{\&}$ is a special case of the operation $\underset{k}{\&}$ at $k = \infty$.

$$\underset{k}{/} A \overset{\text{def}}{=} ((A)_1 / (A)_2 / \ldots / (A)_k) \& \daleth ((A)_{k+1} \& (A)_{k+2} \& \ldots), \quad k \geqslant 1.$$

$$\underset{\infty}{/} A \overset{\text{def}}{=} (A)_1 / (A)_2 / \ldots / (A)_k / (A)_{k+1} / \ldots .$$

Thus, the operation $\underset{\infty}{/}$ is a special subcase of the operation $\underset{k}{/}$ at $k = \infty$.

$$\underset{}{/} A \overset{def}{=} \underset{1}{/}A \lor \underset{2}{/}A \lor \ldots \lor \underset{k}{/}A \lor \underset{k+1}{/}A \lor \ldots \; .$$

In Figure 7 are shown a Petri net and a corresponding net-process which illustrate the meaning of the formula $\underset{k}{\&} a$. The similar nets for the formula $\underset{\infty}{/}a$ are shown in Figure 8.

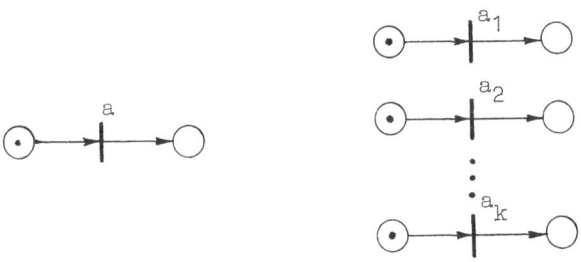

Figure 7

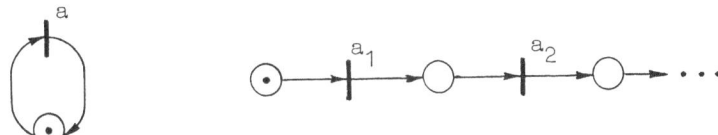

Figure 8

The notions of equivalent formulae, satisfiability and validity for IPL are defined similarly to those of FPL.

The following group of equivalencies can extend the available list of equivalent formula transformations.

7.1. $\underset{k}{\&}A\& \underset{n}{/}A = \eth$, if $k \neq n$

7.2. $\underset{\infty}{\&}A\& \underset{k}{/}A = \eth$

7.3. $/\!\!/ A\& \underset{\infty}{/} A = \eth$

7.4. $/\!\!/ A\&(\underset{\infty}{\&} A) = \eth$

7.5. $\underset{k}{\&}A\& \underset{k}{/}A = \underset{k}{/}A$

7.6. $\underset{\infty}{\&} A\& \underset{\infty}{/} A = \underset{\infty}{/} A$

7.7. $A \; \& \underset{\infty}{/} A = \underset{\infty}{/} A$

7.8. $A\&(\underset{\infty}{\&} A) = \underset{\infty}{\&} A$

7.9. $\mathop{/\!\!/} A\&(\underset{k}{/}A) = \underset{k}{/}A$

7.10. $\mathop{/\!\!/} A\&(\underset{k}{\&}A) = \underset{k}{/}A$

Two new inference rules of partial properties are introduced in addition to those of FPL.

V. $\underset{\infty}{/}\mathcal{P} \Vdash \underset{\infty}{/}A$, if $\quad \mathcal{P} \Vdash A$

VI. $\mathop{/\!\!/} \mathcal{P} \Vdash \underset{i}{/}A$, $\quad i \geqslant 1$, if $\quad \mathcal{P} \Vdash A$.

The following rule is proposed in addition to the inference rules of total properties in FPL.

V. $\underset{\infty}{/}\mathcal{P} \vdash \underset{\infty}{/}A$, if $\quad \mathcal{P} \vdash A$

CONCLUSION

In this paper we develop a unified logical approach to the specification of processes, systems and their properties. The introduced basic set of notions and operations is associated with the basic relations of concurrency, succession and alternative in processes. The inference rules for deduction of partial and total process properties are presented. The generalized operations give the possibility to describe infinite processes with the help of finite formulae.

Both finite and infinite process logics considered here are quite tightly related to "structural" properties of processes. As a result, those properties which are formulated in terms of relations (and their transitive closure) between actions and events can be expressed in these logics. To get the possibility to deduce the properties of more general nature, the logics should be augmented by some additional notions and operations.

In the previous sections we used net representations of processes for informal demonstration of the meaning of the logic formulae. However, we do not give formal rules for the mutual translation of logic and net representations. There can exist different methods of the translation. For example, if a net representation is defined in terms of the net algebra proposed in the paper [8], then a direct correspondence can be established between the operations of the algebra and the operations of FPL or IPL. However, the difference between the structural orientation of the algebra and the behavioural orientation of the logic creates some additional problems. The net representation of processes can contain some superfluous or just contradictory relations and, consequently, does not define properly a reasonable process beha-

viour (as example, occurrence nets which are not K-dense [1] or generalized nets which are not K-dense or L-dense or M-dense [2]). The
net shown in Figure 9 can be defined in the algebra of nets [8]   as
the superposition of concurrent actions   a   and   b   and alternative
actions   a   and   b. This net is not M-dense [2] , as it contains collision between concurrency and alternative.

   An attempt of the direct interpretation of this net in FPL leads
to the formula   (a&b)&(a∨b) which is equivalent to the formula
(a&b&b̄) ∨ (b&a&ā)   and specifies empty process because of the conjuncts
a&ā   and   b&b̄. On the other hand, the net shown in Figure 8 specifies
non-empty process which can be presented by the equivalent M-dense
net shown in Figure 10. For this net there exists a corresponding
formula   a∨b   in FPL.

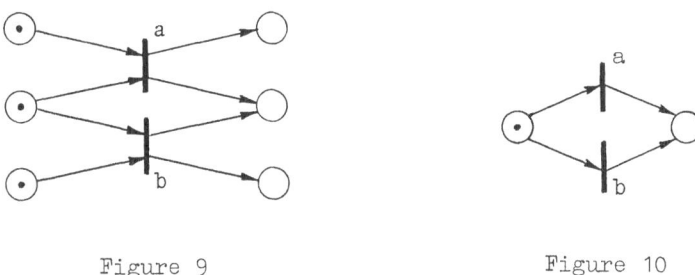

Figure 9   Figure 10

This situation arises because only "well-formed" (dense) net-processes
can be matched with formulae of FPL and IPL. It is possible to modify
the logics introducing "weak" conjunction $\widetilde{\&}$ in the following way: the
semantics of the formula A$\widetilde{\&}$B is defined as a set C_{α}(A$\widetilde{\&}$B) of all maximal common prefixes of the sets C_{α}(A) and C_{α}(B). Then the following
chain of translations can be proposed for the analysis of nets. For a
net (including non-dense nets) a specification in the modified logic
is obtained by direct translation. Then an equivalent transformation
into FPL or IPL formula is fulfilled and the required analysis is done
using appropriate equivalencies and inference rules. The reversing of
this chain makes possible to transform the source net into an equivalent net with desirable properties.

REFERENCES

1. Petri C.A. Non-sequential processes. - ISF-Report-77.05, St.Augustin: Gesellschaft für Mathematik und Datenverarbeitung,1971, 31 p.
2. Kotov V.E., Cherkasova L.A. On structural properties of generalized processes. - Lecture Notes in Computer Science, 188, Springer-Verlag, Berlin, 1984, p. 288-306.
3. Harel D. First-order dynamic logic. - Lecture Notes in Computer Science, 68, Springer-Verlag, Berlin, 1979.
4. Fisher M.J., Ladner R.E. Propositional dynamic logic of regular programs. - J.Comput.on System Science, 18(2), 1979, p. 197-211.
5. Pratt V.R. Process logic. - Proc.ACM Symp.on Principles of Programming Languages, 1979, p. 93-100.
6. Manna Z., Pnueli A. Verification of concurrent programs: temporal proof principles. - Lecture Notes in Computer Science, Springer-Verlag, Berlin, v. 131, 1981, p. 200-252.
7. Best E. The relative strength of K-density. - Lecture Notes in Computer Science, Springer-Verlag, Berlin, v. 84, 1980, p.261-276.
8. Kotov V.E. An algebra for parallelism based on Petri nets. - Lecture Notes in Computer Science, Springer-Verlag, Berlin, 1978, p. 39-55.

Types and Modules for Net Specifications

Bernd Krämer and Heinz-Wilhelm Schmidt

GMD-F2G2, Schloß Birlinghoven, D-5205 St. Augustin 1

Abstract

A specification language for nonsequential systems that unifies algebraic specifications of abstract
data types with high-level Petri net specifications of dynamic behavior is presented. The data structure of
a system, the information content of local states, and static constraints to state changing operations are
specified by sorted Horn clause rules with equality. Behavior is specified by schemes of Predicate-Event
nets together with an initial (distributed) state. Many-sorted algebras provide a standard interpretation
of such specifications in terms of the initial models satisfying the rules, the flow, and the initial state
given. One concern of this paper is to sketch the mathematical semantics of the core language. The
other is to define a notion of abstract system that supports modularity and reusability of specifications
similar to abstract data types.

1 Introduction

Algebraic specifications of abstract data types (ADT's) have been studied for many years as a powerful
tool for the development of software systems (cf. e.g., [10], [6], [9], [5]). Algebraic ADT specifications
provide a formal semantics for many useful programming concepts and are therefore widely accepted for
specifying basically sequential systems. But their value proved to be limited for defining the behavior of
abstract machines in a distributed environment (cf. [8]).

To tackle behavioral issues raised by nonsequential and distributed systems such as concurrency, nondeterminism, communication, and synchronization, a number of dedicated design and specification methods
evolved. Among these we mention Milner's Calculus of Communicating Systems (CCS) [19], temporal logic
methods [21,17], and various Petri net based methods [4]. Petri nets provide a notion of concurrency and
nonsequential processes that is based on partial orders. This makes Petri nets particularly valuable for
distributed application domains in which a clear distinction between concurrency, nondeterminism, and
sequentiality is required [16].

The specification language *SEGRAS* [12,11], whose core concepts are presented here, combines ADT's
and Petri nets in a uniform syntactic and semantic framework. Specifications of nonsequential systems
given in *SEGRAS* generally consist of a) abstract data types describing data structures on which the system
operates, b) Predicate-Event nets [23] describing the dynamic behavior of the system, and c) a description
of the initial (distributed) system state. Such specifications can be considered as a variant of Predicate-
Transition nets (PrT-net) [7]. In our variant a many-sorted partial algebra is used in place of the set
theoretic and logic approach proposed in [7]. A partial algebra provides different data domains and partial
operations on these domains. Their effects are recursively specified by conditional equations which are
equivalent to many-sorted Horn clause logic with equality. This logic was shown to have yet "standard",
i.e., initial models [18].

SEGRAS is the specification language of the ESPRIT project GRASPIN[1]. In this project we are developing techniques and prototypes of tools which are to aid in the construction and verification of nonsequential
systems specifications and their systematic transformation into executable programs. For such a software
engineering approach it is essential that the specification formalism used has computable models. *SEGRAS*

---

[1]GRASPIN is supported in part by the Commission of the European Communities within the ESPRIT program.

seems to be a good compromise between expressive power and semantic strength as there is a strong relationship between initiality and computability.

SEGRAS was designed with the goal to provide an appropriate high-level Petri net language with *data types* and *modularization* capabilities to carry the methodological use of ADT's in programming over to net specifications. This idea was first presented informally in [14] and [13]. A formal treatment of PrE nets combined with many-sorted partial algebras was then given in [24]. Later on it was also proposed to integrate colored nets [26] or Predicate-Transitions nets [1] and ADT's.

Modularity is essential for writing large specifications. Modules comprising only small specifications can be verified and implemented once and for all and can be reused and combined freely to make large specifications. We are going to define a notion of *abstract system* that comprises data abstraction with an interface of named operations and which additionally has a hidden local, possibly distributed state and state changing operations that can actually see and transform parts of the state.

In this paper emphasis is put on the semantics of the core language. As the formalization of the full language (see [12]) is overly complex, a simplified abstract syntax is defined in the next section. This syntax is used in Section 3 to present the mathematical semantics of core concepts of the language. Section 4 introduces concepts for structuring specifications and sketches restrictions on preserved properties of combined specifications. The last section gives a small example of a structured specification.

2 Abstract Syntax of the Language

The abstract syntax of our specification language is based on conditional algebraic specifications of abstract data types, as defined in [15], and on Predicate-Event nets labelled by expressions over a given signature (see [24]). Most of our notions and definitions are kept close to these sources to take advantage of theorems and proofs given there.

This section partly recalls fundamental notions about algebraic specifications adapted to our specific needs. The reader not familiar with the concepts may consult [10,15,5] for detailed information on the topic.

2.1 Algebraic Language

Classically, ADT's consist of sorts, operation symbols, and equations. The sorts serve for naming data domains. The operation symbols name operations on these domains. The operation symbols generate an algebra of terms on which the equations induce a congruence relation. Thus the effect of each operation is abstractly defined in the sense that neither a particular implementation of the operations nor a particular representation of data is referred to.

Signatures

Signatures formalize the notion of a sorted collection of operations available to the user of an abstraction.

Definition 1 *Let S be a countable set, whose elements are called* sorts, *and let S^* be the set of all finite strings over S, including the empty string λ. Then an S-sorted algebra signature Σ is an $S^* \times S$-indexed family of sets $(\Sigma_{u,s})_{u \in S^*, s \in S}$ of operation symbols of arity u, coarity s, and of type $u \to s$.*

Notation. In examples, we write $\sigma : u \longrightarrow s$ to denote $\sigma \in \Sigma_{u,s}$; an S-sorted signature is given as a sequence of operation symbols headed by the keyword **fu**; the sorts indexing a signature are listed after the keyword **sorts**.

As a matter of notational convenience, let S be an arbitrary but fixed set of sorts and Σ an arbitrary but fixed S-sorted signature for the rest of this paper.

271

Terms, Equations, and Rules

A signature determines a language of terms. Terms are used to formulate equations and inequations, from which Horn clause like conditional equations, called rules, are build up.

Definition 2 *Let X be an S-indexed family of sets $(X_s)_{s \in S}$ of variable symbols disjoint[2] from $(\Sigma_{\lambda,s})_{s \in S}$.*
Then the elements of the S-indexed family of sets $T_{\Sigma(X)} = (T^s_{\Sigma(X)})_{s \in S}$ that is minimally defined by the following list of points are called Σ-terms (with variables in X).

1. $X_s \subseteq T^s_{\Sigma(X)}$ *for $s \in S$*
2. $\Sigma_{\lambda,s} \subseteq T^s_{\Sigma(X)}$ *for $s \in S$*
3. $\sigma(t_1, \ldots, t_n) \in T^s_{\Sigma(X)}$ *if $t_i \in T^{s_i}_{\Sigma(X)}$ and $\sigma \in \Sigma_{s_1 \ldots s_n, s}$ for $s_i, s \in S, s_i \neq \lambda (1 \leq i \leq n)$*

A Σ-term is called *ground* if it is without variables. T_Σ denotes the subfamily of ground Σ-terms.

Definition 3 *A Σ-equation (Σ-inequation) is a pair of Σ-terms (t, t') of sort $s \in S$ written $t =_s t'$ (and $t \neq_s t'$ resp.). An equation without variables is called* ground.

Notation. For the remaining part of this text, let X be a fixed family of variable symbols and let $vars(t)$ denote the smallest family $Y \subseteq X$ of variables occurring in term t such that $t \in T_{\Sigma(Y)}$.
We will drop the sort index in equations and inequations, i.e., we write $x = y$ for $x =_s y$.

Definition 4 *A Σ-rule, written $e_0 : e_1, \ldots, e_n$. (for $n \geq 0$), consists of an equation (or inequation) e_0, called* conclusion, *and a possibly empty sequence of equations e_i, called* premises. *A rule is called* positive *if e_0 is an equation and* negative *if e_0 is an inequation $(1 \leq i \leq n)$.*

Partial Specifications of Data Types

Now we show how the ideas of total specification of data types by equations [5] may be modified so as to yield partial specifications. They differ from total specifications by the use of partial operations, i.e., operations that are meaningful only on part of their domain. For example, *tail* is only defined for non-empty lists, or *top* and *pop* are only meaningful for non-empty stacks. In the case of partial specifications, such exceptional situations are simply made undefined (here be means of inequations, so that the problem of error recovery and exception handling is left to the implementor of the data type.

We have chosen partial specifications and partial algebras as their models because we wanted to adequately handle the partiality of operations and relations used in PrE nets in the algebraic framework. Albeit the mathematical theory involved with this approach is more complex than for total specifications, partial specifications have been advocated by other authors, too (e.g., [3]), so that we can adopt some of their results.

Definition 5 *A conditional specification $SPEC$ is a triple (S, Σ, R) where R is a finite set of Σ-rules.*

The example in fig. 1 gives the signature and the rules for the ADT "bintree". We use a mixfix notation for some of the operation symbols. We assume that the sort **Nat** of natural numbers is built-in together with operations on **Nat**. The binary trees have leaves labelled by natural numbers.

In this example, the rules only serve to specify the definedness of terms. Actually, here all terms are defined or, in other words, all "bintree" operations are total. Since this situation occurs so often in partial specifications, in *SEGRAS* we distinguish between total and partial operations on the syntactic level of signatures. For simplicity of the present text we avoid the technicalities required for this distinction. The results however remain the same [15].

Notation. To ease the reading of the examples, equations like $t = t$ are simply written as *literals* of the form "t" in subsequent examples.

---

[2]For A, B S-indexed families of sets, $A \cap B$ means $A^s \cap B^s$ for all $s \in S$.

```
sorts  bintree, Nat

fu  leaf:  Nat  -> bintree.
    tree:  bintree bintree -> bintree.

forall T, T1, T2: bintree; n: Nat.
leaf n = leaf n:  n.
tree(T,tree(T1,T2) = tree(tree(T,T1),T2): T = T,  T1 = T1,  T2 = T2.
```

Figure 1: Abstract Data Type bintree

2.2 Net Language

Petri nets which are labelled by symbols and terms over a given signature and which are accompanied by a set of rules will be used to specify the nonsequential behavior of an abstract system. The rules specify a) data structures the abstract system is operating on, b) the definedness of net elements, c) constraints to event occurrences, and d) the initial state of the system.

This reflects the idea that the various sorts of data involved in the dynamic behavior are considerd integral parts of a system specification and that static constraints to high-level events are formulated by the same axiomatic means as data structures are specified.

PrE Signatures

Now, the notion of algebra signature is extended to a notion of PrE signature in which operation symbols naming net elements are distinguished and two distinct symbols are assumed that capture the flow relation and the initial case in the associated algebra.

Definition 6 *A signature for a Predicate-Event system (PrE signature) over Σ is a triple $\Pi = (\Sigma, \Pi^C, \Pi^E)$ such that*

1. *Σ an algebra signature*

2. *$\Pi^C, \Pi^E \subseteq \Sigma$ are disjoint subsignatures containing the condition and event symbols, resp.*

3. *There are distinctive symbols*
   flow $\in \Sigma_{s_1 s_2, \lambda}$ for $s_1, s_2 \in Sys$, the flow symbol, and
   case $\in \Sigma_{s, \lambda}$ for $s \in Sys$, the initial case symbol,

where the set $Sys \stackrel{def}{=} \{s \mid \sigma \in \Pi^C_{u,s} \cup \Pi^E_{u,s}, u \in S^*, s \in S\}$ of system sorts is formed by collecting the coarities of all condition and event symbols.

Remark. The terms of a sort in *Sys* are used to represent markings and occurrences of events, i.e., they model information related to states and their changes.

Obviously an algebra signature is the degenerate case of a PrE signature with $\Pi^C \cup \Pi^E = \emptyset$.

Notation. We write ac $\sigma : u \longrightarrow s$ to denote $\sigma \in \Pi^E_{u,s}$ and st $\sigma : u \longrightarrow s$ to denote $\sigma \in \Pi^C_{u,s}$.

Π-labelled Nets

We omit fundamental net theoretic definitions here and simply use the terminology and definitions given in [2]. We shall take the symbols of Π^C as the *state predicates* and the symbols of Π^E as the *event schemes* of a *Petri net* $N = (\Pi^C, \Pi^E; F)$. The set of net *elements* is denoted by $N = \Pi^C \cup \Pi^E$. A labeling of the event schemes and of the *flow F* of a net serves for specifying an appropriate transition rule in the following section. In contrast to [2] we admit the empty net because we want an algebraic specification to be a special case of a PrE net specification.

Definition 7 *Let Π be a PrE signature. A Π-labelled net $\mathbf{N}$ is a structure $(\Pi^C, \Pi^E; F, l)$, where $(\Pi^C, \Pi^E; F)$ is a finite (possibly empty) net with $F \subseteq (\Pi^C \times \Pi^E) \cup (\Pi^E \times \Pi^C)$ and $l = (l_E, l_F)$ is a labeling of event schemes and of the flow F such that*

1. *$l_F : F \longrightarrow \mathcal{F}(T_{\Sigma(X)})$ (where $\mathcal{F}(A)$ denotes the finite subsets of A) such that for $(\sigma_1, \sigma_2) \in F$ and forall $t \in l_F(\sigma_1, \sigma_2)$, we have that $t \in T^u_{\Sigma(X)}$ if $\sigma_1 \in \Pi^C_{u,s}$ or $\sigma_2 \in \Pi^C_{u,s}$ (with $s \in S, u \in S^*$).*

2. *$l_E : E \longrightarrow T_{\Sigma(X)}$ such that for $\sigma \in \Pi^E_{u,s}$: $l_E(\sigma) = \sigma(t)$ for some $t \in T^u_{\Sigma(X)}$ (with $s \in S, u \in S^*$).*

PrE Net Specification

A PrE net specification is now introduced as a Π-labelled net together with a set of Σ-rules.

Definition 8 *A PrE specification $SPEC$ is a quadruple $(S, \Pi, R, \mathbf{N})$ where Π is a PrE signature, R is a finite set of Σ-rules, and $\mathbf{N}$ is a Π-labelled net.*

Example

Fig. 2 gives an example of a PrE net specification. We assume the specification bintree of our first example to be included. We also assume that some operation "f" on pairs of natural numbers is specified elsewhere. The task now is to apply this operation recursively to determine the value of a tree based on the values of its left and right subtrees.

We specify an evaluation behaviour in which f is applied concurrently to the subtrees of a given tree t. The tree t is completely split up, and then f is recursively applied to the values computed for sibling subtrees starting from the values on the leaves of t. The splitting and computation processes can be partly concurrent. To preserve the knowledge about the tree structure in the subtrees split up, a data abstraction "treepath" is included. It is used for uniquely referencing the subtrees of a binary tree. According to this specification, a path and hence a subtree can be uniquely determined by a term such as "l(r(r(l(r(l(0))))))". The operation "0" denotes the empty path, i.e., it refers to the main tree; "l" denotes the left and "r" the right subtree of a given tree.

Fig. 3 gives a particular tree "mytree" as a term and its corresponding graphical representation. The result expected from evaluating mytree according to the specified strategy is indicated by the **Nat** term "myval" on which the state predicate _value_ will be defined when the computation has finished.

3 Semantics

The semantic domain for our specifications is based on many-sorted algebras and a many-sorted variant of PrE nets. We call the latter Predicate-Event systems (PrE systems) in analogy to Condition-Event systems. In [24] it is shown that each PrE system abstracts from a possibly infinite Condition-Event system. The definition of partial algebras satisfying a conditional specification is widely standard. Details can be found in [15].

3.1 Many-Sorted Partial Algebras

An algebra signed by a signature Σ, a Σ-algebra for short, provides a set of data elements for each sort in S and (possibly partial) operations for the symbols in Σ.

Definition 9 *A Σ-algebra $\mathbf{A}$ is a structure $(A; (\sigma_{A,u,s})_{s \in S, u \in S^*, \sigma \in \Sigma_{u,s}})$ providing a family of nonempty sets of carriers $(A^s)_{s \in S}$ and a partial operation $\sigma_{A,u,s} : A^u \longrightarrow A^s$ for each σ in $\Sigma_{u,s}$ where $A^{us} \stackrel{\text{def}}{=} A^u \times A^s$ and $A^\lambda \stackrel{\text{def}}{=} \{\emptyset\}$, for $s \in S$ and $u \in S^*$.*

```
sorts   path, e-state

fu   0:            -> path.
     1: path  ->   path.
     r: path  ->   path.

st   _tree_:           path tree    -> e-state.
     _value_:          path Nat     -> e-state.
ac   split up tree__:  tree path    -> e-state.
     get_at_:          Nat path     -> e-state.
     eval__at_:        Nat Nat path -> e-state.

forall T, L, R: tree; p: path; n, n1, n2: Nat.
0:   .                     1:   .
r:   .                     p tree T:   p, T.
p value n:  p, n.          split up tree T p:   T = tree(L,R), p.
get n at p:  n, p.         eval n1 n2 at p:   n1, n2, p.
case ((0) tree T):  T.
```

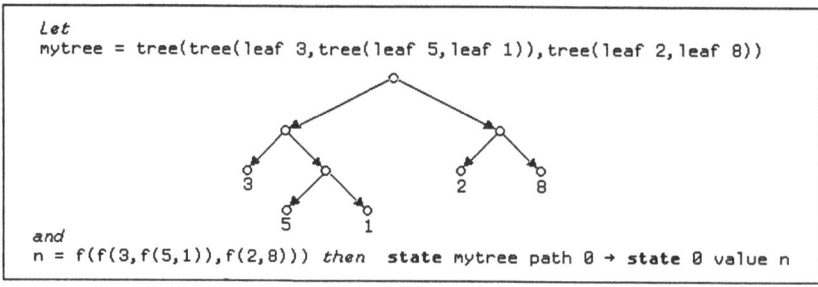

Figure 2: Concurrent evaluation of binary trees

```
let
mytree = tree(tree(leaf 3,tree(leaf 5,leaf 1)),tree(leaf 2,leaf 8))
```

```
                    3        5     1        2     8
and
n = f(f(3,f(5,1)),f(2,8))) then   state mytree path 0 → state 0 value n
```

Figure 3: A binary tree and the expected result of evaluation

Notation. We denote elements of the cartesian product $a \in A^{s_1} \times \cdots \times A^{s_n}$ by tuples $(a_1 \ldots a_n)$ where a_i is the i-th projection of a for $1 \leq i \leq n$.

Σ-algebras form a category with morphism being weak homomorphism, i.e., homomorphisms which preserve definedness but need not preserve undefinedness. We omit the details of such morphism here, although in the theory of ADT's and PrE-systems, the consistency of modular compositions and the correctness of implementation steps consist in showing injectivity and surjectivity of the homomorphisms between certain algebras.

3.2   Models of Algebraic Specifications

A conditional specification $SPEC$ has many-sorted partial Σ-algebras as models. Such models satisfy the given specification in a sense defined below. In particular, we use initial algebras as unique standard models. For defining the notion of satisfaction, we first introduce the notion of assignment of values in the model to variables in the specification. Satisfaction then leads to a class of models, called $SPEC$-algebras, and to a standard model in that class, denoted $\mathbf{P}_{SPEC}$. $\mathbf{P}_{SPEC}$ has as its elements equivalence classes of ground terms under a congruence relation generated by the conclusions of positive rules in R.

We cannot go into the details of these notions here, but refer the interested reader to [10] and the textbook on equational specifications by Ehrig and Mahr [5].

Definition 10 *Let* **A** *be a Σ-algebra. Then an* assignment *is a operation* $\varepsilon : X \longrightarrow A$ *which assigns values in* **A** *to variables in* X. *The following extension of* ε *to a partial operation* $\bar{\varepsilon} : T_{\Sigma(X)} \not\longrightarrow A$ *is called* evaluation of Σ-terms *in* **A**. *It is recursively defined by*

1. $\bar{\varepsilon}(x) = \varepsilon(x)$ *for all variables* $x \in X$

2. $\bar{\varepsilon}(\sigma) = \sigma_A$ *iff* $\sigma \in \Sigma_{\lambda,s}, s \in S$

3. $\bar{\varepsilon}(\sigma(t_1, \ldots, t_n))$ *is defined and equals* $\sigma_A(\bar{\varepsilon}(t_1), \ldots, \bar{\varepsilon}(t_n))$ *iff* $\bar{\varepsilon}(t_1), \ldots, \bar{\varepsilon}(t_n)$, *and* $\sigma_A(\bar{\varepsilon}(t_1), \ldots, \bar{\varepsilon}(t_n))$ *are all defined.*

Remark. In the limit case, where X is the empty set, $\bar{\varepsilon}$ defines a unique *ground term evaluation* in **A** [15], which we also denote ε_A.

For a subfamily $Y \subseteq X$, $T_{\Sigma(Y)}$ is a Σ-algebra and a $\Sigma(Y)$-algebra. As a special case of the above definition, we can therefore consider assignments $\varepsilon : Y \longrightarrow T_{\Sigma(Z)}$ with $Z \subseteq X$. They assign terms with variables to variables and are therefore called *substitutions*.

Definition 11 *Let* $SPEC = (S, \Sigma, R)$ *be a Σ-specification and* **A** *a Σ-algebra. We say*

1. $\mathbf{A} \models t = t'$ *(read "**A** satisfies $t = t'$") for $t, t' \in T_\Sigma^u$ with $u \in S^*$ iff $\varepsilon_A(t)$ and $\varepsilon_A(t')$ are defined and equal in* **A**

2. $\mathbf{A} \models t \neq t'$ *for $t, t' \in T_\Sigma^u$ iff not $\mathbf{A} \models t = t'$ with $u \in S^*$*

3. $\mathbf{A} \models e_0 : e_1, \ldots, e_n$.   *iff forall ground substitutions $\varepsilon : X \longrightarrow T_\Sigma$ we have that $\mathbf{A} \models \bar{\varepsilon}(e_0)$ whenever $\mathbf{A} \models \bar{\varepsilon}(e_i)$ for all $1 \leq i \leq n$*

4. $\mathbf{A} \models R$ *iff $\mathbf{A} \models r_j$ for all $r_j \in R$*

If $\mathbf{A} \models R$ *we also say that* **A** *satisfies $SPEC$ and that* **A** *is a model of $SPEC$.*

Remark. The equality we use in the above definition is also called *weak equality*. Weak equality can be used to specify (conditionally) the definedness of a term t under all assignments of variables in $vars(t)$ by the rule "$t = t : e_1, \ldots . e_n$.". We call it a *definedness rule*.

Hence, the definedness rules given in the previous examples assert the definedness of the bintree and path operations and of the net elements in each Σ-algebra which satisfies the specification.

A key result of the initial algebra approach to data structure specifications is that a particular initial[3] algebra, the quotient term algebra, can be constructed for equational specifications (cf. [10] Theorem 6), and it was shown (by Theorem 8) that it always exists. These results have been adapted in [15] to partial algebras that are specified by rules of our kind and that include multi-valued operations and distinguish between total and partial operations. It was also shown that an initial model exists for a consistent specification of that kind and that it is the initial model for the subset of positive rules of R. Hence we can formulate the following theorem:

Theorem 1 *Let* $SPEC = (S, \Sigma, R)$ *be a conditional specification and let* **PALG**$_{SPEC}$ *be the category of all partial Σ-algebras satisfying the rules in R, together with Σ-homomorphisms between them. Then* **PALG**$_{SPEC}$ *has an initial Σ-algebra.*

The initial algebra provides a framework in which those and only those equations between ground terms are valid which are satisfied by all models of a specification. The idea underlying the proof of this theorem (cf. [15]) is that the carriers are shown to consist of R-induced equivalence classes of Σ-terms. These terms must be consistently defined by the rules in R.

---

[3]Based on the notion of Σ-homomorphisms one can define an *initial* Σ-algebra in a category **PALG** of Σ-algebras to be one which belongs to **PALG** and for which there is a unique Σ-homomorphisms to any other Σ-algebra in **PALG**.

3.3 Dynamic Interpretation of Π-labelled Nets

In [24] the semantics of PrE net specifications was defined by referring explicitly to the Condition-Event system underlying a PrE net specification. It was shown that a net morphism exists from this underlying system to the PrE-net specification. This formalization has the advantage that any concept on the level of Condition-Event systems carries over from the level of the underlying system to the level of PrE-systems. Its disadvantage is that some notions cannot be expressed any more as structural properties on the level of PrE specifications. In the present text, we use many-sorted algebras to model the data associated with a net and provide a dynamic interpretation of PrE net specifications directly.

In Reisig's definition of PrE nets [23], the (one-sorted) value domain and the operations which are labeling the net are given directly by enumeration. In our method, these domains are specified and the specification can be used as a basis for symbolic computations. Their interpretation then is given by a suitable model algebra similar to the pure algebraic case. Moreover, in our PrE nets both conditions and events are labelled as we want the events being the operations that are applicable from the environment of use.

Remark. In the following definitions we speak about markings m instead of cases c to avoid notational confusion.

Definition 12 *Let* $\Pi = (\Sigma, \Pi^C, \Pi^E)$ *be a PrE signature, N be a Π-labelled net, and A be a Σ-algebra. Then*

1. *A* marking of N under A *is a family of operations* $m_{u,s} : \Pi^C_{u,s} \longrightarrow \mathcal{F}(A^u)$ *where* $s \in S, u \in S^*$.

2. *The* initial marking of N under A, m_A, *is the marking defined by*

$$\{(\sigma, a) \mid case_A(\sigma_A(a_1, \ldots, a_n)) \text{ defined where } \sigma \in \Sigma_{u,s}, a \in A^u, s \in S, u \in S^*\}.$$

3. *An* event of N under A *is a pair* (σ, a) *where* $\sigma \in \Pi^E_{u,s}$ *and* $\sigma_A(a_1, \ldots, a_n)$ *defined.*

4. *Let* $e = (\sigma, a)$ *be an event of N under A with $\sigma \in \Sigma_{u,s}, a \in A^u, s \in S, u \in S^*$. Moreover, let $^\bullet\sigma$ be the preset and $\sigma^\bullet$ be the postset of σ in N, and let $l_E(\sigma) = \sigma(t)$. Then*
   $$^\bullet e \stackrel{\text{def}}{=} \{(\sigma', a') \mid \sigma' \in {}^\bullet\sigma, \exists t' \in l_F(\sigma', \sigma) \text{ and } \exists \varepsilon : X \longrightarrow A \text{ an assignment such that } \bar{\varepsilon}(t') = a' \wedge \bar{\varepsilon}(t) = a\}$$
   $$e^\bullet \stackrel{\text{def}}{=} \{(\sigma', a') \mid \sigma' \in \sigma^\bullet, \exists t' \in l_F(\sigma, \sigma') \text{ and } \exists \varepsilon : X \longrightarrow A \text{ an assignment such that } \bar{\varepsilon}(t') = a' \wedge \bar{\varepsilon}(t) = a\}$$
   For a set of events Π^E this notation is extended to $^\bullet\Pi^E$ and $\Pi^{E\bullet}$ in the usual way.

5. *Given two markings m_1, m_2 and a set Π^E of events of N under A. Then we say that m_2 is reachable from m_1 by Π^E in one step, written $m_1[\Pi^E\rangle m_2$, if the following conditions hold for $e, e_1, e_2 \in \Pi^E$:*

   (a) conflict free: $^\bullet e_1 \cap {}^\bullet e_2 \neq \emptyset \Rightarrow e_1 = e_2$ *and* $e_1^\bullet \cap e^\bullet \neq \emptyset \Rightarrow e_1 = e_2$

   (b) concession: $^\bullet e \subseteq m_1$ *and* $e^\bullet \cap m_1 = \emptyset$

   (c) change: $m_2 = (m_1 \backslash \bigcup_{e \in \Pi^E} {}^\bullet e) \cup \bigcup_{e \in \Pi^E} e^\bullet$

Notation. We also use the equivalent set representation of a marking m and write $(\sigma, a) \in m$ to express that $a \in m(\sigma)$, for $\sigma \in \Sigma_{u,s}$.

Note that different state predicates and event schemes give rise to different conditions (elements of a marking) and events. Congruences of Σ-terms lead to congruences of the corresponding conditions and events.

An example for a graphical representation of a marking is depicted in fig. 4.

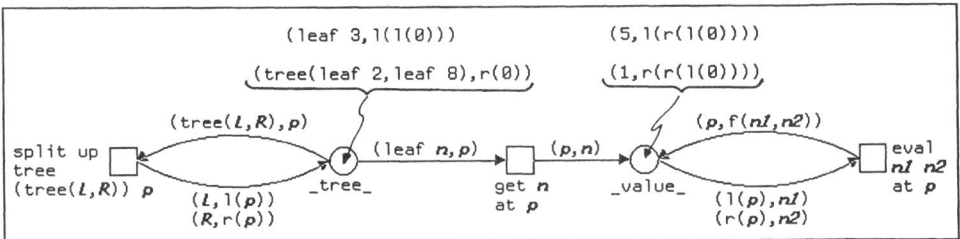

Figure 4: A The Π-labelled net of fig. 2 in some state of computation

Processes of PrE Nets

We now come to defining the behavior of a Predicate-net specification in terms of the processes that may occur on its Π-labelled net. As usual, occurrence nets are taken as domains of nonsequential processes of PrE nets.

Definition 13 *Let* $\Pi = (\Sigma, \Pi^C, \Pi^E)$ *be a PrE-signature,* $(S, \Pi, R, \mathbf{N})$ *a PrE-net specification,* $\mathbf{A}$ *a* Σ-*algebra,* $N_A \overset{\text{def}}{=} \{(\sigma, a) \mid \sigma \in \Pi^C_{u,s} \cup \Pi^E_{u,s}, a \in A^u, s \in S, u \in S^*\}$ *and let* $\mathbf{ON}$ *be an occurrence net (cf. [2]).* *Then the operation* $p : ON \longrightarrow N_A$ *is called a* process *of* $\mathbf{N}$ *iff it satisfies the following conditions:*

1. *for each event* $e \in ON$, $p(e)$ *is an event of* $\mathbf{N}$ *under* $\mathbf{A}$;

2. *for each* Π^C-*cut* B *of* ON : $p \mid_B$ *is injective;*

3. $p(^\bullet e) = \,^\bullet p(e)$ *and* $p(e^\bullet) = p(e)^\bullet$ *for all events* e *of* ON;

4. $p(^\bullet e) \cap p(e^\bullet) = \emptyset$.

ON *is called the* process domain *of* $\mathbf{N}$ *under* p.

Notation. In the graphical representation of a process p we draw the process domain as a net and label its elements n by $p(n)$.

Example

Let s_t denote some event (split up tree$\_,(t,p)$) of $\mathbf{N}$, the net in fig. 2, under some algebra $\mathbf{A}$ which has t as an element of the carrier A^{tree} and $p \in A^{path}$. Similarly, let g_n denote an event (get$\_$at$\_,(n,p)$) and $e_f(n1, n2)$ denote some event (eval$\_$at$\_,(n1, n2, p)$) of $\mathbf{N}$ under $\mathbf{A}$. Then an example of a process of $\mathbf{N}$ is shown in fig. 5.

3.4 Predicate-Event Systems

A model of a PrE net specification is a many-sorted algebra that satisfies elementary net theoretic conditions.

Definition 14 *Let* $\Pi = (\Sigma, \Pi^C, \Pi^E)$ *be a PrE-signature. A* Predicate-Event System *signed by* Π *(*Π-system, for short) is a Σ-algebra $\mathbf{A}$ *satisfying the following two conditions:*

1. *for all* $\sigma \in \Pi^C_{u,s}, \sigma' \in \Pi^E_{v,s}, a \in A^u, a' \in A^v$ *with* $s, s' \in S, u, v \in S^*$: $\sigma_A(a) \neq \sigma'_A(a')$, *i.e., the sets of condition and the set of event instances of the same sort must be disjoint;*

2. *for all* $\sigma \in \Pi^C_{u,s}(\sigma \in \Pi^E_{u,s}), a \in A^u$ *with* $\sigma_A(a)$ *defined* $(s, s' \in S, u \in S^*)$: *there is a* $\sigma' \in \Pi^E_{v,s'}(\sigma \in \Pi^C_{v,s'}), a' \in A^v, v \in S^*$ *such that* $flow_A(\sigma_A(a), \sigma'_A(a'))$ *defined. That is there are no "isolated" condition and event instances.*

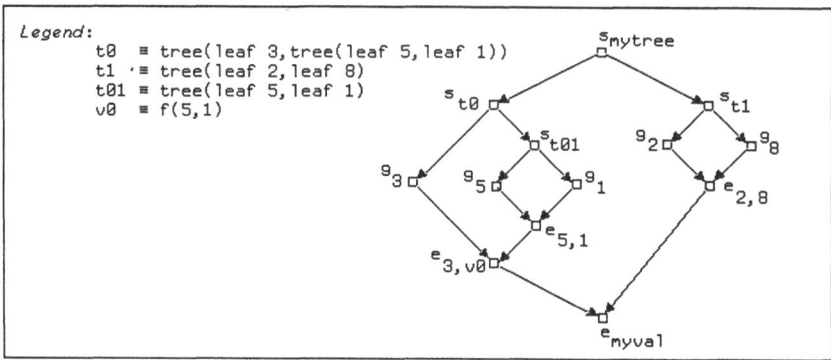

Figure 5: A process of fig. 2

3.5 Models of PrE-Specifications

Now we can explain how the syntactic domain of PrE-net specifications relates to the semantic domain of many-sorted algebras.

Definition 15 *Let* $\Pi = (\Sigma, \Pi^C, \Pi^E)$ *be a PrE signature and* $SPEC = (S, \Pi, R, N)$ *be a PrE net specification. The a PrE-system* **A** *satisfies* $SPEC$ *iff the following conditions hold:*

1. $\mathbf{A} \models R_{SPEC}$

2. *Given an event* $e = (\sigma, a)$ *of* **N** *under* **A**. *Then we have that*
   $(\sigma', a') \in {}^\bullet e$ *iff* $flow_A((\sigma'_A(a'), \sigma_A(a)))$ *defined and*
   $(\sigma', a') \in e^\bullet$ *iff* $flow_A((\sigma_A(a), \sigma'_A(a')))$ *defined.*

Theorem 2 *Let* $SPEC = (S, \Pi, R, N)$ *be a consistent PrE net specification. Then there is an initial model of SPEC in the category of PrE-systems satisfying SPEC.*

In analogy to the pure algebraic case, we denote this initial model by $\mathbf{P}_{SPEC}$.
    The following list of points sketches the proof (cf. [24]):

1. First assume that the preconditions of the theorem are fulfilled, i.e. there exists at least one model of SPEC.

2. The net **N** of SPEC is then translated into a set of rules.

3. Combined with R, the resulting set of rules defines an inital algebra according to theorem 1.

4. This algebra is shown to satisfy the points required for PrE-system by showing that these are implications of the rules.

5. Finally, this algebra is shown to be initial. That is it is demonstrated that any model of SPEC satisfies the rules resulting from the translation.

    As an implication of this proof, the following corollary can be formulated.

Corollary 1 *Let* $SPEC = (S, \Pi, R, N)$ *be a PrE net specification. Then* $SPEC$ *can be translated into an equivalent algebraic specification in the sense that the induced congruences uniquely define the class of PrE systems satisfying SPEC.*

The interesting point about this fact is that PrE systems can be viewed as algebras with some operators distinguished via the PrE signature. Some of the operators are used to identify a variable state of affairs, some are used to identify the changes in states, and the rest describes invariant (structural) properties of the system. Nevertheless all of them can be used in a pure algebraic way to analyze consistency or implementation correctness and to study the use of algebraic verification techniques for concurrent systems.

4 Modularization

For large systems and data structures we may not want to give their entire specification at once. Rather we may find it easier to extend given specifications or to combine them systematically to form a larger system which possibly introduces additional synchronization or extends the freedom of behavioral choice. Synchronization is increased when high level events in the interface of different components combined can be identified, while freedom of choice is increased when high level conditions of different component interfaces can be identified.

SEGRAS offers various features that allow structuring specifications: combination, parameterization, abstract implementation, and scoping. In the framework of ADT's, these concepts have been given an algebraic semantics. As for *SEGRAS* we adopted these concepts and extended them by restrictions on behavioral properties of combined specifications. These behavioral properties are based on the net theoretic notion of processes of PrE-systems and ensure a kind of behavioral consistency and completeness.

Here we confine ourselves to the notion of modularization of PrE-specifications by means of combination and extension.

4.1 Combination and Extension of Specifications

Combinations provide a way of building specifications on existing ones by adding new sorts, operations, rules, and nets to a given specification or by putting existing specifications together. The user of our specification language can view this union as a "glueing" of the labelled nets in the graphical representation. Extension is a specific kind of combination by which the semantics of the extended specifications is preserved. Extendable specifications can be verified once and for all and can be implemented independently from their various extensions as the carrier sets of their models are protected against modifications.

Definition 16 *Let $SPEC = (S, \Pi, R, \mathbf{N})$ and $SPEC' = (S', \Pi', R', \mathbf{N}')$ be two PrE-specifications.*

1. *We call $SPEC$ a subspecification of $SPEC'$ if $S \subseteq S', \Pi \subseteq \Pi', R \subseteq R'$, and $\mathbf{N}$ is a subnet of $\mathbf{N}'$ defined by:*

   (a) $\Pi^C \subseteq \Pi^{C'}$ and $\Pi^E \subseteq \Pi^{E'}$

   (b) $l_E(e) = l'_E(e)$ forall $e \in \Pi^E$ and $l_F(f) \subseteq l'_F(f)$ forall $f \in F$

2. *Let $SPEC$ be a subspecification of $SPEC'$. Then $SPEC'$ is called a combination if $SPEC' = (S \cup S'', \Pi \cup \Pi'', R \cup R'', \mathbf{N} \cup \mathbf{N}'')$ such that $S \cap S'' = \Sigma \cap \Sigma'' = \emptyset$ and $\mathbf{N} \cup \mathbf{N}'' = (\Pi^C \cup \Pi^{C''}, \Pi^E \cup \Pi^{E''}; F \cup F'', l \cup l'')$ where $l_F \cup l''_F(f) \stackrel{\text{def}}{=} l_F(f) \cup l''_F(f)$ forall $f \in F'$. If $SPEC'$ is consistent, there exists a unique morphism $f_{SPEC'} : \mathbf{P}_{SPEC} \longrightarrow \mathbf{P}_{SPEC'}$.*

3. *If $SPEC'$ is a combination and if the initial $SPEC$-algebra $\mathbf{P}_{SPEC}$ is congruent to the Σ-reduct of $\mathbf{P}_{SPEC'}$, written $\mathbf{P}_{SPEC} \cong \mathbf{T}_{SPEC'|\Sigma}$, then $SPEC'$ is called an extension of $SPEC$.*

Remark. In the Σ-reduct of a Σ'-algebra with $\Sigma' = \Sigma \cup \Sigma''$ we forget the sorts in S'', the operation symbols in Σ'', and all data elements which cannot be expressed without these symbols in Σ''.

Note that, if $SPEC'$ is an extension, then the morphism $f_{SPEC'}$ is injective and surjective on the carriers of S.

This definition degrades to the purely algebraic case if $\mathbf{N}' = \emptyset$.

In the categorical treatment of PrE systems and their underlying Condition-Event systems (CE systems), combination corresponds to the disjoint union of possibly infinite CE nets modulo some congruence of net-elements. In an extension, the CE-net underlying $\mathbf{P}_{SPEC}$ is isomorphic to a subnet of the CE-net underlying $\mathbf{P}_{SPEC'}$. For a behavioural extension moreover, the CE-system (i.e. the CE-net plus a selected behaviour in terms of markings and reachability) underlying $\mathbf{P}_{SPEC'}$ additionally "carries" all the processes of $\mathbf{P}_{SPEC}$.

With the above requirements we may ensure consistency and completeness of extensions w.r.t. the terms denoting data and net elements. However, we have not yet ensured similar properties for the processes of $SPEC$ and $SPEC'$. It is still possible that $SPEC$ has processes which are not mapped on processes of $SPEC'$. It is also possible that processes of $SPEC'$ change the marking of the subnet N in N' such that there is no process in $SPEC$ which affects the same change on N.

Therefore we refine our previous definition by two further requirements.

Definition 17 *Let $SPEC' = SPEC + (S'', \Pi'', R'', N'')$ be an extension of $SPEC$. Then $SPEC'$ is called a behavioural extension*

1. *if for every process p with occurrence net ON of $\mathbf{P}_{SPEC}$ can be embedded into some process p' with occurrence net ON' of $\mathbf{P}_{SPEC'}$ by an injective operation $inj : ON \longrightarrow ON'$, such that,*

$$f_{SPEC'}(p(a)) = p(inj(a)).$$

This means, each process in the behaviour specified by $SPEC$ is the Σ-reduct of some process in the behaviour specified by $SPEC'$.

2. *forall markings m_1, m_2 of N and forall processes p of N sucht that $m_1[p\rangle m_2 \Rightarrow m_{1|\Sigma}[p_{|\Sigma}\rangle m_{2|\Sigma}$.*

Notation. $m_1[p\rangle m_2$ denotes the extension from steps to processes.

5 Example: Extension and Combination of PrE-net Specifications

In this section we give some small examples to illustrate the concepts defined.

Storage Cell

Assume an algebraic specification of strings, called string, which provides the sort string and appropriate operations and rules for a monoid over an alphabet. Assume further a specification "length" giving a constant operation "length" of sort **Nat**. Then objects of sort cell each of which may keep arbitrary string values are specified by the PrE net specification cell in fig. 6.

The above specification tells us that an arbitrary system C of sort cell exchanges data of sort string with its environment by means of two actions. This is defined by the terms "put S in C" and "get S from C" where C is a term denoting the data exchanged. The net defines the (distributed) state of C before and after communication has taken place. It also tells us that in case of an occurrence of action "put_in_" S is an input to C because it is undefined in the state before such an occurrence. For an occurrence of "get_from_", however, S is an output of C because it is determined by the state before such an occurrence.

Concurrent Array

A group of cells can be put together to form a kind of array which can be accessed concurrently when different indeces are used at a time (cf. fig. 7).

To illustrate the fact that this specification is a behavioral extension of the first, a typical process is given in fig. 8.

281

```
spec string-cell extends string by

    sorts   cell, c-ac, c-st

    fu      cell:       Nat -> cell.
    ac      put_in_:    string cell -> cell-ac.
            get_from_:  string cell -> cell-ac.
    st      _empty:     cell -> cell-st.
            _in_:       string cell -> cell-st.

    forall C: cell; S: string.
    case (C empty): C.
    S in C:  S, C.                    C empty: C.
    put S in C:  S, C.                get S from C:  S, C.
```

_in_

(S,C)   (S,C)

put S in C   get S from C

S   S

_empty

Figure 6: Specification of a cell able to carry strings

```
spec concurrent-array extends string-cell, length by

    sorts   array

    fu      _sub_:        array Nat -> cell.
    fu      make-array_:  cell -> array Nat.

    forall A: array; i: Nat.
    make-array (cell i) = (A,i):   i < length.
    A sub i = cell i : A.
```

Figure 7: Specification of a concurrent array

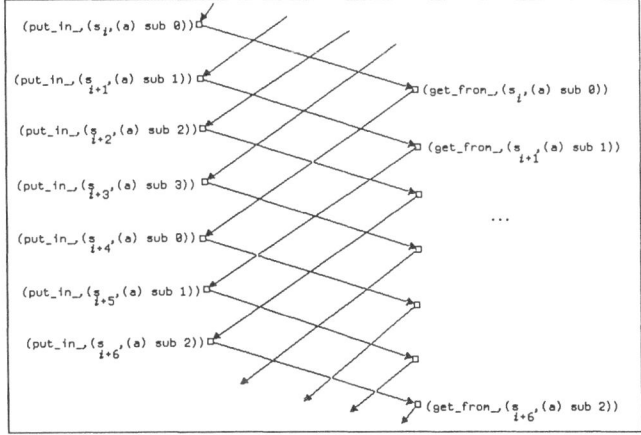

Figure 8: A typical array process

Bounded Ring Buffer

When combining the previous specification with additional behavioral constraints as shown in fig. 9, we arrive at the specification of a ring buffer.

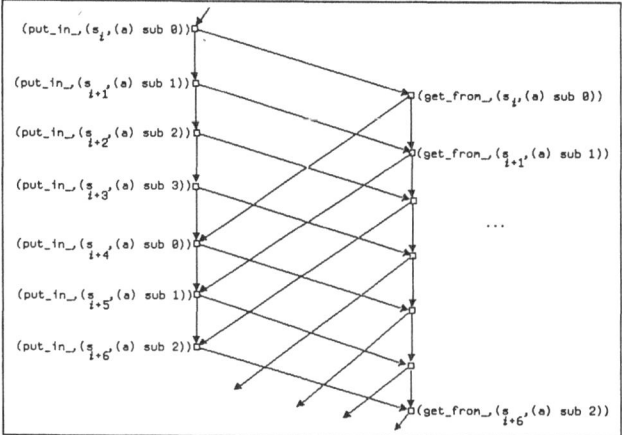

```
spec ring-buffer combines concurrent-array with

    sorts  head-ptr, tail-ptr

    st     _head:  cell -> head-ptr.
    fu     _tail:  cell -> tail-ptr.

    case ((cell 0) head): .
    case ((cell 0) tail): .
    forall A: array; i: Nat.
    (A sub i) tail: A.
    (A sub i) head: A.
    forall array; i: Nat; S: string.
```

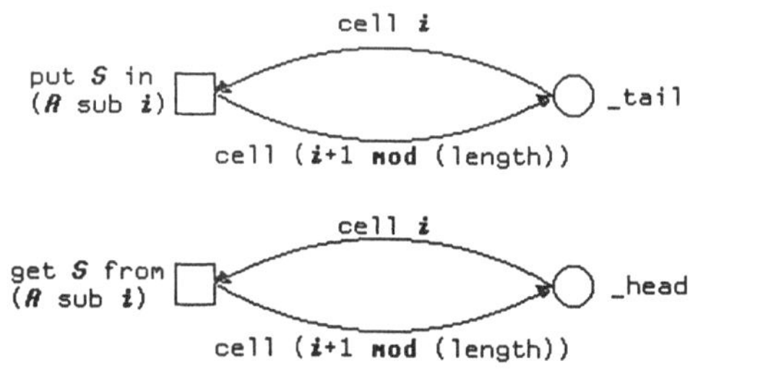

Figure 9: Specification of a concurrent ring buffer

It still allows concurrent put and get operations but all put and all get operations are put in a certain order to achieve FIFO properties. As we also identify the first and the last cell of or buffer by means of the modulo (mod) operation on **Nat**, we have a combination but no extension of concurrent array. Again, a typical process is shown in fig. 10.

Figure 10: A typical ring buffer process

Note that in the full language this example can be simplified by parameterization and it can be made more readable by taking advantage of renaming concepts.

Discrete Pipeline

Finally, we show another example of how FIFO properties can be imposed on a concurrent array. In the shift buffer in fig.11 the put and get operation of any two subsequent cells are identified.

```
spec shift-buffer combines concurrent-array with

    forall A: array; i: Nat; S: string.

    put S in (A sub i) = get S from (A sub (i+1)):
                            put S in (A sub i),
                            get S from (A sub (i+1)),
                            i < ((length) - 1).
```

Figure 11: Specification of a concurrent shift buffer

The idea underlying this specification is that a string produced is put in the last cell of the buffer and passes along the cell sequence to be gotten from a consumer when it arrived in the first cell (see also fig. 12).

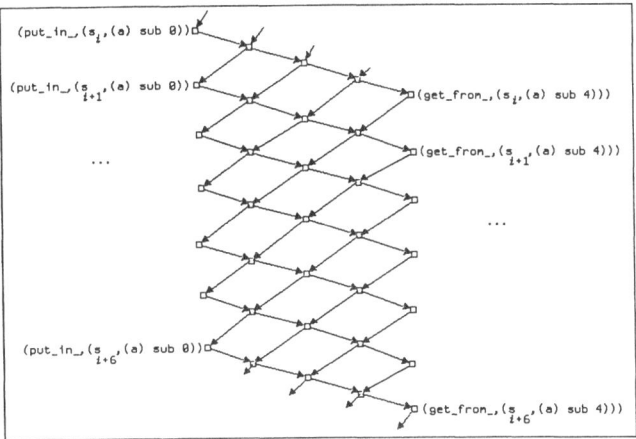

Figure 12: A typical shift buffer process

Conclusions and Future Research

Many practically useful concepts of the S&GRAS language have not been treated in this paper: multi-valued operations (i.e., operations with nontrivial coarity), higher order specifications, polymorphism, subsorts, parameterized specifications, and abstract implementation. The intention was to study our combination of partially specified abstract data types and Predicate-Event nets in the smallest possible framework. However, there are various efforts extending the algebraic theory of abstract data types to handle such concepts. For example, the definition of partial algebras given by Kreowski and Schmidt in [15] already deals with nontrivial coarity of operations. Möller provided a definition of higher order specifications and algebras [20]. Poigné gave a formalization of two-level specifications [22] which allows him to handle polymorphism and higher order types. Our notion of polymorphism relies on multi-leveled specifications which can be considered a generalization of the two-level case. The formal treatment of subsorts, parameterization, and abstract implementation is nearly classics in the algebraic specification framework.

The small examples given in this paper might have shown the potential of the unified framework of many-sorted algebraic and net theoretic specifications. But they might also have raised the demand for a less restrictive methodology and notation to be of practical use. The concrete syntax of the *SEGRAS* specification language [12] is an attempt to provide as much programming power as possible within the framework. The idea is to achieve this by displacing much of the burden of incrementally constructing and manipulating specifications into support tools. Such tools can also supplement omissions by default and may survey the observance of context conditions in a tolerant way.

SEGRAS is designed for interactive use. It is supported by a programming environment, called the *SEGRAS* Lab, which currently consists of a *syntax-directed editor* including the graphical representation of PrE nets[4] a grammar driven database, and a polymorphic typechecker.

Rewrite rules which can be derived from conditional algebraic specifications and the transition rule defined for PrE-systems provide an operational semantics for the language. This semantics can be used to test specifications for adequacy and consistency with their intended initial semantics. In the GRASPIN project we are adapting and integrating a rewrite rule engine into the Lab to support extension checking, consistency and completeness checks for algebraic specifications. An interactive net simulator is under development to provide an interpreter for PrE system specifications. It will permit to observe the token flow through a net on the graphical representation. As tokens are denoted by terms, the rewrite engine will supply its capabilities to rewrite the tokens according to the flow labeling and the behavioral constraints defined.

It is still an open problem how the requirements of behavioral extension can be checked, in general. One solution might be to translate processes into algebraic terms and then to apply the usual algebraic machinery. This was proposed in [25]. Another solution might be to find equivalent definitions of these requirements such that net-theoretic machinery can be applied for checking. This is one of the research topics of the GRASPIN project.

Acknowledgments

We are grateful to Harald Fonio and the 'Advances' referee for their careful reading of this manuscript and valuable suggestions for its improvement.

References

[1] B. Berthomieu, N. Choquet, C. Colin, B. Loyer, J. Martin, and A. Mauboussin. Abstract data nets combining Petri nets and abstract data types for high level specifications of distributed systems. In *Proceedings of the 7th European Workshop on Applications and Theory of Petri Nets*, pages 25–48, Oxford, England, July 1986.

[2] E. Best and C. Fernàndez. Notations and terminology on Petri net theory. *Petri Net Newsletter*, 23:21–46, April 1986.

[3] M. Broy and M. Wirsing. Partial abstract types. *Acta Informatica*, 18:47–64, 1982.

[4] M. Diaz. Modelling and analysis of communication and cooperation protocols using Petri net based models. In C. A. Sunshine, editor, *Protocol Specification, Testing and Verification*, pages 465–510, North-Holland Publishing Company, 1982.

---

[4] All examples in this paper have been produced with this component of the *SEGRAS* Lab.

[5] H. Ehrig and B. Mahr. *Fundamentals of Algebraic Specification 1.* Volume 6 of *EATCS Monographs on Theoretical Computer Science*, Springer-Verlag, Berlin, Heidelberg, New York, Tokyo, 1985.

[6] K. Futatsugi, J. A. Goguen, J. Jouannaud, and J. Meseguer. Principles of OBJ2. In *Conference Record of the Twelfth Annual ACM Symposium on Principles of Programming Languages*, pages 52–66, Louisiana, New Orleans, 1985.

[7] H. J. Genrich and K. Lautenbach. System modelling with high-level Petri nets. *Theoretical Computer Science*, 13(1):109–136, 1981.

[8] S. L. Gerhart, D. Musser, D. Thompson, D. Baker, R. Bates, R. Erickson, R. London, D. Taylor, and D. S. While. An overview of Affirm: a specification and verification system. In S. Lavington, editor, *INFORMATION PROCESSING 80*, pages 343–347, IFIP, North-Holland Publishing Company, 1980.

[9] J. A. Goguen and J. Meseguer. EQLOG: equality, types and generic modules for logic programming. In D. DeGroot and G. Lindstrom, editors, *Functional and Logic Programming*, pages 295–363, Prentice-Hall, 1986.

[10] J. A. Goguen, J. W. Thatcher, and E. G. Wagner. An initial algebra approach to the specification, correctness, and implementation of abstract data types. In R. T. Yeh, editor, *Current Trends in Programming Methodology*, pages 80–149, Prentice-Hall, Englewood Cliffs, New Jersey, 1978.

[11] B. Krämer. *SEGRAS* – a formal language combining Petri nets and Abstract Data Types for specifying distributed systems. In *Proceedings of the 9th Annual International Conference on Software Engineering*, pages 116–125, Monterey, California, March 1987.

[12] B. Krämer. *SEGRAS* – *The GRASPIN Specification Language. GRASPIN Technical Paper*, Sankt Augustin, July 1986.

[13] B. Krämer. Stepwise construction of non-sequential software systems using a net based specification language. In G. Rozenberg, editor, *Advances in Petri nets 1984*, pages 307–327, Springer-Verlag, Berlin, Heidelberg, New York, Tokyo, 1985.

[14] B. Krämer and H. Schmidt. An approach to algebraic specification and stepwise implementation of non-sequential systems. In *Poster Session Proceedings of the 6th International Conference on Software Engineering*, pages 63–64, Information Processing Society of Japan, Tokyo, Japan, September 1982.

[15] H. Kreowski and H. Schmidt. *Some Algebraic Concepts of the Specification Language SEGRAS and their Initial Semantics. GMD-Studien*, Gesellschaft für Mathematik und Datenverarbeitung, October 1984.

[16] L. Lamport. On interprocess communication, part I. *Distributed Computing*, 1:77–85, 1986.

[17] L. Lamport. Specifying concurrent program modules. *ACM Transactions on Programming Languages*, 5(2):190–222, April 1983.

[18] B. Mahr and J. Makowsky. Characterizing specification languages which admit initial semantics. *Theoretical Computer Science*, 31:49–60, 1984.

[19] R. A. Milner. *A Calculus of Communicating Systems. Lecture Notes in Computer Science*, Springer-Verlag, Berlin, Heidelberg, New York, 1980.

[20] B. Möller. Algebraic specification with higher-order operators. In L. Meertens, editor, *Proceedings of the IFIP TC 2 Working Conference on Program Specification and Transformation*, North-Holland, Amsterdam, to appear.

[21] A. Pnueli. The temporal logic of programs. In *Proceedings of the 18th Annual Symposium on Foundations of Computer Science*, pages 46–57, Providence, October 1986.

[22] A. Poigné. On specifications, theories, and models with higher types. *Information and Control*, 68(1–3), January, February, March 1986.

[23] W. Reisig. *Petri Nets*. Volume 4 of *EATCS Monographs on Theoretical Computer Science*, Springer-Verlag, Berlin, Heidelberg, New York, Tokyo, 1985.

[24] H. Schmidt. *Towards a Net-Theoretic Notion of Type based on Predicate-Transition Nets*. Arbeitspapiere der GMD 117, Gesellschaft für Mathematik und Datenverarbeitung, November 1984.

[25] H. Schmidt and M. Papazoglou. *Abstract Implementation of Predicate-Event Systems*. Sankt Augustin, August 1985.

[26] J. Vautherin. Parallel systems specifications with colored Petri nets and algebraic abstract data types. In *Proceedings of the 7th European Workshop on Applications and Theory of Petri Nets*, pages 5–23, Oxford, England, July 1986.

AN INTRODUCTION TO THE MACRO COSY NOTATION

P.E. Lauer and R. Janicki
Department of Computer Science and Systems, McMaster University
Hamilton, Ontario, Canada, L8S 4K1

Abstract

One of the objections to the use of Petri nets or, equivalently, specifications written in the COSY (Concurrent Systems) notation, for modelling realistic systems is that they would grow too large to be of any practical use. Generators for the concise representation of large (possibly infinite) structures in net theory or specifications in COSY are traced to their origins in Carl Adam Petri's thesis [P62]. The generators implemented in the current version of COSY and its accompanying simulation and analysis tools are presented in detail together with design decisions which led to their present form.

1. Introduction

COSY (Concurrent Systems) notation is a formalism for specifying the synchronization properties of concurrent systems by means of collections of synchronized grammars developed at the University of Newcastle upon Tyne by Roy Campbell and Peter Lauer in 1974. The formal semantics of such collections of grammars was modelled by equivalent labelled 1-safe nets and their corresponding labelled occurrence nets by standard net semantics. Subsequently the so-called vector firing sequence semantics due to Mike Shields [S79] was used to define COSY semantics directly without recourse to Petri nets. However, great care was taken never to vitiate the standard net semantics for COSY in all subsequent developments. Hence, there always remained a close relationship between results and developments in net theory and COSY.

This introduction briefly delineates the origin of recursive generators in Petri's thesis [P62] and traces some of their subsequent development by H. Genrich, P. Torrigiani, W. Reisig (GMD) and P. E. Lauer, R. Devillers, J. Cotronis, P. Wong (COSY group). The second chapter gives a short overview of the basic COSY formalism and its vector firing sequence semantics. The bulk of the paper presents the version of generators incorporated in the current implementation of COSY as developed by J. Cotronis [C82] from a set of criteria, requirements and design ideas proposed by P. E. Lauer [L79].

In early applications of <u>path expressions</u>, which form the heart of COSY, to realistic synchronization problems, Roy Campbell and Peter Lauer felt the need to define <u>collective names</u> for collections of events of the same type, where indices served to differentiate names of individual events in a single collection. These collections were thought of initially in two ways: as <u>indexed sets</u>, and as <u>multidimensional arrays</u>.

The latter of the two approaches seemed more appropriate for our purpose of distributing regular connectives over array dimensions to obtain short forms for expressing simple series of synchronization constraints.

For example, having defined a collection of events as a one-dimensional array by stating: **array** series(1:5) **endarray**, writing the <u>distributor</u>:

        ;[series]

is the same as writing the <u>sequence</u>:

        series(1);series(2);series(3);series(4);series(5)

which asserts that the events series(i) must occur strictly in the sequence i=1,2,3,4,5. Similarly when we replace ";" by "," in the above distributor.

Having defined a collection of events as a two-dimensional array by stating: **array** matrix(1:3,1:3) **endarray** and writing the <u>nested distributor</u> ;[,[matrix]] is the same as writing the sequence containing orelements:

        matrix(1,1),matrix(2,1),matrix(3,1)
        ;matrix(1,2),matrix(2,2),matrix(3,2)
        ;matrix(1,3),matrix(2,3),matrix(3,3)

which asserts that the events matrix(i,j) form a process consisting of occurrences of any one event from each column, where the columns are chosen sequentially in the order 1 through 3. In this way one can easily define many regular ascending integral series of event occurrences structured as multidimensional arrays in a concise way.

This was the point Campbell and Lauer had reached in [LC75]. In the same paper the two authors also gave a formal semantics of path expressions by specifying translation rules from path programs to equivalent labeled 1-safe nets. Soon thereafter Carl Adam Petri invited Peter Lauer to visit his group at the GMD to consider some collaborative effort between Lauer at the University of Newcastle upon Tyne and Petri's group. Conversations between Hartmann Genrich, Piero Torrigiani and Lauer concerning the expressive power of the <u>macro path notation</u> using <u>collective names</u> and <u>distributors</u>, led to an initial formulation of the <u>replicator notation</u> which subsequently was developed into the <u>full macro notation</u> which forms the basis of this paper. We will briefly recount our reasoning at that time.

Using the distributor notion together with the ability to define simple ascending sets of integer indices to characterise multidimensional arrays of events does not allow one to specify:

series(5);series(4);series(3);series(2);series(1) or

series(5);series(3);series(1)

in concise shorthand. Hence, we were seeking some device allowing us to express such series which would naturally fit with the notation we already had. Hartmann Genrich suggested that we use a <u>copy operator</u>, perhaps of the form:

[string possibly containing index i $\boxed{i}$ |initial,final,step]

which would be interpreted as "produce a copy of the string with the initial value of the index substituted for every occurrence of i in the string, and concatenate the result with another copy of the string with occurrences of i replaced by the value of the previous value of i plus the step as long as the value of final is not exceeded, otherwise stop."

This idea seems to have been suggested by the <u>recursive copy operator</u> from Petri's thesis [P62,p.24] where he writes, loosely translated: R means "A copy of the shortest ∇-subformula, in whose second sub-formula the R-instance occurs, replaces that instance of R. Interpretation of the result continues with the first symbol of the substituted ∇-formula."

Petri points out that with this rule: *S = ∇S/SR , where *S is the Kleene star prefixed to the nonterminal S, ∇ denotes alternation, and / denotes sequentialization. Actually, according to our reading of this definition it is not a definition of the Kleene star but of the "+".

In this approach the only type of string that can be substituted is a ∇-formula since there is no other way to indicate context. In Genrich's suggestion any string can be indicated as possible substitute due to the use of special parentheses which are associated with a particular occurrence of R (which is now replaced by the box which was called a <u>place-holder</u> at that time). In addition he suggested the for-statement like manner of ensuring termination and the use of the expressions in the for statement like control to generate successive values for the occurrences of indices in the string substituted. Using this initial suggestion it was then possible to write the <u>replicators</u>:

[series(i);$\boxed{i}$|5,1,-1]

[series(i);$\boxed{i}$|5,1,-2]

to express:

series(5);series(4);series(3);series(2);series(1);

series(5);series(3);series(1);

respectively, which are the same as the examples which could not be expressed with the distributors introduced by Campbell and Lauer above, except for the trailing semicolons. To eliminate the trailing semicolons Torrigiani and Lauer used "@" as an operator which strips the last character of a string after the final substitution, allowing them to express the examples which were beyond the power of distributors as the following replicators with the @-operator: [series(i)@; ⬚i |5,1,-1] and [series(i)@; ⬚i |5,1,-2].

In addition to the problem of obtaining succinct shorthand for expressing any ascending and descending series of indices for collections of events, there was the problem of expressing structures of events which in some sense could grow in some regular manner from both sides, as is for instance required for the specification of a bounded stack with 4 cells:

(push(1);(push(2);(push(3);(push(4);pop(4))*;pop(3))*;pop(2))*;pop(1))*.

To express this they had to allow the placeholder "⬚i" to occur anywhere in the string to be copied, and require that exactly one @-operator would occur immediately to the left or the right of the placeholder. This allowed them to express a stack of n cells by the imbricator:

[(push(i)@; ⬚i ;pop(i))* |1,n,1] .

This is essentially the point they reached during their discussions with Hartmann Genrich at that time. Piero Torrigiani and Peter Lauer continued to develop this notation during the rest of Peter's visit with Petri's group and published a number of papers presenting the system notation, an even more powerful macro notation for defining systems as abstract objects instantiated from a SIMULA class like construct [TL77, LTS79]. Furthermore, extended use of the replicator in collective name definitions, and the nesting of paths in replicators yielded considerably more powerful shorthands for path programs. For example a fifo buffer with n cells could be written:

array deposit remove(1:n) **endarray**
[path deposit(i);remove(i) **end** ⬚i |1,n,1]
path ;[deposit] **end**
path ;[remove] **end**

which expands for n=3 to :

path deposit(1);remove(1) **end**
path deposit(2);remove(2) **end**
path deposit(3);remove(3) **end**
path deposit(1);deposit(2);deposit(3) **end**
path remove(1);remove(2);remove(3) **end**

where at the present the reader needs only to see how the expansion takes place. The semantics of the expanded programs will become clear after the section on basic COSY which follows.

We will take up the discussion of further developments of the macro notation by the Newcastle group at a later point. We conclude this historical sketch of the beginnings of this notation by sketching two further developments of these seminal ideas by Genrich and later by Wolfgang Reisig.

Hartmann Genrich used the notion of placeholder in a lecture at the Technical University, Berlin, in November 1975 on the topic of "Logic of Planning, Delegating and Acting". In [G75] he writes:

  □ : is a placeholder for a sub-plan determined by the position of □ .

Thus

  □ : indicates that the execution of an occurrence of □ consists of the substitution and execution of the shortest subplan enclosed by the parentheses "<" and ">" which enclose that occurrence.

  For example he writes : X.<Y.□.Z>.U ≡ X.<Y.<Y.□.Z>.Z>.U .

To terminate the recursion he uses a conditional of the form: [g:X,Y] and defines conditional iteration as : [g*X] := <[g:[X/□],T> where "T" is the trivial plan, and "/" denotes sequential performance. As long as g is satisfied a copy of [g:[X/□],T] is substituted for the placeholder □ . This gives some idea of Genrich's development of Petri's recursive copy operator to allow the place holder to occur anywhere in the string to be copied, and to indicate the range of the copy operator independent of a particular syntactic entity, like the ∇-formula of Petri. Genrich was not particularly concerned with the explicit generation of arbitrary index sets over event types. Hence it remained for the Newcastle group to continue in this direction.

A copy operator applicable directly to nets which resembled the replicator of the Newcastle group was mentioned in [Br80] p.523 and a recursive substitution operator involving interesting ways of constraining the recursion which resemble our so called predicate replicators [LTD80], but also allows a use similar to our distributors was also given.

Wolfgang Reisig also introduced a notion of recursive nets [Re82] to obtain finite characterisations of infinite nets, a problem the Newcastle group tackled in [SL77], by allowing subnets of C/E-systems to be substituted for single events.

Events where substitution is to take place are indicated by a label of the subnet to be substituted, and termination is achieved by reaching a system in which there are no more labeled transitions. This is another approach to copy operators, which is similar to our use of operation names in the system notation [LTS79].

This will have to suffice to give the reader some idea about the relation between the new macro notation presented in this paper and work done by others working in the area of Petri nets and related formalisms. There are similar notions in certain versions of CSP [Ho85] and in Occam [Oc84].

2. Basic COSY Syntax and Semantics

A basic COSY program is a string derived from the production rules given below. The following meta-language conventions have been used in the syntax rules: the symbols "=", "{", "}", "|", "*", "+", "@", are used as meta-symbols. The symbol "=" denotes replacement of its left hand side by the string on its right side. The braces "{}" are used to group items together, "|" indicates alterate productions, "{item}*" indicates production of item zero or more times, "{item}+" indicates production of "item" one or more times. The notation "{item1 @ item2}+" is used as a shorthand for "item1{item2 item1}*".

The syntax of a basic COSY program is given by the following rules:

B1. basicprogram = **program** programbody **endprogram**
B2. programbody = { path | process }+
B3. path = **path** sequence **end**
B4. process = **process** sequence **end**
B5. sequence = { orelement @ ; }+
B6. orelement = { starelement @ , }+
B7. starelement = element | element*
B8. element = event | (sequence)
B9. event = name | name({ integer @ , }+)
B10. name = sequence of letters and digits beginning with a letter

Assume additionally that we may write some comments (usually names) immediately before **program**, **path**, and **process**.

Basic COSY programs consisting of paths only are called COSY path programs, and we restrict ourselves at this moment to this case. The syntactic entities occurring in a path are just regular expressions, the symbols ";" and "," denote sequentialization and arbitrary choice respectively; the symbol "*" is the Kleene star. All regular expressions in paths are considered to be cyclic i.e., the terms **path** and **end** are just an alternative of ( )*.

Synchronization among paths is achieved by the coincident execution of shared events in the sense that if a shared event appears in a number of paths then all these paths containing it must execute this event at the same time. Let us consider an example.

Pr1 = **program**

      P1 : **path** a ; b , c **end**

      P2 : **path** (d ; e)* ; b **end**

   endprogram

This program consists of two paths and there is one shared event "b". The execution of the program is achieved by executing these two paths simultaneously. Since "b" is a shared event, in the combined viewpoint, the first path can execute "b" if and only when the "b" in the second path can be executed, and this achieves the synchronization between these paths.

The formal semantics of basic COSY programs may be defined in terms of vector firing sequences (an approach initiated in [S79]).

With every path P = **path** sequence **end**, we associate its set of events $Ev(P)$. For every regular expression E, let $|E|$ denote the regular language described by E. For every path P = **path** sequence **end** the language $|sequence|$ is called the set of cycles of P and denoted by $Cyc(P)$, i.e.

    $Cyc(P) = |sequence|$.

For example Pr1 we obtain $Cyc(P1)=\{ab,ac\}$, $Cyc(P2)=\{de\}*\{b\}$. A formal (easy) definition of Cyc can be found for example in [S79,LSC81].

From the set of $Cyc(P)$ we construct the set of finite firing sequences of P, denoted by $FS(P)$, as follows:

    $FS(P) = Pref(Cyc(P)*)$.

where for every alphabet A and every $L \subseteq A*$, $Pref(L)=\{x|(\exists y \in A*)\ xy \in L\}$.

With each COSY path-program $Pr = $ **program** $P_1...P_n$ **endprogram** (or simply $Pr = P_1...P_n$) we associate its set of events $Ev(Pr) = Ev(P_1) \cup ... \cup Ev(P_n)$ and the set $VFS(Pr)$, its set of vector firing sequences (or permitted histories). To define $VFS(Pr)$ some additional notions are necessary. A vector $(x_1,...,x_n)$ is a possible history (or behaviour) of $Pr = P_1...P_n$ if each x_i is a possible firing sequence of P_i and furthermore, if the the x_i's agree about the number and order of occurrences of events they share.

Let us define a concatenation on $Ev(P_1)* \chi ... \chi Ev(P_n)*$ as follows:

    $(x_1,...,x_n)\ (y_1,...,y_n) = (x_1y_1,...,x_ny_n)$.

For every i=1,...,n, let $h_i: Ev(Pr)* \longrightarrow Ev(P_i)*$ be an erasing homomorphism given by: $(\forall\ a \in Ev(P))\ h_i(a)=a$ if $a \in Ev(P_i)$ or $h_i(a)=\varepsilon$ if $a \notin Ev(P_i)$,

where ε denotes the empty string.

For every $x \in Ev(Pr)$, let $\mathbf{x} = (h_1(x), \ldots, h_1(x))$, and let

$\quad Vev(Pr) = \{ \mathbf{a} \mid a \in Ev(Pr) \}$.

The set $Vev(Pr)$ is called the set of <u>vector events</u> of Pr. For example Pr1 the vector events are: $Vev(Pr1) = \{a,b,c,d,e\} = \{ (a,\varepsilon),(b,b),(c,\varepsilon),(\varepsilon,d),(\varepsilon,e) \}$. The vector events indicate distribution of events to subsystems, and the sharing of events ("handshake" synchronization).

Vectors of sequences belonging to $Vev(Pr)*$ can be interpreted as <u>partial orders of event occurrences</u>, so that they may be used to model the <u>concurrent behaviour</u> of Pr.

The set of all possible finite behaviours or histories of Pr, the <u>vector firing sequences</u> of Pr, denoted by $VFS(Pr)$, is defined by:

$\quad VFS(Pr) = (FS(P_1) \; X \; \ldots \; X \; FS(P_n)) \cap Vev(Pr)*$.

The set $FS(P_1) X \ldots X FS(P_n)$ in the definition of $VFS(Pr)$ guarantees that each string component x_i of a history $\mathbf{x} = (x_1, \ldots, x_n)$ is a firing sequence of the path P_i, and the set $Vev(Pr)*$ guarantees that all these firing sequences agree about the number and order of occurrences of events they share. This formal model of behaviour allows us to speak formally about dynamic properties of systems specified by a path program $Pr = P_1 \ldots P_n$, like deadlock-freeness, adequacy (liveness) and so on (see [S79,LSC81,B82,L84]).

A number of results have been obtained concerning both the theory and application of vector firing sequence semantics. A review of these results and an almost complete bibliography can be found in [L84].

The semantics of Basic COSY path programs may also be expressed in terms of Labelled Petri Nets [LC75,LSC81,L84,B82,LSB79,LSB80].

In general, a basic program Pr is a string of the form

$\quad$ Pr = **program** $P_1 \ldots P_n Q_1 \ldots Q_m$ **endprogram**

where P_i denote paths and Q_i processes. Although path and processes may be intermixed in a basic COSY program, in the above expression for convenience, we assumed that all paths are collected before processes.

The semantics of a basic COSY program Pr including processes is given by means of the vector firing sequences of an equivalent basic program Pr' involving paths only.

The conversion of Pr into Pr', denoted by Path(Pr), is obtained by the following rules (see [LSB79,LSB80]):

1. For every $a \in Ev(Pr)$ construct a set $I(a)=\{\ i\ |\ a \in Ev(P_i)\ \&\ i=1,\ldots,m\ \}$ and if $I(a)$ is not empty, then:

    a. replace the event "a" in each path it occurs in by the element $(a\&i_1,\ldots,a\&i_k)$

    where k is the cardinality of $I(a)$, and $i_j \in I(a)$ for $j=1,\ldots,k$.

    b. replace the event "a" in process Q_{i_j} by $a\&i_j$ for all $i_j \in I(a)$.

2. Replace all occurrences of "**process**" by "**path**".

Then the semantics of Pr, both in term of vector firing sequences and Petri nets are given in terms of Path(Pr). For example, if

Pr = **program**		Path(Pr) = **program**
path a **end**		**path** (a&1,a&2) **end**
process a **end**	then	**path** a&1 **end**
process a **end**		**path** a&2 **end**
endprogram		**endprogram** .

3. Macro COSY notation

Since the development of the macro COSY notation was considered to be an open-ended effort, i.e. it was permitted to change until it was precisely clear what constitutes a good macro notation, various macro natations and subnotations have been developed, some being extensions of others or, more commonly, differing in many aspects. Results of these experiments as well as examples of use of the macro COSY notation to specify non-trivial distributed and concurrent systems can be found in [TL77,LT78,L79,LTS79,LTS80,LSC81].

The objectives of Cotronis' Ph.D. Thesis [C82] were to re-examine and revise all aspects of the macro notation, its design as a specification language, the formal syntax of macro programs, the expansion rules of macro elements and of complete macro programs, alleviating or eliminating altogether the drawbacks of other notations and grammars; to characterize the strings generated by the expansion of replicators, distributors and of complete macro programs produced by the formal grammar; to investigate some aspects of programming methodology such as when replicators and distributors may be replaced by other replicators and distributors expanding to the same string as the former; and finally to give direct vector firing sequence semantics to macro programs rather than indirectly via the basic programs generated by their expansion.

The guidelines for revising the macro notation and grammar were mainly four:

1. The syntactic well-formedness of a macro program should imply that its expansion is a syntactically well-formed basic program.

2. The notation should allow the generation of a large class of basic programs and their concise representation.

3. The macro grammar should include context-free rules and should be uniform with the grammar of basic COSY.

4. The reading of macro programs should be possible without formal expansion. The structured macro notation defined in [C82] solves this problem but it will not be discussed further in this paper.

In the design of the notation, changes of symbols and of the forms of the collectivisors, replicators and distributors were suggested improving the readability of these constructs and of macro programs as a whole. Some restrictions imposed on what replicators were allowed to generate ensure the readability of unexpanded macro programs. The new replicators generate strings which could not be generated by a single replicator in previous notations. Distributors were extended to generate more strings more economically than replicators. Two new types of replicators were added generating strings which could not be generated by replicators in previous notations. It was precisely specified where distributors and each type of replicator should appear in macro programs.

The new syntax rules for macro programs combine some of the syntax rules of previous grammars, modified to be consistent with changes in the design of the notation. The expansion rules for replicators were modified to deal with their new form and the expansion of distributors was directly defined. The expansion of replicators and distributors was also characterized. The expansion of complete macro programs was formally defined and it was proven that programs permitted by the new grammar generate syntactically well-formed basic programs. Thus, the suggested grammar is not too wide and no meta-restriction rules are needed to eliminate any "unwanted" programs, that is programs not generating basic programs. The conditions under which replicators and distributors may be replaced by other replicators or distributors was also examined.

The macro notation of [C82,W85], called Full (or New) Macro COSY Notation, was implemented using Simula on an IBM/370 (Newcastle upon Tyne) and using C on VAX/780 (McMaster University, Hamilton) [W85,Mi85].

In the next sections we shall present this Full Macro Notation and some of the above mentioned results about the notation.

4. The Macro Program and the Collectivisors

A macro COSY program will consist of collectivisors, paths, processes and body-replicators appearing between the parentheses **program** and **endprogram**. Since after expansion of a macro program all its collectivisors disappear, the macro program should include at least one path, process or body-replicator. The syntax for macro programs is given by rules MN1, MN2 in the Appendix. According to these rules, collectivisors, macro paths, macro processes and body-replicators may appear in any order, with the exception that no collectivisor may appear after all the paths, processes and body-replicators.

Furthermore the following context-sensitive restriction is imposed:

(MPrest) : Collective names should be declared before any path or process involving any of its subscripted operations.

Due to this restriction, among others one pass is sufficient for the expansion of a macro program.

To simplify our considerations, in this and the subsequent sections, we shall frequently associate explicit names with syntactic entities such as collectivisors, replicators, distributors, macro paths etc. Names will be written immediately before entities and will be separated from entities by ":". The names of entities together with ":" do not belong to the macro COSY notation.

Collectivisors are used to declare subscripted events of any finite number of dimensions. Let us consider some examples. For example, the collectivisor C1,

C1 : **array** A(k) **endarray**
     array B C(5) D(3,m) **endarray**

declares the subscripted events

A(1),...,A(k),
B(1),...,B(5),
C(1),...,C(5),
D(1,1),D(2,1),D(3,1),
D(1,2),D(2,2),D(3.2),
...
D(1,m),D(2,m),D(3,m).

If the lower bound in some dimensions of collective names is not 1 but some other fixed integer n we may specify it explicitly, as for instance:

C2 : **array** E(n:k) F(m:n,k) **endarray**

We may also combine the declarations in C1 and C2 into one declaration:

C3 : **array** A(k) B C(5) D(m,3) E(n:k) F(m:n,k) **endarray**

We permit also declaration of subscripted events the indices of which either are not consecutive integers or depend on the index of some other dimension. For example:

```
    S(1,1)
    S(3,1), S(3,2), S(3,3)
    S(5,1), S(5,2), S(5,3), S(5,4), S(5,5).
```
The subscripted events corresponding to S may be declared as follows:

    C4 : **array** #i:1,5,2 [#j:1,i,1[S(i,j)]] **endarray**

The pattern " #i:initial,final,increment " may be read from left to right as "index i takes values from initial to final in steps of increment which upon expansion are replacing index i in each copy of the regularity inside [...]". The same meaning of this pattern will be applied for the body-replicators (section 6) and the sequence-replicators (sections 8,9).

Subscripted events with the same subscripts in all their dimensions may be declared by the same replicators and <u>will not be separated by commas</u>. Also subscripted events with the same subscripts in some of their dimensions may be grouped together in the same replicator, e.g.

    C5 : **array** #i:1,5,2 [T(i) #j:1,i,1 [S(i,j) U(i,j)]] **endarray**

where T corresponds to the events:

    T(1),T(3),T(5)

and U to the events:

```
    U(1,1)
    U(3,1), U(3,2), U(3,3)
    U(5,1), U(5,2), U(5,3), U(5,4), U(5,5)
```
We also permit grouping together subscripted events, indexed by expressions depending on replicator indices indexing other subscripted events, as in C6:

    C6 : **array** #i:1,5,2 [V(i+3) #j:1,i,1[S(i,j)]] **endarray**

where V corresponds to the events V(4), V(6), V(8).

The expressions in collectivisors must satisfy five context-sensitive restrictions (see Appendix, Cresti where i=1,...,5) and the identifiers for replicator indices in collectivisors must satisfy two context-sensitive restrictions (Irest1, Irest2 in the Appendix). All these restrictions, except perhaps Crest4, are obvious and similar to those used in declarations of arrays in programming languages. The restriction Crest4 is imposed to guarantee the independence of the indices of different dimensions of the same collective name and to avoid duplication of declarations of subscripted events. Furthermore it makes the rules for expansion of distributors relatively easy (see the discussion in [C82, section 3.4] and [W85]).

Generally, the shapes of arrays declared by replicators may be characterized as being <u>finite n-dimensional arrays, the indices of which are generated by integer expressions depending on integer variables taking values which form finite arithmetic progressions.</u>

The complete syntax for the collectivisors is given by rules MN3-MN7, restrictions Crest1-Crest5, Irest1, Irest2 (see Appendix).

5. The Body-Replicators, Paths and Processes

Body-replicators generate consecutive regularities of paths and/or processes. We also permit nesting of body-replicators. The syntax is formally given by the rule MN8 (see Appendix). For example the first n paths in an n-frame fifo buffer (see [LTS79] and Section 1) can be specified as:

    #i:1,n,1 [**path** deposit(i);remove(i)] **end**

and m pipelines of size n each associated with a mechanism controlling exits similar to that in the bounded delay priority queues in [LTS79] may be specified as:

    #i:1,m,1

    [#j:1,n,1 [**path** transfer(i,j);transfer(i,j+1)**end**]

      path transfer(i,n+1); csend(i) **end**]

which for m=2, n=3 expands to:

    path transfer(1,1);transfer(1,2) **end**

    path transfer(1,2);transfer(1,3) **end**

    path transfer(1,3);transfer(1,4) **end**

    path transfer(1,4);csend(1) **end**

    path transfer(2,1);transfer(2,2) **end**

    path transfer(2,2);transfer(2,3) **end**

    path transfer(2,3);transfer(2,4) **end**

    path transfer(2,4);csend(2) **end**

We also impose the restriction (BRrest, see Appendix) guaranteeing that body-replicators generate at least one regularity.

The syntax of <u>macro paths and macro processes</u> is similar to the syntax of paths and processes of basic COSY and consists of rules MN9-MN17 in the Appendix. The only difference between these rules and corresponding ones in basic COSY is that here we allow three new types of elements: sequence-replicators and distributors, which <u>cannot</u> be starred, and indexed operations.

6. The Sequence-Replicators

There are two basic types of replicators generating basic COSY sequences:

    <u>the concatenator</u> which generates consecutive regularities and is of the

    form: #i:in,fi,inc [p(i) sep@] and

the imbricator which generates regularities nested within each other and is
of the form: #i:in,fi,inc [p(i) @ t @ q(i)] where p(i), t, q(i) denote
"patterns" and sep is one of the separators ";" or ",".

In the concatenator the "@" strips the separator in front of it in the last copy
of the regularity "p(i) sep". The expansion of the concatenator looks like:

   p(in) sep p(in+inc) sep p(in+2*inc) sep ... sep p(fi')

where fi' denotes the final value of the range of the index (which might be
different from fi, for instance in the case "i:1,6,2" we have fi'=5).

Using the concatenator we can fully specify the n-frame buffer by means of the
macro COSY notation (compare the early macro COSY specification from Section 1):

   MPr1 = **program**
           array deposit remove(n) **endarray**
           #i:1,n,1 [**path** deposit(i);remove(i) **end**]
           path #i:1,n,1 [deposit(i);@] **end**
           path #1:1,n,1 [remove(i);@] **end**
           endprogram

For n=3 MPr1 expands to:

   program
     path deposit(1);remove(1) **end**
     path deposit(2);remove(2) **end**
     path deposit(3);remove(3) **end**
     path deposit(1);deposit(2);deposit(3) **end**
     path remove(1);remove(2);remove(3) **end**
   endprogram

In the imbricator the separators before the first @ and after the second @ in the
last copy of the regularity p(i)q(i) will be replaced by t. The expansion of the
imbricator looks like:

   p(in) p(in+inc) ... p'(fi') t q'(fi') ... q(in+inc) q(in)

where fi' is the same as in the concatenator and p', q' are the same as p, q
respectively but with any trailing separator of p and any leading separator of q
respectively, removed. The reason we have specified a string "t" to be between the
two "@"s instead of just a separator is that we would like our grammar to permit
paths such as the following:

   path empty, #i:1,n,1 [(up(i);@;full*;@;down(i))*] **end**

which specifies a stack of size n with tests for empty and full.

When this path is expanded for n=3 we get:

path empty,(up(1);(up(2);(up(3);full*:down(3))*;down(2))*;down(1))* **end**

in which the starred operation full* appears only once, in the innermost regularity. In general we permit any string to appear at the innermost position as long as it forms a well-formed basic COSY string with the rest of the expansion. The formal syntax for sequence-replicator is given by rules MN18, MN22, MN23 (concatenator), MN24-MN28 (imbricator) and context-sensitive restrictions Irest1, Irest2, Rrest1, Rrest2.

What makes the syntax rules for patterns p(i), q(i) and t of the new macro notation different from earlier rules is that the sequences:

p(in) sep p(in+inc) ... sep p(fi')   for concatenator, and

p(in) p(in+inc) ... p'(fi') t q'(fi') ... q(in+inc) q(in)

for imbricator, must always be a <u>well-formed basic COSY sequence</u>. For the concatenator, roughly speaking, this means that the pattern p(i) itself should be a correct COSY macro sequence. In the case of the imbricator we have a somewhat more complicated situation (see Appendix). More details, including the formal proof of correctness of the syntax rules can be found in [C82].

7. The Left and Right Sequence-Replicators

The sequence-replicators discussed so far have a limitation: they should not expand to empty strings (empty regularities are not permitted by the syntax rules, and the restrictions on in, fi and inc do not permit an empty range). The simplest solution seems to be to forget about the restriction on in, fi, inc but this may result in the string (an incorrect part of the non-starving banker specification of [LT78]):

bnkr(1);#i:k,n+1,-1 [par;rap;@];

which for k<n+1 will be expanded to "bnkr(1);;" which is not a well formed basic COSY string because of the collision of two semicolons. To avoid collision of separators when the replicators generate empty strings, their context should not be the same as in Section 6. A separator should be "missing" either on the left or the right. The expansion has to provide the extra separator. If the separator on the left is missing they will be called <u>left replicators</u>, and if the separator on the right is missing they will be called <u>right replicators</u>. The forms of these replicators are the following:

```
left-concatenator:    #i:in,fi,inc [ sep | p(i) sep' @ ]
left-imbricator:      #i:in,fi,inc [ sep | p(i) @ t @ q(i) ]
right-concatenator:   #i:in,fi,inc [ p(i) sep' @ | sep ]
right-imbricator:     #i:in,fi,inc [ p(i) @ t @ q(i) | sep ]
```

where in, fi, inc, sep, p(i), q(i), t are the same as in Section 8, sep' is either
";" or ",". If the index range is empty the strings generated by the expansion of
these replicators will be empty as well. Otherwise the strings generated by the
expansion of left-concatenator or left-imbricator (right-concatenator or right-
imbricator) will be the same as the strings generated by the expansion of the
concatenator or the imbricator by removing "sep|" ("|sep"), preceded (followed) by
"sep". In this notation the part of the non-starving banker specification can be
written as:

    bnkr(1) #i:k,n+1,-1 [; | par;rap;@];

The formal syntax of these replicators is defined by rules: MN20, MN21 and
MN22-MN28. Let us consider a final example:

    #i:1,n,k [; | (A(i);B(i)), @]

For n=3 the possible expansions would be:

    for k=0 : empty

    for k=1 : ;(A(1);B(1)),(A(2);B(2)),(A(3);B(3))

    for k=2 : ;(A(1);B(1)),(A(3);B(3))

8. The Distributors

Consider the following part of the macro COSY specification MPr1:

    path #i:1,n,1 [deposit(i); @] **end**.

Replicators of the form used above are so regular and occur so frequently that
they might be represented by another simpler mechanism, called the distributors.
The distributor equivalent to the above replicator is simply:

    path ; [deposit] **end**

assuming that the collective name deposit has been previously declared, for
example by the collectivisor **array** deposit remove(n) **endarray**. The distributors do
not generate indices explicitly, like the replicators, but generate indices
defined by the collectivisors.

The simplest distributor, called the simple distributor, has the form:

    sep [mseq]

where sep is either ";" or "," and mseq is a macro sequence in which the events
and indexed events are array-slices. By an array-slice we mean an equivalence
class of indexed events corresponding to the same collective name the indices of
which differ in at least one dimension. Array-slices are represented like indexed
events but with the index fields corresponding to the dimension in which their
elements differ, left blank. We call the dimensions corresponding to blank fields
of an array-slice the distributable dimensions of the array slice. An array slice
could have several distributable dimensions. When all the dimensions of an array
slice are distributable, then these define all the events in the array and are

represented by the collective name itself without any index fields at all. Let us consider the following collectivisor:

C7 : **array** A(0:3) B(4,3) **endarray**

The collective name A has only one dimension and therefore only one array-slice represented by A( ) or A, which defines the following equivalence class of all the events in A: { A(0),A(1),A(2),A(3) }. The collective name B has two dimensions and eight array-slices: B=B(,), B(1,), B(2,), B(3,), B(4,), B(,1), B(,2), B(,3), where for example B(1,) defines the equivalence class { B(1,1),B(1,2),B(1,3) }.

The only syntactic difference between a macro sequence in distributors and a macro sequence anywhere else is that some of the index fields of the indexed events of the former may be empty (see rule MN17 in Appendix). This allows us to define the distributor:

; [A,B(,3)]

where A,B are declared by the collectivisor C7. This distributor expands to:

A(0),B(1,3);A(1),B(2,3);A(2),B(3,3);A(3),B(4,3).

In this case the expansion is possible because the number of distributed A(i)'s and B(i,3)'s is the same. In general, we say that the distributor is well-formed and may be expanded only if it satisfies the so called compatibility criterion, i.e. when all distributable dimensions on which the distributor operates contain the same number of sections, where by a section we mean the equivalence class of events in the array-slice which have the same index in one of the distributable dimensions of that slice.

A distributor in which all the distributable dimensions on which it operates contain n sections may be explained as follows:

"n copies of the macro sequence in the distributor will be concatenated separated by the separator associated with the distributor, and the ith copy (i=1,...,n) of each array-slice will be replaced by the ith section of this slice. The order of the sections is defined to be the order in which their different indices are generated by the array declaration."

In the case of our simple distributor ; [A,B(,3)] , the first (and only) dimension of A and the first dimension of B are distributable dimensions on which this distributor operates. From the collectivisor C7 it follows that A contains four sections: A(0), A(1), A(2), A(3), the first dimension of B has sections: B(1,), B(2,), B(3,), B(4,), and the second dimension of B has sections: B(,1), B(,2), B(,3).

Sometimes it is useful to distribute not only over the whole range of array-slices but over a subrange of them as well. This may be done using a distributor, called the subrange distributor, which is of the following form:

sep #in,fi,inc [mseq]

where sep and mseq are as in the case of the simple distributor and in, fi, inc denote integer expressions representing the subrange over which mseq is to be distributed.

Consider the example: ; #1,3,1 [A] . The string generated by the expansion of ;[A] is: "A(0);A(1);A(3);A(3)", but #1,3,1 between ";" and "[A]" indicates that only copies first, second and third should be generated (from 1 until 3 by 1), thus ;#1,3,1 [A] will expand to: "A(0);A(1);A(2)".

Note that ; #1,3,1 [A] differs from #i:1,3,1 [A(i);@] since the latter will expand to : "A(1);A(2);A(3)".

The distributor ; #1,3,2 [A] will expand to "A(0);A(2)".

In this case the compatibility criterion may be somewhat relaxed, we require only that the slices must contain at least as many sections as specified by the subrange (condition Drest1 in the Appendix). Note that when #in,fi,inc specifies the whole range the above condition is equivalent to the compatibility criterion for the simple case.

From the rules introduced so far it follows that the distributors may be nested. In such a case we require that after the expansion of the outermost distributor, the rest of the distributors must obey the syntax rules (see condition Drest2 in Appendix).

For example:
;[,[A]]
where A is declared by C7, is not valid since after expansion of the outermost distributor a non-valid distributor is generated, namely:
,[A(0)];,[A(1)];,[A(2)];,[A(3)]
The reason for this is that the macro-sequences inside "[ ]" do not consist of array-slices but indexed events. But the distributor
;[,[B]]
where B is declared by C7, is valid, and it expands to:
 B(1,1),B(2,1),B(3,1),B(4,1)
 ;B(1,2),B(2,2),B(3,2),B(4,2)
 ;B(1,3),B(2,3),B(3,3),B(4,3)

with "," applying to the first dimension of B and ";" to the second. In other words we assume the default rule that the outermost distributor applies to the rightmost distributable dimension of each slice; the second outermost to the rightmost not allocated distributable dimensions, etc.

We can also specify explicitly which separator applies to which distributable dimension. We call this kind of distributors to be the selector distributors. For example if we would like to distribute B with ";" applying to the first dimension and "," to the second we should specify:

    ;1[,[B]]

which will expand to:

    B(1,1),B(1,2),B(1,3)
    ;B(2,1),B(2,2),B(2,3)
    ;B(3,1),B(3,2),B(3,3)
    ;B(4,1),B(4,2),B(4,3)

Here it is explicitly specified that ";" applies to the first dimension of B and implicitly that "," applies to the rightmost unallocated dimension of B, according to the default rule. Formally the syntax of distributors is defined by the rule MN19 and the context-sensitive restrictions Drest1–Drest4 in the Appendix.

9.   The Expansion of Macro COSY Programs

In this section we define the formal rules for the expansion of replicators, distributors and complete macro programs.

A macro program denoted by the syntactic variable MPROG is represented by **program** MPBODY **endprogram** . MPBODY has the form: CPPBR1...CPPBRn   , where each CPPBRi denotes a simple path or process or body-replicator possibly headed by collectivisors. If headed by collectivisors CPPBRi may be represented by CL PPBR , where CL denotes a collection of collectivisors and PPBR a single path, or process or a body-replicator. A body-replicator is represented by #i:in,fi,inc [PPBRs] where PPBRs denotes a collection of paths, processes and body-replicators. A body-replicator upon expansion generates a collection of paths, processes and body-replicators represented by: PPBR1...PPBRn ,where each PPBRn denotes a single path or process or body-replicator. A path and a process is represented by: **path** MSEQ **end** and **process** MSEQ **end** respectively, where MSEQ denotes a macro sequence. A macro sequence is represented by: MOR1;...;MORn , where each MORi denotes a macro orelement, which is represented by GEL1,...,GELn where each GELi for i=1,...,n denotes a generalized element.

In general, a generalized element may involve right- and left-replicators and is represented by RR1...RRn M LR1...LRm where RRi and LRi denote right- and left-replicators respectively, and M denotes either a starelement or a sequence-replicator or a distributor. A generalized element may be just a sequence-replicator denoted by SREPL, or a distributor denoted by DISTR, or a starelement represented by EL* or EL, where EL denotes an element which can be either an event or an indexed event, both denoted by EV, or a macro-sequence in parentheses "( )".

The complete expansion of MPROG is given by the function "expand" which is defined recursively as follows:

    expand(MPROG) --> **program** expand(MPBODY) **endprogram**

    expand(CPPBR1...CPPBRn) --> expand(CPPBR1)...expand(CPPBRn)

    expand(CL PPBR) --> expand(PPBR)

    expand(#i:in,fi,inc[PPBRs]) --> expand(Expbrep(#i:in,fi,inc[PPBRs]))

    expand(PPBR1...PPBRn) --> expand(PPBR1)...expand(PPBRn)

    expand(**path** MSEQ **end**) --> **path** expand(MSEQ) **end**

    expand(**process** MSEQ **end**) --> **process** expand(MSEQ) **end**

    expand(MOR1; ... ;MORn) --> expand(MOR1); ... ;expand(MORn)

    expand(GEL1, ... ,GELn) --> expand(GEL1), ... ,expand(GELn)

    expand(RR1...RRn M LR1...LRm) -->

        expand(RR1)...expand(RRn) expand(M) expand(LR1)...expand(LRm)

    expand(RRi) --> expand(Exprrep(RRi))

    expand(LRi) --> expand(Explrep(LRi))

    expand(SREPL) --> expand(Expsrep(SREPL))

    expand(DISTR) --> expand(Expdist(DISTR))

    expand(EL*) --> expand(EL)*

    expand(EV) --> EV'

    expand((MSEQ)) --> (expand(MSEQ))

where EV' is EV with evaluated expressions if EV is an indexed event, Expbrep, Exprrep, Explrep, Expsrep, Expdist are functions that describe the particular expansions of body-replicators, right-replicators, left-replicators, left-replicators, sequence-replicators and distributors respectively.

Now we define these functions in detail.

Let us first define the primitive-recursive operator COPY(j=k,m), where j, k, m are parameters, j is an integer variable, k,m are integers, and which has three string arguments separated by "|" (to distinguish it from "," which is an element of the basic COSY alphabet):

$$
COPY(j=k,m)\{P(j)|T|Q(j)\} = \begin{cases} P(k)\ COPY(j=k+1,m)\{P(j)|T|Q(j)\} & m>k \\ P'(k)\ T\ Q'(k) & m=k \\ T' & m<k \end{cases}
$$

where P(j) and Q(j) are strings in which the integer expressions may depend on j.
The strings P'(k) and Q'(k) are the same as P(k) and Q(k) respectively with the
terminating, and respectively leading separator, if any, removed.
For example:

COPY(j=1,3){A(j-1);|,B,|;C} = A(0); COPY(j=2,3){A(j-1);|,B,|;C} ;C =
= A(0);A(1); COPY(j=3,3){A(j-1);|,B,|;C} ;C;C = A(0);A(1);A(2),B,C;C;C
COPY(j=1,0)){A(j-1);|,B,|;C} = B
COPY(j=1,2){A(j)||} = A(1) COPY(j=2,2){A(j)||} = A(1) A(2)

Let $n = \frac{fi-in}{inc} + 1$ and $f(j) = in + (j-1)*inc$. The expansion of body-

replicator, concatenator and imbricator is defined only if n>0 (which requires
inc≠0) and then is given by the functions: Expbrep, Expcon and Expinb
respectively:

Expbrep(#i:in,fi,inc [body(i)]) = COPY(j=1,n){body(f(j))||},
Expcon(#i:in,fi,inc [p(i) sep @]) = COPY(j=1,n){p(f(j)) sep||},
Expimb(#i:in,fi,inc [p(i) @ t @ q(i)]) = COPY(j=1,n){p(f(j))|t|q(f(j))}.

We point out that Expbrep(BREP) means that only BREP is expanded and not any other
replicator which may be generated by the expansion of BREP! The same remark
concerns also Expcon, Expimb and all expansion functions defined below.

For the left-, right-concatenator and left-, right-imbricator we require only
inc≠0 and the rules are the following: if n>0 then:

Exlcon(#i:in,fi,inc[sep|p(i)sep'@])   = sep COPY(j=1,n){p(f(j))sep'||} ,
Exrcon(#i:in,fi,inc[p(i)sep'@|sep])   = COPY(j=1,n){p(f(j))sep'||} sep ,
Exlimb(#i:in,fi,inc[sep|p(i)@t@q(i)]) = sepCOPY(j=1,n){p(f(j))|t|q(f(j))} ,
Exrimb(#i:in,fi,inc[p(i)@t@q(i)|sep]) = COPY(j=1,n){p(f(j))|t|q(f(j))}sep ,

while for n≤0 Exlcon(...), Exrcon(...), Exlimb(...), Exrimb(...) are simply empty
strings. Formal proofs of the correctness of the above definitions may be found in
[C82]. Now we may define the functions Explrep, Exprrep and Expsrep as follows:

$$Explrep(LR) = \begin{cases} Exlcon(LR) & \text{LR is a left-concatenator} \\ Exlimb(LR) & \text{LR is a left-imbricator} \end{cases}$$

$$Exprrep(RR) = \begin{cases} Exrcon(RR) & \text{RR is a right-concatenator} \\ Exrimb(RR) & \text{RR is a right-imbricator} \end{cases}$$

$$Expsrep(SR) = \begin{cases} Expcon(SR) & \text{SR is a concatonator} \\ Expimb(SR) & \text{SR is an imbricator} \end{cases}$$

The expansion of distributors may be defined using the same operator COPY,
although in this case the precise description of index is somewhat more
complicated.

Let A be a k-dimensional collective name declared by a collectivisor. Let
h(A,i,j) be an integer function which gives the values of the index of the jth
section of A along the dimension i.

Consider the example:

    C8 : **array** A(0:3) B(4,3) C(2,-2:8,5:10) **endarray.**

In this case we have:

    h(A,1,j) = j-1 for j=1,2,3,4 (A has only one dimension),
    h(B,1,j) = j for j=1,2,3,4,
    h(B,2,j) = j for j=1,2,3,
    h(C,1,j) = j for j=1,2,
    h(C,2,j) = j-3 for j=1,...,11,
    h(C,3,j) = j+4 for j=1,...,6.

Consider the well-formed simple distributor of the form:

    sep [p]

where p is a macro sequence of array slices. Since the distributor is well-formed, the compatibility criterion must be obeyed, implying that all the distributable dimensions of array-slices must contain the same number of sections, say m.

Let p(j) be a string derived from p by means of the following rule: for each collective name A occurring in p, the distributable dimension of A applying for sep in p, say i, must be replaced by h(A,i,j).

Now the expansion of sep [p] can be defined as follows:

    Expdist(sep[p]) = COPY(j=1,m){p(j) sep ||}.

The case sep dim [p] (selector distributor), where dim is the number of dimension is almost the same as the simple distributor. The only difference is that in this case we transform p into p(dim,j) instead of p(j), where p(dim,j) is a string derived from p by the following rule: for each collective name A occurring in p, the distributable dimension dim of A in p, is replaced by h(A,dim,j). Thus in this case we have:

    Expdist(sep dim [p]) = COPY(j=1,n){p(dim,j) sep||}.

Let us now obtain a formula for the expansion of a well formed subrange distributor:

$$\text{sep \#in,fi,in [p] .}$$

Here the condition Drest1 is satisfied which means that all the distributable dimensions of array slices on which the distributor applies contain at least m sections where $m = \dfrac{fi-in}{inc} + 1$. Let $f_1(j) = in + (j-1)*inc$. We may define the expansion of this kind of distributor as:

    Expdist(sep #in,fi,in[p]) = COPY(j=1,m){p(f_1(j)) sep||}.

Note that the distributor of the form sep [p] may be considered as a special case of sep #in,fi,inc[p] in which in=inc=1 and fi is the common number of sections in the distributable dimensions in the array slices of p.

Theorem 1.([C82])

The string obtained by the expansion of a distributor of the form

sep #in,fi,inc [p]

may also be generated by a concatenator of the form:

#j:1,n,1 [p(i) sep @]

where $n = \frac{fi-in}{inc} + 1$ and p(j) is obtained from p by substituting the fields of the distributable dimensions of sections in p by h(, ,j) as in the definition of Expdist. □

Let us illustrate the above theorem, for example: ;[C(1,-1,)
where C is declared by C8, is equivalent to: #j:1,6,1 [C(1,-1,j+4) ;@]
since h(C,3,j) = j+4 for j=1,...,6.

Theorem 2.([C82])

The expansion of a macro program given by the function "expand" is a well-formed basic COSY program. □

10. The Semantics of Macro COSY Programs and Final Comments

Theorem 2 allows us to define the semantics of macro-programs in terms of the vector firing sequences of the basic programs obtained by their expansion. The semantics of a macro program MPROG which does not include any macro process will therefore be given in terms of VFS(expand(MPROG)) and that of ones including processes in terms of VFS(Path(expand(MPROG))).

Cotronis [C82] has shown that the equivalent vector firing sequence semantics can also be derived directly from macro-programs.

We may also define the semantics of MPRPG in terms of Petri nets associated to expand(MPROG) or Path(expand(MPROG)) if processes are involved. For more details the reader is advised to refer to [C82].

The essential idea of the above approach was to allow us to specify a system consisting of events like points in any n-dimensional discrete cartesian space, which
is the task of collectivisors combined with the power of replicators. In particular, arithmetic expressions of the usual kind used in cartesian mathematics can be
used to characterize and generate the index sets involved in a concise manner.

Once such a system of independent indexed events has been generated sequential
constraints can be imposed on it by combining subsets of events into sequential
subsystems by means of paths and/or processes. Such subsystems will be
synchronized by means of the events they share with other subsystems.

Appendix

The Syntax of Macro Programs in the Full Macro Notation

MN1.   mprogram = **program** mprogrambody **endprogram**

MN2.   mprogrambody = {collectivisor|mpath|mprocess|bodyreplicator}+

MN3.   collectivisor = **array** {simpleardec|replardec}+ **endarray**

MN4.   simpleardec = {arrayid}+({{iexpr:|}iexpr@,}+)

MN5.   replardec = index_spec[{replardec|arrayid({iexpr @,}+)}+]

MN6.   index_spec = #index:iexpr,iexpr,iexpr

MN7.   arrayid = letter{letter|digit}*

MN8.   bodyreplicator = index_spec[{mpath|mprocess|bodyreplicator}+]

MN9.   mpath = **path** msequence **end**

MN10. mprocess = **process** msequence **end**

MN11. msequence = {morelement @;}+

MN12. morelement = {gelement @,}+

MN12. gelement = {rreplicator}*{starelement|sreplicator|distributor}
             {lreplicator}*

MN13. starelement = element|element*

MN14. element = event|indexedev|(msequence)

MN16. event = letter{letter|digit}*

MN17. indexedev = arrayid{({iexpr@,}+| }

MN18. sreplicator = index_spec[{concseq|imbrseq}]

MN19. distributor = {;|,}{|iexpr}{|#iexpr,iexpr,iexpr}[msequence]

MN20. lreplicator = index_spec[{;|,}|{concseq|imbrseq,q}]

MN21. rreplicator = index_spec[{concseq|imbrseq}|{;|,}]

MN22. concseq = {morelement;}*concor|{morelement;}+ @

MN23. concor = {gelement,}+ @

MN24. imbrseq = imbr_at_seq|{morelement;}* imbor {;morelement}*

MN25. imbor = {gelement,}* imbrstarel {,gelement}*

```
MN26. imbrstarel = imbrel|imbrel*
MN27. imbrel = (imbrseq)
MN28. imbr_at_seq =
          {morelement;}+ {@|at_or1f|at_or1m|at_or1b} {;morelement}*;
          {@|at_or1f|at_or1m|at_or1b} {;morelement}+
          |{morelement;}+ {@|at_or1f|at_or1m|at_or1b} {;morelement}*;
          {at_or1f|at_or1m}
          |{at_or1f|at_or1b}{;morelement}*;{@|at_or1f|at_or1m|at_or1b}
          {;morelement}+
          |{at_or1f|at_or1b}{;morelement}*;{at_or1f|at_or1m}
          |{morelemet;}+{at_or2fb|at_or2fm|at_or2mm|at_or2mb}{morelement}+
          |{morelemet;}+{at_or2fm|at_or2mm}
          |{at_or2mm|at_or2mb} {;morelement}+
          |at_or2mm
          |@ {morelement;}*{at_or1f|at_or1m}
          |@ {morelement;}*{at_or1f|at_or1m|at_or1b}{;morelement}+
          |{at_or1m|at_or1b} {;morelement}+ @
          |{morelement;}+{at_or1f|at_or1m|at_or1b}{;morelement}* @
          |@ msequence @
MN29. at_or1f = @ {,gelement}+
MN30. at_or1m = {gelement,}+ @ {,gelement}+
MN31. at_or1b = {gelement,}+ @
MN32. at_or2fb = @ {,gelement}* , @
MN33. at_or2fm = @ {,gelement}+ , @ {,gelement}+
MN34. at_or2mb = {gelement,}+ @ {,gelement}* , @
MN35. at_or2mm = {gelement,}+ @ {,gelement}* , @ {,gelement}+
MN36. letter = a|...|z|A|...|Z
MN37. digit = 0|...|9
MN38. iexpr = classical integer expression
```

Context-Sensitive Restrictions

MPrest: Collective names should be declared before any path or process
         involving any of its subscripted operations.

Crest1: The upperbound of the dimensions of the collective names to be greather
         than or equal to their corresponding implicit or explicit lowerbound.

Crest2: Each replicator must specify a non empty range for its index.

Crest3: All expressions indexing collective names should yield integers for all
         the values which the indices they involve take.

Crest4: A collectivisor involving nested replicators must be of the form:

$\#k_n:in_n,fi_n,inc_n[...\#k_1:in_1,fi_1,inc_1[Y(h_1,h_2,...,h_m)]...]$ where h_i

for i=1,..,n are expressions involving indices k_j for j=1,...,n such

that:

1. each k_i for i=1,...,n must appear in at least one dimension,

2. an index k_i, i=1,...,n may only appear together with indices k_j for

j>i in a single expression and in at most i-1 expressions with

indices k_j for j>i.

Crest5: An array identifier may only occur in collectivisors.

Irest1: Identifiers for replicator indices should be distinct from any
identifiers used for simple operations.

Irest2: Replicator indices are only defined inside "[ ]" of the replicator with
which they are associated. In the scope of a replicator index no other
replicator index having the same identifier is permitted.

BRrest: The range of the body-replicator indices should be non empty.

Rrest1: inc≠0 and n = $\frac{fi-in}{inc+1}$ > 0 or t' non empty.

Rrest2: The replicators should generate subscripted operations permitted by the
collectivisors.

Drest1: When a subrange is defined the slices will not be required to contain
the same number of sections but at least as many sections as specified
by the subrange.

Drest2: Inside a k-nested distributor there must only be arrays with at least k
dimensions out of which exactly k should be specified as their
distributable dimensions. Furthermore these arrays must be compatible.

Drest3: inc≠0 and n = $\frac{fi-in}{inc+1}$ ≥ 1.

Drest4: 1 ≤ in+(j-1)*inc ≤ m for j=1,...,n, where m is the minimum number of
slices of all the distributable dimensions of the distributor.

References

[B82] Best E., Adequacy Properties of Path Programs, Theoretical Computer
Science 18 (1982), 149-171.

[Br80] Brauer W. (Ed.), Application and Theory of Petri Nets, Lecture Notes in
Computer Science 84, Springer 1980.

[C82] Cotronis J.Y., Programming and verifying asynchronous systems, Ph.D.
Thesis, Report ASM/123, Computing Laboratory, University of Newcastle
upon Tyne, 1982.

[G75]   Genrich H., Handlungslogik: Logik des Planens, Delegierens und Handelns, unpublished memo, 1975.

[Ho85]   Hoare C.A.R., Notes on Communicating Sequential Systems, In: M.Broy (Ed.) Control Flow and Data Flow: Concepts of Distributing Programming, NATO ASI Series Vol. F14, Springer 1985, 123-204.

[L76]   Lauer P.E., Toward a system specification language based on paths and processes, Part 1: The notation, Report ASM/19, Computing Laboratory, University of Newcastle upon Tyne, 1976.

[L79]   Lauer P.E., COSY Subnotations: Replicators and Basic Notation, Part 4, Report ASM/62, Computing Laboratory, University of Newcastle upon Tyne, 1979.

[L84]   Lauer P.E., The COSY approach to distributed computing systems, In: D.A. Duce (Ed.) Distributed Computing Systems Programme, Peter Peregrinus, London 1984, 107-126.

[LC75]   Lauer P.E., Campbell R.H., Formal semantics for a class of high level primitives for coordinating concurrent processes, Acta Informatica 5 (1975), 247-332.

[LSB79]   Lauer P.E., Shields M.W., Best E., Formal Theory of the Basic COSY Notation, Technical Report 143, Computing Laboratory, University of Newcastle upon Tyne, 1979.

[LSB80]   Lauer P.E., Shields M.W., Best E., Design and Analysis of Highly Parallel and Distributed Systems, Lecture Notes in Computer Science 86, Springer 1980, 451-503.

[LSC81]   Lauer P.E., Shields M.W., Cotronis J.Y., Formal behavioural specification of concurrent systems without globality assumptions, Lecture Notes in Computer Science 107, Springer 1981, 115-151.

[LT78]   Lauer P.E., Torrigiani P.R., Toward a system specification language based on paths and processes, Technical Report 120, Computing Laboratory, University of Newcastle upon Tyne, 1978.

[LTD80]   Lauer P.E., Torrigiani P.R., Devillers R., A COSY Banker, Lecture Notes in Computer Science 83, Springer 1980, 223-229.

[LTS79]   Lauer P.E., Torrigiani P.R., Shields M.W., COSY: a system specification language based on path expressions, Acta Informatica 12 (1979), 109-158.

[Mi86]   Milanetti M., A Prototype of the COSY Environment Ported to UNIX. User Manual, Report ASM/135, Department of Computer Science and Systems, McMaster University, Hamilton, Canada, 1986.

[Oc84]   The OCCAM Programming Manual, INMOS Ltd., Prentice Hall, 1984.

[P62]   Petri C.A., Kommunikation mit Automaten, Schriften des IIM Nr.2, Bonn Universitat, 1962.

[Re82]   Reisig W., Recursive Nets, Informatik-Fachberichte 52, Springer 1982, 125-130.

[S79]   Shields M.W., Adequate Path Expressions, <u>Lecture Notes in Computer Science</u> 70, Springer 1979, 249-265.

[SL77]   Shields M.W., Lauer P.E., The equivalence of path expressions and extended semaphore primitives, Report ASM/42, Computing Laboratory, University of Newcastle upon Tyne, 1977.

[TL77]   Torrigiani P.R., Lauer P.E., An object oriented notation for paths and processes, <u>AICA Annual Conference</u>, Vol. 3, pp. 349-371, Pisa, 1977.

[W85]   Wong P., Users introduction to CS: the compiler and expander for the general macro COSY notation, Report ASM/131, Computing Laboratory, University of Newcastle upon Tyne, 1985.

Linear Algebraic Calculation
of Deadlocks and Traps

Kurt Lautenbach

University of Bonn
Dept. of Computer Science
Wegelerstr. 6
D-5300 Bonn

and

GMD-F1P
Postfach 1240
D-5205 St. Augustin 1

ABSTRACT It is shown how deadlocks and traps of a class of place/transition-nets can be calculated as S-invariants of marked graphs.

Key words: S-invariant, deadlock, trap, place/transition-net, Petri net

CONTENTS

1. Introduction

There are two sorts of sets of places which so far have been of some importance for solving liveness problems in P/T-nets. On the one hand these are sets of places on which the token count can vary in a particular way (deadlocks and traps [Ha 73]), on the other hand sets of places on which a weighted token count does not vary (S-invariants [La 73]).

Deadlocks are sets of places which remain empty once they lost all tokens. Traps, on the contrary, are sets of places which remain marked

once they gained at least one token. The variance consists of the possibility to empty marked deadlocks and to mark empty traps.

It is easy to show that under dead markings (no transition enabled)
there is at least one empty deadlock. So it is necessary for the liveness of a marking that all deadlocks remain marked. A reason for this
might be a structural condition; for example, the condition that all
deadlocks contain initially marked traps. Hack has shown that this condition is necessary and sufficient for the liveness in free choice nets
[Ha 73].

In this paper we want to show that deadlocks and traps can be calculated as special S-invariants of some associated net.

In the second section S-invariants, deadlocks and traps are introduced.
Then, for a restricted class of nets, a close relationship between
deadlocks/traps and certain multisets of circuits is proved and demonstrated by means of an example.

A method of calculating deadlocks/traps based on this relationship is
shown in section four.

2. Deadlocks, traps and incomplete circuit systems.

In this section we very shortly introduce some basic concepts in connection with place/transition nets (P/T-nets), (cf. [BF 86],[La 87]).

Then, for a restricted class of nets, a close relationship between deadlocks and traps on the one hand and certain circuit systems on the other
hand is proved and demonstrated by means of an example.

The restriction of the nets is not really necessary with respect to the
method of calculating deadlocks and traps shown in section three.

Without this restriction one can show a more general relationship between deadlocks/traps and path systems.

But this more general relationship is difficult to show and therefore
outside the scope of this paper.

Definition 2.1   (cf. [La 73],[BF 86],[La 87])

Let $N = (S,T;F,K,W)$ be a finite pure P/T-net. For the sets S and T an
arbitrary but fixed order is assumed:

$$S : s_1 < s_2 < \ldots < s_m$$
$$T : t_1 < t_2 < \ldots < t_n \ ,$$

where $m = |S|$ and $n = |T|$.

(1)   A column vector $v : S \to \mathbf{Z}$ indexed by S is called an <u>S-vector</u> of N.

(2)   A column vector $w : T \to \mathbf{Z}$ indexed by T is called a <u>T-vector</u> of N.

(3)   A matrix $N : S \times T \to \mathbf{Z}$ indexed by S and T such that $N(s_i, t_j) := W(t_j, s_i) - W(s_i, t_j)$ is called the <u>incidence matrix</u> of N.

We usually denote a net and its incidence matrix by the same capital letter.

The i-th row and the j-th column of N are denoted by $N(s_i, -)$ and $N(-, t_j)$ respectively.

Let I be an S-vector of N.

(4)   I is called an <u>S-invariant</u> of N iff $I^T \cdot N = 0^T$.

(5) $P_I \subseteq S$ is called the <u>support of I</u> iff $P_I = \{s \in S \mid I(s) \neq 0\}$.

(6)   An S-invariant I of N is called <u>non-negative</u> iff $I \geq 0$.

(7)   An S-invariant $I \gneq 0$ of N is called <u>minimal</u> iff there exists no S-invariant $I' \gneq 0$ of N with $I' \lneq I$.

(8)   A subnet $I_I = (S_I, T_I; F_I, K_I, W_I)$ of N is called the <u>graphical representation</u> of I iff

S_I is the support of I
$T_I := {}^{\bullet}S_I \cup S_I^{\bullet}$
$F_I := F \cap [(S_I \times T_I) \cup (T_I \times S_I)]$
$K_I(s) := K(s) \quad$ for all $s \in S_I$
$W_I(f) := W(f) \quad$ for all $f \in F_I$.

<u>Corollary 2.2</u>
Every integer linear combination of S-invariants is an S-invariant.
□

Before we come to define the central concepts, deadlock and trap, we will restrict the class of nets we want to work with.

<u>Definition 2.3</u>
R is defined as the class of all finite, pure, and strongly connected P/T-nets $N = (S, T; F, K, W)$ with

$$K(s) = \infty \qquad \text{for all } s \in S,$$
$$W(f) = 1 \qquad \text{for all } f \in F.$$

The main reason for this restriction is to keep simple things simple. Since the nets of R are strongly connected they are covered by circuits which are special paths.

This enables us to demonstrate a relation between deadlocks/traps on the one hand and S-invariants on the other hand in a much simpler way than in the general case.

(In the present case we can work with circuit systems whereas in the general case we have to work with path systems.)

Definition 2.4   (cf. [Ha 73])

Let $N = (S,T;F,K,W) \in R$ and $P \subseteq S$.

(1)   P is called a <u>deadlock</u> of N iff $\dot{} P \subseteq P\dot{}$.
(2)   P is called a <u>trap</u> of N iff $P\dot{} \subseteq \dot{} P$.
(3)   Let P be a deadlock (trap). P is called <u>minimal</u> if there is no deadlock (trap) contained in P as a real subset.

The importance of deadlocks/traps will become evident by the following therorems. For the proofs we refer to [Re 85].

Theorem 2.5

Let $N = (S,T;F,K,W) \in R$, $P \subseteq S$, M a marking of N.

(1)   Let P be a deadlock;
           $(\forall p \in P : M(P) = 0) \Rightarrow (\forall M' \in [M> \; \forall p \in P : M'(p) = 0)$
(2)   Let P be a trap;
           $(\exists p \in P : M(P) > 0) \Rightarrow (\forall M' \in [M> \; \exists p \in P : M'(p) > 0)$
(3)   Let P be the support of a non-negative S-invariant; then $\dot{} P = P\dot{}$.
           $\square$

The statements of this theorem are
     (1) Non-marked deadlocks remain non-marked.
     (2) Marked traps remain marked.
     (3) The supports of non-negative S-invariants are deadlocks and traps as well.

In a way theorem 2.5 justifies to understand deadlocks and traps as "semi-S-invariants".

Theorem 2.6

Let $N = (S,T;F,K,W) \in R$ and M a marking of N.

(1) If every deadlock P ≠ Ø of N contains a trap that is marked under
     M then there exists no dead marking in [M>.

(2) If for a marking M' ∈ [M> all deadlocks P ≠ Ø are marked then M' is
     no dead marking.

                                                                      ▯

A remark is in order here. One should note that non-dead does not mean
live. It might be that under a non-dead M the only activated transition
empties a deadlock when it occurs. So M is not a live marking.

Example 2.7

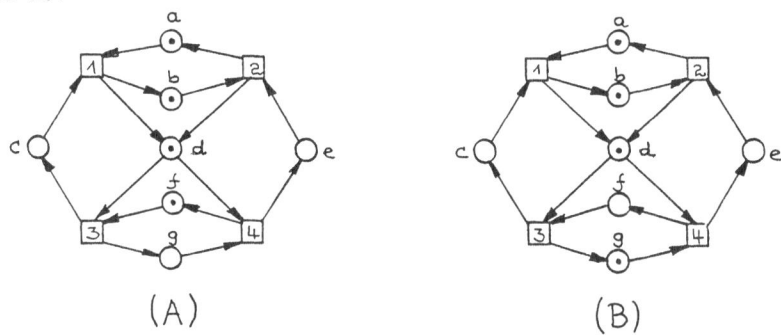

(A)   (B)

Firgure 2.1

The net in fig. 2.1 belongs to the class R. The following sets of places
are minimal deadlocks.

          {a,b}, {c,d,e}, {b,c,d}, {a,d,e}.

The first two deadlocks are also minimal traps. Further minimal traps
do not exist.

Under the marking M_A if fig. 2.1(A) all the deadlocks are marked. The
occurrence sequence 4,2,4 transforms M_A into the marking M_B of fig.
2.1(B) which is dead and under which the deadlock {b,c,d} is non-marked.

Also by the occurrence sequence 3,1,3 M_A is transformed into a dead
marking under which the deadlock {a,d,e} is non-marked.

It is, however, possible to avoid dead markings. One simply has to keep
all deadlocks marked, which can be achieved by the two occurrence sequences 3,1,4,2 and 4,2,3,1 reproducing M_A.

                                                                      ▯

In order to calculate deadlocks and traps we use a method that has been
developed in [La 73] in a similar context.

We will work with circuits. A <u>circuit</u> in a P/T-net $N = (S,T;F,K,W)$ is a graph theoretical circuit in the directed graph $(S \cup T, F)$ with $S \cup T$ as set of nodes and F as set of arcs. A circuit is called <u>simple</u> iff it passes through its nodes exactly once.

Furthermore, we have to work with <u>multisets</u> of circuits. A multiset is a collection of elements that may contain several copies of an element.

<u>Definition 2.8</u>

Let $N = (S,T;F,K,W) \in R$. Let C be a multiset of circuits of N and $C(f)$ denote the number of circuits of C which pass through $f \in F$.

(1)   C is called a <u>D-system</u> iff

$\forall s \in S \; \exists n_s \in \mathbb{N} \; \forall t \in {}^{\bullet}s : C((t,s)) = n_s$.

That is, in a D-system C the same number $n_s \geq 0$ of circuits passes through all <u>input arcs</u> (t,s) of a place $s \in S$. We want to assign these numbers n_s to the places s

$C(s) := n_s$ for all $s \in S$.

(2)   C is called a <u>T-system</u> iff

$\forall s \in S \; \exists n_s \in \mathbb{N} \; \forall t \in s^{\bullet} : C((s,t)) = n_s$.

That is, in a T-system C the same number $n_s \geq 0$ of circuits passes through all <u>output arcs</u> (s,t) of a place $s \in S$.

We want to assign these numbers n_s to the places s

$C(s) := n_s$ for all $s \in S$.

Let C be a D-system or a T-system .

(3)   The set $P_C = \{s \in S \mid C(s) > 0\}$ is called the <u>support</u> of C.

(4)   C is called <u>minimal</u> iff

(a) its support does not contain the support of a circuit system of the same kind as a real subset and

(b) the numbers $C(s)$ are minimal.

In [La 73] D-and T-systems are called "imcomplete" since not necessarily all arcs incidenting with a place are covered.

The relation between deadlocks and D-systems is phrased in the next theorem.

<u>Theorem 2.9</u>

Let $N = (S,T;F,K,W) \in R$ and $P \subseteq S$.

(1)   If C is a D-system the support P of C is a deadlock.

(2)   If P is a minimal deadlock there exists a minimal D-system such
      that P is its support.

For proving this theorem we need the following theorem from real linear
algebra.

Theorem 2.10   (cf. [Ga 60])
Exactly one of the following alternatives holds, where x and y are
column vectors.

Either the equation $Ax = 0$ has a positive solution or the inequality
$y^T A \lessgtr 0^T$ has a solution.

Proof of theorem 2.9
(1)   Let P be the support of a D-system C.
      Let $t \in {}^\cdot P$. Then there exists a place $s \in P$ such that circuits of C
      cover $(t,s) \in F$.

Consequently there exists a place $s' \in {}^\cdot t$ such that $s' \in P$, which implies
$t \in P^\cdot$ and furthermore ${}^\cdot P \subseteq P^\cdot$. P is a deadlock.

(2)   Suppose now that P is a minimal deadlock. First we show that because of the minimality P can be covered by circuits such that the input arcs of all places of P are covered but not necessarily the output
arcs.

We will cover P by a so-called backtrace. Let $s \in P$ and $t \in {}^\cdot s$. Then $t \in {}^\cdot P$
and $t \in P^\cdot$ since ${}^\cdot P \subseteq P^\cdot$. $t \in P^\cdot$ implies that there is a $s_t \in P$ such that
$t \in s_t^\cdot$.

That is, if we start at some place $s \in P$ we may go backwards via an arbitrary input arc (t,s) of s and find then a "next" place $s_t \in P$. By continuing this procedure we obtain a path of the backtrace that exclusively passes through places of P and that ends in a circuit since the
net is finite and covered by circuits.

If there is a place $s' \in P$ with an input arc (t',s') that is not covered
by the backtrace we start a further path at s' such that (t',s') will
be covered.

Since the net is finite this procedure will terminate when all input
arcs of all places of P are covered.

(B)   Let us now assume that the cover does not only consist of circuits.
      Let $N' = (P,T';F',K',W')$ be the subnet of N that is covered. Then

a part of the net N' does not belong to a circuit. By construction, since paths of the backtrace end in a circuit, this part of N' contains the output arc (x,y) of a node x (place or transition) where x belongs to a circuit. By omitting (x,y) we get two subnets of N' which both contain places. We omit also the net y belongs to and decrease the number of places by doing so. Let N" = (P',T";F",K",W") be the subnet of N' the node x belongs to. Then P' is a real subset of P. If N" still contains parts which are not covered by circuits we continue the procedure of decreasing the net until we reach a subnet N'" = (P",T'"; F'", K'", W'") of N" whose set P" of places is a real subset of P and where all input arcs of all places of P" are covered by circuits. Now we can prove as in (1) that P" is a deadlock. This contradicts the fact that P is a minimal deadlock. So, we have proved that minimal deadlocks are covered by circuits in such a way that all input arcs of the places are covered.

(C)   Let us now assume that more than one input place of a transition belongs to P. For example, let $t \in T$ and $\{s,s'\} \subseteq \ ^{\bullet}t \cap P$.

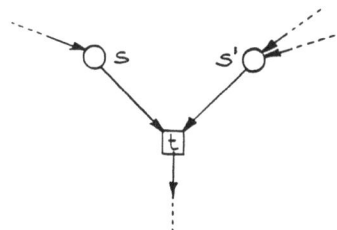

Now we omit the arc (s',t). If the resulting net has parts which are not covered by circuits, we omit also these parts as in (B) and reach the subnet N'" = (P",T'" ; F'" ,K'" , W'") whose set P" of places is a real subset of P and is a deadlock as well. This again contradicts the minimality of P. So, P contains at most one input place of a transition.

(D)   Now we will show that the minimal deadlock P can be covered by a D-system. We follow the corresponding proofs in [Be 74] and [Be 82].

For the subnet N' = (P,T';F',K',W') of N we define the following matrix

$$Z : P \times P \to \mathbb{N}$$
$$Z(s,s') := |s^{\bullet} \cap \ ^{\bullet}s'|.$$

In other words, $Z(s,s') = m \in \mathbb{N}^{+}$ iff there exist m transitions $t_1,...,t_m$ sucht that

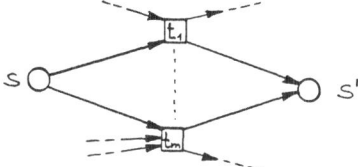

(Recall that the net is pure).

For all $s' \in P$ we find for the column sums

$$\sum_{s \in P} Z(s,s') = |{}^{\cdot}s'| \; .$$

Now we define the matrix

$$A : P \times P \to \mathbf{Z}$$

$$A(s,s') := \begin{cases} Z(s,s') & , \text{ if } s \neq s' \\ -|{}^{\cdot}s'| & , \text{ if } s = s' \; . \end{cases}$$

Let us, for the moment, assume that $A \cdot x = 0$ has a positive solution. Then there exists also a positive integer solution v.

$$A \cdot v = 0$$

yields for all $s \in P$:

$$\sum_{s' \in P} A(s,s') \cdot v(s') = \sum_{s' \in P - \{s\}} Z(s,s') \cdot v(s') - |{}^{\cdot}s| \cdot v(s) = 0$$

Then by

$$C(s) := \begin{cases} v(s) & , \text{ if } s \in P \\ 0 & , \text{ if } s \notin P \end{cases}$$

where $v(s) > 0$ if $s \in P$

we define a D-system C the support of which is P.

Because of (1) it is impossible that there is a D-system whose support is a real subset of P. If C is not minimal with respect to the values $C(s)$ for $s \in P$ then there exists a minimal D-system C' whose support is also P.

To complete the proof we assume that

$$y^T \cdot A \lneq 0^T$$

has a solution w.

Furthermore, let $P := \{s_1, \ldots, s_n\}$, $A_{ij} := A(s_i, s_j)$.

For all $j \in \{1,\ldots,n\}$ we have

$$A_{jj} = - \sum_{\substack{i=1 \\ i \neq j}}^{n} A_{ij}$$

which implies

$$\sum_{i=1}^{n} w_i A_{ij} = \sum_{\substack{i=1 \\ i \neq j}}^{n} (w_i - w_j) A_{ij} \leq 0 \quad \text{for all } j \in \{1,\ldots,n\}$$

and

$$\sum_{i=1}^{n} w_i A'_{ij} = \sum_{\substack{i=1 \\ i \neq j'}}^{n} (w_i - w_j') A'_{ij} < 0 \quad \text{for at least one } j' \in \{1,\ldots,n\}.$$

Let, without loss of generality, $j' = 1$. This yields

$w_1 > w_2 \vee w_1 > w_3 \vee \ldots \vee w_1 > w_n$. We assume $w_1 > w_2$. For $j = 2$ we have

$$(w_1 - w_2) A_{12} + (w_3 - w_2) A_{32} + \ldots + (w_n - w_2) A_{n2} \leq 0 .$$

Since $(w_1 - w_2) A_{12} \geq 0$ we get $w_2 \geq w_3 \vee w_2 \geq w_4 \vee \ldots \vee w_2 \geq w_n$ and we assume $w_2 \geq w_3$.

Continuing that way up to $j = n-1$ we obtain $w_1 > w_2 \geq w_3 \geq \ldots \geq w_{n-1} \geq w_n$.

For $j = n$ we have

$$(w_1 - w_n) A_{1n} + (w_2 - w_n) A_{2n} + \ldots + (w_{n-1} - w_n) A_{n-1,n} \leq 0 .$$

Since $(w_2 - w_n) A_{2n} \geq 0 \wedge \ldots \wedge (w_{n-1} - w_n) A_{n-1,n} \geq 0$ we find $w_n \geq w_1$ which is a contradiction to $w_1 > w_n$.

So
$$y^T \cdot A \not\leq 0^T$$

has no solution which, according to theorem 2.10, means that $A \cdot x - 0$ has a positive solution (cf. example 2.13).

□

Corollary 2.11

Let $N = (S,T;F,K,W) \in R$ and $P \subseteq S$.

(1) If C is a T-system the support P of C is a trap.

(2) If P is a minimal trap there exists a minimal T-system s.t. P is its support.

Proof   Very easy by means of the following lemma.

□

Lemma 2.12

Let $N = (S,T;F,K,W) \in R$ and $N' = (S,T;F^{-1},K,W')$ where $W'(f) = 1$ for all $f \in F^{-1}$.

(1) Every deadlock (trap) of N is a trap (deadlock) of N' and vice versa.

(2) Every D-system (T-system) of N is a T-system (D-system) of N' and vice versa.

Proof   Trivial

                                                                          □

Example 2.13   (cf. [La 73])

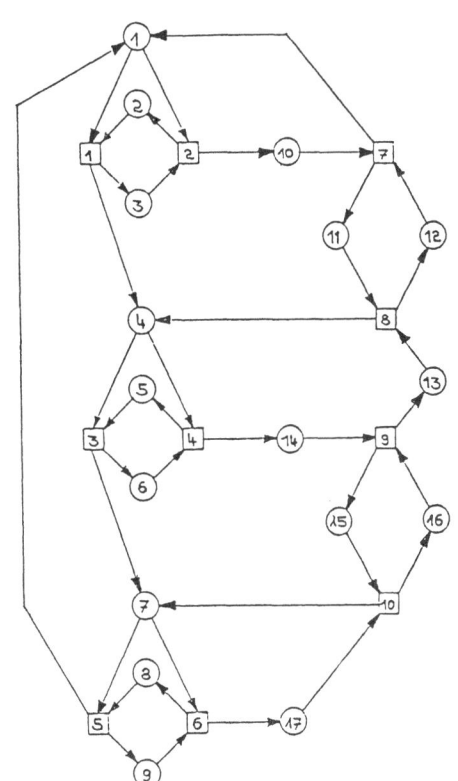

The net N

Figure 2.2

$P = \{1,4,7,12,13,14,15\}$ is a minimal deadlock of the net N in fig. 2.2.

The matrix A is shown in fig. 2.3.

	1	4	7	12	13	14	15
1	-2	1					
4		-2	1			1	
7	1		-2				
A = 12	1			-1			
13		1		1	-1		
14					1	-1	1
15			1				-1

The matrix A

Figure 2.3

v where $v^T = (2,4,1,2,6,7,1)$ is a positive solution of $Ax = 0$. The corresponding minimal D-system C is shown in fig. 2.4.

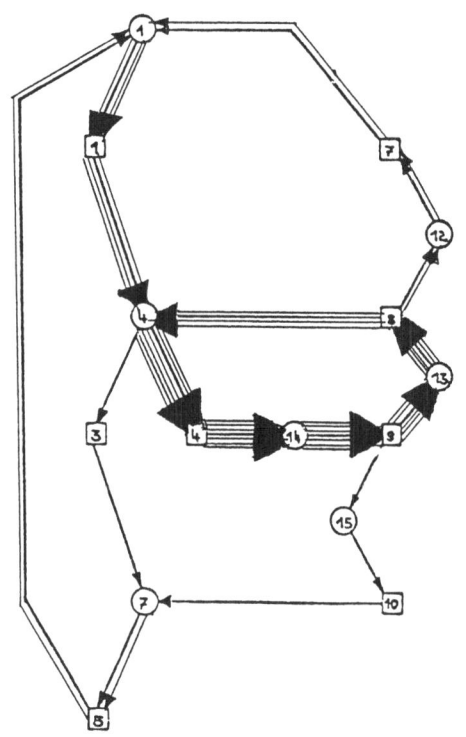

The D-system C

Figure 2.4

Before we show how to calculate D- and T-systems we want to show that unions of deadlocks and traps are also deadlocks and traps, respectively.

Lemma 2.14

Let $N = (S,T;F,K,W) \in R$ and P_1, $P_2 \subseteq S$.

If P_1 and P_2 are deadlocks (traps) of N, also $P_1 \cup P_2$ is a deadlock (trap) of N.

Proof

$$^\bullet P_1 \cup {}^\bullet P_2 = {}^\bullet (P_1 \cup P_2), \quad P_1^\bullet \cup P_2^\bullet = (P_1 \cup P_2)^\bullet$$

$$^\bullet P_1 \subseteq P_1^\bullet \text{ and } {}^\bullet P_2 \subseteq P_2^\bullet \text{ imply } {}^\bullet (P_1 \cup P_2) \subseteq (P_1 \cup P_2)^\bullet$$

$$P_1^\bullet \subseteq {}^\bullet P_1 \text{ and } P_2^\bullet \subseteq {}^\bullet P_2 \text{ imply } (P_1 \cup P_2)^\bullet \subseteq {}^\bullet (P_1 \cup P_2).$$

$\square$

3. Minimal generating systems for deadlocks and traps

In this section first it is shown that the minimal S-invariants in marked graphs are characteristic vectors of minimal circuits. Next, to every net $N \in R$ a marked graph $\hat{N}$ is associated in such a way that the D- and T-systems of N can be calculated as special S-invariants of $\hat{N}$, i.e. as non-negative, integer solutions of a homogeneous linear equation system. Finally, we define minimal generating systems for deadlocks and show that all deadlocks of a net $N \in R$ are marked iff the deadlocks of the generating system are marked.

We want to start by stating that S-invariants "are" deadlocks and traps as well.

Lemma 3.1

Let $N = (S,T;F,K,W) \in R$ and let $I \overset{\geq}{\neq} 0$ be an S-invariant of N.

Then the support P of I is a deadlock and a trap as well, i.e. ${}^\bullet P = P^\bullet$.

Proof

Suppose P is no deadlock, i.e. there exists a transition $t_j \in {}^\bullet P \setminus P^\bullet$

which means
$$\exists s \in P : t_j \in {}^\bullet s$$
$$\forall s \in P : \neg (t_j \in s^\bullet).$$

This can be phrased by means of $N(-,t_j)$:

$$\exists s \in P : N(s,t_j) = 1$$

$$\forall s \in P : N(s,t_j) \neq -1$$

Consequently $I^T \cdot N(-,t_j) > 0$

which contradicts the assumption that I is an S-invariant, so P must be a deadlock.

Similarly, one shows that P is a trap.

□

Remark 3.2

The converse of this lemma is not true in general.

□

The next step towards calculating deadlocks and traps consists of associating a marked graph to every net of the class R.

Definition 3.3

Let $N = (S,T;F,K,W) \in R$. N is called a (stronly connected) marked graph or synchronization graph iff

$$\forall s \in S : |{}^\cdot s| = |s^\cdot| = 1.$$

The subclass of marked graphs in R will be denoted by RG.
For the. nets in RG we can characterize the minimal S-invariants:

Theorem 3.4

Let $N = (S,T;F,K,W) \in RG$ and let $I \ngeq O$ be an S-vector of N.

Then I is a minimal S-invariant of N iff it is the characteristic vector of a simple circuit of N.

Proof

Let I be the characteristic vector of a simple circuit C of N. Let the support of C be $\{s_1,\ldots,s_n\}$.

Without loss of generality we may assume that for C the following holds:

$$s_i^\cdot \cap {}^\cdot s_{i+1} = \{t_i\} \quad \text{for } 1 \le i \le n-1$$

and

$$s_n^\cdot \cap {}^\cdot s_1 = \{t_n\}.$$

Then the upper part of the incidence matrix N is (recall that C is simple)

	t_1	t_2	t_3	$\cdots\cdots$	t_{n-1}	t_n	t_{n-1}	$\cdots\cdots$	$t_{|T|}$
s_1	-1					1			
s_2	1	-1							
s_3		1	-1						
$\vdots$									
s_{n-1}					-1				
s_n					1	-1			

and the fact that the sum of the first n rows equals 0^T can be expressed by

$$I^T \cdot N = 0^T.$$

So I is an S-invariant. Furthermore, I is minimal since none of the first n rows can be omitted.

Now let I be a minimal S-invariant. Because of lemma 3.1 the support P of I is a minimal deadlock such that because of theorem 2.9 P is the support of a minimal D-system. Since $N \in RG$ this minimal D-system is a simple circuit C and, clearly, I its characteristic vector.

□

Corollary 3.5
Let $N = (S,T;F,K,W) \in RG$.
Then there exists a non-negative S-invariant I of N such that N is covered by I, i.e.

$$\forall s \in S : I(s) > 0 .$$

Proof
Since N is strongly connceted there exists for every $s \in S$ a simple circuit C^s through s. Let I^s be the characteristic vector of C^s. Then

$$I := \sum_{s \in S} I^s$$

is, according to corollary 2.2 and theorem 3.4, a non-negative S-invariant covering N.

□

Theorem 3.6
Let $N = (S,T;F,K,W) \in RG$.
(1) Every non-negative S-invariant is representable as a non-negative integer linear combination of minimal non-negative S-invariants.

(2) There exists a basis of minimal non-negative S-invariants for the
     solutions of

$$(3.1) \quad X^T \cdot N = 0^T.$$

<u>Proof</u>

(1) Either a non-negative S-invariant is minimal or it is the sum of
     two non-negative S-invariants, which either are minimal or a sum
of non-negative S-invariants, etc.

(2) Let $V \neq 0$ be an arbitrary solution of (3.1). Then there exists a
     rational number $r \neq 0$ such that rV is an <u>integer</u> solution and, con-
sequently, an S-invariant.

Let E be the (finite) set of minimal non-negative S-invariants of N.
From corollary 3.5 then the existence of a <u>positive</u> S-invariant

$$I = \sum_{J \in E} a_J J \qquad \text{where } a_J \in \mathbb{N} \qquad \text{follows.}$$

That is, for all places $p \in S$ there is at least one $J \in E$ with $J(p) > 0$.
So, one can define a positive S-invariant

$$G = rV + \sum_{J \in E} b_J J \qquad \text{where } b_j \in \mathbb{N} ,$$

and because of (1) there is another representation of G, namely

$$G = \sum_{J \in E} c_J J \qquad \text{where } c_J \in \mathbb{N} .$$

This yields for the arbitrary solution V of (3.1)

$$V = \frac{1}{r} \sum_{J \in E} (c_J - b_J) J .$$

Thus, E is a generating system for the solution of (3.1). Consequently,
E contains a basis.

$\square$

In order to calculate deadlocks and traps in nets of the class R we
want to associate to every $N \in R$ a marked graph $\hat{N} \in RG$. $\hat{N}$ will be deduced
from N by successively replacing shared places by subnets without
shared places.

<u>Definition 3.7</u> (Replacement of shared places)
Let $N = (S,T;F,K,W) \in R$ and let $s \in S$ be a shared place, i.e.
$|{}^{\bullet}s| > 1 \vee |s^{\bullet}| > 1$.

The net $N' = (S',T';F',K',W') \in R$ obtained form N by <u>replacing s</u> is defined as follows (cf. fig. 3.1).

$$S' := (S \smallsetminus \{s\}) \cup \bigcup_{t \in {}^\bullet s} \{(t,s)\} \cup \bigcup_{t \in s^\bullet} \{(s,t)\}$$

$$'T' := T \cup \{s\}$$

$$F' := (F \smallsetminus (\bigcup_{t \in {}^\bullet s} \{(t,s)\} \cup \bigcup_{t \in s^\bullet} \{(s,t)\}))$$

$$\cup \bigcup_{t \in {}^\bullet s} \{(t,(t,s)),((t,s),s)\}$$

$$\cup \bigcup_{t \in s^\bullet} \{(s,(s,t)),((s,t),t)\}$$

$$K'(p) := \infty \qquad \text{for all } p \in S'$$

$$W'(f) := 1 \qquad \text{for all } f \in F'$$

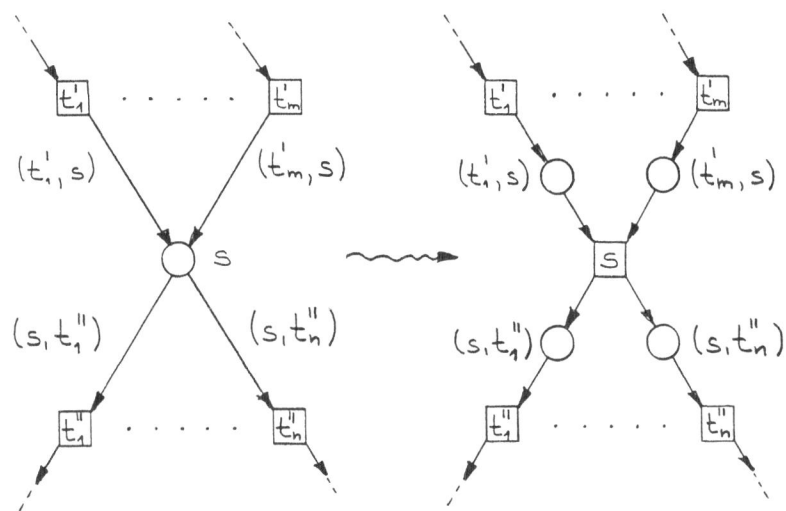

The replacement of $s \in S$

Figure 3.1

Definition 3.8

Let $N \in R$ and $\hat{N} \in RG$. $\hat{N}$ is called the <u>marked graph</u> (or <u>synchronization graph</u>) <u>associated with N</u> iff replacing all shared places in N yields $\hat{N}$.

Theorem 3.9

Let $N = (S,T;F,K,W) \in R$, $H \subseteq S$ be the set of shared places and C be a multiset of circuits in N. Let $\hat{N} = (\hat{S},\hat{T};\hat{F},\hat{K},\hat{W})$ RG be the associate marked graph.

(1)   C is a D-system of N iff

there is a non-negative S-invariant V of $\hat{N}$ such that

$$v^T \cdot N = 0^T \qquad \text{where}$$

(3.2') $\forall s \in H \; \forall t \in {}^\cdot s \; \exists n(s) \in \mathbb{N} : V((t,s)) = C((t,s)) = n(s)$

and   (3.2") $\forall s \in S \smallsetminus H \; \forall t \in {}^\cdot s : V(s) = C((t,s))$.

(2)   C is a T-system of N iff

there is a non-negative S-invariant V of $\hat{N}$ such that

$$v^T \cdot \hat{N} = 0^T \qquad \text{where}$$

(3.3') $\forall s \in H \; \forall t \in s^\cdot \; \exists n(s) \in \mathbb{N} : V((s,t)) = C((s,t)) = n(s)$

and   (3.3") $\forall s \in S \smallsetminus H \; \forall t \in s^\cdot : V(s) = C((s,t))$.

Proof

The proof is based on the fact that the circuits of N and $\hat{N}$ correspond uniquely to each other. We assume without loss of generality that the circuits of C are simple.

(1)   Let C be a D-system of circuits in N and let $\hat{C}$ be the correspond-
ing multiset of circuits in $\hat{N}$. Let k be the characteristic vector of a (simple) circuit in $\hat{N}$ and $\hat{C}(k)$ its multiplicity in $\hat{C}$. Then

$$V := \sum_{k \in \hat{C}} \hat{C}(k) k$$

is a non-negative S-invariant of $\hat{N}$ since theorem 3.4 and corollary 2.2. Obviously, V satisfies (3.2') and (3.2").

Now let V be a non-negative S-invariant of $\hat{N}$ satisfying (3.2') and (3.2"). Since theorems 3.6(1) and 3.4, V is representable as a non-ne-gative integer linear combination of characteristic vectors of simple circuits:

$$V = \sum_{k \text{ simple}} m(k) \cdot k, \quad m(k) \in \mathbb{N} \quad .$$

Let $\hat{C}_V = \{k \mid k \text{ simple circuit} \wedge m(k) > 0\}$ be the corresponding circuit system in $\hat{N}$. Since (3.2') and (3.2"), C is the corresponding circuit system in N and C is a D-system.

(2)   is to be proved in the same way.

                                                                               □

Remark 3.10

We will use the result of theorem 3.9 for calculating deadlocks and traps. Since we then have to do with all D- or T-systems and not with

a special system C the additional restrictions for the linear homogeneous equation system $X^T \cdot \hat{N} = O^T$ are

(3.2) $\forall s \in H \; \forall t \in \text{}^{\bullet}s \; \exists n(s) \in \mathbb{N} : X(t,s) = n(s)$ and

(3.3) $\forall s \in H \; \forall t \in s^{\bullet} \; \exists n(s) \in \mathbb{N} : X(s,t) = n(s)$,   respectively.

It is more convenient for the calculation to express (3.2) and (3.3)
as equations.

Let $s \in H$ and $\text{}^{\bullet}s = \{t_1, \ldots, t_m\}$. Then for s the condition (3.2) is

$$- X(t_1,s) + X(t_2,s) = O$$
$$- x(t_2,s) + X(t_3,s) = O$$
$$\cdots\cdots\cdots\cdots\cdots\cdots$$
$$- X(t_{m-1},s) + X(t_m,s) = O \; .$$

Thus, for calculating solutions satisfying (3.2) we have to augment
the equation system

$$X^T \cdot N = O^T$$

by $\sum\limits_{s \in H} (|\text{}^{\bullet}s| - 1) = (\sum\limits_{s \in H} |\text{}^{\bullet}s|) - |H|$ equations.

For solutions satisfying (3.3) we have to add $(\sum\limits_{s \in H} |s^{\bullet}|) - |H|$ equations.

In the sequel we will denote the augmented equation systems by

(3.5) $X^T \cdot N_D = O^T$ and

(3.6) $X^T \cdot N_T = O^T$,   respectively.

$\square$

Definition 3.11

Let $N \in R$ and let $\hat{N} \in RG$ be its associated marked graph.

Let $\hat{E}_D$ be a set of non-negative integer solutions of (3.5) and E_D the
corresponding set of deadlocks of N.

Let $\hat{E}_T$ be a set on non-negative integer solutions of (3.6) and E_T the
corresponding set of traps of N.

E_D and E_T are called _minimal generating systems_ iff $\hat{E}_D$ and $\hat{E}_T$ are
minimal generating systems for non-negative solutions of (3.5) and
(3.6), respectively.

Corollary 3.12

Let $N \in R$ and let E_D and E_T be minimal generating systems for deadlocks and traps, respectively.

Then every deadlock (trap) is the union of elements of $E_D (E_T)$.

Proof

Let P be a deadlock and let $X_P \gneq O$ be the corresponding solution of (3.5). Then

$$X_P = \Sigma \, a_Q X_Q \quad \text{where } X_Q \in \hat{E}_D, \quad a_Q \in \mathbb{N}.$$

Consequently, P is the union of all those Q where $X_Q \in E_D$ and $a_Q > 0$. These Q, by definition 3.11, are elements of E_D.

The proof is analogous if P is a trap.

$\square$

Theorem 3.13

Let $N = (S, T; F, K, W) \in R$, let E_D and E_T be minimal generating systems for deadlocks and traps, respectively. Let P be a deadlock, Q a trap and M a marking of N.

(1) P is marked under M if all elements of E_D are marked under M.
(2) Q is marked under M if all elements of E_T are marked under M.
(3) P contains a trap marked under M if all elements of E_D contain traps which are marked under M.

Proof

(1) Because of corollary 3.12 every deadlock as a union of elements of E_D is marked if these elements are.
(2) analogous
(3) Since (1) P is the union of deadlocks containing marked traps. The union of these marked traps is a marked trap contained in P (cf. lemma 2.14).

$\square$

Corollary 3.14

Let $N = (S, T; F, K, W) \in R$, let E_D and E_T be minimal generating systems for deadlocks and traps, respectively.

All deadlocks (traps) are marked under a marking M of N iff all $P \in E_D$ ($P \in E_T$) are marked under M .

$\square$

4. Conclusion

We have introduced S- and T-invariants [La 73] and deadlocks/traps
[Ha 73] as structural net components which can be used for analyzing
P/T-nets in a simulation-free way.

Even though there is a considerable difference between S-invariants
and deadlocks/traps it is possible to calculate the latter as S-invariants of associated marked graphs. On the one hand this leads to
generating systems for deadlocks/traps, on the other hand packages for
calculating S-invariants can also be used for calculating deadlocks/
traps.

We did not deal with methods for the calculation itself. Here we refer
to [AT 85], [Ja 85], [MS 82], and [Pa 85].

5. References

[AT 85]   Alaiwan, H.; Toudic, J.M.:
            Recherche des semi-flots, des verrous et des trappes dans
            le réseaux de Petri, Technique et Science Informatiques,
            Vol. 4, 103-112, 1985

[Be 74]   Best, E.:
            Beiträge zur Petri-Netz-Theorie, Diplomarbeit, Universität
            Karlsruhe, 1974, (in German)

[Be 82]   Best, E.:
            A. Theorem on Open Covers in Petri Nets,
            Gesellschaft für Informatik, Special Interest Groups 'Petri
            Nets and Related System Models', Newsletter No. 11, 1982

[BF 86]   Best, E.; Fernandez, C.:
            Notations and Terminology on Petri Net Theory, Gesellschaft
            für Mathematik und Datenverarbeitung, Arbeitspapiere der
            GMD, No. 195, 1986

[CEHP 71]   Commoner, F.; Holt, A.; Even, S.; Pnueli, A.:
            Marked Directed Graphs, JCSS, Vol. 5, 511-523, 1971

[Ga 60]   Gale, D.:
            The Theory of Linear Economic Models, Mc Graw-Hill, NY,
            Toronto, London, 1960

[GL 73]   Genrich, H.J.; Lautenbach, K.:
            Synchronisationsgraphen, Acta Informatica, Vol.2, 143-161,
            1973 (in German)

[GLT 80]   Genrich, H.J.; Lautenbach, K.; Thiagarajan, P.S.:
            Elements of General Net Theory, In: Net Theory and Applica-
            tions, Brauer ed., Springer LNCS 84, 1980

[Ha 73]   Hack, M.H.T.:
            Analysis of Production Schemata by Petri Nets,
            MIT Project MAC, TR 94, 1973

[Ja 85]   Jaxy, M.:
Analyse linearer diophantischer Ungleichungs- und Gleichungs-
systeme im Hinblick auf Anwendungen in der Theorie der
Petri-Netze, Diplomarbeit, Universität Bonn, 1985, (in
German)

[La 73]   Lautenbach, K.:
Exakte Bedingungen der Lebendigkeit für eine Klasse von
Petri-Netzen, Gesellschaft für Mathematik und Datenverarbei-
tung, GMD-Report No. 82, 1973, (in German)

[La 75]   Lautenbach K.:
Liveness in Petri Nets, Gesellschaft für Mathematik und
Datenverarbeitung, GMD-ISF Internal Report 02.1/75-7-29,
1975

[La 87]   Lautenbach, K.:
Linear Algebraic Techniques for Place/transition Nets, To
appear in Springer LNCS, 1987

[MR 80]   Memmi, G.; Roucairol, G.:
Linear Algebra in Net Theory, In: Net Theory and Applications, Brauer ed., Springer LNCS 84, 1980

[MS 82]   Martinez, J.; Silva, M.:
A Simple and Fast Algorithm to Obtain All Invariants of a
Generalized Petri Net, In: Application and Theory of Petri
Nets, Girault, Reisig eds., Springer, Informatik-Fachberichte
No. 52, 1982

[Pa 85]   Pascoletti, K.-H.:
Diophantische Systeme und Lösungsmethoden zur Bestimmung
aller Invarianten in Petri-Netzen, Doctoral Dissertation,
Universität Bonn, 1985, (in German)

[Re 85]   Reisig, W.:
Petri Nets, Springer-Verlag, Berlin, Heidelberg, New York,
Tokyo, 1985

ON DIFFERENT KINDS OF FROZEN TOKENS IN PETRI NETS[+]

Lu,Wei-Ming

Institute of Mathematics, Academia Sinica

Beijing, CHINA

ABSTRACT. Containing a frozen token in a Petri net means that there exists an infinite occurrence sequence in the net such that at least one token in some place is never moved. Frozen tokens are classified here by considering occurrence sequences and processes associated to occurrence sequences. And some applications about frozen tokens are suggested.

1,MOTIVATION.

Petri nets ([6],[7],[8],[9]) have been proposed and widely accepted to be the models of concurrent systems.

A net having a frozen token means that by an infinite run of the net at least one token is superfluous in a place.

For example, the net Fig.1.1 is a system model s.t. when it is executed the light in s is turned on.

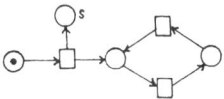

Fig.1.1 A system with a light in s

Of course the net contains a frozen token.

The concept of frozen token was first introduced in [2], and it was suggested to be a way to test distributed systems modeled by Petri nets in [4].

In studying the system behaviour and the relationship between the behav-

+ This is a work supported by the Science Fund of the Chinese Academy
  of Science under Grant M85182

iour and the tokens in a net it was found that a frozen token may be corresponding to some system behaviour, e.g. the just mentioned systems. In order to apply frozen tokens in a modeled system more easily we classify them in the section 3. Afterwards some applications of them are suggested in section 4.

2, PRELIMINARY.

This section is intended to introduce some basic definitions of nets that we make use of. And most of the definitions introduced here come from [1] in order to be commonly accepted.

2.1 Definition A triple $N=(S,T;F)$ is called a net iff
  (a) $S \cap T = \phi$; $S \cup T \neq \phi$;
  (b) $F \subseteq (S \times T) \cup (T \times S)$; $dom(F) \cup cod(F) = S \cup T$.

Graphically an element of the set S is represented by a circle, an element of the set T by a box and an element (x,y) of the flow relation F by an arrow connecting the two nodes x and y. As usual the dot notation is used
  $\cdot x = \{ y \in X \mid (y,x) \in F \}$, $x \cdot = \{ y \in X \mid (x,y) \in F \}$,
the preset and the postset of $x \in X = S \cup T$ respectively.
In the paper the nets in discussion are finite, i.e. $|S| < \infty$, $|T| < \infty$. Note that F can be viewed as a function: $F(x,y) = \begin{cases} 1 & \text{iff } (x,y) \in F \\ 0 & \text{otherwise} \end{cases}$

2.2 Definition A marking is a function: $S \rightarrow N$; a net together with a marking is called a marked net; M_0 will represent the initial marking.

Normally $MN=(S,T;F,M_0)$ is written as a marked net. In diagrams, $M(s)$, the marking of the place s, will be represented by drawing $M(s)$ tokens in the place s, which are black dots.

2.3 Definition Let M be a marking of a net N,
  (a) a transition $t \in T$ is enabled at M and written as $M[t>$ iff $\forall s \in \cdot t$: $M(s) \geq 1$;
  (b) a transition which is enabled at M may occur to yield a follower marking M' as follows:
      $\forall s \in S$: $M'(s) = M(s) - F(s,t) + F(t,s)$;
      it is written as $M[t> M'$

(c) a sequence $\beta = M_0 t_1 M_1 t_2 \ldots t_n M_n$, $n \geq 0$, where $t_i \in T$, and the M_i are markings, $0 \leq i \leq n$, is called an occurrence sequence of the marked net MN iff $M_{i-1}[t_i > M_i$, $1 \leq i \leq n$. A sequence $M_0 t_1 M_1 t_2 \ldots$ is an infinite occurrence sequence.

(d) $[M>$ is the set of marking reachable from M by finite occurrence sequences of MN starting with M.

In the definition the length of β is n and it is written as $|\beta| = n$. So the length of $\beta = M_0$ is zero. If the markings in β are omitted, it is a transition sequence.

2.4 Definition Let $MN = (S, T; F, M_0)$ be a marked net.
  (a) $s \in S$ is n-safe ($n \in \mathbb{N}$) iff $\forall M \in [M_0 >$: $M(s) \leq n$;
  (b) MN is n-safe ($n \in \mathbb{N}$) iff $\forall s \in S$: s is n-safe;
  (c) MN is safe iff $\exists n \in \mathbb{N}$: MN is n-safe.

It is often said that MN is safe or n-safe iff $\exists n \in \mathbb{N} \forall s \in S \forall M \in [M_0 >$: $M(s) \leq n$. Now comes the definition of frozen token and the next section will go deep into it.

2.5 Definition ([2]) A marked net MN will be said to contain a frozen token iff $\exists s \in S \exists$ an infinite occurrence sequence of MN, $\beta = M_1 t_1 M_2 t_2 \ldots$ s.t.

$$\forall i, \quad i \geq 1: M_i(s) \geq \begin{cases} 1 & \text{if } s \notin {}^\bullet t_i \\ 2 & \text{if } s \in {}^\bullet t_i \end{cases}$$

With the definition, a token is 'frozen' in s by β, i.e. at least one token is superfluous.
Note that if $\forall t_i$, $i \geq 1$, $\forall s \in {}^\bullet t_i$: $(t_i, s) \notin F$, then the formula in the above definition can be written as $\forall i$, $i \geq 1$: $M_i(s) \geq 1$. The net in Fig.2.1 (a) does not contain frozen tokens, but the net in (b) contains. Obviously the formula $\forall i: M_i(s) \geq 1$ does not hold for (a) because of the structure $(s, t) \in F \wedge (t, s) \in F$, but it holds for (b).

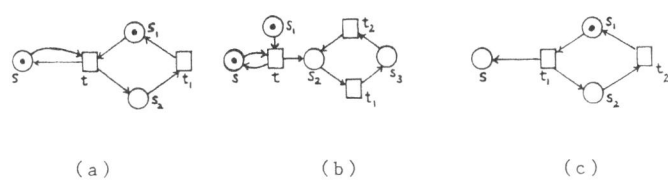

(a)   (b)   (c)

Fig.2.1 which marked net contains frozen token?

For ease to remember, sometimes we say that if a net is pure $(F \cap F^{-1} = \phi)$ then $M_i(s) \geqslant 1$ can be used. Note that the net in Fig.2.1(b) is not pure.

2.6 Proposition ([2]) Suppose that a marked net MN is not
   n-safe; then it contains a frozen token.

The net Fig 2.1 (c) is not n-safe because s can have an unlimited number of tokens. Therefore it does contain a frozen token. In order to study the relationship between a frozen token and system behaviour it is needed to introduce the process definition.

2.7 Definition A net $N=(S,T;F)$ is called an occurrence net iff
  (a) $\forall s \in S:$ $|\cdot s| \leqslant 1 \wedge |s \cdot| \leqslant 1$;
  (b) $\forall x,y \in X:$ $(x,y) \in F^+ \Rightarrow (y,x) \notin F^+$.

The part (b) of the definition above means that any occurrence net is acyclic.
Note that although the marked nets in discussion are finite, the occurrence nets in this paper can be infinite.

2.8. Definition Let $N=(B,E;F)$ be an occurrence net.
  (a) $\leqslant = F^*$, $< = \leqslant \setminus id_x$, $\geqslant = (\leqslant)^{-1}$;
  (b) $li = \leqslant \cup \geqslant$, $co = (X \times X \setminus li) \cup id_x$;
  (c) $l \subseteq X$ is a li-set (chain) iff $\forall x,y \in l:$ $(x,y) \in li$;
  (d) $l \subseteq X$ is a line (maximal chain) iff l is a li-set and
    $\forall z \in X \setminus l \ni x \in l$, $(x,z) \notin li$;
  (e) $L(X;\leqslant)$ or $L(N)$ is the set of lines of $(X;\leqslant)$;
    where $(X;\leqslant)$ is a structure associated to the net N.
  (f) $c \subseteq X$ is a co-set (antichain) iff $\forall x,y \in c:$ $(x,y) \in co$;
  (g) $c \subseteq X$ is a cut (maximal antichain) iff c is a co-set and
    $\forall z \in X \setminus c \ni x \in c:$ $(x,z) \notin co$.
  (h) $C(X;\leqslant)$ or $C(N)$ is the set of cuts of $(X;\leqslant)$.
  (i) a B-cut (E-cut) c is a cut s.t. $c \subseteq B (c \subseteq E,$ respectively).
  (j) $BC(X;\leqslant)$ or $BC(N)$ is the set of B-cuts of N.

Note that the E-elements in N can be considered as events, obviously the events in a line l have sequential behaviour and the events in a cut c have concurrent behaviour, from the above definition. It is assumed that a li-set (a co-set) could be the base of a line (a cut). Then for every li-set (co-set) there is a line (a cut) containing it.

2.9. Definition Let $N=(B,E;F)$ be an occurrence net and $X=B \cup E$.
  (a) $\circ N = \{x \in X \mid \cdot x = \phi\}$, $N^\circ = \{x \in X \mid x \cdot = \phi\}$;

(b) For $A \subseteq X$, $\downarrow A = \{y \in X \mid \exists x \in A: \quad y \leqslant x\}$,

$\uparrow A = \{y \in X \mid \exists x \in A: \quad x \leqslant y\}$;

for $x \in X$, sometimes $\downarrow x$ ($\uparrow x$) is written instead of $\downarrow\{x\}$ ($\uparrow\{x\}$).

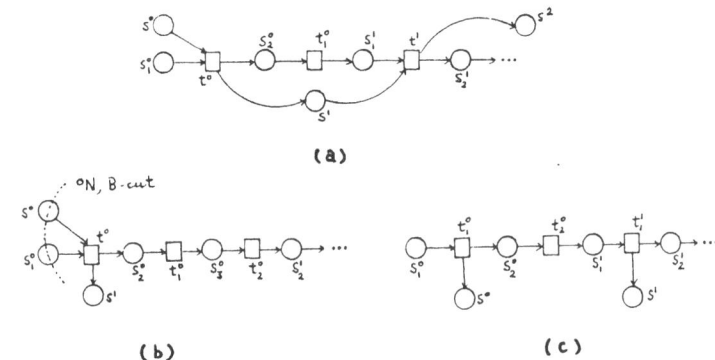

(a)

(b)   **(c)**

Fig.2.2 Three examples of occurrence nets

For the occurrence net of Fig.2.2 (b), $^{\circ}N = \{s^{\circ}, s_1^{\circ}\}$ and it is a B-cut

$\downarrow\{s^{\circ}, t^{\circ}, s^1\} = \{s^{\circ}, s_1^{\circ}, t^{\circ}, s^1\}$

$\uparrow\{s^{\circ}, t^{\circ}, s^1\} = \{s^{\circ}, t^{\circ}, s^1, s_2^{\circ}, t_1^{\circ}, s_3^{\circ}, \ldots\} = X \setminus \{s_1^{\circ}\}$

and N° is well defined. It is the element s^1 only.

2.10 <u>Definition</u> Let $N_1 = (B_1, E_1; F_1)$ and $N_2 = (B_2, E_2; F_2)$ be two occurrence

nets.

(a) ([2]) $N_1 \leqslant N_2$ (N_1 is an initial subnet of N_2) iff $\exists A \in BC(N_2)$:

$B_1 = B_2 \cap \downarrow A$, $E_1 = E_2 \cap \downarrow A$ and $F_1 = F_2 \cap (\downarrow A \times \downarrow A)$.

(b) ([2]) For chains $N_0 \leqslant N_1 \leqslant N_2 \ldots$, it is defined that

$$\overset{\infty}{\underset{i=0}{U}} N_i = (\overset{\infty}{\underset{i=0}{U}} B_i, \overset{\infty}{\underset{i=0}{U}} E_i ; \overset{\infty}{\underset{i=0}{U}} F_i).$$

(c) N_1 ls N_2 (N_1 is a last subnet of N_2) iff $\exists A \in BC(N_2)$: $B_1 = B_2 \cap \uparrow A$,

$E_1 = E_2 \cap \uparrow A$ and $F_1 = F_2 \cap (\uparrow A \times \uparrow A)$.

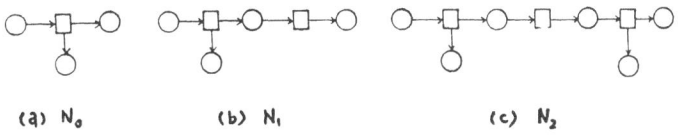

(a) N_0   **(b) N_1**   **(c) N_2**

Fig.2.3 The Nets in order

The nets in Fig.2.3 are in order, i.e. $N_0 \leqslant N_1 \leqslant N_2$. And the net of Fig.2.2 (c)

will be $\overset{\infty}{\underset{i=0}{U}} N_i$.

2.11 Definition Let $MN=(S,T;F,M_0)$ be a marked net ~and let (N,p) be a pair
such that $N=(B,E;F')$ is an occurrence net and $p: B\cup E \to S\cup T$ is a labelling
function. Then (N,p) is a process of MN iff

(a) $p(B)\subseteq S$, $p(E)\subseteq T$;

(b) ${}^\bullet N\subseteq BC(N)$ and $\forall s\in S: M_0(s)=|p^{-1}(s)\cap{}^\bullet N.|$;

(c) $\forall e\in E: p(\bullet e)={}^\bullet p(e)$, $p(e\bullet)=p(e)\bullet$, $|\bullet e|=|{}^\bullet p(e)|$ and $|e\bullet|=|p(e)\bullet|$;

(d) $N= \bigcup_{i\geq 0} N_i$ for some finite occurrence nets N_i with $N_i\subseteq N_{i+1}$, $0\leq i$.

It is easy to see that (the net fig.2.2 (a)/(b),(c)/, a proper
$p_a/p_b,p_c/$) is a process of the net fig.2.1 (a)/(b),(c) respectively/.

2.12 Construction ([2]). Let $MN=(S,T;F,M_0)$ be a marked net and let
$\beta =M_0 t_0 M_1 \cdots$ be an occurrence sequence of MN starting with M_0.
A process (N,p) is associated to β as follows.
First the processes $(N_i,p_i)=(B_i,E_i;F_i,p_i)$ are constructed by induction
on $i\geq 0$.

$\underline{i=0}$. Define $E_0=F_0=\phi$, and B_0 as containing, for each $s\in S$, $M_0(s)$ new
conditions b with $p_0(b)=s$ (this defines p_0 as well).

$\underline{i\to i+1}$. Suppose (N_i,p_i) to be constructed up to M_i. By construction,
N_i^o is a finite B-cut corresponding to M_i,
i.e. $\forall s\in S: M_i(s)=|p_i^{-1}(s)\cap N_i^o|$.
Suppose $\bullet t_i=\{s_1,\ldots,s_q\}$ and $t_i^\bullet=\{s_1',\ldots,s_r'\}$.
Since M_i enables t_i, N_i^o contains conditions $b_1,\ldots,b_q$
with $p_i(b_j)=s_j$ for $1\leq j\leq q$; of course, $b_j^\bullet=\phi$ by the definition of N_i^o.
We choose one such set $\{b_1,\ldots,b_q\}$ and add a new event e_i such
that $b_j F_{i+1} e_i$; also we put $p_{i+1}(e_i)=t_i$.
Further, we add new conditions $b_1',\ldots,b_r'$ such that $e_i F_{i+1} b_j'$ and
$p_{i+1}(b_j')=s_j'$ for $1\leq j\leq r$.
If the occurrence sequence of MN,β, is finite, say $\beta=M_0 t_0 \cdots t_{m-1} M_m$,
then the construction stops at $i=m$, otherwise put $N= \bigcup_{i\geq 0} N_i$ and $P =\bigcup_{i\geq 0} p_i$.

It is important to note that the construction given in 2.12 is non-deterministic
because there may be several different sets $\{b_1,\ldots,b_q\}$ to which a new
event e_i can be added.
As examples, (the net fig 2.2(a)/(b),(c)/, $p_a/p_b,p_c/$) can be derived from
$\beta_a=(tt_1)*/\beta_b=t(t_1 t_2)*, \beta_c=(t_1 t_2)*$ respectively / by 2.12.

3. THE DIFFERENT KINDS OF FROZEN TOKENS.

By 2.12, to an infinite occurrence sequence in 2.5 we may associate
several different processes because of non-determinism.

<u>3.1 Definition</u> Let MN be a marked net and β be an infinite occurrence
sequence of MN. MN is said to contain a frozen token in growing iff
$\forall$ pair $(N,\mathbf{p})$ associated to β by 2.12 $\exists \mathbf{l} \in L(N) \exists c \in C(N): c \cap \mathbf{l} = \phi$.

For example, MN Fig.2.1 (c), $\beta_c = (t_1 t_2)^*$ of MN at the end of the last
section and the unique $(N, \mathbf{p}_c)$ Fig.2.2(c) show that MN contains a frozen
token in growing by 3.1 because there are an $1 = (s_1^0, t_1^0, s_2^0, t_2^0, s_1^1, \ldots)$ and
a $c = (s^0, s^1, s^2, \cdots) $ s.t. $\mathbf{l} \cap c = \phi$. In fact the MN is not n-safe. The phrase 'in
growing' means that if it is concerned to count the number of frozen
tokens in some place, it is growing.

<u>3.2 Proposition</u> Suppose that a marked net MN contains a frozen token in
growing; then it is not n-safe.
Proof. Take an infinite occurrence sequence β of MN s.t. for every pair
$(N,\mathbf{p})$ associated to β by 2.12 there is an $\mathbf{l}$ in $L(N)$, a c in $C(N)$:
$c \cap \mathbf{l} = \phi$.
Consider such an $\mathbf{l}$ as well as a c.
<u>The 1 must be an infinite line:</u>
assume the 1 is finite, $1 = (s_1, t_1, s_2, t_2, \ldots, s_k)$ and $s_1 \in {}^0 N$, $s_k \in N^0$.
Then $\forall x \in \uparrow \mathbf{l} \setminus \mathbf{l} \exists u \in \mathbf{l}: (x, u) \in co$ e.g. $(x, s_k) \in co$ and $\forall y \in \downarrow \mathbf{l} \setminus \mathbf{l} \exists v \in \mathbf{l}: (y, v) \in co$
e.g. $(y, s_1) \in co$ and $\forall c \in C(N) \exists z \in c: z \in \downarrow \mathbf{l} \cup \uparrow \mathbf{l} \Rightarrow$ case 1: $z \in \mathbf{l}$;
case 2: $z \in \uparrow \mathbf{l} \setminus \mathbf{l}$, then $\exists z_1 \in \mathbf{l}: z_1 \in c$; case 3: $z \in \downarrow \mathbf{l} \setminus \mathbf{l}$, then
$\exists z_2 \in \mathbf{l}: z_2 \in c$, i.e. $\forall c \in C(N): c \cap \mathbf{l} \neq \phi$ contradicting the known.

<u>The c must be an infinite cut:</u>
assume the c is finite, $c = (x_1, x_2, \ldots, x_k)$, and $\mathbf{l} = (s, t, \ldots)$, $s \in {}^0 N$.
If $\exists x \in c$ such that $x \in \downarrow \mathbf{l} \setminus \mathbf{l}$, then for $s \in \mathbf{l}$ and $s \in {}^0 N$ it is true that
$(s, x) \in co \Rightarrow \exists y \in \mathbf{l} \forall x_i$, $1 \leq i \leq k$: $y = x_i$ or $(y, x_i) \in co$, i.e. $y \in \mathbf{l} \land y \in c$;
So $\forall x \in c: x \in \uparrow \mathbf{l} \setminus \mathbf{l}$, then because the 1 is infinite $\exists y \in \mathbf{l}$ s.t. (x, y)
$\in co$ and $y \in c$ contradicting $c \cap \mathbf{l} \neq \phi$.
Then it is true that $\exists$ an infinite B cut, $c' \in BC(N)$ s.t. $\forall b \in c': b \in N^0$.
By 2.12 there is an infinite co-set in c', say c_s, satisfying
$\mathbf{p}(b \in c_s) = s \in MN$, because MN is finite. QED.

<u>3.3 Corollary</u> MN contains a frozen token in growing $\Rightarrow$ MN contains a
frozen token.

<u>3.4 Definition</u> Let MN be a marked net which contains a frozen token,
$\bar{s}$ be an S element of MN and β be an infinite occurrence sequence of
MN, $M_1 t_1 M_2 t_2 \ldots$ s.t.
$$\forall i, i \geq 1: M_i(\bar{s}) \geq \begin{cases} 1 & \bar{s} \notin \cdot t_i \\ 2 & \bar{s} \in \cdot t_i \end{cases}$$

MN is said to contain:

(a) a hard frozen token iff $\forall M \in [M_1 > \forall t \in \bar{s}\bullet$: t is not enabled at M.

(b) a soft frozen token iff $\exists t \in T \exists$ infinetely many M in β :
$t \in \bar{s}\bullet \wedge$ t is enabled at $M \wedge$ t is not in β .

(c) an overlay frozen token iff $\exists (N,p)$ associated to β : N consists
of several (≥ 2) infinite connected occurrence nets which are uncon-
nected each other and they are in the relation (ℓs) 2.10 (c).

(d) a coloured frozen token iff $\exists (N_1,P_1)$, (N_2,P_2) associated to
β : N_1 is not connected $\wedge \exists b \in \bullet N_1 : P_1(b)=\bar{s}$ and N_2 is connected $\wedge$
$\forall b \in B(N_2)$: $b\bullet \neq \phi$.

(e) an asynchronous frozen token iff $\exists (N_1,P_1)$, (N_2,P_2) associated to
β : N_1, N_2 are connected $\wedge \exists b \in B(N_1)$: $b\bullet = \dot{\phi} \wedge \forall b \in B(N_2)$: $b\bullet \neq \phi$.

The nets fig.3.1 (a). ((a') too), (b), (c), (d) and (e) contain a hard,
a soft, an overlay, a coloured and an asynchronous frozen token respectively.
Loosely speaking, having a frozen token in a marked net means that there
exists at least one token in the net, which is 'frozen' in a place by
some infinite occurrence sequence. The word 'hard' means that a token,
once is 'frozen', can not be 'unfrozen' afterward; 'soft' means that it
has a lot of chance to move; 'overlay' means that there are at least
two sets of tokens s.t. each set can make the MN running in same way;
'coloured' means that a token is 'frozen' because it is coloured (see
[3]) and 'asynchronous' means that a token is 'frozen' because some
transitions asynchronously occur, i.e. too fast or too slow.

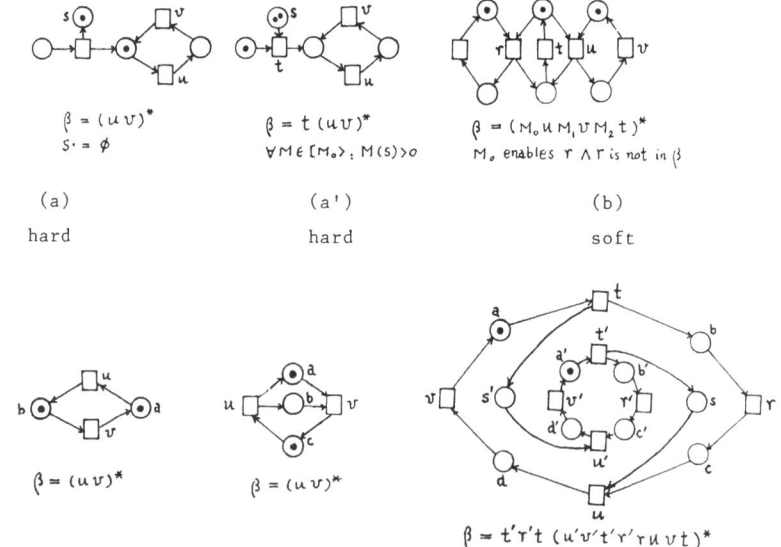

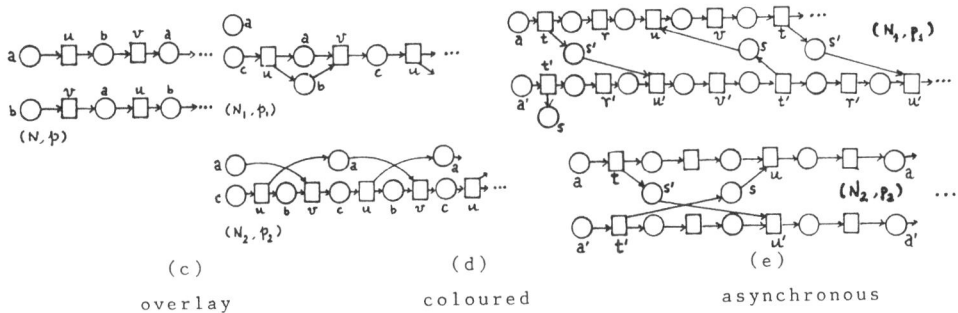

(c)   (d)   (e)

overlay   coloured   asynchronous

Fig.3.1 The different kinds of frozen tokens

3.5 Proposition Let $MN=(N,M_0)$ be a marked net which contains a hard frozen
    token, see 3.4 (a).
    α is a transition sequence of MN starting with M_1 $\Longleftrightarrow$
    α is a transition sequence of MN'
    where $MN'=(N,M_1')$, $\forall s \in S(N) \setminus \{\bar{s}\}$: $M_1(s)=M_1'(s)$, and $M_1'(\bar{s})=M_1(\bar{s})-1$.

Proof. $\Longrightarrow$

    Suppose $\alpha=t_1 t_2 \cdots$ This half of the proposition is proved by induction
    on $i \geq 1$.

    $\underline{i=1}$. t_1 is enabled at M_1 in MN, say $t_1^\bullet=\{s_{11}', \cdots s_{1m_1}'\}$ and
    $\forall s \in \bullet t_1=\{s_{11}, \cdots, s_{1m_1}\}$: $M_1(s) \geq 1$. By 3.4(a) $\bar{s} \notin \bullet t_1$,
    $\forall s \in \bullet t_1$: $M_1'(s)=M_1(s) \geq 1$, i.e. t_1 is enabled at M_1' in MN': After
    t_1 occurs in MN /MN'/; the marking M_2 /M_2'/ is got by 2.3(b). Obvi-
    ously it is true $\forall s \in S(N) \setminus \{\bar{s}\}$: $M_2(s)=M_2'(s)$, and $M_2'(\bar{s})=M_2(\bar{s})-1$.
    $\underline{i \to i+1}$. Suppose $t_1 \cdots t_i$ is a transition sequence of MN starting with
    M_1, then it is a transition sequence of MN' as well. And $\forall s \in S(N)$
    $\setminus \{\bar{s}\}$: $M_{i+1}(s)=M_{i+1}'(s)$, and $M_{i+1}'(\bar{s})=M_{i+1}(\bar{s})-1$. t_{i+1} is enabled at
    M_{i+1} in MN, $t_{i+1}^\bullet=$ $\{s_{i+1,1}', \cdots, s_{i+1,n_{i+1}}'\}$ and $\forall s \in \bullet t_{i+1}=$
    $\{s_{i+1,1}, \cdots, s_{i+1,m_{i+1}}\}$: $M_{i+1}(s) \geq 1$. By 3.4(a), $\forall s \in \bullet t_{i+1}:M_{i+1}'(s)$
    ≥ 1, i.e. t_{i+1} is enabled at M_{i+1}' in MN'. Just the same, after t_{i+1}
    occurs it is got $\forall s \in S(N) \setminus \{\bar{s}\}$: $M_{i+2}(s)=M_{i+2}'(s)$, and $M_{i+2}'(\bar{s})=$
    $M_{i+2}(\bar{s})-1$.

    $\Longleftarrow$

    In a marked net MN, if one increases a marking, then obviously all
    the previous transition sequences remain. That is already known.QED.

In a system modelled by a Petri net, some signal of the system, e.g.
functioning normally, can be indicated by a hard frozen token. For example,
if a nuclear reactor in a power station is fired, then some lights in a
controler of the station are turned on for all time except when the

system is damaged. These lights could be modeled as hard frozen tokens. By 3.5, if one of these lights is broken there is no influence on the system.

4. SOME APPLICATIONS OF FROZEN TOKENS

The tokens in a Petri net can be considered as the resource in a concurrent system modelled by the net. Keeping this thought in mind, some applications of frozen tokens can be suggested. In fact, at the end of the last section it is suggested to have a hard frozen token in a system as a system signal.

By 3.4 (b) having a soft frozen token can be considered as a movable token which is not moved in β . In a system design it is often desired to have an accident processing part but never to use it. Then the power of the part, which is never needed in normal situation, may be designed as a soft frozen token.
For example in a nuclear power station, an emergency cooling system for accidents must be an important part of the station. Although the emergency cooling system does not function in the station's daily running, the power of the system should always keep it ready.

It is known that a powergrid connects several power plants and some of them are reserved e.g. for a periodic overhaul. Then in this system the reserved house may be considered as an overlay frozen token because each powerstation has the same function in the powergrid. A power multi-supply system may be an another example. In many important units, e.g. airport, railway station, wharf, surgical operation room, ... it would be disastrous if the power supply is stopped. To avoid such a case we need multi-supply system, in which each has same function.

For a coloured frozen token in a system, the resource quality being different from others is considered. For example, if the tokens in the place a of the net Fig.3.1 (d) must be used alternatively by some reasons, then this net is a reasonable design and it will run properly, although it does not matter, from a theoretical point of view, to get rid of the token in a at the beginning.
In [4] we have discussed this case theoretically.

Last but not least, in Petri's Beijing Lectures [8] it was mentioned how to synchronize two circuits as shown in Fig.3.1 (e). The way outlined in [8] is to keep the synchronic distance (t,u') being 1 and (t',u) being 1 as well. Now here is another way: to find a frozen token in such system (may be using the algorithm in [4]) and if it exists then to get rid of it, i.e.reinitialize the marking of the system. In the case that for a process (N, ρ) associated to β : $\exists b \in B(N)$: $b \bullet = \phi$ is known, the matter is much easier than the above mentioned methods, and this shown as follows: to find a B-cut in BC(N), say bc, s.t. $\uparrow$bc is an occurrence net N' in which $\forall b \in N': b \bullet \neq \phi$; then to reinitialize the original system with $\forall s \in \rho(bc): M_0(s) = |\rho^{-1}(s)|$ and $\forall s \in S \setminus \rho(bc): M_0(s) = 0$. The fourth possibility to synchronize two circuits is S-complementation or local S-complementation, which of course cannot be treated within the range of this paper.

CONCLUSION.

The frozen tokens in a marked net are considered as the resources,which are reserved for the concurrent system modelled by the net and which are not moved in some occurrence sequence. We have classified 6 kinds of frozen tokens based on occurrence sequences and processes associated to occurrence sequences. On the one hand, if a marked net has no frozen token, then it can be interpreted as a model of a good distributed system in the sense that all the alternative and concurrent functions in the system always use all the available resources. Along this line, some classes of nets have been found, which never contain frozen tokens (see [5]).
On the other hand, a real system normally contains frozen tokens as we have mentioned in the section 4.
What is not contradictory to say is that studying frozen tokens in nets gives us more knowledge about how to design concurrent systems modelled by Petri nets, especially about their resources.

ACKNOWLEDGEMENT. The author would like to thank warmly K. Voss and the colleague of his, who read my paper carefully, for their valuable comments and remarks on this paper.

References

[1]   E.Best and C.Fernandez, "Notations and Terminology on Petri Net Theory", Petri Nets Newsletter 23, 21-46, April 1986.

[2]   E.Best and A.Merceron, "Frozen Tokens and D-Continuity: A Study in Relating System Properties to Process Properties", LNCS Vol. 188, 48-61, 1985.

[3] K.Jensen, "Coloured Petri Nets and The Invariant-Method" Theoret-
     ical Computer Science 14, 317-336, 1981.

[4] W.M.Lu and A.Merceron, "Frozen Tokens: A Way to Test Distributed
     System Modelled by Petri Nets", To appear in Scientia Sinica (Se-
     ries A).

[5] W.M.Lu, A.Merceron and P.S.Thiagarajan, Draft Paper.

[6] J.L.Peterson, PETRI NET THEORY AND THE MODELING OF SYSTEMS, Pren-
     tice Hall, 1981.

[7] C.A.Petri, "Introduction to General Net Theory", LNCS Vol. 84, 1-
     20, 1980.

[8] C.A.Petri, Beijing Lectures, 1981.

[9] W.Reisig, PETRI NETS An Introduction, EATCS Monographs No.4, 1985.

HIGH LEVEL PETRI NETS AND

DISTRIBUTED TERMINATION

Horst Müller

Universität Erlangen-Nürnberg
Institut für Mathematische Maschinen
und Datenverarbeitung (III)
Martensstraße 3
D-8520 Erlangen

Abstract: The paper presents a high level Petri net (HLPN) specification of an algorithm for global termination detection of a system of
parallel processes. The correctness is proved using invariants and an
assertion system.

1. Introduction

One purpose of this paper is to show the usage of high level Petri nets
(HLPN) and assertion systems for the analysis of parallel algorithms
exemplified with the problem of distributed termination.

A scarce introduction of transition systems and assertion systems (see
/Mü/ for details) and a version of HLPN, mostly resting on coloured
Petri nets /Jen/, but also resting on predicate/transition nets /GeLau/
and relation nets /Rei/, is given.

As the main point we give a HLPN-specification of the two counting wa-
ves algorithm for global termination detection and a correctness proof
with help of invariants and an appropriate assertion system.

2. Transition systems, assertion systems and invariants

Without further motivation we define transition systems as the fundamental dynamic system model and assertion systems for verification purposes. For details see /Mü/. An initial transition system TS
$TS = (Q, <[\sigma> \mid \sigma \in \Sigma>, Q_o)$ is composed of a set Q of states, a Σ-in-
dexed family of (recursive) relations $[\sigma> \subseteq Q \times Q$ (transitions, Σ any

nonempty set) and a (recursive) set $Q_0 \subseteq Q$ of _initial states_. For
$\tau = \sigma_1 \ldots \sigma_n \in \Sigma^*$ we define $[\tau> := [\sigma_1 > \circ \ldots \circ [\sigma_n>$ (composition of relations).

For any $X \subseteq Q$ and $\tau \in \Sigma^*$ let $X[\tau> := \{q' \mid \exists q \in X: q[\tau>q'\}$ be the _set_
of τ-successors of X, $[X> := \cup \{X[\tau> \mid \tau \in \Sigma^*\}$ be the set of states
reachable from X (in TS) and call $[Q_0>$ the _reachability set_ of TS (forward reachability).

Every digraph (V,E) with $E \subseteq V \times \Sigma \times V$ may be seen as a TS with
$Q = Q_0 = V$ and $v[\sigma>v' \leftrightarrow (v,\sigma,v') \in E$.

Given an arbitrary transition system TS as above we define an _assertion_
system as a marked digraph $AS = (V,E,\underline{M})$ composed of a vertex set V, a
set $E \subseteq V \times \Sigma \times V$ (of _edges_) and a family $\underline{M} = <\underline{M}_v \mid v \in V>$ of subsets
of Q, which can be thought of as given by unary predicates (_assertions_)
on Q.

TS is _correct_ with respect to AS iff
    $\forall v \in V: (\forall \sigma \in \Sigma: v[\sigma> \neq \emptyset \rightarrow \emptyset \neq \underline{M}_v[\sigma> \subseteq \cup\{\underline{M}_{v'} \mid v[\sigma>v'\})$
This means: if σ is applied to some state in $\underline{M}_v$, a result satisfies
some $\underline{M}_{v'}$.

To express that the assertion system AS is not too small and does not
'forget' some transition enabled in some state of $\underline{M}_v$ we define (partial)
completeness. An assertion system AS (for TS) is _V'-complete_ iff
    $V' \subseteq V$ and $\forall v \in V': (\forall \sigma \in \Sigma : \underline{M}_v[\sigma> \neq \emptyset \rightarrow v[\sigma> \neq \emptyset)$.

AS is _complete_ iff it is V-complete.

A correct and complete assertion system gives partial information on
the reachability set:

Theorem 2.1   If AS is a correct and complete assertion system for a
    transition system TS and $X \subseteq \underline{M}_{v_0}$ for some $v_0 \in V$, then holds
    $[X> \subseteq \cup\{\underline{M}_v \mid v \in [v_0>\}$.

In other words any reachable state (of TS) has to satisfy some assertion $\underline{M}_v$ of AS (if $X = Q_0$ and $V = [v_0>)$.
Expressed conversely any state satisfying no assertion of AS is not
reachable.
Proof(/Mü/): By induction on the path length n of a path
    $q_0[\sigma_1>q_1 \ldots [\sigma_n>q_n$ in TS one shows
    $q_0 \in \underline{M}_{v_0} \rightarrow \exists v_1, \ldots, v_n \in V: v_0[\sigma_1>v_1 \ldots [\sigma_n>v_n$ _and_ $(\forall i: 0<i\leq n \rightarrow q_i \in \underline{M}_{v_i})$

Remark: Assertion systems have some similarity with sets of local in-
    variants defined by Sifakis in /Sif/.
An _invariant_ of a transition system TS is a subset I of Q satisfying
$\forall \sigma \in \Sigma: I[\sigma> \subseteq I$.

Remark: This concept of invariant is more general than the concept of
   S-invariant used in net theory.
Any invariant can be seen as a special (correct and complete) assertion
system with only one vertex v_o, $\underline{M}_{v_o} = I$ and edges $v_o[\sigma>v_o$ for all
$\sigma \in \Sigma$ with $I[\sigma> \neq \emptyset$. Conversely we have:

Theorem 2.2 For any correct and complete assertion system AS to the
   transition system TS $I := \cup\{\underline{M}_v \mid v \in V\}$ is an invariant of TS.
Proof see /Mü/.

3. High level Petri nets

Our formal model for HLPN is based on Jensens coloured PN /Jen/, but
follows as far as possible (and generalizing where needed) the notational proposals of /BeFe/.

Let $\mathbb{Z}$, $\mathbb{N}$, $\mathbb{N}_o$ and $[A \to B]$ denote integers, positive integers, nonnegative integers and total functions from A to B respectively. For
$B \subseteq \mathbb{Z}$ let $[A \to B]_f$ denote the set of total functions for which the
support $\{x \in A \mid f(x) \neq 0\}$ is finite and let <u>singleton</u> be the set of
functions f, for which f(x) = 1 for exactly one x and f(x') = 0 for
all x' $\neq$ x.
For $f \in [A \to \mathbb{Z}]_f$ let $\Sigma f := \underset{x \in A}{\Sigma} f(x)$. $f \in [A \to \mathbb{N}_o]$ is called an <u>A-bag</u>
with f(x) denoting the 'number of occurrences of x in f'.
   $\underline{O}_A = <x \mapsto O> \in [A \to \mathbb{Z}]$ denotes the <u>empty</u> A-bag.
If R is a relation on B and op an operation on B, we lift R and op
from B to $[A \to B]$ by
   f R g $:\leftrightarrow$ $\forall x \in A$: f(x) R g(x) and
   f op g $:= <x \mapsto f(x)$ op $g(x)>$.
For example, let $M,K,G^+ \in [S \to \mathbb{Z} \cup \{\infty\}]$. Then $M \leq K - G^+$ abbreviates
   $\forall s \in S$: $M(s) \leq K(s) - G^+(s)$.
Each $f \in [A \to [B \to \mathbb{Z}]_f]$ has a <u>unique linear extension</u>
   $\tilde{f} \in [[A \to \mathbb{Z}]_f \to [B \to \mathbb{Z}]_f]$ defined by
   $\tilde{f}(g)(b) := \underset{a \in A}{\Sigma} f(a)(b) \cdot g(a)$ for all $g \in [A \to \mathbb{Z}]_f$ and $b \in B$.
Let $f \boxtimes g := \tilde{f}(g)$ for short.

Definition: A <u>high level place/transition system (HLPN)</u> is a tupel N
   $N = (S,T;F,D,K,W,M_o)$ such that
   (1) (S,T;F) is a (Petri) net, i.e.
      $S \cap T = \emptyset \wedge F \subseteq (S \times T) \cup (T \times S) \wedge dom(F) \cup cod(F) = S \cup T \neq \emptyset$

(2) D is a <u>domain function</u> (colour function) from SUT
into nonempty sets, D_x denoting the <u>data domain</u> of $x \in$ SUT.

(3) K is a <u>capacity function</u> with dom(K) = S and
$$\forall s \in S: K(s) \in [D_s \rightarrow \mathbb{N} \cup \{\infty\}]$$

(4) W is a <u>weight function</u> with dom(W) = F and
$$\forall s \in S: \forall t \in T: ((s,t) \in F \rightarrow W(s,t) \in [D_t \rightarrow [D_s \rightarrow \mathbb{N}_o]]) \wedge$$
$$((t,s) \in F \rightarrow W(t,s) \in [D_t \rightarrow [D_s \rightarrow \mathbb{N}_o]])$$

(5) M_o is a   marking of N (see below).

<u>Definition:</u> a) A <u>marking</u> (S-assignment) M of N is a function with
dom(M) = S and $\forall s \in S: M(s) \in [D_s \rightarrow \mathbb{N}_o] \wedge M \leq K$.

b) Let SM_N denote the set of all markings of N.

c) A <u>T-assignment</u> G of N is a function with dom(G) = T and
$\forall t \in T: G(t) \in [D_t \rightarrow \mathbb{N}_o]_f$.

d) For any T-assignment G G^+ and G^- are functions with domain S
and $\forall s \in S: G^+(s) = \underset{t \in T}{\Sigma} W(t,s) \circledast G(t) \wedge G^-(s) = \underset{t \in T}{\Sigma} W(s,t) \circledast G(t)$

<u>Definition</u> (of dynamic behaviour)

a) A T-assignment G is <u>enabled</u> (short M[G>) at the marking M
iff $G^- \leq M \leq K - G^+$.

b) The <u>step-relation</u> is defined on $SM_N \times$ T-assignments $\times SM_N$ by
$M[G>M'$ iff $M[G> \wedge M' = M - G^- + G^+$.

c) A T-assignment G is <u>atomic</u> iff $G(t) \neq \underline{0}_{D_t}$ for exactly one $t = t_o$
and $G(t_o) = \text{char}_{\{d_o\}}$ for some $d_o \in D_{t_o}$. We write $G = d_o \cdot t_o$ in
this case.

d) The step-relation induces a <u>concurrent transition system</u>
$TS_c(N) := (SM_N , <[G> | G \text{ T-assignment}> ,\{M_o\})$.

e) The atomic-step-relation induces a <u>sequential transition system</u>
$TS_s(N) := (SM_N , <[G> | G \text{ atomic T-assignment}> , \{M_o\})$.

By induction on ΣG one shows:

<u>Lemma</u> Any step can be decomposed into a finite sequence of atomic steps.
From this follows:

<u>Lemma</u> The reachability sets of $TS_c(N)$ and $TS_s(N)$ are equal.
For analyzing reachability sets it is therefore sufficient to consider
only atomic steps.

Now we consider <u>describability</u> (representability) of HLPN and assertion
systems in an appropriate data algebra. We consider some usual language
of first order logic with a family $<D_i | i \in I>$ of unary type-predi-
cate symbols (denoting data domains), a family $<op_j | j \in J>$ of opera-
tion symbols and a family $<R_k | k \in K>$ of relation symbols, where to
every operation symbol (and every relation symbol) is associated a

functionalty $(i_1,\ldots,i_{n+1}) \in I^{n+1}$ (resp. $(i_1,\ldots,i_n) \in I^n$) denoting
that op should be interpreted as a partial function from $D_{i_1} \times \ldots \times D_{i_n}$
into $D_{i_{n+1}}$ (resp. relationsymbol R as a subset of $D_{i_1} \times \ldots \times D_{i_n}$).
Fixing some interpretation we get a (partial heterogenious) <u>data alge-</u>
<u>bra</u> $\underline{D}$. For tupels $(x_1,\ldots,x_m)$ of variables (of types $D_{i_1},\ldots,D_{i_m}$) as
usual the family of $(D_i,x_1,\ldots,x_m)$-terms is inductively defined, each
such term denoting a partial function from $D_{i_1} \times \ldots \times D_{i_m}$ to D_i (in $\underline{D}$).

For $X \in \{\mathbb{N}_o, \mathbb{N}_o \cup \{\infty\}, \mathbb{Z}, \mathbb{Z} \cup \{\infty\}\}$ a formal <u>X-sum</u> $\sum\limits_{j=1}^{k} \beta_j \cdot t_j$
with $\beta_1,\ldots,\beta_k \in X$ and $(D_i,x_1,\ldots,x_m)$-terms $t_1,\ldots,t_k$ denotes (in $\underline{D}$)
the function f from $D_{i_1} \times \ldots \times D_{i_m}$ into $[D_i \to X]$ such that
$\quad f(x_1,\ldots,x_m) = \langle y \mapsto \Sigma\{\beta_j \mid t_j = y \land 1 \le j \le k\} \rangle$

A function $f \in [D_j \to [D_i \to \mathbb{Z}]]$ is <u>D-representable</u> iff there is a
formal $\mathbb{Z}$-sum of (D_i,x)-terms (x of type D_j) denoting f.
A relation (set) $R \subseteq D_{i_1} \times \ldots \times D_{i_n}$ is <u>D-representable</u> iff there is a
formula which is true iff $(x_1,\ldots,x_n) \in R$.
The HLPN N is <u>D-representable</u> iff for all $x \in$ SUT D_x is D-representable and for all $(x,y) \in F$ $W(x,y)$ is D-representable.
An assertion system AS is <u>D-representable</u> iff for all $v \in V$
$\underline{M}_v$ is D-representable.

4. Specification and correctness of an algorithm for distributed termination detection by HLPN and assertion systems

We consider the following distributed termination problem: a finite
set of processors communicating by messages are performing a distributed computation. The distributed computation is considered to be globally terminated, if every process is (locally) terminated and no messages are in transit. 'Locally terminated' can be understood to be a
state, in which a process has finished its computation and will not
restart any action unless it receives a message. The problem is to
detect global termination.

In /Ma/ are considered some algorithms for this problem on the base of
an 'atomic model' using a global time concept. In /Ma2/ a legendary
formulation is given for the problem and one of the solving algorithms
(the 'four counter-method'). From this we give the essential translated
parts and corresponding stepwise construction of a HLPN specification.

"Once the time had come for the king of polymicronesia to distribute uncountable islands of his empire fairly among his grandchildren.
He sent messages to the wise men living widely distributed in his
land to get some clever proposal.":

As a data domain we use the (finite) set Wise' = Wise ∪ {king} of wise
men and the king. Messages build a binary relation Mess on Wise ∪ {king}.
A place mw contains the messages on the way, initialy the set
{king}×Wise (or some subset mw_o herefrom with s_o elements). This gives
an intial condition

(1) mw_o = {king}×$Wise_o$ ⊆ {king}×Wise .

"Well known to the king was the willful but steady way of life of
his wise men, who were eating or thinking all the day. If some was
eating, only some message from the king or some of his colleagues
could bring him to thinking. Mostly he took very soon pen, paper,
ink and signet to send some message to some of his colleagues.
Hungry from thinking he returned to the princely meal.":

Two places eat and think together with the invariant

(2) eat + think = Wise

specify the possible states of the wise men. Initialy we suppose them
all eating:

(3) eat_o = Wise .

(This is not essential, any initial state satisfying (2) would suffice)
The possible changes are modeled by three transitions:

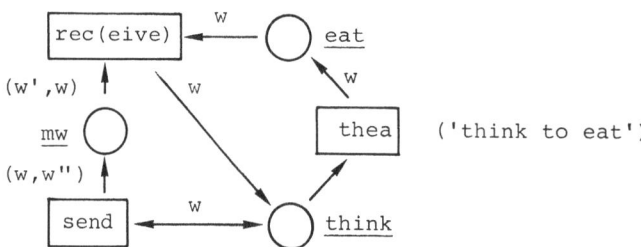

Some further part of the legend give us the information that no messages are lost, that there is no global time available, that the transmission time of messages is finite but unknown and unpredictable and
that the advise for the king is given only in the case of all wise men
eating and no message on the way. Thus the global termination condition
is

(4) eat = Wise and mw = ∅ .

The Privy Councillor proposes "to send emissaries who ask every
philosopher how many messages he has sent and how many he has received and then to test wether the total sum of the number of messages
sent equals the total sum of the number of messages received".

To implement this, we introduce two places emb, ema for emissaries (to
any philosopher corresponds one emissary) before resp. after intervie-
wing a philosopher. A transition interview and two counter places cmr,
cms for the sums of reported numbers of messages received resp. sent.
For being able to reply to the emissaries any philosopher has to count
his messages (on a place mr (messages received) resp. ms (messages sent)
with data domain Wise'× $\mathbb{N}_o$). To get a correct algorithm the emissaries
have to ask only eating philosophers. This gives

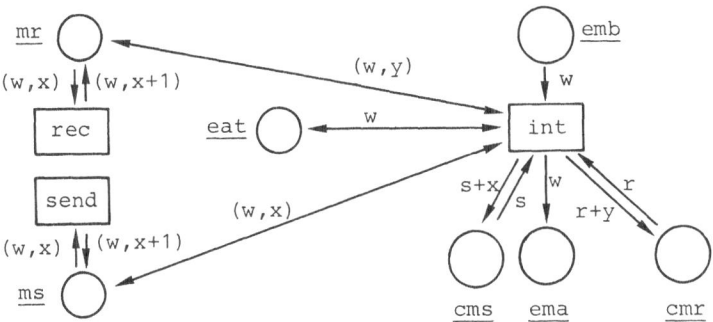

During the interview phase the system should satisfy the local invariant

(5) emb + ema = Wise

and the initial condition

(6) ms_o = { (king,s_o) } ∪ (Wise×{0}) and mr_o = 0 and cms_o = s_o and
    cmr_o = 0 and emb_o = Wise .

The comparison of the sums cms and cmr is implemented by two transitions $st_=$ and $st_{\neq}$, having as a precondition that all emissaries are
back at the court.

> The court-mathematicus shows, that the equality of cms and cmr is
> not - as supposed by the Privy Councillor - sufficient for global
> termination. He proposes to send the emissaries a second time to
> repeat the interviewing once more. If the resulting sums equal the
> sums of the first round, then global termination (4) should hold.

On a place oldsum the sum is recorded (initialized $oldsum_o$ = 0) and
two transitions $sc_=$ and $sc_{\neq}$ (sum comparison) implement this test with
a final place of type Token.

The resulting HPLN is shown in figure 1 resp. table 1.

The partial correctness of this algorithm is expressed by

(7) final = • → eat = Wise and mw = Ø .

For verification of (7) we state at first some invariants, which are
easily proved:

356

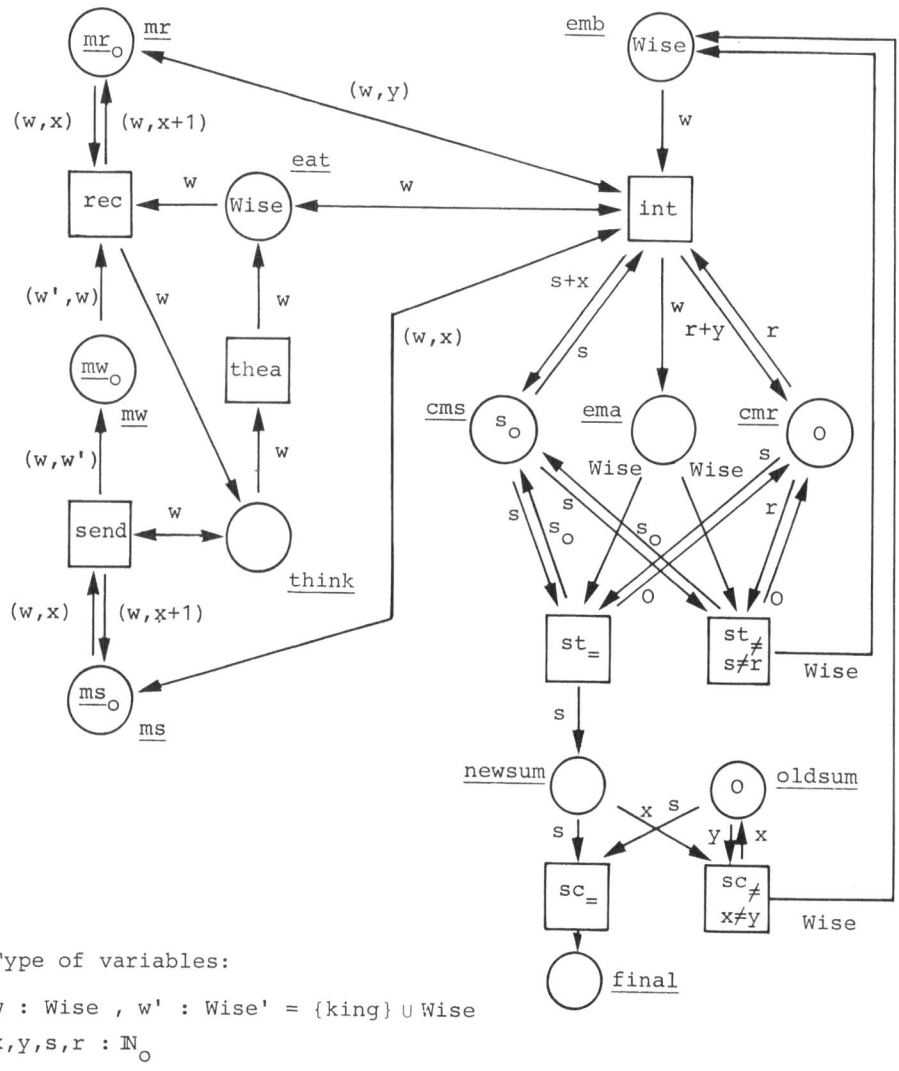

Type of variables:

w : Wise , w' : Wise' = {king} ∪ Wise

$x, y, s, r : \mathbb{N}_O$

Abbreviations for initial marking:

$\underline{mr}_O$ = Wise' × {O} , $\underline{mw}_O$ = {king} × $Wise_O$

$\underline{ms}_O$ = (Wise × {O}) ∪ {(king, s_O)} , s_O = |$Wise_O$|

Figure 1: High Level Petri Net for distributed termination

place s	data domain D_s	initial marking $M_0(s)$	receive $Mess \times \mathbb{N}$ (w',w,y)		send $Mess \times \mathbb{N}$ (w,w',x)	thea Wise w	interview $Wise \times \mathbb{N}^4$ (w,x,y,r,s)	$st_=$ $\mathbb{N}_0$ s_0	$st_{\neq}$ $\mathbb{N}_0^2-Id$ $(s,r)_0$	$sc_=$ $\mathbb{N}_0$ s_0	$sc_{\neq}$ $\mathbb{N}_0^2-Id$ $(x,y)_0$
			$W(s,rec)$	$W(rec,s)$							
eat	Wise	Wise	w			w	w				
think	Wise	Wise	w		w	w	w				
mw (mess. on the way)	Mess	$\{king\}\times Wise_0$		(w',w)	(w,w')						
mr (mess. received)	$Wise'\times\mathbb{N}_0$	$Wise'\times\{0\}$	(w,y)	$(w,y+1)$			(w,y) (w,y)				
ms (mess. sent)	$Wise'\times\mathbb{N}_0$	ms_0			(w,x) $(w,x+1)$		(w,x) (w,x)				
emb (emiss. before int)	Wise	Wise					w		Wise		Wise
ema (emiss. after int)	Wise						w	Wise	Wise		
cms (count ms)	$\mathbb{N}_0$	s_0					s $s+x$	s s_0	s s_0		
cmr (count mr)	$\mathbb{N}_0$	0					r $r+y$	s 0	r 0		
newsum	$\mathbb{N}_0$							s	s	s	x
oldsum	$\mathbb{N}_0$	0								s	y
final	$\{\bullet\}$	0								$\bullet$	.

$ms_0 = Wise\times\{0\}\cup\{(king,s_0)\}$, $s_0 = |Wise_0|$, $Mess = Wise'\times Wise - Id_{Wise}$

Table 1 : High Level Petri Net for global termination detection

(I1) $\Sigma\underline{mr} + \Sigma\underline{mw} = \Sigma\underline{ms}$

(the number of messages sent equals the sum of received messages and the messages on the way)

(I2) $\underline{eat} + \underline{think} = \text{Wise}$

(I3) ($\underline{emb} + \underline{ema} = \text{Wise } \underline{and} \underline{newsum} = \emptyset = \underline{final}$) $\underline{or}$

    ($\underline{emb} + \underline{ema} = \emptyset \underline{and} \underline{newsum} \in \mathbb{N}_o \underline{and} \underline{final} = \emptyset$) $\underline{or}$

    ($\underline{emb} + \underline{ema} = \emptyset \underline{and} \underline{newsum} = \emptyset \underline{and} \underline{final} = \bullet$)

(I4) $\underline{cmr}, \underline{cms} \in \text{singleton}$

These invariants and the initial conditions allow to postulate that all places with the only exception of $\underline{mw}$ have capacity $K(s) = \underline{1}$ and $\underline{mw}$ is unbounded (allowing multiple messages from w' to w being on the way). Moreover we have

(I5) $\forall w \in \text{Wise'}: (\exists x \in \mathbb{N}_o : \underline{mr}(w,x) = 1 \underline{and} \forall x' \in \mathbb{N}_o - \{x\}: \underline{mr}(w,x')=0)$

                $(\exists y \in \mathbb{N}_o : \underline{ms}(w,y) = 1 \underline{and} \forall y' \in \mathbb{N}_o - \{y\}: \underline{ms}(w,y')=0)$

(functionality of $\underline{mr}$, $\underline{ms}$)

By (I5) $\underline{mr}$, $\underline{ms}$ are considered as functions from Wise' to $\mathbb{N}_o$.

We define $\Sigma\underline{mr}/A := \Sigma\{x \mid \underline{mr}(w,x) = 1 \underline{and} w \in A\}$, $\Sigma\underline{mr} := \Sigma\underline{mr}/\text{Wise'}$ (for $\underline{ms}$ likewise). That the reported numbers of messages do not exceed the actual numbers is expressed by the invariant

(I6) $\Sigma\underline{mr} \geq \Sigma\underline{mr}/\underline{ema} \geq \underline{cmr} \underline{and} \Sigma\underline{ms} \geq \Sigma\underline{ms}/\underline{ema} \geq \underline{cms}$

(This follows immediately from monotonicity of $\underline{mr}$ and $\underline{ms}$)

In the following we consider the assertion system in figure 2.

<u>Lemma</u> The initial marking satisfies $\underline{M}_1$.
Proof : trivial checking.

<u>Lemma</u> The assertion system AS of figure 2 is correct and complete with respect to the HLPN in figure 1.
Proof: For all assertions $\underline{M}_i$ (i = 1,...,5) and all atomic steps $M[d_t \cdot t> M'$ we have to show

a) $\underline{M}_i[d_t \cdot t> \subseteq \cup\{\underline{M}_i, \mid i[t>i'\}$ (correctness) and

b) $(\exists M \in \underline{M}_i: M[d_t \cdot t>) \rightarrow \exists i': i[t>i'$ (in AS) (completeness) .

Let s' denote M'(s) for all $s \in S$. We make case analysis on transitions.

<u>Case t = rec</u>, $d_t = (w',w,y)$: Occurrence of rec modifies only $\underline{eat}$, $\underline{mr}$ and $\underline{mw}$ such that $\Sigma\underline{mw'} = \Sigma\underline{mw}-1$, $\underline{eat'} \subseteq \underline{eat}$, $\Sigma\underline{mr'}/A \geq \Sigma\underline{mr}/A$, $w \in A \rightarrow \Sigma\underline{mr'}/A = \Sigma\underline{mr}/A + 1$ and $w \notin A \rightarrow \Sigma\underline{mr'}/A = \Sigma\underline{mr}/A$.

This shows the inheritance of $\underline{emb} + \underline{ema}$, $\underline{oldsum} \leq \Sigma\underline{ms}$, $\underline{oldsum} \leq \Sigma\underline{mr}$, $\underline{oldsum} < \Sigma\underline{ms}/\underline{emb} + \underline{cms} \leq \Sigma\underline{ms}$, $\underline{oldsum} \leq \Sigma\underline{mr}/\underline{emb} + \underline{cmr} \leq \Sigma\underline{mr}$, $\underline{oldsum} < \underline{newsum} \leq \Sigma\underline{ms}$, $\underline{newsum} \leq \underline{mr}$, $\underline{M}_3$ and $\underline{M}_5$. For $w \in \underline{emb}$ furthermore $\underline{ema} \subseteq \underline{eat}$ and $\underline{cmr} + \Sigma\underline{mr}/\underline{emb} + \Sigma\underline{mw} = \underline{cms} + \Sigma\underline{ms}/\underline{emb}$ are inherited.

This gives $\underline{M}_1[d_t \cdot rec> \subseteq \underline{M}_1$ for $w \in \underline{emb}$. For $w \in \underline{ema}$ from $\underline{M}_1$
and $M[d_t \cdot rec>$ we get $\Sigma\underline{mw} > 0$ and
$\underline{oldsum}' = \underline{oldsum} \leq \Sigma\underline{mr}/\underline{emb} + \underline{cmr} < \Sigma\underline{mr}/\underline{emb} + \underline{cmr} + \Sigma\underline{mw}$
$\qquad = \Sigma\underline{ms}/\underline{emb} + \underline{cms} = \Sigma\underline{ms}'/\underline{emb}' + \underline{cms}'$.
This shows $\underline{M}_1[d_t \cdot rec>\underline{M}_3$ for $w \in \underline{ema}$. Because of $\Sigma\underline{mw} = 0$
rec is not enabled in $\underline{M}_2$, $\underline{M}_4$.

The other cases are similar or even simpler.

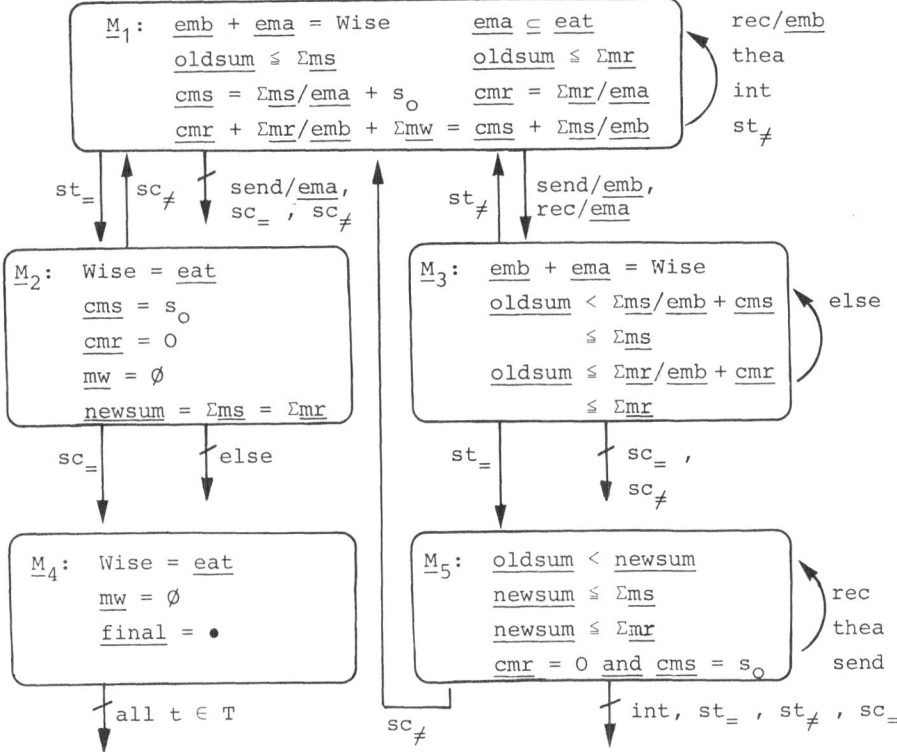

Figure 2: Assertion system,
remarks: a) $\underline{M}_i \not\xrightarrow{t}$ denotes: $d_t \cdot t$ is not enabled for all $d_t \in D_t$ in $\underline{M}_i$;
$\qquad$ b) $\underline{M}_i \xrightarrow{t/A} \underline{M}_j$ denotes: $M \in \underline{M}_i$ and $M[d_t \cdot t>M'$ and $d_t \in A \to M' \in \underline{M}_j$.
Last part of the correctness proof is to show (7):
Let M be a rechable marking satisfying $\underline{final} = \bullet$. The premisses of
theorem 2.1 (with $X = \{M_0\}$ and $v_0 = 1$) have been shown above. So we
conclude $\exists i (1 \leq i \leq 5$ $\underline{and}$ $M \in \underline{M}_i)$. By (I3) we get $\underline{emb} + \underline{ema} = \emptyset$, yielding $M \notin \underline{M}_1 \cup \underline{M}_3$, and $\underline{newsum} = \emptyset$, yielding $M \notin \underline{M}_2 \cup \underline{M}_5$ (because
$\underline{newsum} \in \mathbb{N}_0$ in $\underline{M}_2 \cup \underline{M}_5)$. The only remaining possibility is $M \in \underline{M}_4$,
wich states $\underline{eat} = Wise$ $\underline{and}$ $\underline{mw} = \emptyset$.

5. How to get an appropriate assertion system

As in the case of verification of sequential programs the problem of discovering appropriate assertions is difficult and should be combined with the specification process. From the given problem hints should be found for different phases of the system which form a partition or covering of the state space (or the reachable part of it). In the distributed termination system (I3) gives such a partition into three phases:

1) Interview-phase,

2) directly after positive counter comparison(st_) and

3) termination phase (final = •).

Blocks of such partitions may be refined by considering subsets satisfying (or not satisfying) further essential system properties. In the distributed termination problem such properties/questions are:

a) The counted messages equal the number of messages sent/received by the interviewed wise men (formal: $cms = \Sigma ms/ema + s_o$ and $cmr = \Sigma mr/ema$) .

b) How are the counted values oldsum, newsum and Σmr, Σms related?

These properties and analysing enabled transitions yields breaking of the interview-phase into M_1 and M_3 and breaking phase 2 into M_2 and M_5.

6. Remark on the verification procedure

Verifying a system with help of an assertion system has the advantage that no concurrent occurrences of transitions (or transition sequences) have to be considered. It is sufficient to consider only any atomic step (independent of all other steps). This checking of any atomic step (verification condition) is done in most cases by some simple calculations in the underlying data structure. A machine may he helpful to make these calculations and assure that all atomic steps have been considered.

References

/BeFe/ E. Best, C. Fernández : Notations and Terminology on Petri Net Theory. Petri Net Newsletter 23 (1986)

/BeMa/ C. Beilken, F. Mattern, M. Reinfrank : Verteilte Terminierung – ein wesentlicher Aspekt der Kontrolle in verteilten Systemen. Bericht Nr. 41/85 Uni Kaiserslautern, FB Informatik SFB124 (1985)

/GeLa/ H.J. Genrich, K. Lautenbach : System Modelling with High Level
        Petri Nets. Theor. Comp. Sc. 13, 109-136 (1981)
/Jen/   K. Jensen : Coloured Petri Nets and the Invariant-Method.
        Theor. Comp. Sc. 14, 317-336 (1981)
/Kos/   S.R. Kosaraju : Decidability of reachability in vector addi-
        tion systems. Proc.14th Ann.Symp. on Theory of Comp. (1982)
/Ma/   F. Mattern : New Algorithms for Distributed Termination Detec-
        tion in Asyncronous Message Passing Systems. Report 42/85
        Uni Kaiserslautern, FB Informatik SFB124 (1985)
/Ma2/   F. Mattern : Das Märchen von der verteilten Terminierung.
        Informatik-Spektrum 8, 342-343 (1985)
/Min/   L. Minsky : Computation, Finite and Infinite Machines.
        Prentice Hall (1967)
/Mü/   H. Müller : Reachability Analysis with Assertion Systems.
        In: Theoretical Computer Science, 5th GI-Conference
        Karlsruhe 1981. Ed.: P. Deussen. Springer Lect. Notes in
        Comp. Sc. 104, 214-223 (1981)
/Rei/   W. Reisig : Petri Nets, EATCS Monographs on Theoretical Comp.
        Science, Vol. 4, Springer (1985)
/Sif/   J. Sifakis : Global and Local Invariants in Transition Systems.
        Information and Control 53, 91-107 (1982)

COMMUNICATION AND DATABASE ORIENTED MODELLING

OF MULTILATERAL COOPERATION -

A COMPARISON BASED ON PETRI NETS

Volker Obermeit, Ralf Steinmetz
Technische Hochschule Darmstadt

Bernd Baumgarten, Heinz-Jürgen Burkhardt,
Peter Ochsenschläger, Rainer Prinoth
GMD Darmstadt

Abstract

The design of distributed sytems requires integrating viewpoints of various disciplines into a common modelling approach. Distributed applications are characterized by the need to correlate different processing states and data, a task which comprises communication and database aspects.

The present paper deals with the question of how the modelling of distributed applications is influenced by the viewpoint taken by the modeller. To answer this question we model an example from the area of human cooperation first from the standpoint of a communication systems specialist and then from the standpoint of a database systems expert. In both cases, we begin by identifying substructures and their interrelations as suggested by the respective concepts. Expressing both models as higher level Petri nets we are then able to investigate the relation between the communication and database aspects.

Topologically, the two nets differ mainly in their degree of explicitly modelling local states and message passing. The tasks of coordination and synchronization between the participants, however, are so strictly determined by the problem that their net representations are very similar. The major difference lies in the field-specific interpretations of the nets, which induce dissimilar partitions into subnets, namely into participants and message types on one hand and into transactions on the other hand.

1. Introduction

Distributed applications generally include aspects of both communication and databases. In this paper, we investigate the relation between these aspects by comparing modelling concepts from the two fields. The latter are presented by means of an example from the area of human cooperation, namely the contract phase of a car purchase financed by a bank loan.

We start by structuring the application problem into application substructures and their interrelation. As is to be expected, this is done partly by reasoning at the general problem level and partly with particular regard to the modelling concepts of the communication and database fields. Then, by mapping these substructures and their relations on the concepts dedicated to the two areas, we obtain a database- and a communication-oriented model. As Petri nets are particularly suited for the modelling of complex concurrent systems /Petri73,77/, these models are both finally formalized as product nets in order to allow a systematic comparison of the concepts.

In section 2 a natural language description of the contract phase is given. In section 3 a communication oriented model is derived from this description and presented as a product net. Similarly, section 4 derives a database oriented model, which is then also described as a product net. Both nets serve in section 5 for a comparison of the structures and concepts applied under the aspects of communicating systems on one hand and databases on the other hand. Finally some hints are given how to bridge the gap between the two areas.

A more detailed version of this paper has appeared as /OSBBOP86/.

Our thanks are due to Mrs. Gallinat for preparing the illustrations.
For their encouragement and support we are indebted to Prof. Dr. Schek, head of the data base systems section in the Department of Computer Science of the Technische Hochschule Darmstadt, to Dr. Raubold, head of the Institute for Systems Technology in the GMD, as well as to all the other participants of our joint seminar on synchronization.

2. Negotiating the Contract of a Car Purchase Financed by a Loan - Natural Language Description

A customer who wishes to acquire a car contacts a licensed dealer of a certain manufacturer. He informs the dealer about his personal data, financial situation, and the model he is interested in.

The dealer is contractually connected not only to the manufacturer but also to a bank. He functions as a loan broker and applies on behalf of the customer for a loan with his partner bank. The latter checks the customer's credit rating. When the check proves positive, the bank notifies the dealer of its approval, making a specific loan offer. Otherwise the loan is refused. The manufacturer finds out whether the required model is available. If this is the case, he assures the dealer of delivery; otherwise he informs the dealer that delivery is not possible.

When both loan and supply are assured, the dealer makes the customer an offer; otherwise he gives the customer a negative reply. If bank and manufacturer give one positive and one negative reply then the dealer responds to the positive reply with a cancellation.

In the case he gets an offer the customer can either decline or he can accept by placing an order. The customer's order is transformed by the dealer into an order to the bank and another to the manufacturer. As soon as both orders are confirmed the dealer confirms the order to the customer.

In comparison with a real car purchase, the following simplifications are assumed:

The customer gives the dealer the full information, as to what kind of car he is willing to buy under which conditions, in one single step. There are no further steps of negociation between customer and dealer before the dealer's offer or his negative reply. Similar assumptions hold for the dialogues between the dealer on one side and his partner bank and the manufacturer, respectively, on the other side.

The bank disposes of unlimited financial resources and decides randomly whether it grants a loan or not. If it does, then the loan offer depends only on the customer's financial situation.

The manufacturer assures delivery whenever the wanted model is on stock. He has limited resources, which are, however, increased randomly.

3. Communication Oriented Model

The interplay described informally in section 2 is an example of a cooperation of autonomous partners for the purpose of reaching a common goal. Autonomy means in this context that the partners dispose of their own resources, that they are capable of concurrent actions and possess a certain freedom of decision. The cooperation between them is based on a common initial understanding, which is then perpetuated consistently by the actions of the single partners. Due to their spatial distribution and the independence of their memories, the only form of interaction between them is the exchange of messages. Thus we denote by a cooperation the establishment and perpetuation of a common understanding (context) of communication partners by the exchange of data.

The simplest form of cooperation is the bilateral one, where we have a communication association between two partners. A multilateral cooperation can be described as a correlation of bilateral cooperations, where the vehicle is a multitude of two-party communication associations.

In our example, the participating communication partners play the roles of clients and contractors. Each client/contractor-relation constitutes a two-party cooperation. Each client enters direct relations to one or more contractors. Each contractor on his part can be a (sub-)client, namely if he cannot satisfy the order alone. Consequently, the client/contractor-relations and thus the multilateral cooperation are structured as a tree.

The nodes of the tree represent clients and/or contractors. The root of the tree represents the main client, while its leaves comprise those (sub-)contractors, who do not also play the role of a client:

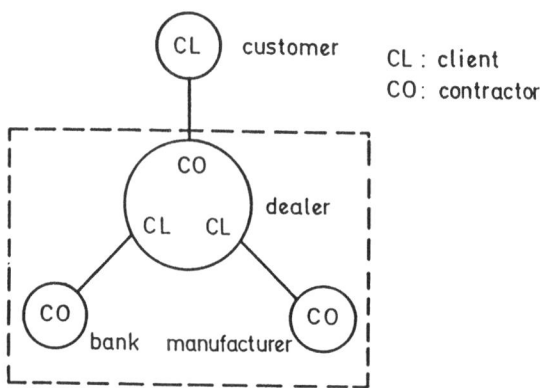

Fig. 1: Client-contractor-tree

In fig. 1, the dealer can be regarded as representing towards the customer the complete service to be rendered (dashed rectangle).

The common initial context and the consistent perpetuation of the common understanding can be considered and modelled as the correlation of local processing states to global cooperation states. We call this correlation synchronization.

The tool for maintaining the common context is the dialogue structure shown in fig. 2. In /OSBBOP86/, we show how this structure can be derived from an interplay of 'handshakes'.

Wanting to describe multitudes of car purchases, we base our model of this application on the notion of roles played by individual participants. In particular we introduce and interrelate the roles

customer C,   dealer   D,
bank   B,   manufacturer M.

In these roles the characteristical behaviour of the participants of this distributed application is laid down. In the context of a particuler car purchase these roles will be played by particular persons and institutions, e.g.

the customer Will Knotpaigh,   the dealer   Kay Owe & Co.,
the bank   Lown & Sharques,   the manufacturer Russ T. Tincan.

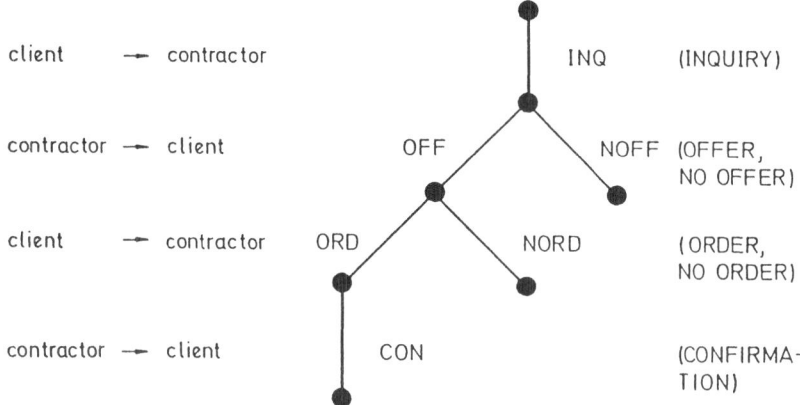

client   → contractor   INQ   (INQUIRY)

contractor → client   OFF   NOFF   (OFFER, NO OFFER)

client   → contractor   ORD   NORD   (ORDER, NO ORDER)

contractor → client   CON   (CONFIRMA-TION)

Fig. 2: Dialogue structure

The formal tool we use in our model are product nets, a form of higher level Petri nets with individualized tokens (whose identity is referred to in arc labels and transition inscriptions) as well as inhibitor arcs and erase arcs. Product nets were introduced and formalized in 1984 /EcPr84, EcPr85/ for the purpose of modelling communication systems in the PROSIT project /BEP85/.

In the course of our work with formal models for communication systems, under particular consideration of standardization activities regarding Open Systems Interconnection, certain patterns of communicative behaviour crystallized. In constructing our models, we treated these patterns both as building blocks and as strategies for connecting them, in the style of a design method.

On the Petri net level these building blocks are essentially subnets bounded (relative to the remainder) by transitions. Generally, building blocks obey the following demands: Omitting labels reduces them (often) to place-transition-nets modelling single activities in single roles. In this form they can only work for one activity at a time. Their state of progression in a single activitiy is recorded in interior places. Each of their transitions 'serves' (is neighbour of) at most one 'interface' (place outside the building block). Interface places are often drawn as triangles in order to emphasize the direction of information flow.

In our example, the entire net consists of four subnets (building blocks) describing the roles C, D, B, M, connected by 'communication' ('interface') places. CUSTOMER, BANK and MANUFACTURER essentially consist of a CLIENT or CONTRACTOR kernel topology (figs. 3, 4) which is responsible for producing the dialogue structure shown above.

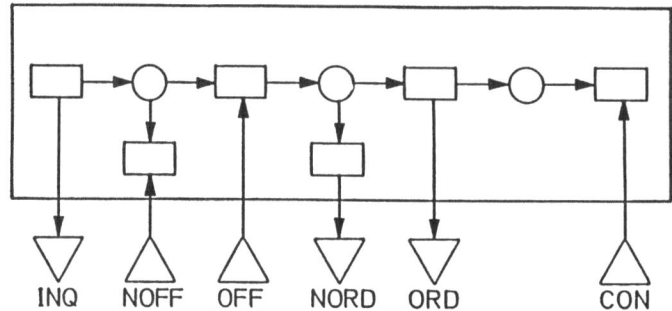

Fig. 3: Client

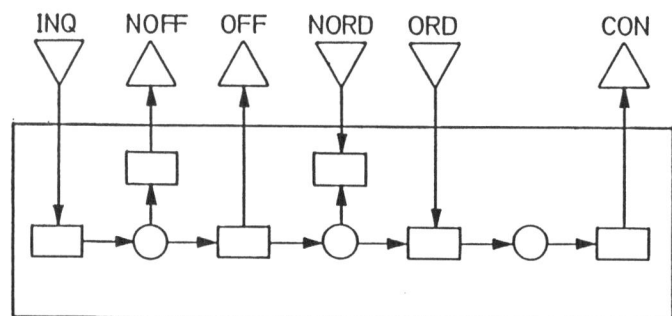

Fig. 4: Contractor

The DEALER building block models the dealer's task of coordinating the dialogues to the customer, bank and manufacturer. In a more
detailed model the dealer contains additionally two CLIENT and one
CONTRACTOR building blocks, such that concurrent activites within the
dealer (-organization) may be represented, too /BBOP86/.

In the full model (fig. 5), the kernel topologies are enhanced
with further topological details:

- counters x.REF which permit to distinguish locally between concurrent car purchases. These references accompany the messages exchanged between the participants (like 'our/your reference').
- B.LOANOFF models how the bank determines a possible loan offer from
  the customer's informations. References to loans actually granted
  are created by means of B.LOANREF.
- The available manufacturer's stock in M.STOCK is increased by the
  production of additional cars (M.PROD) whose potential model
  description is taken from M.MODELS. They are identified uniquely
  relative to their manufacturer by use of M.CARREF. M.NOFF will
  only occur if no suitable car is on stock (inhibitor arc from
  M.STOCK). Making an offer (M.OFF) will reserve the car, a cancellation (M.NORD) will make it available again.

On the dealer's part, D.BANK, D.MANUF and D.PRICE model the associations

dealer	$\rightarrow$		partner bank,
dealer	$\rightarrow$		manufacturer,
dealer, model	$\rightarrow$		nrice.

When the answers from bank and manufacturer do not lead to an offer to the customer, the B- and M-tags in the D.NOFF-transitions indicate by whose negative reply this was caused.

Due to space limitations we confine the list of arc labels to the neighbourhood of the transitions M.INQ, M.OFF, M.NOFF. The interested reader will probably arrive at a labelling very similar to ours if he or she just fills in the informations to be passed around according to the problem description; otherwise he or she is referred to the full listing in /OSBBOP86/.

partial list of arc labels:
51: ⟨D_INQ_M, d_name, d_ref, m_name, car_spec⟩
52: ⟨M_NOFF_D, m_name, m_ref, d_ref, d_name⟩
53: ⟨M_OFF_D, m_name, m_ref, d_ref, d_name⟩

61: ⟨m_name, m_ref⟩
62: ⟨m_name, m_ref+1⟩
63: ⟨m_name, m_ref, d_ref, d_name, car_spec⟩
64: ⟨m_name, car_spec, car_ref, FREE⟩

4. Database Oriented Model

The same scenario as in chapter 2 can also be modelled under the aspect of using a database management system. This leads to a radically different view of our model, as it is heavily influenced by the available tools:
- Static aspects will be modelled by the data base, consisting of raw data and its logical structure as defined in the conceptual schema. This schema describes the structure of the observed miniworld, and the data contents comprise the actual state.
- Dynamic aspects are modelled by transactions, embedded in application programs and executed by the database management system. These transactions implement the consistent state changes of the data base.

We will constitute our model by fitting our miniworld, consisting of customers, dealers, manufacturers and banks, into the framework of a conceptual schema and by defining transactions which simulate the actions in our scenario. Distribution and autonomy of the actors is of

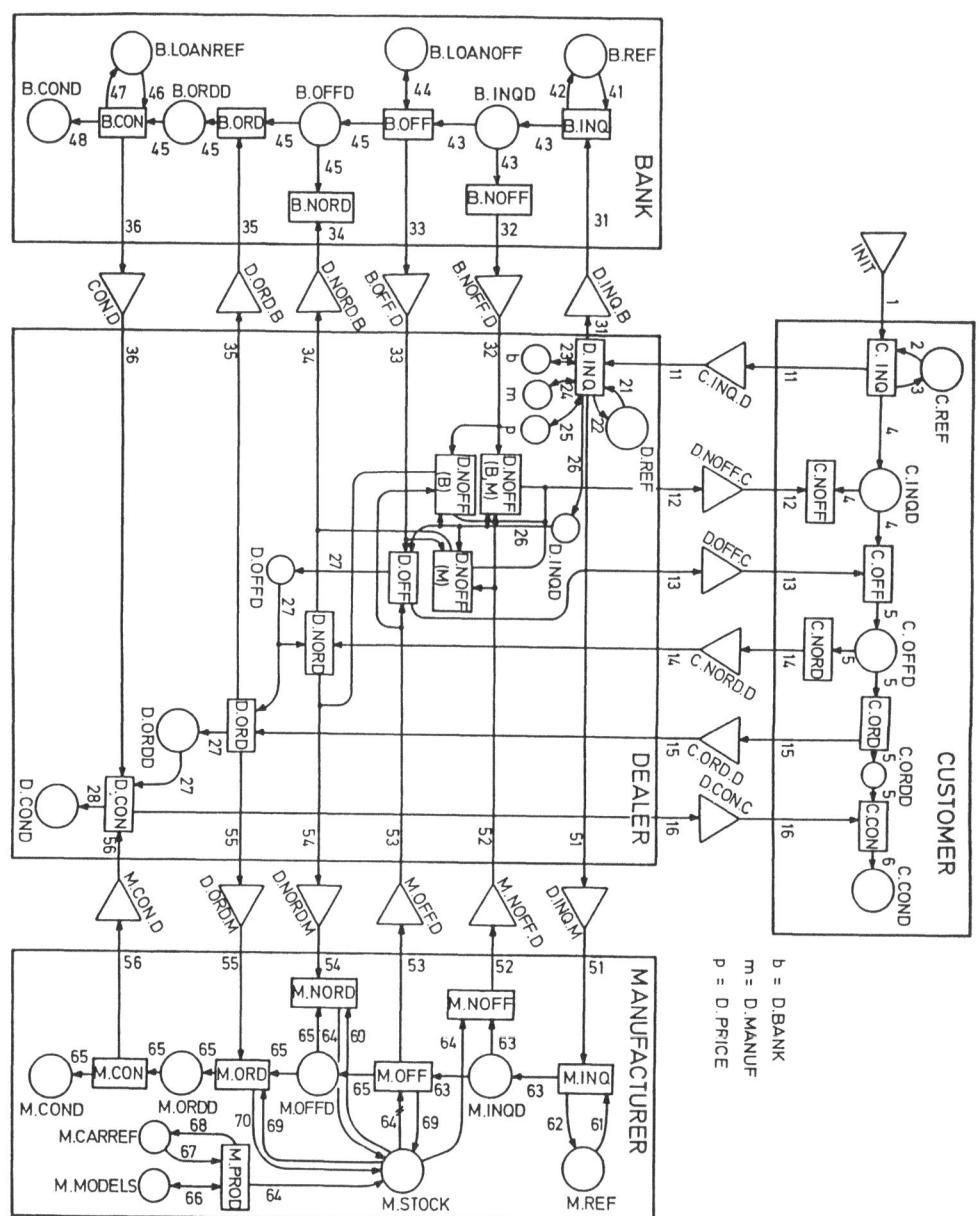

Fig. 5: The Communication Oriented Petri Net Model

minor importance, as the database management system acts as a global centralised coordination instance and hides the communication oriented aspects from the actual user. This can be done through artificial centralisation of the distributed process or through the use of a distributed data base management system.

Our first step consists of an informal problem description and a collection of the essential facts to isolate the miniworld concerned. Afterwards we model the relevant objects and their relationships as a conceptual schema based on the relational model of Codd /CODD70/. Here all data reside in tables (relations), which consist of identically structured records (tuples) consisting of atomic data fields (attributes). Relationships are implicitly contained in the data, e.g. through equality of corresponding attribute values. Our nomenclature will be 'relation-name (attribute-name-1, ..., attribute-name-n)'. When defining a conceptual schema one should take into account that certain aspects of the miniworld can possibly be modelled outside the data base (i.e. as external agents interacting with it). In our case we decided to use this open approach: The customer is actually modelled as the user of the database management system, and his interactions with the dealer comprise the I/O-interface of the system. The remaining data are partially grouped, corresponding to the actors to which they belong. This grouping could be used as a base for a partitioning in the case of a distributed database management system.

The following conceptual schema shows all the necessary relations, each one described by its name (in capital letters) and its list of attributes (enclosed in parentheses, attribute names in lower case). Attribute domains (e.g integer) are omitted for the sake of simplicity. The groups are separated by short explanations (enclosed in **).

CONCEPTUAL SCHEMA:

STOCK (manufacturer, model, status, car-id)
   ** Each tuple represents a car available from a certain
     manufacturer. **
LOANOFF (bank, price, finances, loan offer)
OFFERED LOANS (bank, customer, price, loan offer, loan-id)
   ** LOANOFF models an oracle which decides whether a loan is offered
     or not. Offers actually made are stored in OFFERED LOANS. **
PARTNER BANK (dealer, bank)
LICENSED DEALER (dealer, manufacturer)
PRICE CATALOG (model, price)

OFFER FILE (offer-id, dealer, customer, car-id, model, price,
          manufacturer, bank, loan offer, loan-id)
   ** All offers to customers are held in the OFFER FILE for further
      use.**
LOAN CONTRACT (dealer, customer, car-id, price, loan offer, bank)
PURCHASE CONTRACT (dealer, customer, car-id, model, manufacturer)

      The last step is to define transaction programs to simulate the
actual proceeding in our scenario. As the semantics of transactions
guarantee atomicity, consistency, integrity-preservation and durabili-
ty /HäRe83/, we see that our program must be divided into several
(sub-) transactions, since in reality there exist observable intermediate states of the process (i.e. offer made to customer, but not yet
answered). Our particular example is written in SQL /IBM83/ embedded
in Pseudo-PASCAL. Due to space considerations only the main program's
body is shown in table 1. The reader is recommended to consult
/OSBBOP86/ for refinements of the sub-transactions and a full listing.

4.1 Representation of the Database Oriented Model as a Petri Net
      For the purpose of comparing the two viewpoints of this 'car pur-
chase' example (given in sect. 3 and 4) the database 'implementation'
will be transformed into a Petri net of the type used for the communication style representation /EcPr84, EcPr85/. We distinguish two
layers:
1. the user-interface including the user himself (customer),
2. the database-layer.

4.1.1 The User Interface including the Customer
      The relevant attitudes and reactions of persons intending to buy
a car have to be incorporated into the Petri net. This layer (fig. 6)
is established for the very purpose of distinguishing all surrounding
activities from the kernel, the database system.

4.1.2 The Database Layer
      From the entire model (fig. 7), we selected the transaction
'manufacturer inquiry', on which we demonstrate the transformation
method as well as the results achieved.
      The following rules characterize the transformation:
 - A transaction (subtransaction) is transformed into a set of transitions, together with the corresponding places and arcs.
      E.g. the transaction 'manufacturer inquiry' is composed of: the
      transitions 'positive manufacturer' and 'negative manufacturer'

the places 'manufacturer-input', 'manufacturer-output' and
'stock', the arcs no. 8, 30, 53, 54, 55 and '56.
- Every relation is represented by a place.
    E.g. the relation STOCK corresponds to the place 'stock'.
- The attributes of the relations constitute the components (in the
  Petri net context) of the tokens of the place.
    E.g. the components 'manufacturer, model, status, car-id' of the
    relation STOCK make up the components of the tokens at the place
    'stock'.
- The manipulation of data will show up as the dynamics of the Petri
  net (the respective markings of the places and the occurrences of
  the transitions).

```
program    LOAN_PURCHASE ;
  begin
    read customer inquiry ;
    begin_transaction  OFFER ;
        select  bank, manufacturer, price
        from    partner bank, licensed dealer, price catalog
        where   "specifications of customer inquiry apply" ;
        call transaction  MANUFACTURER_INQUIRY ;
        call transaction  BANK_INQUIRY ;
        case concat(manufacturer_answer,bank_answer) of
          '++' : begin
                    insert  offer  into offer file ;
                    return ('positive')
                 end
          '+-' : begin
                    call transaction CANCEL_BANK_OFFER ;
                    return ('negative')
                 end
          '-+' : begin
                    call transaction CANCEL_MANUFACTURER_OFFER ;
                    return ('negative')
                 end
          '--' :    return ('negative')
        end;
    end_transaction  OFFER ;
    if offer positive
    then begin
            display offer ;
            read customer decision ;
            if decision = 'buy'
            then begin
                    call transaction PURCHASE ;
                    print purchase contract ;
                    print loan contract ;
                 end
            else call transaction CANCEL_OFFER ;
         end;
end.
```

Table 1: LOAN_PURCHASE (top level)

In performing these transformations, we observed that many sequences of operations on relations which often appear in real implementations can be transformed into simple transitions with adequately labelled arcs.

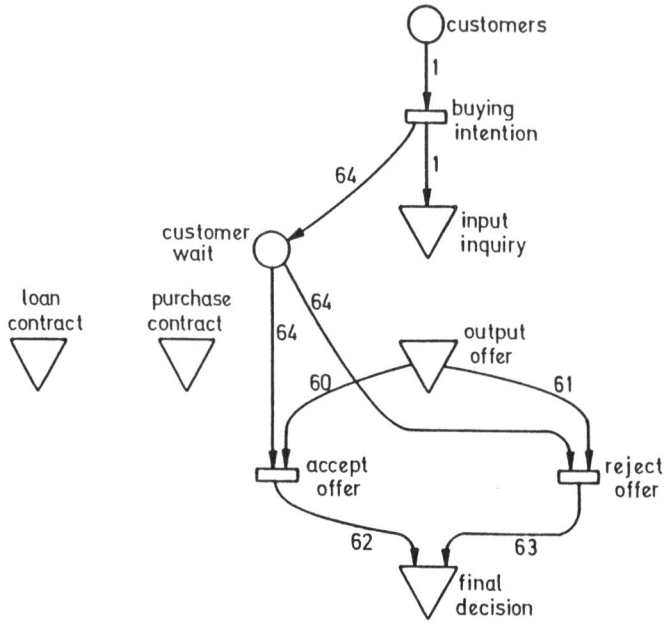

Fig. 6: Database oriented Petri net model - user layer

4.1.3 Places and arcs of the transaction 'manufacturer inquiry'

Stock: This is the place corresponding to the relation STOCK, which represents the supply of cars. Every car is identified by the 'car-id'. Its status can be 'sold', 'available' or 'reserved'. The place 'stock' can thus hold tokens of the following types:

⟨manufacturer,model,AVAILABLE,car-id⟩
⟨manufacturer,model,SOLD,car-id⟩
⟨manufacturer,model,RESERVED,car-id⟩

Variable names are written lower case while constants - here the different status values - are written in capital letters. At the beginning we need no tokens in this place, because the production of cars is explicitly modelled by a different transaction 'production'.

Manufacturer-input: This place and the manufacturer-output represents parameter passing between the dealer and the manufacturer:

⟨transaction-id,customer,dealer,manufacturer,model⟩

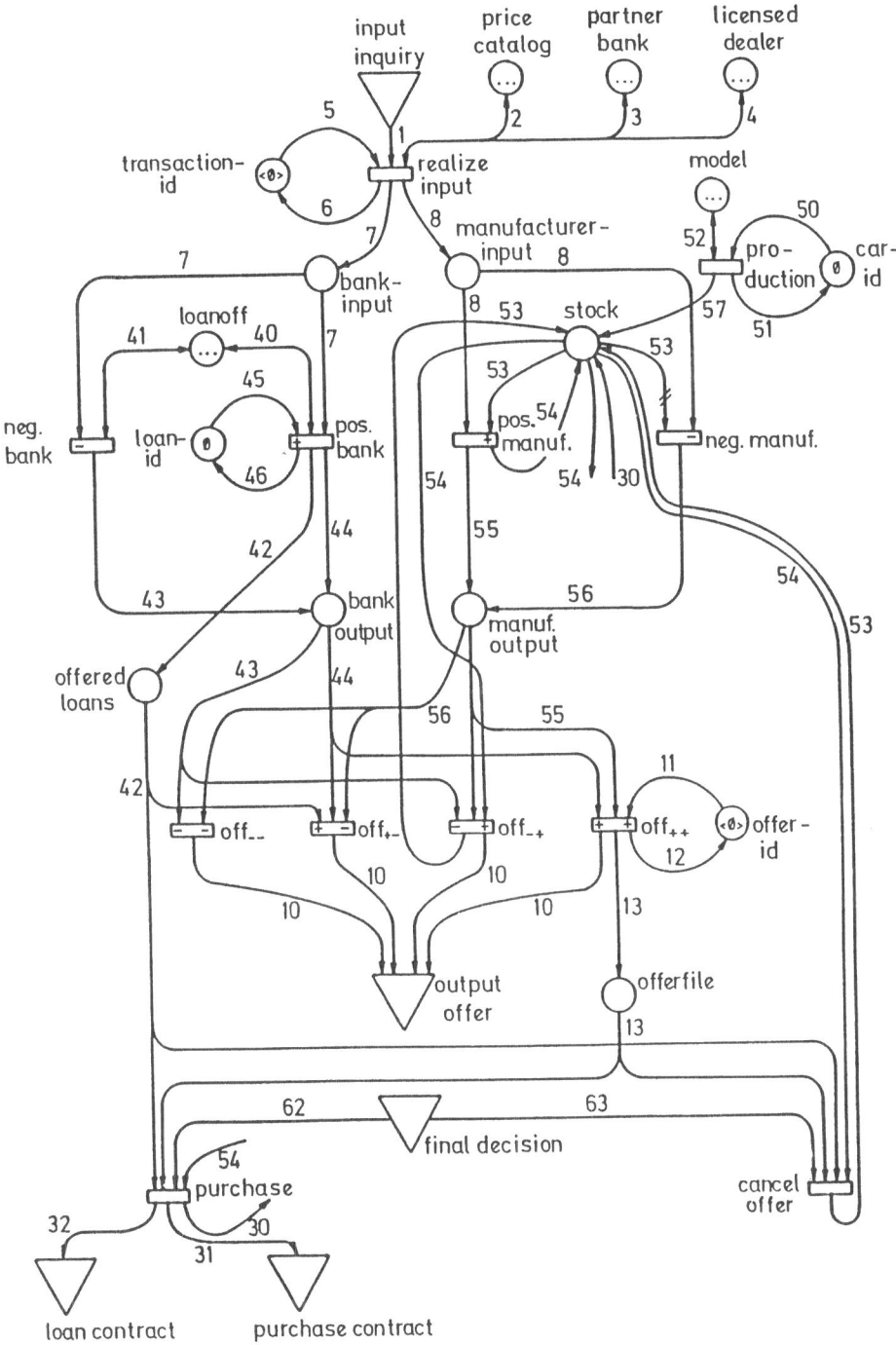

Fig. 7: Database Oriented Petri Net Model – Database Layer

Manufacturer-output: The outcomes of the inquiries to the manufacturers show up as the tokens in this place. They are of the types

〈transaction-id,NO-CAR-OFFER,dealer,customer,car-id,model,
                                                    manufacturer〉
〈transaction-id,RESERVED,dealer,customer,car-id,model,
                                                    manufacturer〉

The arcs attached to this transaction bear the following labels:
 8: 〈transaction-id,customer,dealer,manufacturer,model〉
30: 〈manufacturer,model,SOLD,car-id:
53: 〈manufacturer,model,AVAILABLE,car-id〉
54: 〈manufacturer,model,RESERVED,car-id〉
55: 〈transaction-id,NO-CAR-OFFER,dealer,customer,car-id,model,
                                                    manufacturer〉
56: 〈transaction-id,RESERVED,dealer,customer,car-id,model,manufacturer〉

5. A Comparison of the Models

In the following, we compare the two net models, viz. the communication oriented one ('COMM') and the database oriented one ('DABA').

Both models decompose the total task given by the problem description (section 2) into subtasks of different extents (bilateral coordination steps, transactions). In both cases, the interplay of the respective subactivities in the context of each model achieves the full aim of the cooperation.

A comparison of the two models reveals that the database standpoint of the action constitutes a higher level (more abstract) view than the communication standpoint:

- an occurrence sequence of COMM beginning with D.ORD and ending with D.CON (including occurrences of B.ORD, B.CON, M.ORD, M.CON) corresponds to an occurrence of the transition 'purchase' in the DABA model.
- an occurrence sequence in the COMM model of D.NORD, followed by B.NORD and M.NORD in any order, corresponds to an occurrence of the 'cancel offer' transition in the DABA model.
- etc.

It turns out that the ways of modelling the synchronization mechanisms
- for the coordination of subactivities to a total task
- for the separation of different total tasks (several car purchases) during the concurrent access to common resources (e.g. manufacturer's stock)

essentially differ only in that the COMM model explicitly addresses the spatial separation of the systems.

The DABA model can be considered as a requirement specification for arbitrary implementations. If a centralized realization is intended, no further efforts are needed with respect to design logic. In the case of a distributed realization, non-local transactions may arise, which necessitate communication in order to ensure their atomicity.

The task treated in this paper nicely demonstrates an example of how these communication aspects are to be dealt with in such a case:

Table 2 contains corresponding partitions of the transitions of both models; it is obtained by superposing the DABA transaction structure given in section 4 and the COMM net (fig. 5) and DABA net (figs. 6, 7), respectively. In the net models each transaction consists of the subnet spanned by the transitions listed.

transaction .....sub-transaction	COMM transitions	DABA transitions
OFFER	D.INQ D.NOFF(B,M) D.OFF	realize input off-- off++
.....BANK_INQUIRY	B.INQ B.OFF B.NOFF	neg. bank pos. bank
.....MANUFACTURER_INQUIRY	M.INQ M.OFF M.NOFF	neg. manuf. pos. manuf.
.....CANCEL_BANK_OFFER	B.NORD D.NOFF(M)	off+-
.....CANCEL_MANUFACTURER_OFFER	M.NORD D.NOFF(B)	off-+
PURCHASE	D.ORD D.CON B.ORD B.CON M.ORD M.CON	purchase
CANCEL_OFFER	D.NORD B.NORD M.NORD	cancel offer
PRODUCTION	M.PROD	production

Table 2: Transaction structure in the net models

Obviously, in a realisation by a distributed database system there will be 'local' transactions (BANK_INQUIRY, MANUFACTURER_INQUIRY, PRODUCTION), as well as 'distributed' transactions (OFFER, CANCEL_BANK_

OFFER, CANCEL_MANUFACTURER_OFFER, ˙PURCHASE, CANCEL_OFFER). The former
are transactions of the type met in non-distributed databases.

Of course, different partitions are conceivable, including such
that lead exclusively to local transactions. The particular choice of
a partition will generally neither facilitate nor complicate the task
of modelling: Different coverings by transactions will only shift the
complexity of implementation between the database user and the database system.

Distributed transactions are user-friendly, as they guarantee per
se the integrity of the distributed databases they manipulate. In
their case it is up to the (invisible) interplay of the local database
managers concerned to ascertain this integrity (e.g. by use of the
CCR algorithm specified in ISO OSI standards for tree-structured
transactions /BOP85b, BOP87/). If only local transactions are used,
the user has to take care that the interplay of the transactions will
produce the intended global results.

References

/BBOP86/ B.Baumgarten, H.J.Burkhardt, P.Ochsenschläger, R.Prinoth:
The Signing of a Contract - a Tree-Structured Application
Modelled with Petri Net Building Blocks, in: Advances in Petri
Nets 1985, G. Rozenberg (ed.), Springer LNCS 222, 1986

/BEP85/ H.J.Burkhardt, H.Eckert, R.Prinoth: Modelling of OSI-Commu-
nication Services and Protocols using Predicate/Transition
Nets, in: Protocol Specification, Testing, and Verification,
IV, Y.Yemini et al.(ed.), North-Holland, 1985

/BOP85a/ B.Baumgarten, P.Ochsenschläger, R.Prinoth: Building Blocks
for Distributed System Design, in: Protocol Specification,
Testing, and Verification, V, M. Diaz (ed.), North-Holland,
1986

/BOP85b/ B.Baumgarten, P.Ochsenschläger, R.Prinoth: A Formal Model of
the CCR Algorithm, Arbeitspapiere der GMD Nr.186, 1985

/BOP87/ B.Baumgarten, P.Ochsenschläger, R.Prinoth: Synchronization
in Tree-Structured Transactions - a Case Study, to appear in:
Protocol Specification, Testing, and Verification, VI,
North-Holland

/Codd70/ E.F. Codd: A Relational Model for Large Shared Data Banks,
Communications of the ACM Vol.13 , no.6, June 1970.

/EcPr84/ H.Eckert, R.Prinoth: Produktnetze - Definition eines PROSIT-
Beschreibungsmittels, Arbeitspapiere der GMD Nr.92, 1984

/EcPr85/ H.Eckert, R.Prinoth: Grundsätzliche Betrachtungen und
Bemerkungen zu Produktnetzen, Studien der GMD Nr.106, 1985

/HäRe83/ T.Härder, A.Reuter: Principles of Transaction-Oriented Database Recovery, ACM Computing Surveys Vol.15 , no.4, December
1983

/IBM83/   IBM Manual: SQL Application Programming, no.SH24-5018-2 , 1983

/OSBBOP86/ V.Obermeit, R.Steinmetz, B.Baumgarten, H.J.Burkhardt,
          P.Ochsenschläger, R.Prinoth: Datenbank- und kommunikations-
          orientierte Modellierung einer mehrseitigen Kooperation - ein
          Vergleich auf der Basis von Netzen, Studien der GMD Nr.115,
          1986

/Petri73/ C.A.Petri: Concepts of Net Theory, in: Mathematical Founda-
          tions of Computer Science - Proceedings, High Tatras, 1973

/Petri77/ C.A.Petri: Communication Disciplines, in: Computing System
          Design - Proceedings, Newcastle upon Tyne, 1977

THE STRUCTURE OF FACTS IN OCCURRENCE NETS

Helmut Plünnecke

Gesellschaft für Mathematik und Datenverarbeitung
Institut für Methodische Grundlagen (F1)
Postfach 1240, Schloß Birlinghoven
D-5205 St. Augustin 1

Abstract

A fact in a C/E-system $\Sigma:=(B,E;F,C)$ is a conceivable "dead" pure transition t which could be added to the system Σ without changing its behaviour. The transition t is completely described by the corresponding
pair $(^{\cdot}t,t^{\cdot})$; and therefore facts are essentially pairs (I,J) with
$I,J \subseteq B$. Facts with minimal (I,J) are called *basic* facts; and it is immediately clear that a basic fact (I_0,J_0) implies all facts (I,J) with
$I_0 \subseteq I \wedge J_0 \subseteq J$.

Processes of Σ can be conceived as mappings from appropriate occurrence nets into Σ. Then every fact of Σ can be derived from the facts
in the corresponding occurrence nets. Therefore, knowledge of the general structural properties of facts in occurrence nets may be helpful
to understand how facts can be realized by C/E-systems. The aim of
this paper is to show that facts in occurrence nets have a rather simple structure provided some finiteness conditions hold true. Obviously, to this end it is suffucient to investigate the *basic* facts of occurrence nets; it will turn out that for all basic facts (I_0,J_0) of an
occurrence net we have $|I_0| \leq 2$; and that the set of all of these basic
facts can be partitioned into six classes each of which contains facts
of a special structure.

1 Introduction

Facts play an important role in net theory[1]; they can be used to specify and to describe the behaviour of systems (cf. [5], [6], [9]). In a sense, their expressive power is "complete": Every sentence about a C/E-system which *only* concerns the holding and not-holding of conditions is a logical consequence of those sentences expressed by facts; and the set of facts uniquely determines the set of cases of the C/E-system.[2]

The facts of a contact-free C/E-system $\Sigma := (B,E;F,C)$ are established by the possible processes of this system; these processes can be conceived as mappings from corresponding occurrence nets into the underlying system net preserving the "local" structure (cf. [1], [2]). We consider an occurrence net $\Sigma_0 := (S,T;F)$ a special non-empty C/E-system $\Sigma_0^c := (S,T;F,C)$ where C is the set of *all* S-cuts[3] of Σ_0; we call the corresponding poset $O := (S,T;\leq)$ an occurrence poset, and denote the set of all *finite* S-cuts of Σ_0 by C-fin.

By definition, a fact in a C/E-system $\Sigma := (B,E;F,C)$ is a conceivable "dead" pure transition t which could be added to the system Σ without changing its behaviour (because t would be enabled in no case). The transition t is completely described by the corresponding pair $(\dot{t},t\dot{})$. In this sense, obviously, a pair (I,J) corresponds to a fact of Σ iff

(1) $\qquad\qquad I,J \subseteq UC \wedge I \cap J = \emptyset \wedge (\forall H \in C)(I \subseteq H \Rightarrow J \cap H \neq \emptyset).$

Pairs (I,J) with the property (1) are called facts with respect to C; by Fact(C) we denote the set of *all* of these pairs. The sets Fact$_0(C)$ and Fact$_{00}(C)$ are subsets of Fact(C); the consist of all (I,J) with $|I| < \infty$ and with $|I| < \infty \wedge |J| < \infty$, respectively, and we call the respective facts "finite" and "totally finite". Because the definition of Fact(C) depends on nothing else than the set C, we can immediately apply (1) for any subset C' instead of C thus defining *facts* of $\Sigma := (B,E;F,C)$ with respect to C', the sets of which we denote by Fact(C'), Fact$_0(C')$, and Fact$_{00}(C')$, respectively; and again we have that every

---

[1] See [9] from where most of the net-related definitions have been taken.

[2] This is also true in the infinite case: One only has to replace conjunctions and disjunctions by approriate expressions of predicate logic with universal or existential quantifiers, respectively.

[3] S-cuts are those cuts of Σ_0 which contain *only* elements of S (unbranched places).

sentence about the C/E-system which *only* concerns the holding and not-holding of conditions *in the cases* C' is a logical consequence of those sentences expressed by Fact(C').

On the set Fact(C) we define a partial order by

(2) $\qquad\qquad (I,J) \subseteq_2 (I',J') \iff I \subseteq I' \wedge J \subseteq J'$;

and for any $C' \subseteq C$ the minimal elements in the poset (Fact(C'), $\subseteq_2 \cap (C' \times C')$) are called basic facts (with respct to C'), and by basic-Fact(C'), basic-Fact$_0$(C'), and basic-Fact$_{00}$(C') we denote the set of all basic facts in Fact(C'), Fact$_0$(C'), and Fact$_{00}$(C'), respectively.[4] It is clear that a fact (I,J) implies every fact (I',J') which is greater than or equal to (I,J) (i.e. $(I,J) \subseteq_2 (I',J')$). Hence, provided every fact contains a basic fact, every fact can be trivially derived from the basic facts[5].

The set of cases of Σ is uniquely determined by the cases of the corresponding ("processes defining") occurrence nets; and hence the facts of these occurrence nets uniquely determine the facts of Σ. To describe the relation between the facts of Σ and the facts of the corresponding occurrrence nets in more detail we introduce the following concept:

A fact (I,J) of Σ is called an *occurring* fact iff I is subset of at least one case of Σ; and we call a fact (I',J') in a corresponding occurrence net an *occurrence* of the occurring fact (I,J) iff I' is subset of at least one S-case of the occurrence net, $p(I')=I$, and $p(J') \subseteq J$, where p is the process defining mapping from the occurrence net into Σ.

Then obviously the following holds:

If (I,J) is an occurring fact of Σ, then there is at least one corresponding occurrence net with a fact (I',J') which is an occurrence of (I,J).

Therefore, knowledge of the general structural properties of facts in occurrence nets can be used to understand how facts can be realized by C/E-systems. The aim of this paper is to show that facts in occurrence nets have a rather simple structure provided some finiteness conditions hold true.

---

[4] It is easy to see that basic-Fact$_0$(C')=basic-Fact(C')$\cap$Fact$_0$(C') and basic-Fact$_{00}$(C')=basic-Fact(C')$\cap$Fact$_{00}$(C').

[5] This connection between arbitrary facts and basic facts is the reason for the name "basic".

In order to get an idea what kind of finiteness conditions should
be reasonably assumed imagine we observe a C/E-system and try to find
out how it behaves. Then at any time we can have observed only a
finite number of occurrences of condition holdings (and events); especially, infinite cases never can have been observed as a whole. This
suggests the following two restraints for the corresponding occurrence
posets: Facts should only relate to *finite* cuts, if (I,J) is a fact
then I should be finite, and all "intervals" between any two points
should be finite[6].

The first of these "finiteness restraints" can now be expressed in
the following way: Instead of investigating the set of *all* facts we
(only) will consider the set $Fact_0(C\text{-fin})$ and describe the structure
of its basic facts. This is less restrictive than it may seem at first
glance: Obviously, *all* S-cuts of occurrence nets decribing processes
of *finite* C/E-systems are finite (and hence $C=C\text{-fin}$); and there is a
large class of occurrence posets[7] (with possibly infinite S-cuts)
where we at least have[8] $Fact_{00}(C)=Fact_{00}(C\text{-fin})$.

The other finiteness conditions we need are expressed by length-finiteness, interval-finiteness, and degree-finiteness: An occurrence
poset is called length-finite [interval-finite] iff all chains
[intervals] between any two points are finte, and it is called degree-
finite iff all elements only have a finite number of immediate predecessors and successors.[9] Our main results can now be statet as follows:

1.1 Theorem. In an occurrence poset every finite fact (I,J) with respect to $C\text{-fin}$ contains a finite basic fact with respect to
$C\text{-fin}$[10]. ∎

1.2 Theorem. Let $O:=(S,T;\leq)$ be a interval-finite occurrence poset
which has at least one finite S-cut, and let $(I,J)\in basic\text{-}Fact_0(C\text{-fin})$.
Then $|I|\leq 2$; moreover, (I,J) belongs to one of six classes *a-f* each of

---

[6] A formal definition is given below.

[7] See 1.4.

[8] Obviously, $Fact_{00}(C)\subseteq Fact_{00}(C\text{-fin})$.

[9] It is clear that interval-finiteness implies lenght-finiteness.

[10] I.e. $(\exists(I',J')\in basic\text{-}Fact_0(C\text{-fin}))(I',J')\subseteq_2(I,J)$.

which contains facts of a special structure[11] according to the following list:

   a) $|I|=0$ $|J|>0$: J "unlimited"   − Cf. Fig. 1
   b) $|I|=1$ $|J|>0$: J "limited" above   − Cf. Fig. 2
   c) $|I|=1$ $|J|>0$: J "limited" below   − Cf. Fig. 3
   d) $|I|=1$ $|J|>0$: J twosidely "limited" − Cf. Fig. 4
   e) $|I|=2$ $|J|>0$: J twosidely "limited" − Cf. Fig. 5
   f) $|I|=2$ $|J|=0$: J empty   − Cf. Fig. 6

Examples for the different structures are depicted in Fig. 1-6; the
elements of I and J are represented by ⊙ and ⊖ , respectively; and
continuous lines [————], broken lines [------], and dotted lines
[···········] denote the neighbourhood relation, the ≤-relation and the
<-relation, respectively.   ■

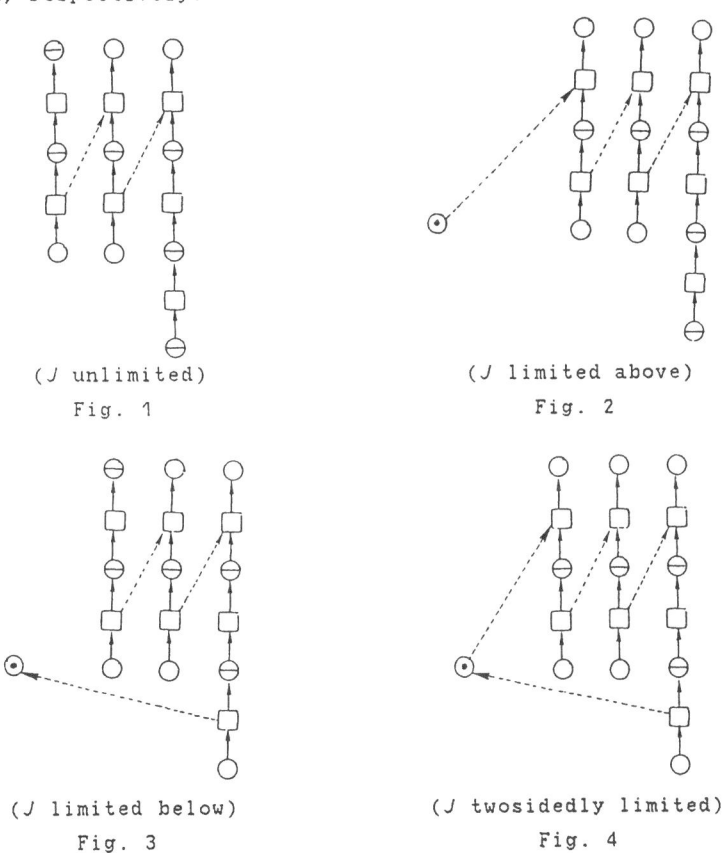

(*J* unlimited)   (*J* limited above)
Fig. 1   Fig. 2

(*J* limited below)   (*J* twosidedly limited)
Fig. 3   Fig. 4

---

[11] A formal description of the different structures is given in sec-
     tion 3.

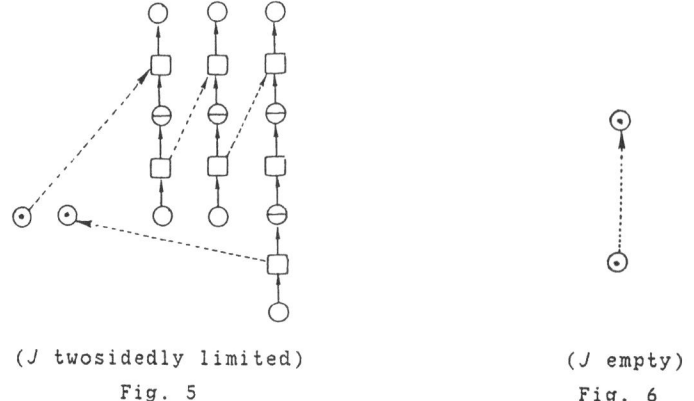

(*J* twosidedly limited)   (*J* empty)

Fig. 5   Fig. 6

In the next two theorems conditions for the interval-finiteness of
occurrence nets are stated (so that theorem 1.2 can be applied).

1.3 Theorem. If all S-cuts of an occurrence poset $O:=(S,T;\leq)$ are finite, then O is interval-finite.   ∎

1.4 Theorem. Let $O:=(S,T;\leq)$ be a length-finite occurrence poset; let
there be at least one finite S-cut. and let at least one of the following properties (*a*), (*b*) hold true:

  (*a*) All elements of *S* are contained in at least one finite S-cut
    (i.e. $S = \mathbf{U}C$-fin).

  (*b*) *Q* is degree-finite.

Then both properties (*a*) and (*b*) hold, O is interval-finite, and,
moreover, we have $Fact_{oo}(C) = Fact_{oo}(C\text{-fin})$.   ∎

If $O:=(S,T;\leq)$ is an occurrence poset, then the induced poset $(S,\leq)$
has two properties[12]: It is combinatorial (i.e. $\leq$ is the reflexive
and transitive closure of the "neihgbourhood relation") and N-dense
(i.e. certain "N-shaped" figures are not contained in the poset). It
will turn out that the following proofs work in general combinatorial
N-dense posets and that the special structure of occurrence posets is
only used to prove these two properties of $(S,\leq)$. In the following, we
therefore no longer refer to the special structure of occurrence nets,
but we will investigate arbitrary combinatorial N-dense posets defining its "facts" according to (1); the straightforward (re-)translation
of our results to the original problem of occurrence nets mainly is
left to the reader.

---

[12] A formal definition and a proof of these properties will be given
in section 2.

This paper is organized as follows: In section 2 the basic definitions are given, and some propositions concerning finiteness assumptions, and, especially, some theorems about lattice properties of posets of finite cuts are derived. The main results about the structure
of basic facts are proved in section 3. Section 4 contains some con-
cluding remarks about the application of these results.

2 Preliminary Considerations

We start with some well-known concepts concerning partial orders.[13]

2.1 Definition. Throughout the following let $P:=(X,\leq)$ be a *non-empty*
partially ordered set (poset), i.e. a non-empty set together with a
reflexive, antisymmetric, and transitive relation $\leq$ on it.
$$x\widetilde{\sim}y \;:\Longleftrightarrow\; x,y\in X \,\wedge\, \neg x<y \,\wedge\, \neg y<x$$
defines the *co-relation* of P; and by $x<y$ and $x\sim y$ we denote $(x\leq y \wedge x\neq y)$
and $(x\widetilde{\sim}y \wedge x\neq y)$, respectively. Let $x,y\in X$ and $Q\subseteq X$. Then:

$x\triangleleft y \;:\Longleftrightarrow\; x<y \,\wedge\, (\forall z\in X)(x\leq z\leq y \Rightarrow x=z\vee z=y) \;:\Longleftrightarrow\; x,y$ are neighbours.

$\;\;\;\;\;\;\;\;^{\cdot}Q := \{z\,|\,z\in X \,\wedge\, (\exists q\in Q)z\triangleleft q\}, \;\;\;\; Q^{\cdot} := \{z\,|\,z\in X \,\wedge\, (\exists q\in Q)q\triangleleft z\};$

$\;\;\;\;\;\;\;\;\downarrow Q := \{x\,|\,x\in X \,\wedge\, (\exists q\in Q)x\leq q\}, \;\;\;\; \uparrow Q := \{x\,|\,x\in X \,\wedge\, (\exists q\in Q)q\leq x\};$

if $Q:=\{q\}$ is a singleton, then we simply write $^{\cdot}q$, $q^{\cdot}$, $\downarrow q$, and $\uparrow q$ instead of $^{\cdot}\{q\}$ etc.; and $(Q,\leq)$ is an abbreviation for the induced sub-
poset $(Q,(\leq)\cap(Q\times Q))$.

$L\subseteq X$ is called a chain iff $(\forall x,y\in L)(x\leq y\vee y\leq x)$, $A\subseteq X$ is called a antichain
iff $(\forall x,y\in A)(x\widetilde{\sim}y)$; lines are maximal chains, cuts are maximal anti-
chains. C is the set of cuts, C-fin is the set of finite cuts. (Note
that *here* C and C-fin relate to the the *whole* set X on which the partial order is defined whereas in an occurrence net $O:=(S,T;\leq)$ C and
C-fin only consist of those [finite] cuts that are subsets of S.) For
$I,J,I',J'\in C$ and $C'\subseteq C$ we set
$$(I,J)\subseteq_2(I',J') \;:\Longleftrightarrow\; I\subseteq I' \,\wedge\, J\subseteq J';$$

$\;\;\;\;\;$Fact$(C')\;\;\; := \{(I,J)\,|\,I,J\subseteq UC' \,\wedge\, I\cap J=\emptyset \,\wedge\, (\forall H\in C')(I\subseteq H \Rightarrow J\cap H\neq\emptyset)\},$

$\;\;\;\;\;$Fact$_0(C')\;\; := \{(I,J)\,|\,(I,J)\in\text{Fact}(C') \,\wedge\, |I|<\infty\},$

$\;\;\;\;\;$Fact$_{00}(C') := \{(I,J)\,|\,(I,J)\in\text{Fact}(C') \,\wedge\, |I|<\infty \,\wedge\, |J|<\infty\};$

and we call the respective elements of Fact(C'), Fact$_0(C')$, and
Fact$_{00}(C')$ the facts, the finite facts, and the totally finite facts
(of P with respect to C'). $\subseteq_2$ defines a partial ordering on $C\times C$; the
respective minimal elements in (Fact$(C'),\subseteq_2$), in (Fact$_0(C'),\subseteq_2$), and

---

[13] Cf. [8].

in $(Fact_{oo}(C'), \subseteq_2)$ are called basic facts, finite basic facts, and totally finite basic facts (with respect to C); and the respective sets of basic facts[14] are basic-$Fact(C')$, basic-$Fact_o(C')$, and basic-$Fact_{oo}(C')$. ∎

The following definitions deal with finiteness properties and N-density.

2.2 Definition.

P combinatorial $:\Longleftrightarrow (\forall a, b \in X)(a<b \Rightarrow (\exists x_1, \ldots, x_n \in X)(a = x_1 \blacktriangleleft x_2 \blacktriangleleft \ldots \blacktriangleleft x_n = b)$.

P length-finite $:\Longleftrightarrow (\forall x, y \in X)(\forall L)($Line $L \Rightarrow |\{z|z \in L \wedge x \leq z \leq y\}| < \infty$.

P interval-finite $:\Longleftrightarrow (\forall x, y \in X)|\{z|z \in X \wedge x \leq z \leq y\}| < \infty$.

P degree-finite $:\Longleftrightarrow (\forall x \in X)(|{}^{\cdot}x| < \infty \wedge |x^{\cdot}| < \infty)$.

P N-dense $:\Longleftrightarrow (\forall a, b, c, d \in X)(a<b \wedge c<b \wedge c<d \wedge a \sim c \wedge a \sim d \wedge b \sim d$

$$\Rightarrow (\exists e \in X)(c<e \wedge e<b \wedge a \sim e \wedge e \sim d)). \quad ∎$$

Next we define occurrence posets direcly without deriving them from occurrence nets; it can easily be seen that either ways lead to the same notion.

2.3 Definition. $O:=(S,T;\leq)$ is called an occurrence poset iff (3)-(6) hold:

(3) $(S \cup T, \leq)$ is a combinatorial poset;

(4) $S \cap T = \emptyset$,

(5) $(\blacktriangleleft) \subseteq (S \times T) \cup (T \times S)$,   [Two neighbouring elements cannot belong both to S or to T.]

(6) $(\forall s \in S)(|{}^{\cdot}s| \leq 1 \wedge |s^{\cdot}| \leq 1)$.

An S-cut H in the occurrence poset $O:=(S,T;\leq)$ is a cut H in $(S \cup T, \leq)$ with $H \subseteq S$; the set of all S-cuts of an occurrence poset is denoted by C, and C-fin is the set of all finite S-cuts. ∎

Connections between the different finiteness conditions (and occurrence posets) are stated in the next theorems.

2.4 Proposition. If P is interval-finite, then P is length-finite; if P is length-finite, then P is combinatorial.

Proof. Clear (using [8]:3.2).

---

[14] Cf. footnote 4.

2.5 Theorem. If $O:=(S,T;\leq)$ is an occurrence poset, then O and $(S,\leq)$ are combinatorial and N-dense; moreover, if O is length-finite or interval-finite, then $(S,\leq)$ has the respecitve property.

Proof. The N-density can be proved using [8]:4.6, the rest is clear. ∎

2.6 Theorem. Let $O:=(S,T;\leq)$ be an occurrence poset with at least one S-cut. Then H is an S-cut in O if and only if H is a cut in $(S,\leq)$.

Proof. Let H be a cut in $(S,\leq)$. If H is no S-cut, then there is $t \in T$ such that $H \cup \{t\}$ is an antichain. There is a S-cut Q in O; and hence there is $q \in Q$ with $q < t$ or $t < q$. Therefore $s \in {}^\cdot t \cup t^\cdot$ exists. We get $s \notin H$; and the special structure of an occurrence poset implies that $H \cup \{s\}$ is an antichain. Hence H is no cut in $(S,\leq)$, contradiction. This proves "if"; the converse direction is trivial. ∎

Due to the preceding theorems, in the following we can restrict ourselves to the investigation of arbitrary combinatorial N-dense posets.

2.7 Theorem. If P is combinatorial and if all cuts are finite, then P is interval-finite.

Proof. See [8]:3.7 and [8]:3.2(9). ∎

2.8 Lemma. Let P be N-dense, $S \in C$, and
$$u \in \downarrow S \wedge u = x_1 \lessdot x_2 \lessdot \ldots \lessdot x_n = v \wedge v \in \uparrow S.$$
Then there is m such that $1 \leq m \leq n$ and $x_m \in S$.

Proof. We assume that no such m exists. Then there are m, a, and d such that $a, d \in S$, $a < x_{m+1}$, $x_m \lessdot x_{m+1}$ and $x_m < d$. Then $a \sim x_m$, $a \sim d$ and $x_{m+1} \sim d$. Hence by N-density we get the existence of e with $x_m < e < x_{m+1}$ contradicting $x_m \lessdot x_{m+1}$. ∎

2.9 Lemma. Let P be length-finite and N-dense, and let $p, q \in X$. Then $|\uparrow p \cap \downarrow q \cap UC\text{-fin}| < \infty$, i.e. between any two points there are only finitely many points which are contained in finite cuts.

Proof. We set $Y := UC\text{-fin}$. Then
(7) $(\forall x, y \in X)(|\uparrow x \cap \downarrow y \cap Y| = \infty \Rightarrow$
$(\exists z \in X)(x < z < y \wedge (|\uparrow x \cap \downarrow z \cap Y| = \infty \vee |\uparrow z \cap \downarrow y \cap Y| = \infty)))$.
To prove (7) let $x, y \in Y$, $Q := \uparrow x \cap \downarrow y \cap Y$, and $|Q| = \infty$; then there are $w \in Y$ and

$H \in C$-fin such that $x < w < y$ and $w \in H$; using 2.8 we see that for every $q \in Q$ there is $h \in H \cap Q$ with $q \leq h \vee h \leq q$; and hence because of $|H| < \infty$ there is $z \in H$ with the desired property. ■(7)

The proof of 2.9 is done by an indirect argument: We assume $|\uparrow p \cap \downarrow q \cap Y| = \infty$; and by repeately applying (7) we get an infinite chain between p and q contradicting the length finiteness. ■

2.10 Lemma. Let P be length-finite and N-dense, $Y \subseteq UC$-fin and $x \in X$. Then $|\dot{x} \cap Y| < \infty$ and $|x\dot{} \cap Y| < \infty$.

Proof. We use an indirect argument and assume $|x\dot{} \cap Y| = \infty$. There are h and H with $H \in C$-fin and $h \in x\dot{} \cap H$. For every $z \in x\dot{}$ we have $z \in \downarrow H$ (otherwise: $z \in \uparrow H$, $x \in \downarrow H$, $x \triangleleft z$, $z \in H$ due to 2.8). Hence by 2.9 we get $|x\dot{} \cap Y| \leq |\uparrow x \cap \downarrow H \cap Y| < \infty$, contradiction. ■

2.11 Theorem. Let P be length-finite and degree-finite, and let C-fin$\neq \emptyset$. Then P is interval-finite; and we have $X = UC$-fin,

(8) $\qquad (\forall A)(A$ antichain $\wedge |A| < \infty \Rightarrow (\exists A' \in C$-fin$)A \subseteq A')$

and, moreover, $Fact_{oo}(C) = Fact_{oo}(C$-fin$)$.

Proof. The interval-finiteness and (8) follow from [8]:6.4; $X = UC$-fin is an immediate consequence of (8). Obviously, $Fact_{oo}(C) \subseteq Fact_{oo}(C$-fin$)$. Let

(9) $\qquad (I,J) \in Fact_{oo}(C$-fin$) \backslash Fact_{oo}(C)$.

Then I, J are finite, and there is $B \in C$ with $I \subseteq B$ and $J \cap B = \emptyset$. Hence $J' \subseteq B$ exists such that $J \subseteq \downarrow J' \cup \uparrow J'$ and $|J'| < \infty$. Because of (8) there is $B' \in C$-fin with $I \cup J' \subseteq B' \cap B$; and (9) implies the existence of $j \in J \cap B' \subseteq \downarrow J' \cup \uparrow J'$, hence $J' \subseteq B'$, $j \in J' \subseteq B$ contradicting $J \cap B = \emptyset$. ■

2.12 Theorem. Let P be length-finite and N-dense, and let C-fin$\neq \emptyset$. Then

(10) $\qquad (\forall A \subseteq UC$-fin$)(A$ antichain $\wedge |A| < \infty \Rightarrow (\exists A' \in C$-fin$)A \subseteq A')$,

i.e. every finite antichain that can be "covered" by finite cuts is contained in a finite cut.

Proof. We consider the poset $P' := (UC$-fin$, \leq)$; then by 2.10 we see that P' is degree-finite. Hence 2.11 can be applied, and (8) yields (10). ■

2.13 Corollary. Let $O := (S,T;\leq)$ be a length-finite occurrence poset with C-fin$\neq \emptyset$. Then

$\qquad (\forall A \subseteq UC$-fin$)(A$ antichain $\wedge |A| < \infty \Rightarrow (\exists A' \in C$-fin$)A \subseteq A')$;

i.e. every finite antichain that can be "covered" by finite S-cuts is contained in a finite S-cut.

Proof. Direct consequence of 2.5, 2.6, and 2.12 (applied to $(S, \leq)$). ∎

Proof of 1.3. Indirect: Let $x, y \in S \cup T$ and $I := |\uparrow x \cap \downarrow y| = \infty$. Because of 2.5 and 2.7 there must be an infinite antichain A in I. Then $A \cap S$ is finite; hence $J := A \cap T$ must be infinite, and therefore $J \subseteq I \cap S$ is an infinite antichain, contradiction. ∎

Proof of 1.4. If O has the property (a), then 2.10 with $X := S \cup T$ and $Y := S$ yields the degree-finiteness of O; hence (b) is true. If O has the property (b), then we get (a) and $Fact_{oo}(C) = Fact_{oo}(C\text{-fin})$ by application of 2.11 to the poset $(S, \leq)$ using 2.6; moreover, applying 2.11 to $(S \cup T, \leq)$ proves that O is interval-finite. ∎

For our proofs in the next section we need that using a certain order relation the (finite) cuts form a lattice.

2.14 Definition. For $A, B \in C$ let
$$A \sqsubseteq B :\iff \downarrow A \subseteq \downarrow B.$$
∎

It is immediately clear that $(C, \sqsubseteq)$ and $(C\text{-fin}, \sqsubseteq)$ are posets and that[15]
$$A \sqsubseteq B :\iff \uparrow B \subseteq \uparrow A.$$
But in the next section we need more: We have to use that $(C\text{-fin}, \sqsubseteq)$ is a lattice[16].

2.15 Definition. Let $(Q, \leq)$ be an arbitrary poset and D a subset of Q. Then D is called below [above] finitely bounded iff there is a *finite* subset F of Q such that
$$(\forall d \in D)(\exists f \in F) f \leq d \quad \text{and} \quad (\forall d \in D)(\exists f \in F) d \leq f, \text{ respectively;}$$
and D is called finitely bounded iff it is below finitely bounded *and* above finitely bounded. $(Q, \leq)$ is called a *boundedly* complete lattice iff for every *non-empty* and finitely bounded subset D of Q the greatest lower bound $\sqcap D$ and the least upper bound $\sqcup D$ exist. Obviously, a *finite* subset D of Q always is finitely bounded (choose $F := D$). For (finite) $D := \{x, y, \ldots\}$ we will also write $x \sqcap y \sqcap \ldots$ instead of $\sqcap D$ etc. ∎

[15] Cf. [4]:Proposition 1.2.
[16] Lattice properties of $(C, \sqsubseteq)$ extensively were investigated in [7], and for the special case of occurrence nets, in [4].

2.16 Theorem. Let P be combinatorial and N-dense. Then $(C, \sqsubseteq)$ is a boundedly complete and distributive lattice. And for every non-empty and below [resp. above] finitely bounded subset D of C we have

(11) $\qquad \sqcap D = \mathrm{Min}(UD)$ and $\sqcup D = \mathrm{Max}(UD)$, respectively;

(12) $\sqcap D = \mathrm{Max}(\cap\{\downarrow H \mid H \in D\})$ and $\sqcup D = \mathrm{Max}(\cup\{\downarrow H \mid H \in D\})$, respectively.

(13) $\sqcap D = \mathrm{Min}(\cup\{\uparrow H \mid H \in D\})$ and $\sqcup D = \mathrm{Min}(\cap\{\uparrow H \mid H \in D\})$, respectively.

Proof. Let D be a non-empty and below finitely bounded subset of C; then there is a finite subset F of C such that $(\forall d \in D)(\exists f \in F) f \sqsubseteq d$. We set $M := \mathrm{Min}(UD)$. Obviously, M is a antichain. Let $p \in X$. If $p \notin \uparrow UD$ then there are $q \in UD$ and $y_1, \ldots, y_n$ such that $p = y_1 \blacktriangleleft \ldots \blacktriangleleft y_n = q$; hence there is m such that $y_m \in UD$ and $y_r \notin UD$ for $r < m$; and using 2.8 it is not difficult to show $y_m \in M$; therefore $p \in \downarrow M$. If $p \in \uparrow UD$ then there are $q \in \mathrm{Min}(UF)$ and $y_1, \ldots, y_n$ such that $q = y_1 \blacktriangleleft \ldots \blacktriangleleft y_n = p$; hence there is m such that $y_m \in \uparrow UD$ and $y_r \notin \uparrow UD$ for $r < m$; and again using 2.8 we get $y_m \in M$ (which is also true if $m = 1$); therefore $p \in \uparrow M$. We get $M \in C$ and $UD \subseteq \uparrow M$, an immediate consequence of which is $\sqcap D = \mathrm{Min}(UD)$. The proof of $\sqcup D = \mathrm{Max}(UD)$ is dual. (12) and (13) are immediate consequences of (11). Finally, the mapping $H \longmapsto \downarrow H$ for $H \in C$ establishes an isomorphism between $(C, \sqsubseteq)$ and $(\{\downarrow H \mid H \in C\}, \subseteq)$, and this yields because of (12) and (11) the the distributivity of $(C, \sqsubseteq)$. ∎

2.17 Theorem. Let P be length-finite and N-dense. Then $(C\text{-fin}, \sqsubseteq)$ is a boundedly complete and distributive lattice. And for every non-empty and below [resp. above] finitely bounded[17] subset D of C-fin the greatest lower [resp. least upper] bound exist in C-fin, and we have
$$\sqcap D = \mathrm{Min}(UD) \quad \text{and} \quad \sqcup D = \mathrm{Max}(UD) , \quad \text{respectively.}$$

Proof. Let D be a non-empty and below finitely bounded subset of C-fin and $M := \mathrm{Min}(UD)$. Hence there is a *finite* subset F of C-fin such that $(\forall U \in D)(\exists G \in F) G \sqsubseteq U$. P is combinatorial; hence we can apply 2.16. Therefore $M \in C$, and M is the greatest lower bound of D in C; and we have $H := \mathrm{Min}(UF) \in C$, $|H| < \infty$, and $(\forall U \in D) H \sqsubseteq U$. There is $V \in D$, hence $H \sqsubseteq M \sqsubseteq V$. Because H and V are fintite, 2.9 yields the finiteness of M, and, therefore, $M \in C$-fin. This proves $\sqcap D = \mathrm{Min}(UD)$, $\sqcup D = \mathrm{Max}(UD)$ is dual; the distributivity follows from 2.16. ∎

2.18 Corollary. If 2.16 or 2.17 can be applied, then for $A, B, Z \in C$ we have $A \cap B \subseteq (Z \cap A) \cup B$ and $A \cap B \subseteq (Z \cup A) \cap B$.

---

[17] Of course, the "bounding" set F from definition 2.15 is a subset of C-fin in this case.

Proof. By application of 2.16 and 2.17, respectively. (Use: If $U, V \in C$ and $x \in U \cap V$ then $x \in U \sqcap V$ and $x \in U \sqcup V$.) ∎

3 Basic facts

This section contains the main results about basic facts. The next theorem deals with a trivial case that we will exclude later on.

3.1 Proposition. For arbitrary posets P we have
$$\text{Fact}(\emptyset) = \text{basic-Fact}(\emptyset) = \{(\emptyset, \emptyset)\},$$
and the same applies for $\text{Fact}_o(\emptyset)$ and $\text{Fact}_{oo}(\emptyset)$.

Proof. Clear. ∎

Next we prove that the finite basic facts in C-fin are really "basic".

3.2 Theorem. Let P be an arbitrary poset. Then
$$(\forall I, J \subseteq X)((I, J) \in \text{Fact}_o(C\text{-fin}) \iff$$
$$(\exists (I', J') \subseteq_2 (I, J))(I', J') \in \text{basic-Fact}_o(C\text{-fin})).$$

Proof. "$\Longleftarrow$" is clear. Let $(I, J) \in \text{Fact}_o(C\text{-fin})$. We consider the set K of facts $(I', J') \subseteq_2 (I, J)$. Let (A_n, B_n) be a non-empty descending chain (whith respect to $\subseteq_2$) of these facts, and we set $(A, B) := (\cap\{A_n\}, \cap\{B_n\})$. If $(A, B) \notin \text{Fact}_o(C\text{-fin})$, then there is $U \in C\text{-fin}$ such that $A \subseteq U$ and $B \cap U = \emptyset$. Since A_n and $B_n \cap U$ are finite for all n and because of $B \cap U = \cap\{B_n \cap U\}$ there is m such that $A = A_m$ and $B \cap U = B_m \cap U$; hence $A_m \subseteq U$, $B \cap U = B_m \cap U \neq \emptyset$, $(A, B) \in \text{Fact}_o(C\text{-fin})$, contradiction. Hence by Zorn's Lemma there is a minimal fact (I', J') in K; and obviously, (I', J') has the desired properties: $(I', J') \subseteq_2 (I, J)$ and $(I', J') \in \text{basic-Fact}_o(C\text{-fin})$. ∎

Proof of 1.1. By application of 3.2, 3.1, and 2.6. ∎

We now will consider basic facts of the form $(I, \emptyset)$, and after that we are going to study the structure of facts (I, J) with $J \neq \emptyset$.

3.3 Theorem. Let P be a length-finite and N-dense poset and C-fin$\neq \emptyset$, and let $I \subseteq X$. Then (cf. 1.2, class f – Fig. 6)
$$(I, \emptyset) \in \text{basic-Fact}_o(C\text{-fin}) \iff (\exists a, b \in \cup C\text{-fin})(a < b \wedge I = \{a, b\}).$$

Proof. "⟸" is clear. We now assume

(14) $(I,\emptyset)\in$ basic-Fact$_0$ $(C$-fin).

Hence $I\subseteq UC$-fin and $|I|<\infty$. If $(\forall x,y\in I)\neg x<y$, then due to 2.12 there is
$H\in C$-fin with $I\subseteq H$; and by (14) we get $H\cap\emptyset\neq\emptyset$, contradiction. Therefore
there are $a,b\in I$ with $a<b$; and (14) implies $I=\{a,b\}$.   ∎

The concepts of F-chain and F-structure defined in the following
will be used to describe the structure of finite basic facts.

3.4 Definition. Let $x,y\in X$ and $Y,Z\subseteq X$. Then:
$x\triangleleft y :\Longleftrightarrow (\exists u,v\in X)(x\triangleleft u \wedge v<u \wedge v\triangleleft y)$.
$(Y,\leq_v)$ is an F-chain iff $(Y,\leq_v)$ is a total length-finite order[18] and
$$(\forall a,b\in Y)(a\triangleleft_v b \Rightarrow a\triangleleft b),$$
where $\triangleleft_v$ is the neighbour relation[19] of $\leq_v$.
$(Z,Y;\leq_v)$ is an F-structure iff the following holds: Z is an antichain
with $|Z|\leq 2$ and $(Y,\leq_v)$ is an F-chain such that for all finite cuts H
with $Z\subseteq H$ there are $a,b\in Y$ with $a\leq_v b$, $a\in\downarrow H$, and $b\in\uparrow H$.
On the set of F-structures we define a partial order (analogous to the
partial order on the set of facts) by
$$(Z',Y';\leq_w) \subseteq_{\exists} (Z,Y;\leq_v) \Longleftrightarrow Z'\subseteq Z \wedge Y'\subseteq Y;$$
and the minimal elements in this partial order are called irreducible
F-structures.   ∎

The figures $7(a)$–$8(b)$ illustrate $x\triangleleft y$: Fig. $7(a)$ and Fig. $7(b)$ re-

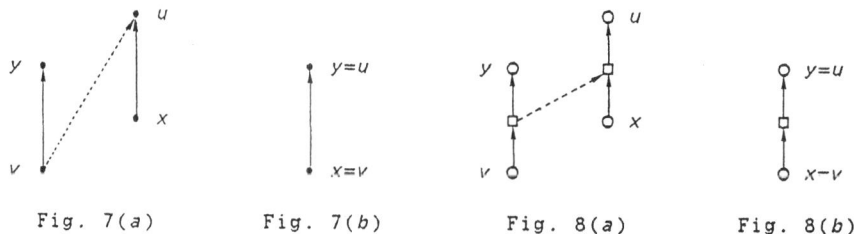

Fig. 7(a)   Fig. 7(b)   Fig. 8(a)   Fig. 8(b)

late to arbitrary posets $(X,\leq)$, whereas in Fig. $8(a)$ and Fig. $8(b)$ examples of the $\triangleleft$-relation in the subposet $(S,\leq)$ of an occurrence poset
$(S,T;\leq)$ are depicted. (Continuous lines [———], broken lines [------],
and dotted lines [·········] denote the neighbourhood relation, the
$\leq$-relation and the $<$-relation, respectively.)

---

[18] I.e. Y is a chain with respect to the order relation $\leq_v$ and the
    number of elements between any two elements is finite.
[19] I.e. $p\triangleleft_v q$ iff $p<_v q \wedge (\forall r)(p\leq_v r\leq_v q \Rightarrow p=r \vee r=q)$.

We now are prepared to state our main result:

3.5   Theorem.   Let P be an interval–finite and N–dense poset which has at least one finite cut. Let $I, J \subseteq UC\text{-fin}$, $I \cap J = \emptyset$, and $J \neq \emptyset$. Then

(15) $(I, J) \in \text{basic-Fact}_0 (C\text{-fin})$

iff there is an order relation $\leq_v$ such that $(I, J; \leq_v)$ is an irreducible F–structure.

Moreover, if $(I, J; \leq_v)$ is an irreducible F–structure, then there are I_1, I_2, L_1, L_2 such that

(16) $L_1 \subseteq (\text{Max}(J; \leq_v))^{\cdot} \ \wedge \ |L_1| \leq 1 \ \wedge \ L_2 \subseteq^{\cdot} (\text{Min}(J; \leq_v)) \ \wedge \ |L_2| \leq 1,$

(17) $|I_1| \leq 1 \ \wedge \ |I_2| \leq 1 \ \wedge \ I_1 \cup I_2 = I,$

(18) $(I_1 \neq \emptyset \iff L_1 \neq \emptyset) \ \wedge \ (I_2 \neq \emptyset \iff L_2 \neq \emptyset),$

(19) $(\forall i \in I_1)(\forall l \in L_1) i < l \ \wedge \ (\forall i \in I_2)(\forall l \in L_2) l < i,$

(20) $|I| = 2 \implies I_1 \cap \uparrow L_2 = \emptyset \ \wedge \ I_2 \cap \downarrow L_1 = \emptyset.$

Proof. The proof is carried out in several steps.

I)   If $(I, J; \leq_v)$ is an F–structure, then $(I, J) \in \text{Fact}_0 (C\text{-fin})$.

   Proof. Indirect: There is $H \in C\text{-fin}$ with $I \subseteq H$ and $H \cap J = \emptyset$. There are $a, b \in J$ with $a \leq_v b$, $a \in \downarrow H$, and $b \in \uparrow H$. Hence $a' \in J \cap \downarrow H$ and $b' \in J \cap \uparrow H$ exist with $a' \blacktriangleleft b'$. Hence $a' \vartriangleleft b'$; and therefore u and v exist such that $a' \blacktriangleleft u$, $v < u$, and $v \blacktriangleleft b'$. Because of $H \cap J = \emptyset$, applying 2.8 we get $u \in \downarrow H$ and $v \in \uparrow H$ contradicting $v < u$.   ∎ I

II)   Let (15). Then we set $C' := \{H | H \in C\text{-fin} \ \wedge \ I \subseteq H\}$; and for every $u \in J$ there is $H_u \in C'$ such that $H_u \cap J = \{u\}$ (otherwise: $(I, J \setminus \{u\})$ would be a fact).   ∎ II

III)   If (15), we define for $u, v \in J$

$$u \leq_r v \ :\iff \ u \in \downarrow H_v.$$

   Then

$$u \leq_r v \ \iff \ v \in \uparrow H_u.$$

   Proof. "⇒", indirect: There are $u, v \in J$ with $u \in \downarrow H_v$ and $v \in (\downarrow H_u) \setminus H_u$. Then $v \neq u$, $u \notin H_v$,

$$K := H_u \sqcup H_v = \text{Max}(H_u \cup H_v) \in C\text{-fin}$$

   by 2.17; hence (by 2.18) $I \subseteq K$ and $K \cap J = \emptyset$, contradiction.
   "⇐" is the dual proposition.   ∎ III

IV)   If (15) then $(J, \leq_r)$ is a total length–finite order.

   Proof. Let $u, v, w \in J$.

   Obviously, $u \in \downarrow H_v$ or $u \in \uparrow H_v$; hence by III we get $u \leq_r v$ or $v \leq_r u$.
   $u \leq_r u$ is clear. If $u \leq_r v$ and $v \leq_r u$, then we get (using III) $u \in \downarrow H_v$, $v \in \downarrow H_u$, $u \in \uparrow H_v$, $u \in H_v$, $u = v$.

Let $u \leq_r v$ and $v \leq_r w$, and we assume $\neg u \leq_r w$. Then we get $u \neq v$, $v \neq w$, $u \neq w$, $u \in \downarrow H_v$, $v \in \downarrow H_w$, $w \in \downarrow H_u$; and using 2.17 it follows $(H_u \sqcup H_v) \cap J = \{v\}$, $(H_v \sqcup H_w) \cap J = \{w\}$, $(H_w \sqcup H_u) \cap J = \{u\}$, $K := H_u \sqcup H_v \sqcup H_w$,

$$K \cap J \subseteq ((H_u \sqcup H_v) \sqcup H_w) \cap J \subseteq ((H_u \sqcup H_v) \cap J) \cup (H_w \cap J) \subseteq \{v,w\},$$

analogously $K \cap J \subseteq \{w,u\}$ and $K \cap J \subseteq \{u,v\}$; hence $K \cap J = \emptyset$. But we have $I \subseteq H_u$, $I \subseteq H_v$, $I \subseteq H_w$, hence (by 2.18) $I \overset{\cap}{\subseteq} K$ and $K \cap J \neq \emptyset$ because of (15); contradiction.

Therefore $(J, \leq_r)$ is a total order. If $u \leq_r v \leq_r w$, then using III we get $v \in (\uparrow H_u \cap \downarrow H_w) \cap J =: Q$; Q is finite due to 2.9, and this implies the length-finiteness of $(J, \leq_r)$.   ∎IV

V)   Let (15), $K \in C$-fin, $u,v \in J$, $u \leq_r v$, $u \in \downarrow K$, and $v \in \uparrow K$. Then $K \cap J \neq \emptyset$.
Proof. Indirect: Let $K \cap J = \emptyset$. We get (using 2.17 and III) $u \notin K$, $v \notin K$, $v \neq u$, $K' := K \sqcup H_u$, $K' \cap J = \emptyset$, $v \in \uparrow H_u$, $v \in (\uparrow K') \setminus K'$, $K'' := K' \sqcap H_v$, $K'' \cap J = \emptyset$. But 2.18 implies $I \subseteq K''$; therefore $K'' \cap J \neq \emptyset$, contradiction.   ∎V

VI)   Let (15). Then $(J, \leq_r)$ is an F-chain.
Proof. Indirect: Then due to IV there are $m,n \in J$ with $m \triangleleft_r n$ and $\neg m \triangleleft n$. We set $K_1 := H_m \sqcap H_n$, $K_2 := H_m \sqcup H_n$, and $R := (\uparrow K_1) \cap (\downarrow K_2)$. Then $K_1 \subseteq K_2$, $K_1 \cap J = \{m\}$, $K_2 \cap J = \{n\}$, and (by 2.18) $I \subseteq K_1 \cap K_2$; and $u,v \in R$ exist with $m \triangleleft u$ and $v \triangleleft n$. We now are going to prove $v < u$ which implies $m \triangleleft n$ contradicting our supposition.
If $\{u,v\}$ is an antichain, then there is a maximal antichain Q contained in R such that $u,v \in Q$; if $u < v$, then there is a maximal antichain Q contained in R such that $v \in Q$. It is easy to see that in any case Q is a cut in $(X, \leq)$; R is finite due to the interval-finiteness of P and the finiteness of H_m and H_n; hence $Q \in C$-fin. It is $m \in \downarrow Q$ and $n \in \uparrow Q$; hence using V we get $Q \cap J \neq \emptyset$. But on the other hand: $m \notin Q$ and $n \notin Q$; and for $j \in J$ with $j \neq m$ and $j \neq n$ we get $j <_r m \vee n <_r j$,

$$(j \in \downarrow H_m \wedge j \in \downarrow H_n) \vee (j \in \uparrow H_m \wedge j \in \uparrow H_n),$$

hence $j \notin R$; and therefore $Q \cap J \subseteq (R \setminus \{m,n\}) \cap J = \emptyset$, contradiction. ∎VI

VII) Let (15). If
(21) $(\exists i \in I)(\exists m_1, m_2, l_1, l_2)(\text{Max}(J; \leq_r) = \{m_1\} \wedge \text{Min}(J; \leq_r) = \{m_2\} \wedge$
$$m_1 \triangleleft l_1 \wedge i < l_1 \wedge l_2 \triangleleft m_2 \wedge l_2 < i),$$
then we set $L_1 := \{l_1\}$, $L_2 := \{l_2\}$, and $I_1 := I_2 := \{i\}$.

VIII) Let (15) and not (21), $A \in C$-fin, and $u \in J$ such that (cf. III)

397

(22) $(\forall v \in J)(u \leq_r v \Rightarrow v \in (\downarrow A)\backslash A)$.

Then there are $i_1 \in I$, m and l_1 such that $Max(J;\leq_r)=\{m\}$, $m \blacktriangleleft l_1$, and $i_1 < l_1$; and we set $I_1 := \{i_1\}$ and $L_1 := \{l_1\}$.

Proof. Due to the interval finiteness, $(\uparrow H_u)\cap \downarrow A$ is finite. Hence there are m and l_1 with $Max(J;\leq_r)=\{m\}$ and $m \blacktriangleleft l_1$.

If $I' := I \cup \{l_1\}$ is an antichain, then: I' is finite, due to 2.12 there is $K \in C$-fin with $I' \subseteq K$; we get $K' := K \cup H_m$, $l_1 \in K$, $m \notin K'$, and $I \subseteq K'$; if $u \in J$ and $u <_r m$ then $u \in (\downarrow H_m)\backslash H_m$ and (hence) $u \notin K'$; $K' \cap J = \emptyset$ contradicting $I \subseteq K'$ and (15).

Hence there is $i_1 \in I$ such that $i_1 < l_1$ or $l_1 \leq i_1$; and the latter can be excluded because of $i_1 \in H_m$ and $l_1 \in (\uparrow H_m)\backslash H_m$.   ∎VIII

IX) Dual to *VIII* is following proposition: Let (15) and not (21), $A \in C$-fin, and $u \in J$ such that

(23) $(\forall v \in J)(v \leq_r u \Rightarrow v \in (\uparrow A)\backslash A)$.

Then there are $i_2 \in I$, m and l_2 such that $Min(J;\leq_r)=\{m\}$, $l_2 \blacktriangleleft m$, and $l_2 < i_2$; and we set $I_2 := \{i_2\}$ and $L_2 := \{l_2\}$.   ∎IX

X) Let (15) and not (21). If no $A \in C$-fin and no $u \in J$ exist such that (22) holds, then

(24) $(\forall A \in C\text{-fin})(\forall u \in J)(\exists v \in J)(u \leq_r v \wedge v \in \uparrow A)$;

and we set $I_1 := \emptyset$ and $L_1 := \emptyset$.   ∎X

XI) Let (15) and not (21). If no $A \in C$-fin and no $u \in J$ exist such that (23) holds, then

(25) $(\forall A \in C\text{-fin})(\forall u \in J)(\exists v \in J)(v \leq_r u \wedge v \in \downarrow A)$;

and we set $I_2 := \emptyset$ and $L_2 := \emptyset$.   ∎XI

XII) Let (15). We define I_1, I_2, L_1, and L_2 according to *VII-XI*. If $K \in C$-fin and $I_1 \cup I_2 \subseteq K$, then in any case (24) and (25) hold for $A=K$ (in the cases *VII*, *VIII*, and *IX* because of 2.8 the maximal and the minimal element of J with respect to $\leq_r$ are v's with the desired property, respectively); hence due to V we get $K \cap J \neq \emptyset$. This shows that $(I_1 \cup I_2, J) \in Fact_0(C\text{-fin})$; and therefore and since (I,J) is a *basic* fact we have $I=I_1 \cup I_2$.

To prove (20) we assume $|I|=2$. This can only happen if *VIII* and *IX* were used. Then $I=\{i_1, i_2\}$, and assuming

$$I_1 \cap \uparrow L_2 \neq \emptyset \vee I_2 \cap \downarrow L_1 \neq \emptyset$$

we get that (21) holds with $i := i_1$ or $i := i_2$, respectively, contradicting the construction in *VIII* and *IX*.

This proves together with *VI* that $(I, J; \leq_r)$ is an F-structure and that (16)-(20) hold.   ∎XII

XIII) If $(I,J)\in\text{basic-Fact}_0(C\text{-fin})$, then $(I,J;\leq_r)$ is an irreducible F-structure.

Proof. Let

(26) $(I,J)\in\text{basic-Fact}_0(C\text{-fin})$.

Due to *XII*, $(I,J;\leq_r)$ is an F-structure. If $(I,J;\leq_r)$ is not irreducible, then there is $I'\subseteq I$ and $J'\subseteq J$ such that $(I'\ne I \lor J'\ne J)$ and $(I',J';\leq_r)$ is an F-structure; and due to *I* we get $(I',J')\in\text{Fact}_0(C\text{-fin})$ contradicting (26).   ∎*XIII*

XIV) If $(I,J;\leq_v)$ is an irreducible F-structure, then

$(I,J)\in\text{basic-Fact}_0(C\text{-fin})$.

Proof. Assume that

(27) $(I,J;\leq_v)$ is an irreducible F-structure.

Due to *I*, $(I,J)\in\text{Fact}_0(C\text{-fin})$. If (I,J) is no basic fact, then there is a fact (I',J') such that $I'\subseteq I$ and $J'\subseteq J$ and $(I'\ne I \lor J'\ne J)$; and due to *XII* we get that $(I',J';\leq_r)$ is a F-structure contradicting (27).   ∎*XIV*

The theorem follows immediately from *XII*, *XIII*, and *XIV*.   ∎

Using the preceding theorem we get the different structure of basic facts (I,J) (with $J\ne\varnothing$) corresponding to the classes *a-e* from theorem 1.2:

3.6 Theorem. Let P be an interval-finite and N-dense poset which has at least one finite cut, and let $(I,J)\in\text{basic-Fact}(C\text{-fin})$ with $J\ne\varnothing$. Then using the notions from 3.5 we get the existence of I_1 and I_2 such that (16)-(20) hold and that exactly one of the following propositions *A-E* is true corresponding to *a-e* from 1.2:

A) $I_1 = I_2 = \varnothing$.

B) $I_1 \ne \varnothing \land I_2 = \varnothing$.

C) $I_1 = \varnothing \land I_2 \ne \varnothing$.

D) $I_1 = I_2 \ne \varnothing$.

E) $I_1 \ne \varnothing \land I_2 \ne \varnothing \land I_1 \ne I_2$.   ∎

Proof of 1.2. Because of 2.5 we can apply 3.3 and 3.6 to the poset $(S,\leq)$, and this yields the theorem due to 2.6.   ∎

4 Concluding remarks

Let us consider the following problem: Given six conditions a, b, c, x, y, and z corresponding to "local" states in some real system, we want to construct a device which ensures that

(28) $a \wedge b \wedge c \Rightarrow x \vee y \vee z$

always holds. We assume that our construction can be described by a C/E-system and that a, b, and c can hold concurrently. Then (28) de-fines the occurring[20] fact

(29) $(\{a,b,c\},\{x,y,z\})$;

and the above results tell us something about the "logical" structure we have to realize for any special concurrent occurrence of a, b, and c (i.e. any special holding of $a \wedge b \wedge c$):

We consider a special occurrence of a, b, and c; and in order to shorten our discussion we assume that before and after this occu-rence there are "not-holdings" of $x \vee y \vee z$ and that during this special holding of $a \wedge b \wedge c$ the conditions x, y, and z are "used" at most once and are really "needed" (i.e. during this special holding of $a \wedge b \wedge c$: For each of the conditions x, y, and z there is at most one begin of holding and at most one end of holding; and each of $x \vee y$, $x \vee z$, and $y \vee z$ is *not* true at some point of time.) Then we know by 3.5 (for this special occurrence of $a \wedge b \wedge c$):

There is a total oder among x, y, and z, say $x < y < z$, and there are $i_1, i_2 \in \{a,b,c\}$ such that the following holds (cf. Fig. 8(a) and Fig. 8(b)):

   The begin of the holding of x
        precedes or equals the begin of the holding of i_2.
   The begin of the holding of y
        precedes or equals the end of the holding of x.
   The begin of the holding of z
        precedes or equals the end of the holding of y.
   The end of the holding of i_1
        precedes or equals the end of the holding of z.

It is immediately clear that a device which works in this way solves our problem; the interesting result is that there are essentially no other solutions (cf. 3.6 and 1.2).

Applying the preceding example to the case that x, y, and z denote the burning of different candles X, Y, and Z we can exclude the possi-bility that the begin of the burning of Y *equals* the end of the burn-

---

[20] Cf. the introduction.

ing of X etc; for it seems to be impossible that lighting one candle and extinguishing another one is *one* "atomic" indistinguishable action. Hence it cannot be avoided that two candles burn concurrently. A similar problem arises if we have a series of lamps on a scale where a burning lamp shall indicate that a quantity to be measured has a special value. Then it should be guaranteed that always *exactly one* lamp is burning; but from the above considerations one may conclude that this is impossible: Either we must allow that two lamps are burning at the same time, or, that sometimes no lamp is burning. In other words: If we insist that always at least one lamp is burning, than reading the scale defined by these lamps can lead to an ambiguous value.

Summarizing our reasoning we can say: Uncertainty seems to be an inherent property of measurement which not always can be avoided. One may object to the assumptions from which this result was derived; especially an observation is a process which is *different* from (but connected with) the process to be observed and should be considered explicitly. The point I wanted to make was that there is a connection between facts and uncertainty; it will be subject of further research to get a better understanding of these problems.

Acknowledgement

I am very much obliged to the referees of this paper who have studied it in detail. Their remarks helped me to eliminate minor errors and to improve the representation of some concepts.

4 References

[1] Best, E.; Devillers, R.: Concurrent Behaviour: Sequences, Processes and Programming Languages. – St. Augustin : Gesellschaft für Mathematik und Datenverarbeitung, GMD-Studien Nr. 99 (1985)

[2] Best, E.; Fernandez, C.; Plünnecke, H.: Concurrent Systems and Processes: Final Report on the Foundational Part of the Project BEGRUND. – St. Augustin : Gesellschaft für Mathematik und Datenverarbeitung, GMD-Studien Nr. 104 (1985)

[3] Fernandez, C.; Nielsen, M.; Thiagarajan, P. S.: A Note on Observable Occurrence Nets. In: Lecture Notes in Computer Science Vol. 188 : Advances in Petri Nets 1984, pp. 122–138. – Berlin : Springer-Verlag (1985)

[4] Fernandez, C.; Thiagarajan, P. S.: A Lattice Theoretic View of K-density. – In: Lecture Notes in Computer Science Vol. 188 : Advances in Petri Nets 1984, pp. 139–153. – Berlin : Springer-Verlag (1985)

[5] Genrich, H. J.;Lautenbach, K.: Facts in Place/Transition-Nets. –
In: Mathematical Foundations of Computer Science 1978,
pp. 213-231 – Berlin : Springer-Verlag (1978)

[6] Genrich, H. J.;Lautenbach, K.; Thiagarajan, P. S.: Elements of
General Net Theory. In: Lecture Notes in Computer Science Vol. 84
: Net Theory and Applications, pp. 21-163. – Berlin :
Springer-Verlag (1980)

[7] Plünnecke, H.: Schnitte in Halbordnungen. – St. Augustin :
Gesellschaft für Mathematik und Datenverarbeitung, ISF-Report
81.09 (1981)

[8] Plünnecke, H.: K-density, N-density and Finiteness Properties. –
In: Lecture Notes in Computer Science Vol. 188 : Advances in
Petri Nets 1984, pp. 392-412. – Berlin : Springer-Verlag (1985)

[9] Reisig, W.: Petri Nets : An Introduction. – Berlin :
Springer-Verlag (1982)

OBSERVING NET BEHAVIOUR

Lucia Pomello

Dipartimento di Scienze dell'Informazione

Universita' degli Studi di Milano

via Moretto da Brescia 9

20133 Milano – Italy

ABSTRACT

Net system models are discussed from the point of view of the observations an observer can make on their behaviour. It is shown that while the behaviour of a contact-free C/E system, that can be considered as a "completely" specified system model, does not exhibit any difference depending on the kind of the observations the observer makes on it, the same is not true for safe net systems and for labelled safe net systems (considered as "incompletely" specified system models).

1. INTRODUCTION

It is widely recognized that any model of a system depends on the observer point of view and is therefore partial (see for example [Ash64]). In fact the cutting between the system and its environment, the choice of the events taken into consideration, and of that ones considered as atomic, and so on are all observer depending restrictions of the phenomenology taken into account. In addition, once a specific point of view is assumed, the various system models can be observed and compared by means of different modalities (by means of different kinds of observation) each one of them taking into account different characteristics of the system behaviour. Moreover, different classes of models characterize different specification levels. It is interesting to analyse how different modalities of observation generate different classes of distinction for the various classes of models.

In this paper such a comparison is performed among three classes of models defined inside Net Theory [Bra80], [ACPN86], namely contact-free C/E systems, safe net systems and labelled safe net systems.

C/E systems can be considered as <u>completely specified</u> models of concurrent systems. C/E systems are in fact based on the "extensionality principle": 'a repeatable event is fully characterized by the <u>extension of the change in conditions</u> effected by each of its occurrences. Thus, an event e may occur singly whenever its preconditions •e are holding and its postconditions e• are not holding' [Pet79]. As a consequence, each event (condition) is completely specified by its pre- and post-conditions (events) and then the underlying net is <u>simple</u>, and pre- and post-conditions (events) of an event (a condition) are disjoint, the underlying net is <u>pure</u>.

In this paper we shall consider only <u>contact-free</u> C/E systems. Contact-freeness implies that the concession for an event in a case is determined solely by the validity of its pre-conditions and is independent from the validity of its post-conditions.

Safe net systems do not satisfy the "extensionality principle , since an event (a condition) is not univocally determined by its pre- and post-conditions (events), and pre- and post-conditions (events) of an event (a condition) are not necessarily disjoint, i.e. the underlying net can be not simple and not pure. This means: that event (or that condition) is not at the same atomic level as the other elements of the model, or, in other words, some events (or conditions) are not enough different one from another to have different extensions; then a safe net system is a <u>not extensionally specified</u> model of a concurrent system.

Labelled safe net systems are safe net systems in which the transitions are labelled either by labels in an alphabet of "observable events", which correspond to interactions with the observer/environment, or by a special simbol (τ) denoting an "unobservable event".

Transition labelling is performed in accordance to a design discipline in which the presence of transitions with the same label is due to an incomplete knowledge about the system: the structure and behaviour of the system is determined through the knowledge of its interactions with a given environment.

Labelled safe net systems do not respect the "extensionality principle" not only because the underlying net can be impure and not simple, but also because two or more events may have the same name without having the same pre- and post-conditions. Therefore it may happen that an observable/unobservable event is not univocally determined by its pre- and post-conditions, and dually, that it does not univocally determine its pre- and post-conditions. Then, a labelled safe net system is an <u>incompletely specified</u> model of a concurrent system.

Beside labelled safe net systems we shall consider particular subclasses obtained by different transition labelling restrictions: the subclass in which no event is unobservable (not-τ-labelled systems), and that one in which the labelling is strict, i.e. in which observable events have different names (strictly-labelled systems). In spite of these labelling restrictions, these labelled system subclasses still are incompletely specified models and moreover, as we shall show, the results presented here are independent from such restrictions.

A deep discussion on the interpretation of the labelling can be found in [Vos86] where motivations in favour of the strict labelling are also given.

We consider how the differences among the three mentioned classes of net systems influence the observations about the experiments an observer can make on a concurrent system model.

We assume that the behaviour of a system can be observed or experimented by an observer only by "interacting" with it, the observer can be considered as a possible environment in which the system is embedded and with whom the system interacts; we assume the interaction system-observer to be synchronized through shared events only.

An observation of a system may (or may not) take into account concurrency, distinguishing (or identifying) real concurrency from its sequential non-deterministic simulation. An observation can consist only in verifying which sequences of events (or which sequences of sets of concurrent events) belong to the system behaviour; or in perceiving and keeping trace of the intermediate states, reachable by means of a subsequence of events (or of sets of events), considering the set of events which have concession after the subsequence considered. An observation can also take into account other characteristics of the system behaviour (e.g.: if it diverges, which kind of choices it contains, and the like) taking trace of the appropriate information. Therefore different observers can make different observations on the basis of which different distinction criteria between system models are assumed.

During the last years many observation equivalence notions for concurrent systems have been defined each one of which implicitly assumes the observer making a specific kind of observation.

Some of such notions have been compared and generalized to capture concurrency in [Pom86]. In this paper Failure equivalence [HBR84], Exhibited-Behaviour equivalence [DDPS85], Observation equivalence [Mil80], Behaviour equivalence [And82] and String equivalence (which is based only on the occurrence sequences) are taken into account together with the corresponding notions based on sequences of sets of concurrent events as defined in [Pom86] and with the equivalence notion, (γ,ϵ)-equivalence, defined for C/E systems in [GLT80] and in [Rei85].

In this paper divergence is not taken into account, therefore Testing-equivalence [DH84], Failure&Divergence-equivalence [Bro85] and Readiness [OH84] are not considered.

In this paper we ascertain whether and how different distinctions between models performed through different modalities of observation, as captured by the various considered equivalence notions, depend on the incomplete specification of the observed models, i.e. on the net class modelling the systems. In particular we show that while the behaviour of a contact-free C/E system does not exhibit any difference under the different modalities of observation, the same is not true for safe net systems and for labelled safe net systems. For contact-free C/E systems all the considered equivalence notions coincide; for safe net systems the equivalence notions which distinguish real concurrency from its sequential simulation are more restrictive than the other ones which identify concurrency with its sequential simulation; while for labelled safe net systems the various equivalence notions show different degrees of restrictiveness.

This paper is organized as follows: in section 2.1 some basic definitions on nets are recalled, the definition of labelled safe net system is given, two transition rules for sets of observable transition images which may concurrently occur called in the following set image transition rules are defined; in section 2.2 (γ,ϵ)-equivalence and the Concurrency-equivalence notions are defined. The equivalence notions based on event sequences are not defined since they are particular cases of the previous ones. Finally, in section 3.1 the relationships among the equivalence notions for the different net system models are discussed and sketched in table 1, while in section 3.2 they are formally proved.

The work on equivalence notions carried out at the Dipartimento di Scienze dell'Informazione together with Fiorella De Cindio, Giorgio De Michelis and Carla Simone and that one on the comparison of the equivalence notions and their generalization to capture concurrency developed by the author during her stay at GMD-F1 in the course of 1984 have deeply influenced and contributed to the work reported here. In particular a first analysis of the different observations in dependence on the incomplete specification of the observed model was presented in [DPS85].

2. BASIC DEFINITIONS AND EQUIVALENCE NOTIONS.

In the following we recall some basic definitions of Net Theory [Bra80], [Rei85], [ACPN86] and introduce some equivalence notions for concurrent systems modelled by safe net systems giving an intuitive interpretation to the observations they are based on.

2.1. Basic Definitions.

Definition 2.1.

● A net is a triple N=(S,T;F) such that:
  (i) $S \cup T \neq \emptyset$ and $S \cap T = \emptyset$;
  (ii) $F \subseteq (P \times T) \cup (T \times P)$;
  (iii) $dom(F) \cup codom(F) = S \cup T$.

S is the set of S-elements, T is the set of T-elements and F is the Flow relation of N; $X = S \cup T$ is the set of elements of N.

We denote: $^\bullet x = \{y \in X \ / \ (y,x) \in F\}$, $x^\bullet = \{y \in X \ / \ (x,y) \in F\}$ respectively the pre-set and the post-set of $x \in X$ and, in general, for $Y \subseteq X$ $^\bullet Y = \bigcup_{x \in Y} {}^\bullet x$ and $Y^\bullet = \bigcup_{x \in Y} x^\bullet$

● A net N=(S,T;F) is connected iff $\forall x_1, x_2 \in X : x_1 (F \cup F^{-1})^* x_2$

● A net N=(S,T;F) is pure iff $\forall x \in X : {}^\bullet x \cap x^\bullet = \emptyset$.

● A net N=(S,T;F) is simple iff $\forall x,y \in X : ({}^\bullet x = {}^\bullet y \ \wedge \ x^\bullet = y^\bullet) \Longrightarrow x = y$.

● Let N=(S,T;F) be a net and $T' \subseteq T$. The subnet generated by T' is the net N'=(S',T';F') with : $S' = {}^\bullet T' \cup T'^\bullet)$ and $F' = F/ ((S' \times T') \cup (T' \times S'))$.

Definition 2.2.

Let N=(S,T;F) be a net.

(i) $T' \subseteq T$ is a set of independent transitions (Ind(T')) iff
$$\forall t_1, t_2 \in T' : \quad t_1 \neq t_2 \ \Rightarrow \ ({}^\bullet t_1 \cup t_1{}^\bullet) \cap ({}^\bullet t_2 \cup t_2{}^\bullet) = \emptyset$$

(ii) Let $U \subseteq T$ and $S' \subseteq S$; U is a step enabled at S' (or U has concession at S')
(S'[U>) iff:
$$Ind(U) \quad \underline{and} \quad {}^\bullet U \subseteq S' \quad \underline{and} \quad (U^\bullet - {}^\bullet U) \cap S' = \emptyset.$$
If S'[U> then the occurrence of U yields $S'' \subseteq S$ (S'[U>S'') such that:
$$S'' = (S' - {}^\bullet U) \cup U^\bullet.$$

(when U is a singleton {t} then we say that the transition t is enabled, or has concession, at S' yielding S'': S'[t>S'')

A sequence $V = U_1, ..., U_n$ of sets of independent transitions $U_i \subseteq T$, i=1,...,n , is an occurrence sequence from S' to S'' (S'[V>S'') iff
$$\exists S_1, ..., S_{n-1} \subseteq S : \quad S'[U_1>S_1 \ ..... \ S_{n-1}[U_n>S''.$$
With $T_{(V)}$ we shall denote the set of transitions involved by the sequence V.

Remark

The transition rule defined here enables a transition even if its pre-set and post-set are not disjoint.

That is, this rule differs from that one given for Elementary Net systems in [RT86] and in [ACPN86], where a transition is enabled under the further condition that its pre-set and post-set are disjoint.

For pure nets the two rules coincide.

Definition 2.3.

A <u>Net System</u> $(N;M_0)$ is a net $N=(S,T;F)$ with an <u>initial marking</u> $M_0 \subseteq S$ associated to it. $[M_0\rangle \subseteq P(S)$ denotes the <u>set of markings</u> reachable from M_0 by finite occurrence sequences.

Definition 2.4.

Let $(N;M_0)$ be a net system. Then it is a <u>safe net system</u> iff
$$\forall \ t \in T, \ \forall M \in [M_0\rangle : \ ^\bullet t \subseteq M \ \Rightarrow \ (t^\bullet - {}^\bullet t) \cap M = \emptyset$$

Definition 2.5.

$(N;C)$ is a <u>Condition/Event system</u> (C/E system) iff the following conditions hold:

(i) $N=(S,T;F)$ is a <u>simple</u> and <u>pure</u> net in which:

S-elements are interpreted as <u>conditions</u> and usually denoted by B;

T-elements are interpreted as <u>events</u> and usually denoted by E;

(ii) $C \subseteq P(B)$, the <u>case class</u>, is an equivalence class of the <u>full reachability relation</u> $R=(r \cup r^{-1})^*$, where $r \subseteq P(B) \times P(B)$ is :
$$c_1 \ r \ c_2 \ \langle = \rangle \ \exists \ G \subseteq E : c_1 [G\rangle c_2;$$

(iii) $\forall \ e \in E \ \exists \ c \in C$ such that e has concession in c.

Definition 2.6.

Let $(N;C)$ be a C/E system. Then it is <u>contact-free</u> iff
$$\forall \ e \in E, \ \forall \ c \in C : \ ^\bullet e \subseteq c \ \Rightarrow \ e^\bullet \cap c = \emptyset$$

Remark.

A safe system net $(N;M_0)$ with underlying net N simple and pure and such that each transition is not dead at M_0 (i.e. such that: $\forall \ t \in T, \ \exists \ M \in [M_0\rangle : M[t\rangle$) corresponds to a contact-free C/E system with "initial case" M_0 and in which only cases forward reachable from M_0 are considered.

In safe net systems as well as in contact-free C/E systems it holds: $\forall \ t \in T : \ ^\bullet t \neq \emptyset$. Here we require also: $\forall \ t \in T: t^\bullet \neq \emptyset$, i.e.: $T \subseteq \text{dom}(F) \cap \text{codom}(F)$; therefore the underlying nets are such that pre-set and post-set of each transition are not empty.

Labelling the transitions allows the observer to distinguish observable actions from the unobservable ones (the latter ones identified by a unique special symbol τ), and not to distinguish different transitions corresponding to the same observable action (the ones labelled with the same symbol different from τ). In order to avoid the use of multi-sets rather than of sets to handle steps, the only restriction is that transitions which can occur concurrently must not be labelled with the same non-τ label.

Definition 2.7.

Let L be a finite alphabet denoting the set of observable actions and $\tau \notin L$ a special symbol denoting a (hidden) unobservable action, a __labelling__ h of a safe net system $(N;M_0)$ is a total function

$$h : T \rightarrow L \cup \{\tau\} \quad \text{such that:}$$

$$\forall t_1, t_2 \in T \ (t_1 \neq t_2), \ \forall M \in [M_0\rangle : \quad h(t_1) = h(t_2) \neq \tau \ \Rightarrow \ M \ [\{t_1, t_2\}\rangle$$

h can be extended to a homomorphism h: $T^* \rightarrow L^*$.

ϵ denotes the empty word over L^* and can be either the image of the empty word over T^* or the image of a sequence of unobservable actions, i.e. of a sequence of transitions τ-labelled by h.

Definition 2.8.

● A __labelled safe net system__ $(\langle N;M_0;h\rangle)$ is a safe net system $(N;M_0)$ with an associated labelling h : $T \rightarrow L \cup \{\tau\}$.

According to different restrictions on the labelling or on the structure of the underlying net, we get the following further definitions:

● A __strictly-labelled safe net system__ is a labelled safe net system $\langle N;M_0;h\rangle$ in which h is such that $\forall t_i, t_j \in T \ (t_i \neq t_j) : \quad h(t_i) = h(t_j) \Rightarrow h(t_i) = \tau$.

● A __non-τ-labelled safe net system__ is a labelled safe net system $\langle N;M_0;h\rangle$ in which h is such that $\forall t \in T : h(t) \neq \tau$.

Remark

A labelled safe net system $(\langle N;M_0;h\rangle)$ is a __safe net system__ iff h is a bijection such that $\forall t \in T : h(t) \neq \tau$. If in addition N is simple and pure and each transition is not dead at M_0 then it is a __contact-free C/E system__ in which only the forward behaviour from the initial case M_0 is considered.

The equivalence notions considered here below are based on the forward system behaviour. Therefore also in the case of contact-free C/E systems, where a forward and a backward behaviour is defined (see definition 2.5 (ii)), we shall consider only the forward behaviour from an initial case.

In order to consider only behavioural aspects that can be inferred by observation, on labelled safe net systems can be defined two transition rules taking into account occurrences of sets of observable transition images which may concurrently occur: the one is the "set-image transition rule" and the other the "EB-set-image transition rule". The latter one is due to Exhibited-Behaviour equivalence requirement to take into account only markings able to contribute to the occurrence of transitions whose images are observable (see definition 2.17).

In the case of equivalence notions based on sequences of single events instead of on sequences of event sets, these two transition rules can be easily reduced by considering singletons respectively to "image transition rule" and to "EB-image transition rule".

First we extend the labelling to sets of transitions:

<u>Definition 2.9.</u>
Let $ST=2^T - \emptyset$ and $SL=2^L - \emptyset$, then $h^\wedge : ST \rightarrow SL \cup \{\tau\}$ is defined by means of h as follows:

$$h^\wedge(\{t_1,...,t_k\}) = \begin{cases} \tau & \text{if } \forall t \in \{t_1,...,t_k\} : h(t) = \tau \\ \{ \partial \in L \, / \, \exists 1 \, 1 \leq i \leq k \, : \, h(t_i) = \partial \} & \text{otherwise} \end{cases}$$

$h^\wedge$ can be extended to a homomorphism $h^\wedge: ST^* \rightarrow SL^*$.
λ denotes the empty word over SL^* and
- $\lambda = h^\wedge(\lambda)$ where λ is the empty subset-word over ST^* <u>or</u>
- $\lambda = h^\wedge(U_1....U_n)$ if $\forall i=1...n \, : \, h^\wedge(U_i) = \tau$

<u>Definition 2.10.</u> (set-image transition rule)
A set of observable transition images $A \in SL$ <u>may occur in one step</u> at M yielding M' ($M(A \gg M'$) iff
$\exists W \in ST^+ (W = U_1 ... U_n$, with $U_i \in ST$ and $i=1...n)$:
$\quad \exists j=1...n : h^\wedge(U_j)=A$ <u>and</u> $\forall k=1...n \, (k \neq j): \, h^\wedge(U_k) = \tau$
<u>and</u> $M[W \gg M'$
A sequence of sets of observable transition images $Z \in SL^*$ has concession at M yielding M' ($M(Z \gg M'$) iff
$Z=\lambda \Rightarrow \exists W \in ST^*: h^\wedge(W)=\lambda$ <u>and</u> $M[W \gg M'$
$Z=A_1...A_n \Rightarrow \exists M_1,...,M_{n-1}: \, M(A_1 \gg M_1(A_2 \gg ... M_{n-1}(A_n \gg M'$

In the EB-set-image transition rule we shall use the following auxiliary boolean functions:

<u>Definition 2.11.</u>
$E : [M_0 \rangle \rightarrow BOOL$ associates the value true either to the reachable markings in which only sequences of τ-labelled transitions have concession, or to that ones in which the τ-labelled transitions having concession are in conflict with non-τ-labelled transitions; it is defined as follows:
$E(M')= ([\forall v \in T^*: M'[v\rangle, h(v)= \epsilon]$ <u>or</u> $[(\exists t \in T: h(t)=\tau$ <u>and</u> $M'[t\rangle) ==\rangle$
$\qquad\qquad\qquad\qquad (\forall t_i \in T: h(t_i)=\tau$ <u>and</u> $M'[t_i\rangle, \, \exists t_2 \in (\bullet t_i)\bullet: h(t_2) \neq \tau)] \,)$

<u>Definition 2.12.</u>
$B : ST^+ \rightarrow BOOL$ associates the value true to the sequence of steps whose transitions can be partitioned in sets in such a way that each set generates a connected

subnet containing at least one transition whose image is observable; it is defined as follows:

$$\forall W \in ST^+ \quad B(W) = \exists T_1,...,T_m \in ST: (T_1 \cup...\cup T_m = T_{(W)} \text{ and } T_1 \cap...\cap T_m = \emptyset \text{ and}$$
$$\forall i=1...m \ (N_{Ti} \text{ is connected and } \exists t \in T_i: h(t) \neq \tau))$$

(where $T_{(W)}$ denotes the set of transitions involved by W and N_{Ti} denotes the subnet generated by T_i)

This function allows to check that a non-τ-labelled transition occurrence can involve only occurrences of τ-labelled transitions belonging to the same system component, preventing the behaviour evolution of independent concurrent system components from interfering with each other.

Definition 2.13. (EB-set-image-transition rule)

(i) The empty subsetword $\lambda \in SL^*$ has EB-concession at M yielding M' ($M((\lambda \gg M')$ iff
   $\exists W \in ST^*: h^\wedge(W)=\lambda \text{ and } M[W \gg M' \text{ and } E(M')=\text{true}.$

(ii) A set of images of observable transitions $A \in SL$ has EB-concession at M yielding M' ($M((A \gg M')$) iff
   $\exists W \in ST^+(W=U_1 ... U_n, \text{ with } U_i \in ST \text{ and } i=1...n):$
   $\qquad \exists j=1...n : h^\wedge(U_j)=A \text{ and } \forall k=1...n \ (k \neq j) : h^\wedge(U_k) = \tau$
   $\qquad \text{and } M[W \gg M' \text{ and } B(W)=\text{true and } E(M')=\text{true}.$

2.2. Equivalence Notions.

In the following we shall introduce the definitions of the equivalence notions for concurrent systems modelled by labelled safe net systems.

To each equivalence notion introduced below corresponds a class of experiments of which the observer takes trace in different ways. The general form of an experiment is the following: the observer interacts with the system by means of a keyboard with as many keys as the cardinality of the powerset of the observable event names. The keys with the names of the observable events which can concurrently occur are lit up. The observer lets a set of observable events occur by pushing the corresponding key.

In the case of the equivalence notions based on sequences of single events the keyboard keys are in correspondence one to one with the names of the observable events.

The first equivalence notion we present is derived from that one introduced for comparing the behaviour of C/E systems [Bra80], [Rei85], which considers two C/E systems equivalent if there exists a bijection between their cases and a bijection between their events in such a way that their behaviours correspond each other. Here we give a modified definition for safe net systems with no labelling associated to them. Labelled safe net systems are therefore in general not comparable by means of (γ,ϵ)-equivalence. In the next section we shall show that (γ,ϵ)-equivalence corresponds to Concurrency-String-equivalence.

Definition 2.14. ((γ,ϵ)-equivalence)

Let $\langle N_1;M_{01}\rangle$ and $\langle N_2;M_{02}\rangle$ be safe net systems and for $i = 1, 2$ $T_i' = \{ t\epsilon T_i \, / \, \exists \, M\epsilon[M_{0i}\rangle : M[t\rangle \}$, $\langle N_1;M_{01}\rangle$ and $\langle N_2;M_{02}\rangle$ are $\underline{(\gamma,\epsilon)\text{-equivalent}}$ $(\langle N_1;M_{01}\rangle \approx^{(\gamma,\epsilon)} \langle N_2;M_{02}\rangle)$ iff there exist two bijections $\gamma : [M_{01}\rangle \rightarrow [M_{02}\rangle$ and $\epsilon : T_1' \rightarrow T_2'$ such that:

(i) $\gamma(M_{01}) = M_{02}$

(ii) $\forall \, W\epsilon ST_1'$, $\forall \, M_1,M_1'\epsilon [M_{01}\rangle$, $M_1[W\rangle M_1' \, \Longleftrightarrow\, \gamma(M_1) \, [\epsilon(W)\rangle \, \gamma(M_1')$

The previous definition differs from (γ,ϵ)-equivalence definition for C/E systems ([Bra80], [Rei85]) in requiring bijections only between the forward reachable markings and only between non-dead transitions at the initial markings, and in requiring condition (i). These differences are due to the previously discussed restriction to forward behaviour.

Let for $i=1,2$ $\langle N_i;M_{0i}; h_i : T_i \rightarrow L \cup \{\tau\}\rangle$ be labelled safe net systems, then we have the following definitions:

Definition 2.15. (CS-equivalence)

$\langle N_1;M_{01};h_1\rangle$ and $\langle N_2;M_{02};h_2\rangle$ are $\underline{\text{Concurrency-String-equivalent}}$ ($\langle N_1;M_{01};h_1\rangle \approx^{CS} \langle N_2;M_{02};h_2\rangle$) iff

$$\forall \, Z\epsilon SL^* : \quad M_{01}(Z\rangle\rangle \, \Longleftrightarrow\, M_{02}(Z\rangle\rangle.$$

The observer takes only trace of the sequences $Z\epsilon SL^*$ of event sets he can activate.

Definition 2.16. (CF-equivalence)

$\langle N_1;M_{01};h_1\rangle$ and $\langle N_2;M_{02};h_2\rangle$ are $\underline{\text{Concurrency-Failure-equivalent}}$ ($\langle N_1;M_{01};h_1\rangle \approx^{CF} \langle N_2;M_{02};h_2\rangle$) iff $\forall Z\epsilon SL^*$, $\forall X\subseteq SL$:

$\exists \, M_1 : (M_{01}(Z\rangle\rangle M_1 \, \underline{\text{and}} \, \forall A\epsilon X \, M_1(A\not\rangle\rangle) \, \Longleftrightarrow\, \exists \, M_2 : (M_{02}(Z\rangle\rangle M_2 \, \underline{\text{and}} \, \forall A\epsilon X \, M_2(A\not\rangle\rangle).$

The observer takes trace, for each sequence $Z\epsilon SL^*$ of event sets he can activate, of the sets of event sets that cannot be activated after it.

Definition 2.17. (CEB-equivalence)

$\langle N_1;M_{01};h_1\rangle$ and $\langle N_2;M_{02};h_2\rangle$ are $\underline{\text{Concurrency-Exhibited-Behaviour-equivalent}}$ ($\langle N_1;M_{01};h_1\rangle \approx^{CEB} \langle N_2;M_{02};h_2\rangle$) iff $\forall Z\epsilon SL^*$:

● $\underline{\text{if}}$ $Z= \lambda$ $\underline{\text{then}}$:

- $\forall \, M_{01}((\lambda\gg M_{11} \, \Longrightarrow\, \exists \, M_{02}((\lambda\gg M_{12} : \, \forall A\epsilon SL \, M_{11}((A\gg \, \Longleftrightarrow\, M_{12}((A\gg$

- $\underline{\text{and}}$ vice versa

- if $Z = A_1....A_n$ ($n \geq 1$, $\forall i=1...n : A_i \in SL$) <u>then</u> :

  $- \forall M_{01}((\lambda \gg M_{11} ((A_1 \gg M_{21} .... ((A_n \gg M_{n+1\ 1} ==> \exists M_{02}((\lambda \gg M_{12}((A_1 \gg M_{22} ... ((A_n \gg M_{n+1\ 2}:$

    $\forall i=1...n+1, \forall A \in SL : M_{i1}((A \gg <=> M_{i2}((A \gg$

- <u>and</u> vice versa

The observer takes trace, for each sequence $Z \in SL^*$ of event sets he can activate, of the sets of event sets which can occur after each event set A_i belonging to Z.

<u>Definition 2.18.</u> (CO-equivalence)

$\langle N_1;M_{01};h_1 \rangle$ and $\langle N_2;M_{02};h_2 \rangle$ are <u>Concurrency-Observation-equivalent</u>

($\langle N_1;M_{01};h_1 \rangle \approx^{CO} \langle N_2;M_{02};h_2 \rangle$) iff

$\quad \forall n \geq 0 : \langle N_1;M_{01};h_1 \rangle \approx_n^{CO} \langle N_2;M_{02};h_2 \rangle$ where :

- $\langle N_1;M_{01};h_1 \rangle \approx_0^{CO} \langle N_2;M_{02};h_2 \rangle$ is always true
- $\langle N_1;M_{01};h_1 \rangle \approx_{n+1}^{CO} \langle N_2;M_{02};h_2 \rangle$ iff $\forall Z \in SL^*$

  ($\forall M_1 : M_{01}(Z \gg M_1 ==> \exists M_2 : M_{02}(Z \gg M_2$ <u>and</u> $\langle N_1;M_1;h_1 \rangle \approx_n^{CO} \langle N_2;M_2;h_2 \rangle$)

  <u>and</u>

  ($\forall M_2 : M_{02}(Z \gg M_2 ==> \exists M_1 : M_{01}(Z \gg M_1$ <u>and</u> $\langle N_1;M_1;h_1 \rangle \approx_n^{CO} \langle N_2;M_2;h_2 \rangle$)

As it has been pointed out by Darondeau [Dar82], O-equivalence (and therefore also CO-equivalence) does not correspond to a "realistic" class of experiments. In fact CO-equivalence is the limit of a sequence of equivalence notions, each one of which is based on a class of experiments. W.r.t. each equivalence notion in the sequence, the observer takes trace of the experiments he makes on the basis of the distinctions performed by the previous equivalence in the sequence.

<u>Definition 2.19.</u> (CB-equivalence)

$\langle N_1;M_{01};h_1 \rangle$ and $\langle N_2;M_{02};h_2 \rangle$ are <u>Concurrency-Behaviour-equivalent</u>

($\langle N_1;M_{01};h_1 \rangle \approx^{CB} \langle N_2;M_{02};h_2 \rangle$) iff

- $\langle N_1;M_{01};h_1 \rangle \approx^{CS} \langle N_2;M_{02};h_2 \rangle$ <u>and</u>
- $\forall Z \in SL^*$, ($\forall M_1,M_2 : M_{01}(Z \gg M_1$ <u>and</u> $M_{02}(Z \gg M_2$):

  $\forall A \in SL : M_1(A \gg <==> M_2(A \gg$.

The observer has to verify that the systems are deterministic, i.e. that after each sequence of event sets he can activate, it is always possible to activate the same set of event sets, and that he can activate in both cases the same set of sequence of event sets $Z \in SL^*$.

The equivalence notions based on sequences of single events instead of on sequences of sets of concurrent events, i.e.: String-/ Failure-/ Exhibited-Behaviour-/ Observation-/ Behaviour-equivalence, are derivable as particular cases of the previous ones when considering singletons; furthermore as well as the previous ones they will be shortened by their initial letters.

3. COMPARING THE EQUIVALENCE NOTIONS.

In the following we shall prove the relationships among the equivalence notions for systems modelled by the three different considered net classes, showing how the different distinction power depends from the incomplete specification of the models. In section 3.1 we shall informally explain the results that will be proved in section 3.2.

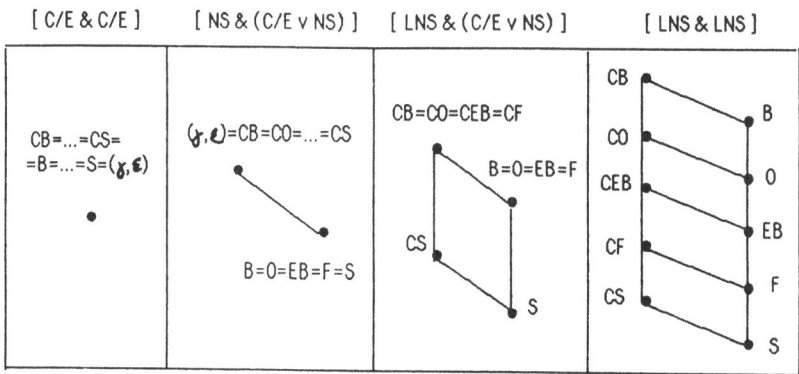

(Abbreviations: C/E := contact-free C/E system;
NS := safe net system;
LNS := labelled / strictly-labelled /
non-τ-labelled safe net system)

Table 1.

3.1

The different distinctions between models by means of the different modalities of observation depend from the incomplete specification of the observed models as well as sketched in Table 1, which can be read as follows.

[C/E & C/E]
Any modality of observation of completely specified models generates the same class of distinction, i.e. if the compared models are contact-free C/E systems then all the considered equivalence notions coincide. (See Theorem 2.)

[LNS & LNS]
On the contrary, observing incompletely specified model is dependent on the modalities of observation; i.e. if the compared models are labelled safe net systems

then the considered equivalence notions show growing degrees of restrictiveness. (See Theorem 1.)

The relationships among the considered equivalence notions are maintained even if the labelling is such that the nets contain either unobservable actions (belong to the class of strictly-labelled nets) or actions with identical names (belong to the class of non-τ-labelled nets). (See Theorem 5 and Theorm 6.)

[NS & (C/E v NS)]

It is sufficient that one of the compared models is, instead of a completely specified model, a not extensionally specified model in order to have different distinctions when observing sequences of set of concurrent events instead of event sequences; i.e. if one (or both) of the two compared systems is not a contact-free C/E system but is modelled by a safe net system, (in particular by a system with a not pure underlying net) then the equivalence notions distinguishing real concurrency from its sequential simulation are more restrictive than that ones identifing concurrency with its sequential simulation; while in each one of the two classess the different notions still coincide.

This is due to the fact that in an impure net if two events can be activated in any order then they are not necessarily concurrent. And this can be easily verified comparing the models $N_1=<N_1;M_{01};h_1>$ and $N_2=<N_2;M_{02};h_2>$ in Fig.1. If in a model an event is assumed to be not atomic, because the net is not pure, (the transitions labelled by a and b in N_2) then the possibility to experiment the same event sequences (a followed by b and b followed by a) on the two compared models does not guarantee that on them it is possible to experiment also the same sequences of concurrent event sets ({a,b}). (See Theorem 3)

[LNS & (C/E v NS)]

If one of the compared models is a completely specified model (or also a not extensionally specified model) and the other one is an incompletely specified model then the equivalence notions based only on sequences (String-equivalence and Concurrency-String-equivalence) are less restrictive than that ones based on observation. (See Theorem 4)

The possibility of different observations in case of safe net systems instead of contact-free C/E systems is as previously discussed due to impurity (see [NS & (C/E v NS)]); it seems, and it is under investigation, that observation depends from simplicity in the case of equivalence notions in which also conditions are observable (e.g.: (γ,ε)-equivalence [GLT80], Exhibited-Functionality equivalence [Ber86], interface equivalence [Vos86]).

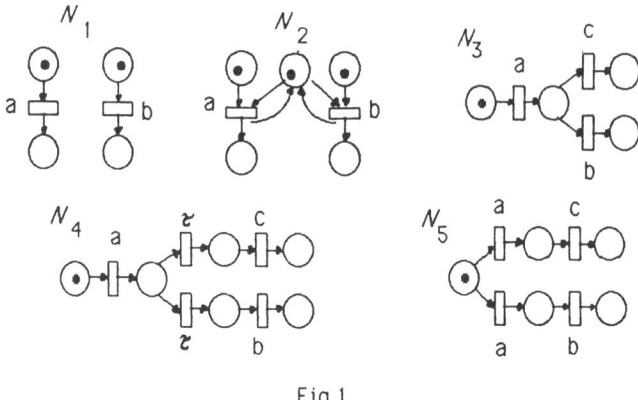

Fig. 1

3.2.

<u>Theorem 1.</u>
Let $\langle N_1;M_{01};h_1\rangle$ and $\langle N_2;M_{02};h_2\rangle$ be labelled safe net systems, then the equivalence notions introduced in the previous section result partially ordered w.r.t. their strength as well as represented in the graph drawn in column [LNS & LNS] of Table 1 (in which the strongest equivalences are at the top).

<u>Proof.</u>
These relationships have been proved in [Pom86].

<u>Theorem 2.</u>
Let $\langle N_1;M_{01};h_1\rangle$ and $\langle N_2;M_{02};h_2\rangle$ be contact-free C/E systems, then <u>all</u> the equivalence notions introduced in the previous section <u>coincide</u> $((\gamma,\epsilon)$-equivalence too).

<u>Proof.</u>

(a) $\approx S = \approx CB$
we prove: (a_1) $\approx S = \approx CS$, showing that $\approx CS \supset \approx S$ ($\approx S \supset \approx CS$ is trivial); and then
(a_2) $\approx CS = \approx CB$, showing that $\approx CB \supset \approx CS$ ($\approx CS \supset \approx CB$ is trivial).

● (a_1) $\approx CS \supset \approx S$ is proved by contradiction:
∃ $N_1 = \langle N_1;M_{01};h_1\rangle$, $N_2 = \langle N_2;M_{02},h_2\rangle$: $N_1 \approx^S N_2$ <u>and</u> $N_1 \neq^{CS} N_2$
=> ∃ Z∈SL*: $M_{01}(Z\rangle\rangle$ <u>and</u> $M_{02}(\not{Z}\rangle\rangle$ (or vice versa)
=> ∀ z∈seq(Z): $M_{01}(z\rangle\rangle$ <u>and</u> (since $N_1 \approx^S N_2$ also) $M_{02}(z\rangle\rangle$ but then, being N_2 a contact-free C/E system, in which it holds that: ∀ M∈[$M_{02}\rangle$, ∀$t_1,t_2\in T_2$: $M[t_1\rangle M'[t_2\rangle$ <u>and</u> $M[t_2\rangle M''[t_1\rangle$ <=> $M[\{t_1 t_2\}\rangle$, it should be also $M_{02}(Z\rangle\rangle$ that is a contradiction.

(seq(Z) denotes the set of sequences of observable events obtained by interleaving the sets of concurrent events in the sequence Z, i.e.: if $Z=U_1...U_n$, $U_i \in SL$ then $seq(Z)=\{w_1...w_n$ / $\forall$ i=1,..,n : $w_i \in$ interleaving(U_i) } where for $U \in SL$ interleaving$(U)=\{ u_1...u_k$ / $k=card(U)$ and $\forall$ i=1...k : $u_i \in U$ and $\forall$ i,j=1...k $(i \neq j)$: $u_i \neq u_j$ }; if $Z=\lambda$ then $seq(Z)=\{\epsilon\}$))

● (a_2) $\approx^{CB} \supset \approx^{CS}$ is proved by contradiction:

$\exists$ $N_1=\langle N_1;M_{01};h_1\rangle$, $N_2=\langle N_2;M_{02};h_2\rangle$: $N_1 \approx^{CS} N_2$ and $N_1 \not\approx^{CB} N_2$

=> $\exists Z \in SL^*$, $\exists M_1,M_2$: $(M_{01}(Z\gg M_1$ and $M_{02}(Z\gg M_2)$ and $\exists A \in SL$: $(M_1(A\gg$ and $M_2(A\gg$) but then $M_{01}(ZA\gg$ and $M_{02}(ZA\gg$ since, being N_2 a contact-free C/E system, M_2 is the only marking reachable from M_{02} by means of Z and therfore we get the contradiction: $N_1 \not\approx^{CS} N_2$.

(b) $\approx(\gamma,\epsilon) = \approx^S$

We show $\approx(\gamma,\epsilon) = \approx^{CS}$ as from (a) we get $\approx(\gamma,\epsilon) = \approx^S$.

● (b_1) $\approx^{CS} \supset \approx(\gamma,\epsilon)$

Let $\langle N_1;M_{01}\rangle \approx(\gamma,\epsilon) \langle N_2;M_{02}\rangle$ then $\exists$ h_1, h_2 : $\langle N_1;M_{01};h_1\rangle \approx^{CS} \langle N_2;M_{02};h_2\rangle$ if h_1 and h_2 are constructed in such a way that: $\forall t_1,t_2$ $(t_1 \in T_1', t_2 \in T_2')$: $t_2=\epsilon(t_1)$ <=> $h_1(t_1)=h_2(t_2)$ then it is immediate to derive CS-equivalence definition from (γ,ϵ)-equivalence.

$(T_i'$ denotes the set of transitions that are no-dead under the initial marking as in definition 2.14.)

● (b_2) $\approx(\gamma,\epsilon) \supset \approx^{CS}$

Let $\langle N_1;M_{01};h_1\rangle \approx^{CS} \langle N_2;M_{02};h_2\rangle$ then, being $\langle N_1;M_{01};h_1\rangle$ and $\langle N_2;M_{02};h_2\rangle$ contact-free C/E systems and therefore h_1 and h_2 bijections, the bijection $\epsilon: T_1' \rightarrow T_2'$ is constructed by requiring: $\epsilon(t_1)=t_2$ <=> $h_1(t_1)=h_2(t_2)$ for $t_1 \in T_1'$ and $t_2 \in T_2'$; while $\gamma: [M_{01}\rangle \rightarrow [M_{02}\rangle$ is constructed by putting $\gamma(M_1)=(M_2)$ <=> $\exists$ $W \in ST_1^*$: $M_{01}[W\rangle M_1$ and $M_{02}[\epsilon(W)\rangle M_2$. Being the systems contact-free C/E systems, for each sequence of steps M_1 and M_2 are univocally determined and therefore both the conditions (i) and (ii) of definition 2.14 are guaranted.

Theorem 3.

        Let $\langle N_1;M_{01};h_1\rangle$ be a contact-freee C/E system and $\langle N_2;M_{02};h_2\rangle$ a safe net system, then the equivalence notions based on sequences of single events coincide and are strictly weaker than that one based on subset sequences; (γ,ξ)-equivalence coincide with the latter ones.

        The same relationships are valid if $\langle N_1;M_{01};h_1\rangle$ and $\langle N_2;M_{02};h_2\rangle$ are both safe net systems. (See column [NS & (C/E v NS)] in Table 1)

Proof.

        The systems $N_1=\langle N_1;M_{01};h_1\rangle$ and $N_2=\langle N_2;M_{02};h_2\rangle$ of fig.1, that are respectively a contact-free C/E system and a safe net system, are equivalent w.r.t. each equivalence notion based on sequences of single events, but not w.r.t. each equivalence notion based on subset sequences, i.e.:

$$\forall X \in \{S, F, EB, O, B\} \quad N_1 \approx^X N_2 \quad \text{but} \quad N_1 \not\approx^{CX} N_2.$$

Since in a safe net system the marking reachable from a given marking by means of a sequence of events (or of sets of concurrent events) is univocally determined even if the underlying net is not pure or not simple, by means of reasoning similar to that ones used in proving theorem 2., it is immediate to prove: $\approx^S = \approx^B$ and $\approx^{CS} = \approx^{CB} = \approx(\gamma, \epsilon)$.

Theorem 4.

Let $\langle N_1; M_{01}; h_1 \rangle$ be a labelled safe net system and $\langle N_2; M_{02}; h_2 \rangle$ a contact-free C/E system or a safe net system, then:

a) $\quad \forall X \in \{S, F, EB, O, B\} \quad \approx^X \supset \approx^{CX}$

b) $\quad \approx^F = \approx^{EB} = \approx^O = \approx^B \quad \underline{and} \quad \approx^{CF} = \approx^{CEB} = \approx^{CO} = \approx^{CB}$

c) $\quad \approx^S \supset \approx^F \quad \underline{and} \quad \approx^{CS} \supset \approx^{CF}$

i.e. we get the partial order represented by the graph as drawn in column [LNS & (C/E v NS)] of Table1.

The same relationships are valid if $\langle N_1; M_{01}; h_1 \rangle$ is a strictly-labelled or a non-τ-labelled safe net system.

Proof.

a) is proved by means of the same counterexample of theorem 3.

b) is proved by contradiction showing that if two systems $\langle N_1; M_{01}; h_1 \rangle$ and $\langle N_2; M_{02}; h_2 \rangle$ satisfy the hypothesis and are not B-equivalent (and therefore not CB-equivalent), then they are also not F-equivalent (CF-equivalent); in the proof is essential the fact that in $\langle N_2; M_{02}; h_2 \rangle$, being h_2 a bijection, the marking reachable from M_{02} by means of a sequence of events (or of set of concurrent events) is univocally determined.

c) is proved by means of a counterexample which shows two models that are CS-equivalent (and therefore also S-equivalent) but not equivalent w.r.t. the other equivalence notions: (e.g.: $N_3 = \langle N_3; M_{03}; h_3 \rangle$ and $N_4 = \langle N_4; M_{04}; h_4 \rangle$ of fig.1, that are respectively a contact-free C/E system and a labelled (nay a strictly-labelled) safe net system, are such that: $N_3 \approx^{CS} N_4$ but $N_3 \not\approx^F N_4$.

Theorem 5.

Let $\langle N_1; M_{01}; h_1 \rangle$ be a labelled safe net system and $\langle N_2; M_{02}; h_2 \rangle$ a strictly-labelled safe net system or a non-τ-labelled safe net system, then the relationships among the equivalence notions w.r.t. their strength are not changed w.r.t. theorem 1. (See column [LNS & LNS] in Table 1).

Proof.

The theorem is proved by means of counterexamples which show models that are equivalent w.r.t. an equivalence notion but not w.r.t. that one immediately stronger than the previous one; e.g., considering the systems in fig.1. and fig.2., $\forall$ X∈{S, F, EB, O, B } $N_1 \approx^X N_2$ but $N_1 \neq^{CX} N_2$; $N_3 \approx^{CS} N_4$ but $N_3 \neq^F N_4$; $N_6 \approx^{CF} N_4$ but $N_6 \neq^{EB} N_4$ or $N_7 \approx^{CF} N_4$ but $N_7 \neq^{EB} N_4$; $N_5 \approx^{CEB} N_4$ but $N_5 \neq^O N_4$; N_4 (N_5) is CO-equivalent but not B-equivalent with a system isomorphic to itself.

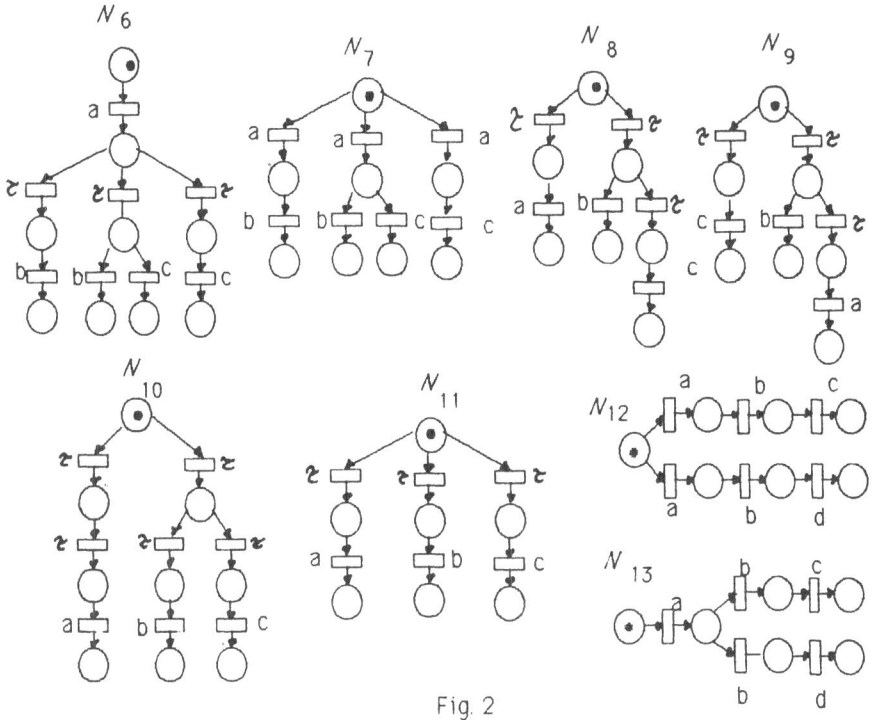

Fig. 2

Theorem 6.

Let $N_1 = \langle N_1; M_{01}; h_1 \rangle$ and $N_2 = \langle N_2; M_{02}; h_2 \rangle$ be both strictly-labelled (or non-τ-labelled) safe net systems, then the relationships among the equivalence notions w.r.t. their strength are not changed w.r.t. theorem 1. (See column [LNS & LNS] in Table 1.).

Proof.

The theorem is proved by means of counterexamples which show models that are equivalent w.r.t. an equivalence notion but not w.r.t. that one immediately stronger than the previous one; e.g., considering the systems in fig.1. and fig.2., $\forall$ X∈{S, F, EB, O, B }

$N_1 \approx^X N_2$ but $N_1 \not\approx^{CX} N_2$; $N_3 \approx^{CS} N_4$ but $N_3 \not\approx^F N_4$, and $N_3 \approx^{CS} N_5$ but $N_3 \not\approx^F N_5$; $N_8 \approx^{CF} N_9$ but $N_8 \not\approx^{EB} N_9$, or $N_7 \approx^{CF} N_8$ but $N_7 \not\approx^{EB} N_5$; $N_{10} \approx^{CEB} N_{11}$ but $N_{10} \not\approx^O N_{11}$, or $N_{12} \approx^{CEB} N_{13}$ but $N_{12} \not\approx^O N_{13}$; N_4 (N_5) is CO-equivalent but not B-equivalent with a system isomorphic to itself.

4. REFERENCES.

[ACPN86]   Proc. of Advanced Course on Petri Nets, Bad Honnef 8-19 Sept.'86, (1986).

[And82]   C. Andre', Behaviour of a Place-Transition Net on a Subset of Transitions, IFB 52, Springer Verlag, (1982).

[Ash64]   W.R. Ashby, An Introduction to Cybernetics, Univ. Paperbaks, Methuen, (1964).

[Ber86]   G. Berthelot, Transformations and Decompositions of Nets, in [ACPN86], (1986).

[Bra80]   W. Brauer ed., Net Theory and Applications, LNCS 84, Springer-Verlag, (1980).

[Bro85]   S.D. Brookes, A Semantics and Proof System for Communicating Processes, in K.R. Apt (ed.), Logics and Models of Concurrent Systems, Springer-Verlag, (1985).

[Dar82]   P. Darondeau, An enlarged Definition and complete Axiomatization of Observational Congruence of Finite Processes, LNCS 137, Springer-Verlag, (1982).

[DDPS85]   F. De Cindio, G. De Michelis, L. Pomello, C. Simone, Exhibited-Behaviour Equivalence and Organizational Abstraction in Concurrent System Design, Proc. 5th Int. Conf. on Distributed Computing, IEEE (1985).

[DPS85]   G. De Michelis, L. Pomello, C. Simone, Observing Nets, Proc. Int. Symp. on Circuits and Systems ISCAS '85, Kyoto, (1985)

[DH84]   R. De Nicola, M. Hennessy, Testing equivalences for processes, TCS 34, 83-134, (1984).

[GLT80]   H.J. Genrich, K. Lautenbach, P.S. Thiagarajan, Elements of General Net Theory, in [Bra80], (1980).

[HBR84]   C.A.R. Hoare, S.D. Brookes, A.W. Roscoe, A theory of Communicating Sequential Processes, J.ACM Vol.31/3, (1984).

[Mil80]   R. Milner, A Calculus for Communicating Systems, LNCS 92, Springer Verlag, (1980).

[OH84]   E.R. Olderog, C.A.R. Hoare, Specification-oriented semantics for communicating processes, Tech. monograph PRG-37, Prog. Research Group, Oxford Univ., (1984), to appear also in Acta Inform.

[Pet79]   C.A. Petri, Concurrency as a Basis of Systems Thinking, Proc. 5th Scandinavian Logic Symposium, Aalborg University Press, (1979).

[Pom86]   L. Pomello,   Some Equivalence Notions for Concurrent Systems, An Overview, in G. Rozemberg (ed.) Advances in Petri Nets'85, LNCS 222, Springer-Verlag, (1986).

[Rei85]   W. Reisig, Petri Nets – An Introduction, EATCS Monographs on Theoretical Computer Science, Vol. 4, Springer-Verlag, (1985).

[RT86]   G. Rozenberg, P.S. Thiagarajan, Petri nets: basic notions, structure, behaviour, in J.W. de Bakker, W.P. de Roever and G. Rozenberg (eds.) Current Trends in Concurrency, LNCS 224, Springer-Verlag, (1986).

[Vos86]   K. Voss, System Specification with Labelled Nets and the Notion of Interface Equivalence, Arbeitspapiere der GMD Nr. 211, (1986).

This research has been developed with the financial support of the Italian Ministero della Pubblica Istruzione.

ALGEBRAIC MODELS OF PARALLELISM
AND NET THEORY

M.W. Shields
Electronic Engineering Laboratories
The University of Kent at Canterbury
Great Britain

ABSTRACT

In this note we present a comparison between Net Theory and Algebraic languages such as Milner's CCS [2]. We argue that a mild generalisation of three postulates from which Condition/Event systems may largely be derived, gives rise to a class of algebraic models at this level of abstraction. As C/E systems provide the basic level of interpretation in Net theory, so too do these C/E algebraic systems provide a level of interpretation for general algebraic models, and one in which concurrency may be formally defined.

1. Introduction

In this paper we shall examine a relationship between topological models of parallelism, such as Net Theory, and algebraic models, such as CCS.

Such an exercise appears to us to be useful, given the acknowledged importance of the two approaches.

The most obvious way of establishing a formal relationship between the two is to define a method of translating objects of one kind into objects of another.

For example, in CCS there is a rule that if E is a behaviour expression (describing some possible pattern of communication) and α is an action (possible communication), then $\alpha.E$ is a behaviour expression (in which α occurs and then anything allowable by E.)

A translation rule might say that if the net of figure 1 corresponds to E under the translation, then the net of figure 2 corresponds to $\alpha.E$.

This approach is intuitively appealing, but lacks justification in terms of the semantics of the two models.

We suggest, however, that a comparison on the semantic level must be made before we argue about the legitimacy of translations from one model to the other. Furthermore, we claim that it is important that the comparison should take place at the same level of abstraction. From the point of view of nets, this must be that of Condition/Event Systems, which provide the basic level of interpretation for Nets. We accordingly seek a similar level of abstraction in CCS and in algebraic specification languages generally.

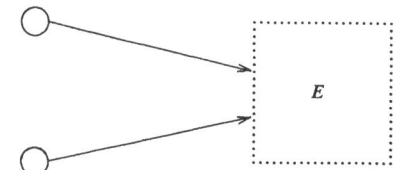

Figure 1

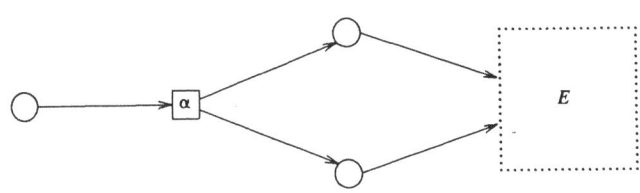

Figure 2

2. Three Basic Principles of Net Theory

Let us first consider three basic principles upon which Net Theory may, with some slight additions, be erected ([2],p.25). These principles are:

N1 States are distributed.

N2 In general, changes to states are caused by event occurences which are local.

N3 Each event determines a unique local state change.

Let us consider the consequences of these postulates. Firstly, we must have a collection B of local states. Each 'global' state is a collection of these. We thus have a set $C \subseteq \underline{P}(B)$[1] of *cases*. Each of the local states $b \in c \in C$ will be said to hold in c. We must also have a set E of events. Each event determines a unique state change, that is; there is a unique set $\cdot e$ of states which cease to hold after e has occurred and a unique state $e\cdot$ of states which begin to hold after e has occurred.

---

[1] $\underline{P}(B)$ denotes the power set of B.

Very little more is required to yield the Condition/Event system model of concurrent systems. We have already our triple (B,E,F), where

$$F = \{(p,e) \in B \times E \mid p \in \dot{e}\} \cup \{(e,p) \in E \times B \mid p \in e\dot{}\}$$

If we add to our postulates the 'common sense' requirements that $B \cap E = \emptyset$ or that $B \cup E = \emptyset$ or that isolated elements may be neglected - and hence that $domain(F) \cup range(F) = E \cup B$ - then (B,E,F) satisfies the net axioms.

Whether we do or not, these three postulates give us a model of systems which *specialises* to that of Condition/Event Systems.

3. The Basic Principles Extended

In order to compare algebraic languages with Net Theory, we shall attempt to found the former on an extension of the three principles that began the previous section.

A1 States are *structured*

A2 In general, changes of state are local changes to structure

A3 Each event determines a unique local state change.

Let us elaborate on these.

States are structured

The main difference between the two sets of principles is the replacement of *distributed* by *structured*. Again we have a set B of local states, but global states are now no longer merely subsets of B. They are organised in some way which determines how changes to individual local states may be grouped together to make event occurrences.

We might think of each local state as a possible state of a sequential process, part of a system, and structure as indicating how a given set of active processes are permitted to interact at that time.

We illustrate this with an example. Consider the following expression.

$$(p_1 \mid (p_2 \mid p_3) \backslash A) \backslash B \mid p_4 \qquad \text{(E)}$$

which is a behaviour expression in CCS. Each p_i is associated with a set of possible actions (which in CCS are communications which may be made with other p_j or with an environment).

Informally, we may consider the expression to be describing the instantaneous state of a system comprising four sequential processes organised as in the picture in figure 3.

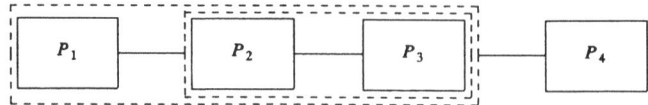

Figure 3

Each process P_i may be considered to be carrying the local state p_i. By analogy with C/E systems, $\{p_1,...,p_4\}$ may be considered as a case or global state of the system. However, things are a little more complicated because of the way in which the system is structured. Whether or not an event may occur depends not only on its pre- and post-conditions - which processes must take part in its occurrence and their local states - but also on the way in which the processes are structurally interrelated.

In example (E), the two processes P_2 and P_3 are composed in parallel and may interact via message passing. But this composite has some of its possible actions hidden by the restriction operator \A which prevents them being used to communicate with any other process in the system.

Any given global state will be a combination of local states (of the sequential processes that are currently active, say). We propose that such combinations be represented syntactically, as in example (E).

We expect, therefore, that in general we would have

(1)   A set B of local states;

(2)   For each $n > 0$, a set (possibly empty) Σ_n, of *operators* of arity n.

Thus, we have a Σ-algebra signature, in which B is the set of constants (operators of arity 0).

The *terms* of the algebra are defined as follows. Each element of B is a term. If op is an operation of arity n and $t_1,...,t_n$ are terms then so is $op(t_1,...,t_n)$.

We propose

Global states are represented by terms in a Σ-algebra.

Changes are Local

We shall say that each t_i is a subterm of $op(t_1,...,t_n)$ and extend the notion of subterm to a partial order on the terms of the algebra. We shall write $t \leq t'$ to denote that t is a subterm of t'.

Given that we are representing global states by terms, we may now interpret our second principle by saying that in general

Event occurrences bring about local changes to terms representing global states. These changes are the result of substitutions of subterms for subterms.

We explain what we mean by the substitution of a term t'' for a term t' in a term t. We write the resulting term as $t[t'' \setminus t']$. If t' isn't a subterm of t, then $t[t'' \setminus t'] = t$ and of course if $t = t'$ then $t[t'' \setminus t'] = t''$. Otherwise, we have $t = g(t_1,\ldots,t_n)$ and $t' \le t$ and $t \ne t'$. In that case we define $t[t'' \setminus t'] = g(t_1[t'' \setminus t'],\ldots,t_n[t'' \setminus t'])$.

Events Determine Unique Local Changes

We translate this as meaning that each event e is associated with a unique pair $\cdot e$ and $e\cdot$. If t' and t'' are terms representing global states, then if e occurs transforming t' to t'', then we must have $\cdot e \le t'$ and $t'' = t'[e\cdot \setminus \cdot e]$.

Write $t'\text{-}e \rightarrow t''$ to mean that occurrence of event e transforms term t' to term t''.

This assignment of terms to e cannot be an arbitrary matter. Somehow, this must reflect (a) the relationship between e and the local states in the terms and (b) the meaning of the term-building operations.

This latter will be given by rules, at least one per operation, of the general form:

$$\frac{t_{i_1} - e_1 \rightarrow t'_{i_1} \,\&\, \cdots \,\&\, t_{i_k} - e_k \rightarrow t'_{i_k}}{g(t_1,\ldots,t_n) - e \rightarrow t} \tag{R}$$

where t is a term and e is an action and both depend functionally upon the t_i, t'_j and e_k.

For example, in CCS, we have a rule of the form

$$\frac{E_1 - e \rightarrow E'_1}{(E_1 + E_2) - e \rightarrow E_1}$$

In order to start things going, we also need a set of rules for the individual local states and events, of the form $b\text{-}e \rightarrow t$, where t is a term. There may be several such rules associated with b. Collecting them together we get a rule that looks like

$$b = \sum_i e_i \cdot t_i \tag{D}$$

In general, then, an algebraic specification language consists of a Σ-*algebra* and at least one rewrite rule of the form (R) for each operation. A *specification* in the language consists of a single term (the initial term, representing the initial 'global' state), and a set of definitions of the form (D).

For a given set of definitions (D) and rules (R), we may determine all possible legitimate *derivations* $t-e{\rightarrow}t'$. In accordance with our three postulates, we consider the following type of subclass.

(α) The global states, members of a set C are terms in which no $b \in B$ occurs more than once. (This is akin to safety in C/E systems).

(β) For each event e, there are unique minimal terms $\cdot e$ and $e \cdot$ such that $\cdot e-e{\rightarrow}e\cdot$, that is, if $t-e{\rightarrow}t'$, then $\cdot e \le t$ and $t' = t'[e\cdot\diagdown\cdot e]$.

(γ) If $t-e{\rightarrow}t'$ and either of t and t' belong to C then so does the other.

We call these, C/E systems.

Within a C/E system, we may define a relation *ind* on events, of *independence*, interpreted as follows. If e ind e' and $t-e{\rightarrow}t_1$ and $t-e{\rightarrow}t_2$, then the occurrences of e and e' are *concurrent*.

If t is a term, let $\bar{t}$ denote the set of atomic terms occurring in t.

Define e_1 ind e_2 iff $(e_1\cdot \cup \cdot e_1)\cap(e_2\cdot \cup \cdot e_2) = \emptyset$

It is easy to show that *ind* is irreflexive and symmetric, that is, *ind* is an *independence relation* [3].

4. Asynchronous State Machines and Concurrent Behaviour

From notions of structured or distributed state and local change, we may construct a behavioural semantics for nets or C/E algebraic systems in the same way. In both cases we have a set of 'global states', Q, a set of events E, a transition relation $R \subseteq Q \times E \times Q$ and a relation *ind* $E \times E$. Let us give a name to this type of object.

4.1 Definition

An *asynchronous state machine* is a quadruple (Q,E,R,ind) where
(1) Q is a set (of global states)
(2) E is a set (of events)
(3) $R \subseteq Q \times E \times Q$ is a transition relation. If $(q,e,q') \in R$, then we write $q-e{\rightarrow}q'$.
(4)   *ind* $\subseteq E \times E$ is an independence relation
and the following hold:
(A)   Suppose $q,q_1,q_2 \in Q$ and $e \in E$ with $q-e{\rightarrow}q_1$ and $q-e{\rightarrow}q_2$, then $q_1 = q_2$.
(B)   Suppose $q,q_1,q_2 \in Q$ and $e_1,e_2 \in E$ with $q-e_1{\rightarrow}q_1$ and $q-e_2{\rightarrow}q_2$. Then, if e_1 ind e_2 then there exists $q' \in Q$ such that $q_1-e_2 \rightarrow q'$ and $q_2 -e_1 \rightarrow q'$.

If $S = (\mathbf{N}, C)$ is a C/E system, with $\mathbf{N}$ as the underlying net and C as the set of cases, then S determines an **ASM**, $ASM(S)$, with $Q=C$, $E=T$ (the transitions of $\mathbf{N}$), R defined by $c\text{-}t{\rightarrow}c'$ iff $c[t > c'$ and ind given by t ind t' iff t co t' and $t \neq t'$.

If t is the initial term and D is the set of definitions of the form (D) determining a specification of a C/E system, then we may define an **ASM**, $ASM(t,D)$ with Q as the set of terms that may be derived from t, E as the set of actions occurring in the definitions D, R as the obvious relation and ind as defined at the end of section 3.

The main point about ASM's is that they 'accept' trace languages. This means, as a consequence of theorem 5.7, which we shall meet later, that an **ASM** determines a set of label partial orders. Each such partial order described a period of asynchronous behaviour in the well-known way.

From an independence relation ind on A we may construct a relation $\equiv_{ind}$ on A^*. Define $x \equiv_{ind} {}^{(1)} y$ if there exist $u,v \in A^*$ and $a,b \in A$ such that $x = u.a.b.v$ and $y = u.b.a.v$ and a ind b. Let $\equiv_{ind}$ be the equivalence relation generated by $\equiv_{ind}{}^{(1)}$.

4.2 Proposition

$\equiv_{ind}$ is a monoid congruence. Let $\langle x \rangle_{ind}$ denote the $\equiv_{ind}$ class of x and let $A_{ind}{}^*$ denote the quotient monoid.

$A_{ind}{}^*$ is right and left cancellable and has no non-trivial zero dividers, that is, if $\langle x \rangle_{ind}.\langle y \rangle_{ind} = \langle \varepsilon \rangle_{ind}{}^2$ then $x = y = \varepsilon$.

It follows that the relation $\leq$ defined - $\langle x \rangle_{ind} \leq \langle y \rangle_{ind}$ iff there exists $z \in A^*$ such that $\langle x.z \rangle_{ind} = \langle y \rangle_{ind}$ - is a partial ordering.

With every **ASM** M and $q \in Q$, we may associate a language $\mathbf{L}(M,q) \subseteq E_{ind}{}^*$ in the same sort of way that a transition system plus initial state determines a language of strings.

If $a \in E$ and $x \in E^*$, we define

$q' - \langle a \rangle_{ind} \rightarrow q''$ iff $q' - a \rightarrow q''$

$q' - \langle a.x \rangle_{ind} \rightarrow q''$ iff $q' \langle a \rangle_{ind} \rightarrow q'''$ and $q''' - \langle x \rangle_{ind} \rightarrow q''$ some q'''

Note that these relations are well defined as a consequence of (B) of 4.1.

We may then set $\mathbf{L}(M,q) = \{ \langle x \rangle_{ind} | q - \langle x \rangle_{ind} \rightarrow q'$ some $q' \}$.

The use of quotient monoids for representing concurrent behaviour was pointed out in [3]. In this paper its author explains that a single trace ($\equiv_{ind}$ class) determines a labelled partially ordered set.

---

^{2} ε denotes the null string

We may say a good deal more than this, however, as we shall see in
the next section.

5. Traces and Posets

This section sketches machinery using which we may considered
ASM's as acceptors for certain types of labelled poset. Each such set
describes a period of asynchronous behaviour in the obvious way. If
e,e' are two elements of such a poset, labelled by a and b and $e < e'$ in
this poset, then we would say that in that behaviour, the occurrence
e of the event a preceded the occurrence b of the event e'. Likewise,
if e and e' are incomparable with respect to $\leq$, then we say that that
in that behaviour, the occurrence e of the event a was concurrent to
the occurrence b of the event e'.

If X is a poset, then we write $\overline{X}$ for the underlying set and $\leq_x$ for
its order relation.

The set of labelled posets accepted by an **ASM** is more than a set.
It is partially ordered with respect to the relation defined as
follows.

5.1 Definition

Let X,Y be posets, then $X[Y(X$ is a prefix of $Y)$ <=>

1) $\overline{X} \subseteq \overline{Y}$
2) $\forall x \in \overline{X} \, \forall y \in \overline{Y}: y \leq_Y x$ <=> $y \leq_X x$

If P is a set of posets, then $(P,[)$ is obviously itself a poset.

We concern ourselves with certain types of set of posets, namely
the *behaviour systems* defined below. They are related to the *event
structures* of [5] in the same way that a language of finite strings
(for more details, see [4]) is related to its closure as a language
of finite and infinite strings.

5.2 Definition

A behaviour system is a set **B** of posets such that
1) $\forall X \in$ **B**: $Y[X = > Y \in$ **B** (left closure)
2) $\forall X,Y \in$ **B** $\forall x \in X \cap Y = > (y \leq_X x$ <=> $y \leq_Y x)$(event consistency)

A labelled behaviour system (**B**,λ) is a behaviour system **B** together
with a function

$$\lambda: Env(\mathbf{B}) \to A \, ,$$

where A is some set of labels and $Env(\mathbf{B})$ is the set $\bigcup_{X \in \mathbf{B}} \overline{X}$. As we have
said, $\lambda(e) = a$ is to be read 'e is an occurrence of a'. We restrict
ourselves to labellings which determine a concurrency structure on the
poset with respect to some independence relation.

5.3 Definition

Let X be a poset, $\lambda:\overline{X} \to A$ and ind be an independence relation on A. X is an (A, ind)-poset w.r.t. λ if

1) If $x, y \in \overline{X}$ with $x\ co_X\ y$ then $\lambda(x)\ ind\ \lambda(y)$
2) If $x, y \in \overline{X}$ with $\lambda(x)\ ind\ \lambda(y)$ then $x\ co_X\ y$ or for some z, $x <_X z <_X y$ or $y <_X z <_X x$.

We may now define the class of labelled posets that we will be working with.

5.4 Definition

$(\mathbf{B}, \lambda)$ is conservatively labelled if and only if

1) There exists a set A and an independence relation ind such that every element of $X \in \mathbf{B}$ is an (A, ind)-poset w.r.t. $\lambda|X$.
2) If $X, Y \in \mathbf{B}$ and there exists $\phi:\overline{X} \to \overline{Y}$ such that ϕ is a poset iso-morphism with $\lambda(\phi(x)) = \lambda(x)$ all x then $X = Y$.

We now show how to map (A, ind) posets to sets of strings.

5.5 Definition

Let X be a finite (A, ind) poset. Let $T(X)$ denote the set of all totally ordered posets $(\overline{X}, \leq)$ with $\leq_X \subseteq \leq$.

Let $\rho(X)$ denote the subset of A^* obtained as follows. $x \in \rho(X)$ iff there exists a poset $Y \in T(X)$ such that if $e_1, \ldots, e_n$ are the elements of $\overline{X}$ and $e_1 <_Y \ldots <_Y e_n$ then $x = \lambda_X(e_1) \ldots \lambda_X(e_n)$.

We may prove:

5.6 Lemma

With the above notation, $\rho(X) \in A_{ind}{}^*$.

The representation theorem (5.7) shows that ρ is actually an iso-morphism of partial orders between $(\mathbf{B}, [)$ and $(\rho(\mathbf{B}), \leq)$. First, let $\underline{L} \subseteq A_{ind}{}^*$. We say that $\underline{L}$ is left closed in $A_{ind}{}^*$ iff whenever $<x>_{ind} \leq <y>_{ind}$ and $,y>_{ind} \in \underline{L}$, then $<x>_{ind} \in \underline{L}$.

5.7 Representation Theorem ([4])

Let $(\mathbf{B}, \lambda)$ be a (A, ind) linguistic, then ρ is a poset isomorphism

$$\rho: \mathbf{B} \to \rho(\mathbf{B}) \subseteq A_{ind}{}^*$$

and $\rho(\mathbf{B})$ is left-closed in $A_{ind}{}^*$.

Conversely, if $\underline{L}$ is a left-closed subset of $A_{ind}{}^*$, then there exists an (A, ind)-linguistic behaviour system $(\mathbf{B}, \lambda)$ such that $\rho(\mathbf{B}) = \underline{L}$.

Finally, if $(\mathbf{B}, \lambda)$ and $(\mathbf{B}', \lambda')$ are (A, ind) linguistic, then

$$\rho(\mathbf{B}) = \rho(\mathbf{B}') => \mathbf{B} \approx \mathbf{B}'$$

Here $\approx$ indicates that the two systems are isomorphic, where 'iso-morphic' has the obvious definition.

From this theorem, we may see how to provide any C/E algebraic
specification language with a behavioural semantics in terms of
partially ordered sets.

Form the **ASM**, $ASM(t,D)$. It is easy to show that $\mathbf{L}(ASM(t,D),t)$ is
left-closed in $E_{ind}{}^*$. We may now exhibit the totality of potential
asynchronous behaviour by a unique (up to isomorphism) linguistic
behaviour system $(\mathbf{B},\lambda)$ such that $\rho(\mathbf{B}) = \mathbf{L}(ASM(t,D),t)$.

6. General Languages

We have seen that it is possible to view algebraic methods as being
based on a 'basic level of interpretation' in which they generalise
the case/transition approach to C/E systems. In this way, one might
even see algebraic methods as generalising Net methods on the basic
level.

Let us now turn to the problem of providing general algebraic
specification languages with a behavioural semantics.

We take our inspiration from General Net Theory. C/E systems have
a direct behavioural semantics in terms of processes nets ([1],p.100).
Higher level nets may be interpreted in terms of lower level nets, and
hence, ultimately, C/E systems, using net *morphisms* exhibiting higher
level entities as composites of entities on the lower level.

We therefore propose that general algebraic specifications be
interpreted in terms of C/E specifications through relabellings.

Consider a specification consisting of an initial term t_o and a set
of definitions of the form (D) above.

Let T denote the set of all terms that may be derived from t_o by
application of the rewrite rules to appropriate subterms and let E
denote the set of all actions occurring in rewrites. Let B be the set
of all atomic terms occurring in elements of T.

Suppose that t_C, T_C, E_C and B_E represent the corresponding entities
in a C/E specification. A *folding* of the system specified by t_C onto
the system specified by t_o is a pair of mappings $f:B_C \to B$ and $g:E_C \to E$,
such that if f is extended to an algebra homomorphism f^* - by

$$f^*(op(t_1,\ldots,t_n)) = op(f^*(t_1),\ldots,f^*(t_n))$$

and $f^*(b) = f(b)$ for atomic terms - then

(1) $f^*(t_C) = t_o$
(2) f^* is a bijection, $f^*:T_C \to T$
(3)   If $t - e \to t'$ for $t,t' \in T_C$ and $e \in E_C$, then $f^*(t) - g(e) \to f^*(t')$
(4)   If $t - e \to t'$ for $t,t' \in T$ and $e \in E$, then there exists $e' \in E_C$
       such that $g(e') = e$ and $f^{*-1}(t) - e' \to f^{*-1}(t')$.

A folding establishes an isomporphism between the two transition systems determined by t_O and t_E, which allows one to distinguish between multiple copies of atoms or events with the same name.

We suggest that, intuitively, the two systems have the same behaviour - apart from the names of actions and local states. If this is correct, then, we may obtain an asynchronous behavioural semantics for the t_O system from that of the t_e system by relabelling events of the latter using g.

Given a folding, one may finally provide the t_O system with a concurrent semantics. Given the **ASM** M for the t_E system, we construct the trace language $L(M, t_E)$. Using the representation theorem, one may then construct a labelled behaviour system, $(\mathbf{B}, \lambda)$. Finally, we relabel this behaviour system to give a set of partial orders giving the asynchronous behaviour of the t_O system, $(\mathbf{B}, \lambda')$, where $\lambda'(e) = g(\lambda(e))$ for each e belonging to $\mathbf{B}$.

REFERENCES

[1] Brauer (ed), Advanced Course on General Net Theory of Systems and Processes, Hamburg, 1979. LNCS 90

[2] Milner, A.R.J.G.: A Calculus of Communicating Systems, LNCS 92, Springer 1980

[3] Mazurkiewicz, A: Concurrent program schema and their interpretations. Proc. Aarhus Workshop on Verification of Parallel Programs, 1977.

[4] Shields, M.W.: Concurrent Machines, The Computer Journal, Vol.28, No. 5, 1985

[5] Nielsen, M., Plotkin, G., Winskel, G.: Petri Nets, Event Structures and Domains. Proc. Symposium on Semantics of Concurrent Computation, Evian-les-Bains, 1979.

TOWARDS A SYNCHRONY THEORY FOR P/T NETS

M. Silva
Dpto. Automática, E.T.S.I.I.Z.
María Zambrano, 50 (Actur)
50015 ZARAGOZA (SPAIN)

Abstract . Synchrony theory is a branch of General Net Theory devoted to the study of transition firing dependences. Considering place/transition nets (P/T nets), these dependences will be studied by means of four related kinds of synchronic concepts: *synchronic lead (SL)* , *synchronic distance (SD)* , *deviation bound (DB)* and *fairness bound (FB)* . The main relations among them are presented, giving an unified view. The boundedness of any *synchronic function* for two subsets of transitions is characterized by a *synchronic relation* . Particular attention is given here to algebraic characterizations of synchronic relations.

1. INTRODUCTION

Synchrony theory is a branch of General Net Theory devoted to the study of transition firing dependences.

Synchronic distance is a fundamental concept introduced by C.A. Petri in [PETR 75]. It is a *metric* (hence the term "distance") by which it is possible to describe some behavioural aspects of a net model with *quantitative assertions* . Finite synchronic distances represent mutual interdependences between the firing of transitions; i.e. they represent a given degree of "synchronization" in the firing of transitions. A more precise measure is obtained by introducing a weight in the original concept [SIFA 79], [GENR 80], [GOLT 82], [SUZU 83]. In the sequel we will directly give the name synchronic distance to a modified weighted generalization (formally introduced in definition 2.4 as W-synchronic distance).

Nevertheless, the (weighted) synchronic distance may be infinite even if there exist strong synchronic interdependences among the firing of transitions. In other words, the synchronic distance does not capture many important synchronic dependences. It can be viewed as a computationally *linearly-based* concept related to a broader concept for characterizing synchronic interdependences: the *fairness bound.* Synchronic lead and firing deviation bound are the basic properties on which synchronic distance and fairness bound are constructed.

Among the different properties considered in synchrony theory, synchronic distance turns out to be the most restrictive. *Asynchrony* between two subsets of transitions (events) (in the sense of the existence of firing independence), $\{T_i, T_j\}$, will be characterized by the unbounded firing deviation of T_i with respect to T_j, and viceversa.

Synchronic distance (see: [PETR 75], [ANDR 79], [SIFA 79], [GENR 80], [GOLT 82], [SUZU 83], [MURA 85], [CHON 86]) and fairness (see: [BEST 84], [CARS 85], [MURA 85], [LEU 86], [MERC 86], [SILV 86]; for free-choice nets [THIA 84] and for simple nets [SUZU 85]) are the most studied synchronic concepts. Nevertheless, it is very important to note that the definitions of a given concept do not usually coincide in the literature and they can differ considerably (see, for example, the k-fairness concept in [BEST 84] and the B-fairness concept in [SILV 86]). In any case and as a starting point, our definitions consider *all possible firing sequences* in the P/T net model. This draws an important distinction between the original synchronic distance concept (defined on C/E systems by considering processes; for a didactical presentation see [REIS 85]) and our W-synchronic distance concept. In particular, Petri argues that concurrency between two transitions always yields at least synchronic distance 2; but this will be true in our case if there exists a cyclic behaviour that contains both transitions.

Four synchronic concepts are studied in this paper. *Deviation bound* (*DB*) and *fairness bound* (*FB*) express relative *finite delay properties*. *Bounded deviation* and *bounded fairness* are *relations* between subsets of transitions for which the corresponding property is bounded. *Synchronic lead* (*SL*) and *synchronic distance* (*SD*) appear as computationally linearly-based concepts related to *DB* and *FB*. *Lead-synchronization* and *distance synchronization* are relations that express the respective boundedness. Distance-synchronization and bounded-fairness relations characterize *interdependences* (they are *symmetric*). Lead-synchronization and bounded-deviation relations characterize the existence of *dependences* in the firing of some transitions *with respect to* another subset. The four synchronic relations are *reflexive* and *transitive*. Bounded deviation and lead synchronization relations are, after some elements (transitions subsets) classifications, *partial order relations*, while bounded fairness and distance synchronization are *equivalence relations*. Surprisingly, the first two turn out to be alternative definitions of the same relation, but the value of the deviation bound and the synchronic lead can be different.

For each basic concept two kind of relations are introduced:

(1) *behavioural* (two types can be introduced, because the computation can start at the initial marking, M_O, or at any marking reachable from M_O).

(2) *structural* (the function's boundedness is considered for any initial marking).

Using behavioural and/or structural synchronic relations a fine characterization of the *synchronic relations based structure* of the net is obtained. Synchronic relations are strongly related to *T-invariants*. Basically, behavioural synchronic relations are related to the firing count vectors associated to the directed circuits of the coverability graph, while structural synchronic relations are related to the non-negative right annullers of the token flow matrix. In particular, generalizing the presentation in [SUZU 83], the weights that make the synchronic distance bounded for *any* initial marking belong to a *linear space*.

The paper is structured as follows: Basic concepts and some properties are introduced in §2. Necessary and sufficient conditions for boundedness in synchronic concepts are presented in §3. In §3.1 *structural* synchronic relations are considered, while in §3.2 *behavioural* synchronic relations are characterized. For particular classes of nets, some interesting corollaries allow the computations of synchronic relations to be done in *polynomial* time and space. In §4 the synchronic relations based struture of a net is considered.

2. BASIC SYNCHRONIC CONCEPTS AND PROPERTIES

This section presents some fundamental concepts on which to base an analysis of the dependences of transition firings in a place/transition net. The importance of this is clear because in P/T nets any atomic activity is represented by the firing of a transition. In the sequel it will be assumed that:

1) any atomic activity takes *finite time* .

2) a *complete abstraction* is made of the way in which conflicts (when effective) are solved. In other words, our analysis ignores the way in which *arbiters* or *schedulers* work.

The firing Deviation Bound (*DB*) is the most basic concept. A related linearly-based property is the Synchronic Lead (*SL*). The Fairness Bound (*FB*) and the Synchronic Distance (*SD*) are constructed by making *DB* and *SL*, respectively, symmetric. Because the basic activity of a net can be easily summarized in a linear form (by its *state equation),* the analysis of the *SL* and the *SD* can be connected with other more classic nets analyses (in particular, with linear marking invariants, like implicit places) [SIFA 79] [SILV 87].

Any of the above concepts, {*DB, FB, SL, SD* }, is defined for the net model (i.e. for any firable sequence or computation) and not for a particular firing sequence (or computation).

To characterize the existence of dependences or interdependences between the firings of transition subsets, *synchronic relations* are introduced. Two subsets of transitions are in a synchronic relation iff the corresponding synchronic function is bounded. The synchronic relations are defined in a *behavioural* sense, for a marked net $<N,M_0>$, or in a *structural* sense, for any initial marking on N.

The material presented in this section is structured in three parts. Although the reader is assumed to be familiar with P/T net theory (see, for example, [BRAU 80], [SILV 85]), section 2.1 states the main notations (in accordance with [SILV 85]). In §2.2 finite delay properties are introduced, while in §2.3 linearly-based synchronic properties are considered.

2.1 Preliminaries

Let N be a place/transition net, $N=<P,T,Pre,Post>$, where:

* P is the set of places $(n=|P|)$.

* T is the set of transitions $(P \cap T = \emptyset)$ $(m=|T|)$.

* *Pre* (and *Post*) is the pre (post) incidence function or input (output) function:

$Pre : PxT \rightarrow N$ (it represents arcs going from places to transitions)

$Post : PxT \rightarrow N$ (it represents arcs going from transitions to places)

where $N = \{0,1,2,3,...\}$

The tuple $<N,M_0>$ represents a marked net, obtained by considering the initial marking M_0 on the structure described by N.

Structural analysis of P/T nets is based on considering their *state equation* (see, for example, [BRAU 80], [SILV 85]). If the firing of transition t_j from marking M_{k-1} allows marking M_k to be reached, $M_{k-1} [t_j > M_k$, then:

$$M_k = M_{k-1} + C \cdot U_j = M_{k-1} + C(t_j) \quad \text{(state equation)}$$

where: * $C = [C_{ij}]$, $C_{ij} = Post(p_i ,t_j) - Pre(p_i ,t_j)$

* $U_j (g) = 0 \quad \forall g \neq j, U_j(j) = 1$

The j-th column of C, $C(t_j)$, represents the flow of tokens caused by the firing of transitions t_j . The i-th row of C, $C(p_i)$, represents the flow of tokens for place p_i: $M_k(p_i) = M_{k-1}(p_i) + C_{ij}$. Matrix C will be called the *token flow matrix* [SILV 85] of the net(*flow matrix,* for short). If the net is *pure* (self-loop free, i.e. $Pre(p,t) \cdot Post(p,t) = 0$) the flow matrix may be interpreted as a place-transition *incidence matrix..*

Let: * σ be a firing sequence,

* $\overline{\sigma}$ be σ's *characteristic vector* (i.e. the Parikh mapping of σ) whose i-th component, $\overline{\sigma}(t_i)$, is the number of ocurrences of transition t_i in σ.
* $T_i \in T$, be a subset of transitions. Its *firing count* in σ is $\overline{\sigma}(T_i)$, the number of times that the transitions belonging to T_i are fired in σ: $\overline{\sigma}(T_i) = \sum_{t_j \in T_i} \overline{\sigma}(t_j)$

* $\|V\|$ be the support of vector V .

* $M_i [\sigma > M_j$ express that σ is applicable from M_i , M_j being the reached marking.

* $R(N,M_0)$ be the set of all markings reachable from M_0 .

* $L(N,M_0)$ be the set of all firing sequences from M_0..

* θ be a set of transition subsets of T : $\theta = \{T_i \mid T_i \subset T \}$.

A marked net $<N,M_0>$ is said to be (marking) *bounded* iff for any place p_i there exists a $k_i \in N$ such that $M(p_i) \leq k_i$ for any reachable marking $M,M \in R(N,M_0)$. Marking boundedness in a net characterizes the finiteness of its state space. A marked net $<N,M_0>$ is said to be *live* iff in any reachable marking M and $\forall t_i \in T$, a sequence σ_i , $M[\sigma_i > M_i$, is applicable such that M_i enables transition t_j . Liveness in nets guarantees the possibility of an infinite activity of all the transitions in the net. If a net is live, it is deadlock free. A global cyclic behaviour is characterized by reversibility. A marked net $<N,M_0>$ is said to be *reversible* iff the initial marking can always be returned to from whichever other reachable marking .

By summing the state equations for $k=1,2,...,q$ we obtain the following matrix equation:

$$M_q = M_0 + C \cdot \overline{\sigma}, \quad \text{where } M_0 \ [\ \sigma > M_q$$

A *consistent component* is any $X \in N^m$ such that $C \cdot X = 0$. If there exists a sequence σ applicable from $M \in R(N,M_0)$ such that $\overline{\sigma} = X$, then $M[\sigma > M$ (*marking reproduction*). A *net is consistent* iff there exists an $X > 0$ $(\forall i \in (1,m) : X(i) \in N^+)$.

A *conservative component* is any $Y \in N^n$ such that $Y^T \cdot C = 0$. Then $Y^T \cdot M_q = Y^T \cdot M_0$ (*token conservation law*). A *net is conservative* iff there exists an $Y > 0$ $(\forall j \in (1,n) : Y(j) \in N^+)$.

2.2 On basic finite delay synchronic concepts

The firing deviation bound and fairness bound are introduced in this section in a somewhat different form from our previous work [LEU 86] [SILV 86].

Before going on to the more formal definitions, let us try to introduce, very informally, the firing deviation bound and the fairness bound between two transitions, t_i and t_j . Let the firing of t_i mean that a certain *user i (active subsystem)* takes a given *resource (passive element)* and the firing of t_j means the same, but for *user j* . Assuming that resources are taken for finite time, if t_i may only fire finitely often without firing t_j , user i will not prevent user j from taking the common resource. So we say that t_i is in *bounded firing deviation relation* with respect to t_j , $(t_i,t_j) \in BD$. This characterizes a *relative* (from t_i to t_j) and *finite delay* property. A quantitative bound of the delay introduced on user j by user i is given by the *firing deviation bound, DB(t_i ,t_j)*.

If the above relation is made *symmetric* (i.e. if t_i may only fire finitely often without firing t_j , and viceversa), t_i and t_j are said to be in *bounded fairness relation*, $(t_i,t_j) \in BF$. This characterizes a *relative* (between t_i and t_j) and *symmetric finite delay* property. A quantitative bound of both delays (that introduced by i on j and viceversa) is given by the *fairness bound, FB(t_i ,t_j) = max\{DB(t_i ,t_j), DB(t_j ,t_i)\}*.

In practice the activity of a given user (or resource) is represented by a subset of transitions. The extension of the above basic concepts from two transitions, t_i and t_j , to two subsets of transitions, T_i and T_j , leads to the following definitions. From these definitions, many interesting properties can be directly shown. Because of the lack of space they are presented without formal proof. The properties relative to symmetry or non-symmetry are considered in the examples.

Definitions 2.1. FIRING DEVIATION CONCEPTS

1) The *firing deviation bound* (*deviation bound*, for short) of T_i with respect to T_j in $<N,M_0>$ is:

$$DB(T_i,T_j) = sup \ \{\overline{\sigma}(T_i) \mid \sigma \in L(N,M) \ , \ M \in R(N,M_o) \ \text{and} \ \overline{\sigma}(T_j) = 0\}$$

2) T_i is in *bounded (firing) deviation relation* with respect to T_j in $<N,M_o>$, $(T_i,T_j) \in BD$, iff $\exists k \in N$ such that $DB(T_i,T_j) \leq k$.

3) T_i is in *structural bounded deviation relation* with respect to T_j in N, $(T_i,T_j) \in SBD$, iff they are in bounded deviation relation for any initial marking, $M_o \in N^n$. ◆

Properties 2.1.

1) Non-negativity: $DB(T_i,T_j) \geq 0 \ \ \forall T_i,T_j \in \theta$.

2) Non-symmetry: $\exists <N,M_o>$ with $T_i,T_j \in \theta$ such that $DB(T_i,T_j) \neq DB(T_j,T_i)$.

3) $(T_i,T_j) \in SBD \Rightarrow (T_i,T_j) \in BD$. ◆

Examples 2.1. Let us consider the net in figure 1.1.

a) $DB(t_1,t_2) = \infty$ and $DB(t_2,t_1) = \infty$ {i.e. there is no synchronic dependence}.

b) $DB(t_3,t_4) = 1$ {$\exists$ synchronic dependence}; $DB(t_4,t_3) = \infty$.

c) $DB(t_1,t_3) = 1$ and $DB(t_3,t_1) = 1$ {synchronic dependence in both directions}.

For the net in figure 1.2 we can write:

d) $(t_1,t_3) \in SBD \Rightarrow (t_1,t_3) \in BD$.

e) $(t_1,t_4) \notin SBD$ but $(t_1,t_4) \in BD$.

f) For $T_i = \{t_i\}$ any pair (t_i,t_k) is in BD, any pair of even (odd) transitions are in SBD.

From the firing deviation bound, let us now introduce the fairness bound as a symmetric concept. From the bounded deviation relation, the bounded fairness relation and their structural counterparts will be introduced. Just as liveness properties are useful for characterizing the absence of total or partial deadlocks, bounded fairness is interesting for characterizing the absence of total or partial *starvation* .

Definitions 2.2. FAIRNESS CONCEPTS

1) The *fairness bound* of T_i with respect to T_j in $<N,M_o>$ is:

$$FB(T_i,T_j) = max \ \{DB(T_i,T_j), DB(T_j,T_i)\}$$

2) T_i and T_j are said to be in *bounded fairness relation* in $<N,M_o>$, $(T_i,T_j) \in BF$, iff $\exists k \in N$ such that: $FB(T_i,T_j) \leq k$.

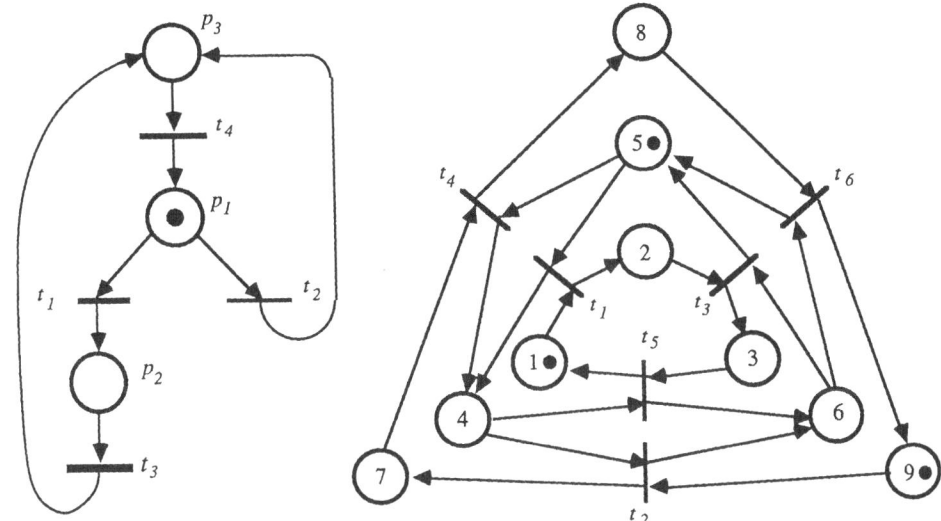

(1) t_1 and t_2 are not synchronized, while between t_1 and t_3 there exits a synchronic dependance.

(2) Bounded-fair net, but not structurally bounded fair.

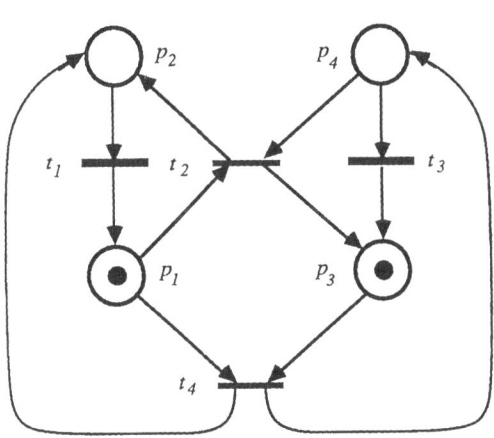

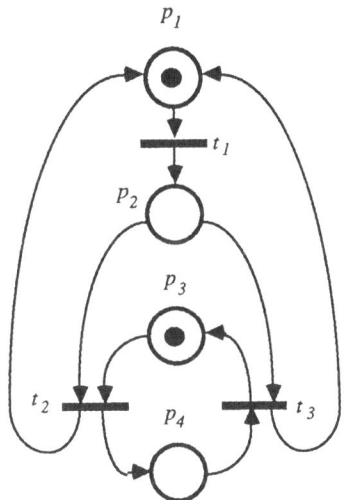

(3) t_1 and t_4 are in bounded-fairness relation, but $\nexists$ finite synchronic distance.

(4) T_o be finite, the synchronic distance between t_1 and t_2 must be weighted $(t_1, 2t_2)$.

Figure 1. Four live, binary and reversible nets.

3) T_i and T_j are said to be in *structural bounded fairness relation* in N, $(T_i, T_j) \in SBF$, iff they are in bounded fairness relation for any initial marking, $M_o \in N^n$ ♦

Properties 2.2.

1) Non-negativity: $FB(T_i, T_j) \geq 0$ $\forall T_i, T_j \in \theta$.

2) Symmetry: $FB(T_i, T_j) = FB(T_j, T_i)$ $\forall T_i, T_j \in \theta$.

3) $(T_i, T_j) \in SBF \Rightarrow (T_i, T_j) \in BF$. ♦

Examples 2.2. Let us consider the net in figure:

a)   Fig. 1.1: $FB(t_3, t_4) \equiv FB(t_4, t_3) = \infty$ b) Fig. 1.1: $FB(t_1, t_3) \equiv FB(t_3, t_1) = 1$

c)   Fig. 1.3: $FB(t_1, t_4) \equiv FB(t_4, t_1) = 2$ d) Fig. 1.3: $FB(t_1, t_3) \equiv FB(t_3, t_1) = \infty$

2.3. On linearly-based synchronic properties

Synchronic lead and synchronic distance are said to be linearly-based because their computations are based on the dot product of two vectors: $\overline{\sigma}$ and W (a weight vector). The invariants expressed by synchronic leads and distances can be shown to be very strongly related to the bounds of some (implicit) places [SIFA 79] [SILV 87]. For the *quantitative* definition of these concepts, distinctions will be made for the computations realized: (1) from the *initial* marking, *ISL* and *ISD*, or (2) from any marking *reachable* from the initial one, *SL* and *SD*. As in §2.2, many interesting properties are directly introduced without formal proofs. In general, they are technically very simple.

The basic definition of synchronic lead and distance are not precise enough to characterize many phenomena of practical interest. Let us directly consider the weighted concepts as the basic one.

Let: * W_i and $W_j \in N^m$ be two vectors which express the weight associated with the transitions of the subsets T_i and T_j [i.e. $\|W_i\| = T_i$ and $\|W_j\| = T_j$]

     * $W_{ij} = W_i - W_j$ [$\Rightarrow W_{ji} = -W_{ij}$]

Definitions 2.3. SYNCHRONIC LEAD CONCEPTS

1) The *initial synchronic lead* of W_i with respect to W_j (W_{ij} -initial synchronic lead, for short) in $<N, M_o>$ is:

$$ISL(W_{ij}) = \sup \{W_{ij}^T \cdot \overline{\sigma} \mid \sigma \in L(N, M_o)\}$$

2) The *synchronic lead* of W_i with respect to W_j (W_{ij} -synchronic lead, for short) in $<N, M_o>$ is:

$$SL(W_{ij}) = \sup \{W_{ij}^T \cdot \overline{\sigma} \mid \sigma \in L(N, M) \text{ and } M \in R(N, M_o)\}$$

3) T_i is *lead-synchronized* in $<N,M_o>$ with respect to T_j, $(T_i,T_j) \in LS$, iff $\exists W_i \neq W_j \in N^m$ and $\exists k \in N$ such that $SL(W_{ij}) \leq k$.

4) T_i is *structurally lead-synchronized* in N with respect to T_j, $(T_i,T_j) \in SLS$, iff they are in LS - relation for any initial marking, $M_o \in N^n$.   ◆

At this point it is very important to observe that synchronic lead relations (definitions 2.3.3 and 4) are not defined for particular weights. Synchronic lead relations can be defined *for a given couple of weights*, $\{W_i, W_j\}$, in the obvious way. These additional relations, *WLS* and *WSLS* may be of interest in some cases.

Clearly the synchronic lead is *not* a linear function, although it is based on $W^T \cdot \overline{\sigma}$, a linear computation. The next properties deal with non-linearity.

Properties 2.3.

1) If $\lambda \geq 0$ then $SL(\lambda \cdot W, M_o) = \lambda \cdot SL(W, M_o)$

         else   if $SL(W,M_o) \neq 0$ then $SL(\lambda \cdot W, M_o) \neq \lambda \cdot SL(W, M_o)$

2) Let $\alpha, \beta \geq Q^+$, and $W \in Z^m$

$$SL(\alpha \cdot W_1 + \beta \cdot W_2, M_o) \leq \alpha \cdot SL(W_1, M_o) + \beta \cdot SL(W_2, M_o)$$

The same properties also holds for *ISL*.   ◆

Observe that for $\alpha = \beta = 1$, $W_1 = W_i - W_j$ and $W_2 = W_j - W_k$, the property 2.5.2 express a sort of *triangle inequality*.

Properties 2.4. For any $W_{ij} \in Z^m$ $(W_{ji} = -W_{ij})$:

1) Non-negativity: $SL(W_{ij}) \geq ISL(W_{ij}) \geq 0$.

2) Non-symmetry: $\exists <N,M_o>$ such that $ISL(W_{ij}) \neq ISL(W_{ji})$ or $SL(W_{ij}) \neq SL(W_{ji})$.

3) $SL(W_{ij}) \leq ISL(W_{ij}) + ISL(W_{ji})$.

4) If $ISL(W_{ji}) = 0$, then $SL(W_{ij}) = ISL(W_{ij})$.

5) If $(T_i,T_j) \in SLS$ for W_{ij}, then $(T_i,T_j) \in LS$ for W_{ij}.   ◆

It may be interesting to prove Property 2.4.3. Let $\sigma = \alpha \cdot \beta \in L(N,M_o)$ be a sequence such that $SL(W_{ij}) = W_{ij}^T \cdot \overline{\beta}$. By definition, for $\sigma_1 = \alpha$ and $\sigma_2 = \alpha \cdot \beta$ we can write: $ISL(W_{ij}) \geq W_{ij}^T \cdot (\overline{\alpha} + \overline{\beta})$ and $ISL(W_{ji}) \geq -W_{ij}^T \cdot \overline{\alpha}$. By adding the two initial synchronic leads: $ISL(W_{ij}) + ISL(W_{ji}) \geq W_{ij}^T \cdot \overline{\beta} = SL(W_{ij})$. Moreover, it can be proved that for reversible nets the equality holds [i.e. $ISL(W_{ij}) + ISL(W_{ji}) = SL(W_{ij})$].

Examples 2.3. Let us consider the net in:

a) Fig. 1.4: $ISL(t_1,t_2) = SL(t_1,t_2) = \infty$

b) Fig. 1.4: $ISL(t_1,2t_2) = 1$; $SL(t_1,2t_2) = 2$, from $M(p_1) = M(p_4) = 1$.

c) Fig. 1.3: $ISL(t_1,2t_4) = 0$; $ISL(2t_4,t_1) = \infty$, $SL(t_1,2t_4) = 2$; $SL(2t_4,t_1) = \infty$.

d) Fig. 1.3: $ISL(t_1,t_4) = \infty$; $ISL(t_4,t_1) = 1$; $SL(t_1,t_4) = \infty$ $SL(t_4,t_1) = 1$.

Definitions 2.4 SYNCHRONIC DISTANCE CONCEPTS

1) The *initial synchronic distance* of W_i with respect to W_j (W_{ij}-*initial synchronic distance*, for short) in $<N,M_o>$ is: $ISD(W_{ij}) = max\{ISL(W_{ij}), ISL(W_{ji})\}$.

2) The *synchronic distance* of W_i with respect to W_j (W_{ij}-*synchronic distance*, for short) in $<N,M_o>$ is:

$$SD(W_{ij}) = max \{SL(W_{ij}), SL (W_{ji})\}.$$

3) T_i and T_j are in *distance synchronization relation* in $<N,M_o>$, $(T_i,T_j) \in DS$, iff $\exists W_i \neq W_j \in N^m$ and $\exists k \in N$ such that $SD(W_{ij}) \leq k$.

4) T_i and T_j are in *structural distance synchronization relation* in N, $(T_i,T_j) \in SDS$, iff they are in *DS*-relation for any initial marking.

Properties 2.5. For any $W_{ij} \in Z^m$:

1) $0 \leq ISD(W_{ij}) \leq SD(W_{ij}) \leq ISL(W_{ij}) + ISL(W_{ji})$.

2) If $<N,M_o>$ is *reversible* : $SD(W_{ij}) = ISL(W_{ij}) + ISL(W_{ji})$.

3) If $(T_i , T_j) \in SDS$ for W_{ij}, then $(T_i ,T_j) \in DS$ for W_{ij}. ◆

Examples 2.4. All the nets in figure 1 are reversible, then:

a) Fig. 1.1: $ISL(t_1,t_3) = 1$, $ISL(t_3,t_1) = 0 \Rightarrow SD(t_1,t_3) = 1$.

b) Fig. 1.3: $ISL(t_3,t_4) = 1$, $ISL(t_4,t_3) = \infty \Rightarrow SD(t_3,t_4) = \infty$.

c) Fig. 1.4: $ISL(t_1,2t_2) = 1$, $ISL(2t_2,t_1) = 1 \Rightarrow SD(t_1,2t_2) = 2$.

For the net in figure 1.3 there does not exist *DS* -relation between t_1 and t_4 .

Properties 2.6. If any transition is firable at least once in $<N,M_o>$, *SD* is a *distance* :

1) $SD(W,M_o) \geq 0$; $SD(W,M_o) = 0 \Leftrightarrow W = 0$ [$W = W_1-W_2 = 0 \Rightarrow \|T_1\| = \|T_2\|$].

2) Symmetry: $SD(W,M_o) = SD(-W,M_o)$.

3) Triangle inequality: $SD(W_i - W_j, M_o) \leq SD(W_i - W_k, M_o) + SD(W_k - W_j, M_o)$.

The same property holds for ISD and SSD .   ♦

3. ALGEBRAIC CHARACTERIZATION OF SYNCHRONIC RELATIONS

In this section we are not interested in the precise value of the different synchronic functions, but in their boundedness. For each synchronic function, there exists a *synchronic dependence* between two subsets of transitions iff the synchronic function is bounded. In other words, only *qualitative* (and not *quantitative*) aspects will be considered here (by means of synchronic relations). The results to be presented allow a very precise algebraic characterization of the behavioural and structural relations.

3.1. Structural synchronic relations

The first result presented in this section shows that the transformation of a structurally marking unbounded net into a structurally marking bounded one preserves the structural synchronic relations. A theorem on the characteristic vectors of infinite sequences in structurally marking bounded nets appears below. Later, necessary and sufficient conditions are given to characterize the four basic structural synchronic relations, {SBD, SBF, SLS, SDS } . Theoretically the algorithms we can deduce are of exponential complexity. Some corollaries of the basic theorems allow polynomial computations for net subclasses or general nets with particular properties (e.g. conservativeness and consistency).

LEMMA 3.1.[SILV 86]. For any structurally marking unbounded net there exist finite initial markings and associated firing sequences such that every structurally marking unbounded place will contain an arbitrarily large number of tokens.   ♦

This lemma can be easily proved by taking an M_o large enough. For denotational convenience, the next theorem is numbered 3.0.

THEOREM 3.0. The removal of structurally marking unbounded places, together with their incident arcs, from a P/T net preserves all the structural synchronic relations: {SBD, SBF,SLS,SDS} .   ♦

Proof. The places of a net are restrictions on the firing possibilities of their output transitions. Therefore, if N^* is a net obtained by removing some places and their incident arcs: $L(N,M_o) \subseteq L(N^*,M_o^*)$, and any synchronic firing relation for N^* implies the corresponding one for N (e.g. SBD^* $\Rightarrow SBD$). By considering Lemma 3.1, it is easy to conclude that the structurally unbounded places cannot represent any restriction on the firing of their output transitions, for arbitrarily long firing sequences. Then structural synchronic relations in the original net, N , and in the net obtained by removing its structurally marking unbounded places coincide.   ♦

The above theorem allows degenerate nets without places to be obtained. Clearly, this means that transitions are firable in a completely independent way (i.e. there is no synchronic relation!).

Theorem 3.0 permits the reduction of a structurally marking unbounded net to a given structurally marking bounded one. By this we can concentrate the algebraic characterization study on the structurally marking bounded nets. In the rest of this section let N be a structurally marking bounded net.

THEOREM 3.1. N is structurally marking bounded iff for any initial marking, M_O, and for any sequence or subsequence, σ, applicable in $<N,M_O>$ we can write:

$$\overline{\sigma} = \Sigma_i \lambda_i.R_i + \mu$$

where: * R_i is a reproduction vector (i.e. an elementary consistent component)

   * $\lambda_i \in Q^+$

   * $\mu \geq 0$ is a bounded vector such that $C.\mu \ngeq 0$ is impossible.   ◆

Proof. It follows from the structural marking boundedness hypothesis :

($\Leftarrow$) Premultiplying σ by C ($C \cdot \overline{\sigma}=M-M_O$ and $C \cdot R_i = 0$) :

$$C \cdot \overline{\sigma} = M\text{-}M_O = \Sigma_i \lambda_i \cdot C \cdot R_i + C \cdot \mu = C \cdot \mu$$

As is known [BRAU 80], a P/T net is structurally marking bounded iff $\nexists \mu \geq 0$, such that $C \cdot \mu \ngeq 0$.

($\Rightarrow$) Let $<N,M_O>$ be a k-bounded net (i.e. $M(p) \leq k$ $\forall p \in P$). Then the number of reachable markings is less than $(k+1)^n$. If σ is "infinite", it can be decomposed as: $\overline{\sigma} = R^* + \mu$, where R^* is a consistent component ($C.R^* = 0$, $R^* \geq 0$) and $\mu \geq 0$ is such that $\Sigma_j \mu(j) < m(k+1)^n$ (because this is a bound of the number of edges in the reachability graph, then of the length of any cycle). Therefore $\mu \geq 0$ is a bounded vector such that $C \cdot \mu \ngeq 0$ is impossible.   ◆

Because in the sequel we will be interested only in boundedness properties, the existence of μ can be disregarded.

Let :   * $R = (R_1 R_2 ... R_q)$ be the matrix in which the reproduction vectors are the columns. Matrix R can have an exponential number of columns [MART 82]. It can be partitioned into row vectors as follows:

$$R = \begin{pmatrix} r_1 \\ r_2 \\ ... \\ r_m \end{pmatrix} , \quad \text{where } m = |T|$$

   * $\theta = \{T_i \ |T_i \subset T \}$, be a set of transitions subsets .

$*$ $T_i \in \theta$. The vector r^i expresses how many times the transitions belonging to T_i are fired in each reproduction vector. For example, if $T_i = \{t_1,t_2\}$, $r^i = r_1 + r_2$

THEOREM 3.2. STRUCTURAL BOUNDED-DEVIATION

Let N be a structurally marking bounded net. $(T_i , T_j) \in SBD \Leftrightarrow \|r^i\| \subseteq \|r^j\|$. ◆

Proof. According to theorem 3.1, the "infinite" part of any characteristic vector σ can be expressed as: $\overline{\sigma}^* = \Sigma_k \, \lambda_k \cdot R_k$. But for M_o large enough any R_k can be executed infinitely often without executing any other R_i.

Now it is easy to observe that $\|r^i\| \subseteq \|r^j\|$ is equivalent to $\forall R_k \in R$ $r^i(k) > 0 \Rightarrow r^j(k) > 0$. Then the number of firings of transitions in T_i is bounded between two firings of transitions from T_j (sufficiency). Let us suppose now that $\exists R_d$ such that $r^i(d) > 0$ and $r^j(d) = 0$. Then the transitions of T_i can be fired infinitely often in R_d without firing transitions of T_j (necessity). ◆

THEOREM 3.3. STRUCTURAL BOUNDED-FAIRNESS

Let N be a structurally marking bounded net. $(T_i , T_j) \in SBF \Leftrightarrow \|r^i\| = \|r^j\|$. ◆

Proof. $(T_i , T_j) \in SBF \Leftrightarrow (T_i,T_j) \in SBD$ and $(T_j,T_i) \in SBD$. Applying theorem 3.2, $\|r^i\| = \|r^j\|$ is a necessary and sufficient condition. ◆

Examples 3.1. Places p_2 and p_4 are structurally marking unbounded in the net in figure 2.1. By removing it, the net in figure 2.2 is obtained. Trivially we can write for N^*:

$$R^* = \begin{pmatrix} 1 & 1 & 0 \\ 1 & 0 & 0 \\ 0 & 0 & 1 \\ 0 & 1 & 0 \end{pmatrix} \begin{matrix} r_1 \\ r_2 \\ r_3 \\ r_4 \end{matrix}$$

Considering $\theta = \{t_1,t_2,t_3,t_4\}$ (then $r^i = r_i$), there is no couple of transitions in structural bounded fairness relation. If $\theta = \{t_1,\{t_2,t_4\},t_3\}$, then $(t_1,\{t_2,t_4\}) \in SBF$. ◆

Let us now consider the linearly-based synchronic relations. "A priori" it is expected that their characterization represent particular cases of the above results.

THEOREM 3.4 . STRUCTURAL LEAD-SYNCHRONIZATION

Let N be a structurally marking bounded net. T_i is in structural LS-relation with respect to T_j in N, $(T_i , T_j) \in SLS$, iff there exists weights W_i , $W_j \in N^m$, such that $(W_i - W_j)^T \cdot R \le 0$, where $\|W_i\| = T_i$ and $\|W_j\| = T_j$.

448

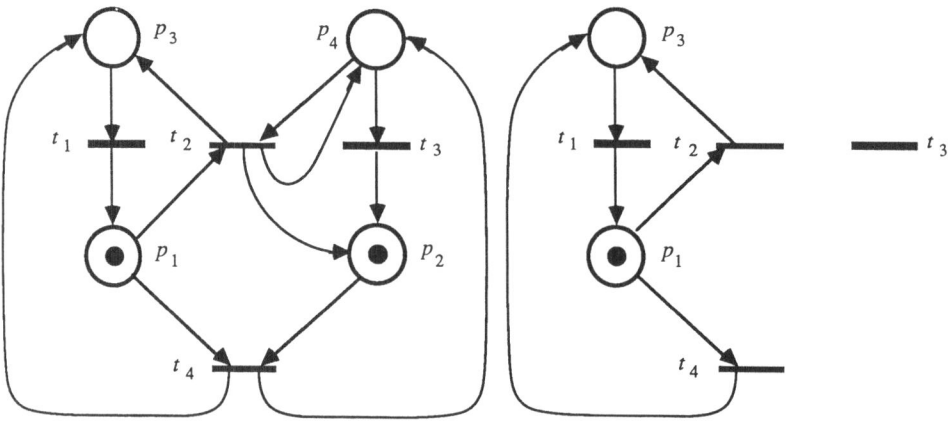

(1) Original marked net, $< N,M_o >$.

(2) Net obtained by removing the structural unbounded places, N^*.

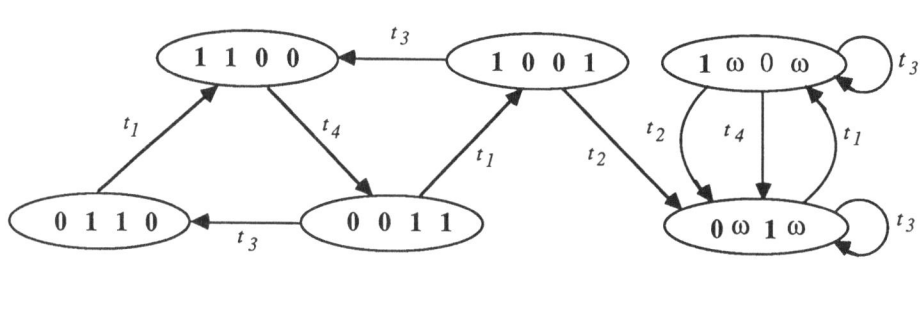

(3) $CG(N,M_o)$

Figure 2. Structurally (marking) unbounded net and the computation of synchronic relations.

Proof. If N is structurally marking bounded, according to theorem 3.1 we can write for any sequence $\sigma : \overline{\sigma} = \Sigma_i \lambda_i R_i + \mu$. Substituting σ in the $SL(W)$ definition ($\forall M_o$, $W_{ij} = W_i - W_j$):

$$SL(W_{ij}) = sup_{\sigma} (W_{ij}^T.\overline{\sigma}) = sup_{\sigma} (\Sigma_i \lambda_i W_{ij}^T R_i + W_{ij}^T \cdot \mu).$$

Then, $SL(W) = \infty \Leftrightarrow \exists R_k$ such that $W_{ij}^T \cdot R_k > 0$. Therefore $(T_i, T_j) \in SLS \Leftrightarrow W_{ij}^T \cdot R \leq 0$. ♦

The next corollary introduces a result that at a first glance may be surprising, because synchronic lead has been defined as a linearly-based property related to the deviation-bound concept.

COROLLARY 3.1. Structural lead-synchronization and structural bounded deviation are *the same relation* :

$$(T_i, T_j) \in SLS \Leftrightarrow (T_i, T_j) \in SBD$$ ♦

Proof. ($\Rightarrow$) $\exists W_i, W_j \in N^m$, $\|W_i\| = T_i$ and $\|W_j\| = T_j$ such that:

$$(W_i - W_j)^T \cdot R \leq 0 \Leftrightarrow W_i^T \cdot R \leq W_j^T \cdot R \Rightarrow \|r^i\| \subseteq \|r^j\|.$$

($\Leftarrow$) Let us see how a W_i and W_j can be constructed such that if $\|r^i\| \subseteq \|r^j\| \Rightarrow (W_i - W_j)^T \cdot R \leq 0$.

Let: $* W_i(f) := $ **if** $t_f \in T_i$ **then** 1 **else** 0; $V_i = R^T \cdot W_i$

$* W_j^o(g) := $ **if** $t_g \in T_j$ **then** 1 **else** 0; $V_j = R^T \cdot W_j^o$

$V_i(k)$

$*$ **If** $V_j \neq 0$ **then** $u = max_k$ ---------- **else** $V_i = 0$ {i.e. $\|r^j\| = \|r^i\| = \emptyset$ } and any finite

$V_j(k)$ u can be used

If $\|r^i\| \subseteq \|r^j\|$, then $V_i \leq u \cdot V_j \Leftrightarrow W_i^T \cdot R \leq u \cdot (W_j^o)^T \cdot R$.

By defining $W_j^T = u \cdot (W_j^o)^T$, two weights, W_i and W_j, have been constructed such that $(W_i - W_j)^T \cdot R \leq 0$; i.e. T_i and T_j are in SL-relation. ♦

Example 3.2. For the net in figure 1.3, we can write:

$$R = \begin{pmatrix} 1 & 2 \\ 0 & 1 \\ 1 & 0 \\ 1 & 1 \end{pmatrix} \begin{matrix} r_1 \\ r_2 \\ r_3 \\ r_4 \end{matrix} \quad \text{If } W_1 = \begin{pmatrix} 1 \\ 0 \\ 0 \\ 0 \end{pmatrix}, \quad W_4 = \begin{pmatrix} 0 \\ 0 \\ 0 \\ 1 \end{pmatrix} \text{ then we have: } V_1 = \begin{pmatrix} 1 \\ 0 \\ 1 \end{pmatrix} \text{ and } V_4 = \begin{pmatrix} 1 \\ 1 \\ 2 \end{pmatrix}$$

Then, according to the approach outlined in the above proof:

$$W_{14} = W_1 - 2W_4 \text{ is such that } W_{14}^T \cdot R \leq 0 \Rightarrow (t_1, t_4) \in SLS$$

$$W_{41} = W_4 - W_1 \text{ is such that } W_{41}^T \cdot R \leq 0 \Rightarrow (t_4, t_1) \in SLS$$

Obviously, $(t_1,t_4) \in SBF$ and $(t_4,t_1) \in SBF$.

The above result, corollary 3.1, can be easily understood because the *SLS*-relation does not consider "a priori" any set of weights vectors, $W = \{W_i \,/\, W_i \in N^m \text{ and } //W_i// = T_i \}$.

THEOREM 3.5. STRUCTURAL DISTANCE-SYNCHRONIZATION

Let N be a structurally marking bounded net. T_i and T_j are in structural *DS*-relation in N,$(T_i,T_j) \in SDS$, iff there exist weights W_i ,$W_j \in N^m$ such that $(W_i - W_j)^T \cdot R = 0$, where $\| W_i \| = T_i$ and $\|W_j\| = T_j$.   ◆

Proof. Because $(T_i ,T_j) \in SDS$ for $W_{ij} \Leftrightarrow (T_i ,T_j) \in SLS$ for W_{ij} and $(T_j ,T_i) \in SLS$ for W_{ij} , according to theorem 3.4 it is necessary and sufficient to have $(W_i - W_j)^T \cdot R \le 0$ and $(W_j-W_i)^T \cdot R \le 0$. Therefore $(W_i - W_j)^T \cdot R = 0$.   ◆

Clearly the $W_{ij} = W_i - W_j$ vectors that make the *SD* bounded for any initial marking form a *linear space*. A basis of this can be computed easily by using standard linear algebra on matrix R. In particular, if the net is conservative and consistent a basis of the linear space that makes the synchronic distance bounded can be formed by maximal subsets of *linearly independent* rows asociated with places in the token flow matrix C.

Example 3.3. All the nets in figure 1 are conservative and consistent. For nets 1,3 and 4, the rows associated with p_1 and p_3 form a basis of their W -spaces.   ◆

The next corollary points out an interesting result.

COROLLARY 3.2. $(T_i ,T_j) \in SDS \Rightarrow (T_i ,T_j) \in SBF$. The converse is not true.   ◆

Proof. $(W_i - W_j)^T \cdot R = 0 \Leftrightarrow W_i^T \cdot R = W_j^T \cdot R \Rightarrow \|r^i \| = \|r^j \|$.

The converse is not true because we can have support equality without vector equality.   ◆

Example 3.4. For the net in figure 1.3, we can write: $(t_1,t_4) \in SBF$ but $(t_1,t_4) \notin SDS$.

As was mentioned before, the computation of R is exponential [MART 82]. In the following interesting particular cases the conclusion on structural distance synchronization relation can be drawn in polynominal time and space.

Let $B = (B_1 B_2 ... B_n)$ be a *basis* of right annullers of C $(C \cdot B_i = 0)$. The computation of B is a classical polynomial complexity problem. Matrix B can be partitioned into row vectors as follows:

$$B = \begin{pmatrix} b_1 \\ b_2 \\ ... \\ b_m \end{pmatrix}, \quad \text{where } m = /T/$$

451

Theorem 3.6. Let N be a structurally marking bounded net and the following couples of structural synchronic property and algebraic condition:

$$(T_i, T_j) \in SLS \qquad \exists W_i, W_j \in N^m \quad \text{such that } (W_i\text{-}W_j)^T \cdot B \leq 0,$$

$$\text{where: } \|W_i\| = T_i \text{ and } \|W_j\| = T_j.$$

$$(T_i, T_j) \in SDS \qquad \exists W_i, W_j \in N^m \quad \text{such that } (W_i\text{-}W_j)^T \cdot B = 0,$$

$$\text{where: } \|W_i\| = T_i \text{ and } \|W_j\| = T_j.$$

For each structural relation, the corresponding algebraic condition is:

1) a sufficient condition.

2) a necessary and sufficient condition if the net N is also consistent. ◆

Proof. a) $(W_i\text{-}W_j)^T \cdot B \leq 0 \Rightarrow (W_i\text{-}W_j)^T \cdot R \leq 0$ because $\forall R_i \in R$ there exists a real vector γ_i such that $R_i = B \cdot \gamma_i$. Then (1) follows from theorems 3.4 and 3.5.

b) If N is consistent $(W_i\text{-}W_j)^T \cdot R \leq 0 \Rightarrow (W_i\text{-}W_j)^T \cdot B \leq 0$, because R contains a basis of the right-annullers space of C. Then (2) holds obviously from "a" and "b". ◆

Do not forget that for consistent nets, structural marking boundedness and conservativeness coincide. The computation of consistency and conservativeness is a theoretically polynomial complexity problem (although the "usually" more efficient computation schema is based on the *simplex method!*).

Theorem 3.7. Let N be a structurally marking bounded net. If there exists a $\lambda \neq 0$ such that $W_i^T \cdot B = \lambda \cdot W_j^T \cdot B$, then T_i and T_j are in structural distance-synchronization relation and in structural bounded fairness relation. ◆

Proof. $W_i^T \cdot B = \lambda \cdot W_j^T \cdot B \Rightarrow W_i^T \cdot R = \lambda \cdot W_j^T \cdot R$. By taking $W_i = \lambda \cdot W_j$, $(W_i - W_j)^T \cdot R = 0$ and T_i and T_j are in structural distance synchronization relation for W_i -W_j. By applying corollary 3.2, T_i and T_j are also in structural bounded fairness relation. ◆

Several corollaries can be deduced from this theorem; these state sufficient conditions which are fast to compute. Among them we can note the following one.

Corollary 3.3. If B is a 0-1 defined matrix, then $b_i = b_j \Leftrightarrow (t_i, t_j) \in SDS \Rightarrow (t_i, t_j) \in SBF$ ◆

Structural synchronic relations have been studied for structurally marking unbounded nets by transforming them into others which are structurally marking bounded. Let us end this section with a very compact theorem that for any net expresses necessary conditions to verify the corresponding structural synchronic relations.

The *elementary repetitive components* [BRAU 80] [SILV 86] of a net N with flow matrix C are the minimal support vectors $A_i \in N^m$ such that $C \cdot A_i \geq 0$.

THEOREM 3.8.

1) Only *some* sequences σ of N satisfy:

$$\bar{\sigma} = \Sigma_i \lambda_i . A_i + \mu$$

where: $\lambda_i \in Q^+$ and $\mu \geq 0$ is a bounded vector.

2) If $(T_i, T_j) \in SBD \Rightarrow \|a^i\| \subseteq \|a^j\|$

3) If $(T_i, T_j) \in SBF \Rightarrow \|a^i\| = \|a^j\|$

4) If $(T_i, T_j) \in SLS \Rightarrow \exists W_i, W_j$ such that $(W_i - W_j)^T . A \leq 0$.

5) If $(T_i, T_j) \in SDS \Rightarrow \exists W_i, W_j$ such that $(W_i, W_j)^T . A = 0$.

In cases 4 and 5: $\|W_i\| = T_i$ and $\|W_j\| = T_j$. ◆

The proof of this theorem is now straightforward. *Non-sufficiency* is rooted in theorem 3.8.1, because not all sequences satisfy the expressed condition. This is easy to check: if we have $M(p) = \omega$ (arbitrarily big) and there exists $t \in T$ such that p is its unique input place and p is not an output place of t (Fig. 2.1, transition t_3), in this case $\sigma_o = t$ can be fired consecutively ω times, and $\sigma_o \notin A$. On the other hand it is interesting to realize the connections between theorems 3.i and theorems 3.8.i (i = 1,5). Theorems 3.8.4 and 3.8.5 where introduced in [SIFA 79], while theorem 3.8.3 was introduced in [SILV 86].

3.2 Behavioural synchronic relations

The results to be presented in this section are somewhat analogous to those introduced in §3.1. In this case our interest will be focused on the synchronic relations in a marked net, $<N, M_o>$.

Let us consider directly the case where $<N, M_o>$ is not necessarily bounded. $CG\ (N, M_o)$ denotes its *coverability graph* (i.e. the graph in which each node explicitly represents a reachable marking or a covering of a set of reachable markings). The coverability graph of a net $<N, M_o>$ (see, for example, [BRAU 80], [REIS 85]) can be obtained from the *coverability tree* (they are considered, for example, in [PETE 81] [SUZU 83]) by merging all the nodes having the same label (markings). Figure 2.3 represents the CG of the net in Fig.2.1. If the net $<N, M_o>$ is bounded, the coverability graph becomes a simple *reachability graph* .

An *elementary cycle* C_i of $CG\ (N, M_o)$ is a *directed circuit* and the characteristic vector of the represented sequences will be denoted by D_i .

In the CG of Fig. 2.3 we have elementary cycles with:

$$D_1 = (1\ 0\ 1\ 1)^T \qquad\qquad D_3 = (1\ 0\ 0\ 1)^T$$

$$D_2 = (0\ 0\ 1\ 0)^T \qquad\qquad D_4 = (1\ 1\ 0\ 0)^T$$

LEMMA 3.2. [SILV 86]. Let $k \in N$. Given an elementary cycle C_i in CG, there always exists a reachable marking M_u, $M_o\ [\sigma_o > M_u$, from which the firing of the transitions labeling arcs in C_i can be repeated k times. ◆

Let us consider the net in Fig. 2.1 and $b \geq k$. Because $M = (0\ \omega\ 1\ \omega)^T$ appears in the CG (Fig. 2.3), the marking $M_u = (0\ a\ 1\ b)^T$, is reachable. Therefore, t_3 can be fired k-times.

Let $D = (D_1\ D_2\ ...\ D_q)$ be the matrix in which the characteristic vectors of the elementary cycles of $CG(N,M_o)$ are the columns. D can be partitioned into row vectors as follows:

$$D = \begin{pmatrix} d_1 \\ d_2 \\ ... \\ d_m \end{pmatrix}, \qquad \text{where } m = |T|$$

Unfortunately, the computation of D is doubly exponential because:

1) The computation of $CG(N,M_o)$ is exponential [BRAU 80].

2) The computation of the elementary cycles of a graph is also exponential.

Given this theoretical complexity, the structural analysis appears doubly interesting because:

1) Conclusions hold for any initial marking.

2) Computations are theoretically much less complex.

The limitations of structural analysis techniques reside, basically, in the fact that the state equation only partially represents the activity of the net ($\overline{\sigma}$ does not represent σ) and partially represents its structure (recall self-loops).

Now let θ be a set of transitions subsets. The vector d^i expresses how many times the transitions belonging to T_i are fired in each elementary sequence.

THEOREM 3.9. BEHAVIOURAL SYNCHRONIC RELATIONS

1) For *any* sequence or subsequence σ applicable in $<N,M_o>$ we can write:

$$\overline{\sigma} = \Sigma_i\ \lambda_i \cdot D_i + \mu$$

where: * D_i is the ith characteristic vector of the elementary cyclic sequences.

454

* $\lambda_i \in Q^+$

* $\mu \geq 0$ is a bounded vector.

2) $(T_i ,T_j) \in BD \quad \Leftrightarrow \quad \| d^i \| \subseteq \| d^j \|$

3) $(T_i ,T_j) \in BF \quad \Leftrightarrow \| d^i \| = \| d^j \|$

4) $(T_i ,T_j) \in LS \quad \Leftrightarrow \exists W_i ,W_j$ such that $(W_i - W_j)^T \cdot D \leq 0$.

5) $(T_i ,T_j) \in DS \quad \Leftrightarrow \exists W_i ,W_j$ such that $(W_i - W_j)^T \cdot D = 0$.

In cases 4 and 5 we suppose: $\| W_i \| = T_i$ and $\| W_j \| = T_j$.   ◆

The proof of theorem 3.9.i is absolutely analogous to that of theorem 3.i. Thus they are omitted. Observe that for unbounded nets, theorem 3.9 directly gives a necessary and sufficient condition (theorem 3.8 only gives necessary structural conditions). Theorem 3.9.5 was previously presented in [SUZU 83]. The W_{ij} vectors that makes the *ISD* and the *SD* finite form a *linear space*.

COROLLARY 3.4. Let us consider any marked P/T net $<N,M_o>$.

1) The bounded deviation relation and the synchronic lead relation are the *same*.

2) $(T_i ,T_j) \in SD \Rightarrow (T_i ,T_j) \in BF$. The converse is not true.   ◆

The strong analogy between corollaries 3.i and 3.4.i is obvious.

4. SYNCHRONIC RELATIONS BASED STRUCTURE

In this section we are interested in the study of the infinite synchronic structure of P/T nets. These structures can be based on the behavioural or on the structural synchronic relations introduced in §2.

It can be shown that the elements of θ are *classified* according to the bounded fairness equivalence relations, while the quotient set appears to be *partially ordered* by the deviation bound relations. In turn, the elements of each bounded fairness equivalence class are classified according to the distance synchronization equivalence relations which are more restrictive. All this determines the behavioural and the structural synchronic relations based structure of a P/T net.

4.1 Basic structure

It is not difficult to directly prove that the eight basic synchronic relations (four behavioural and four structural) are *reflexive* and *transitive* (see [SILV 86] for bounded fairness relations). Bounded fairness and distance synchronization are also *symmetric* , and therefore *equivalence* relations. Instead of producing direct proofs of this, the identification of the different relations will be done as corollaries of the main theorems (§3).

Let $\theta/\equiv$ be the *quotient* set (θ modulo $\equiv$) defined by the equivalence relation $\equiv$ on the set θ.

THEOREM 4.1.

1) Let θ be a set of transitions subsets of T. The BF and SBF relations are *equivalence* relations.

2) Let W be a set of weight vectors associated to T. The DS and SDS relations are *equivalence* relations.

3) If W is consistent with θ (i.e. $T_i = ||W_i||$), the distance synchronization relations refine the classification of the bounded fairness relations. In other words, each element of θ/BF (or of θ/SBF) is a disjoint subset of elements of θ/DS (or of θ/SDS).

Proof. 1) Support equality is an equivalence relations, then the equivalence of: (a) SBF follows from theorems 3.0 and 3.3; (b) BF follows from theorem 3.9.3.

2) Vector equality is an equivalence relation, then the equivalence of: (a) SDS follows from theorems 3.0 and 3.5 ($W_i^T \cdot R = W_j^T \cdot R$); (b) DS follows from theorem 3.9.5 ($W_i^T \cdot D = W_j^T \cdot D$).

3) Vector equality implies support equality (see Corollary 3.2). [In fact theorem 4.1.3 is a restatement of corollary 3.2, given that both distance synchronization and bounded fairness are equivalence relations]. ♦

THEOREM 4.2. Let θ be a set of transition subsets of T. The following quotient sets are *partially ordered*:

1) θ/SBF by the SBD relation (i.e. $<\theta/SBF$, $SBD>$ is a *poset*).

2) θ/BF by the BD relation (i.e. $<\theta/BF$, $BD>$ is a *poset*).

Proof. Clearly in the quotient set θ/SBF there is no pair of elements with equal support. According to theorems 3.0 and 3.2, the SBD relation acts as the usual strict inclusion relation between the elements (subsets of transitions) of θ/SBF. Then $<\theta/SBF$, $SBD>$ is a poset.

The same reasoning holds for the behavioural relations by using theorem 3.9.2. ♦

Example 4.1. Consider the net in figure 1.3. For the initial marking we have:

$$R = D = \begin{pmatrix} 1 & 2 \\ 0 & 1 \\ 1 & 0 \\ 1 & 1 \end{pmatrix} \begin{matrix} t_1 \\ t_2 \\ t_3 \\ t_4 \end{matrix} \quad \text{[i.e. behavioural and structural relations coincide in this case]}$$

Figure 3 despicts the synchronic structure for the case $T_i = \{t_i\}$. In the next case all the subsets of transitions are in fairness relations:

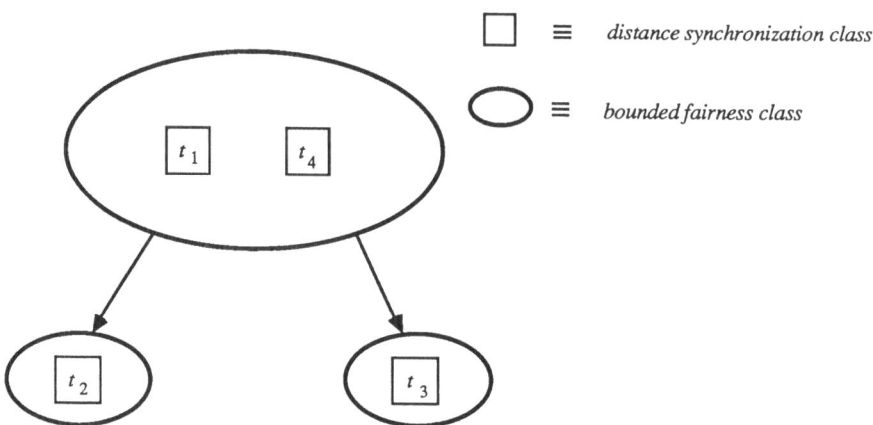

Figure 3. Synchronic relations based structure of the net in figure 1.3

(1) θ/SBF is a partially ordered set (poset) for the relation $SBD \equiv SLS$

(2) The elements of θ, θ_i, can be classified according to any of the following two equivalence relations: {SBF, SDS}.

$(T_i,T_j) \in SDS$	$\Rightarrow$	$(T_i,T_j) \in SLS$	and	$(T_j,T_i) \in SLS$
$\Downarrow$		$\Updownarrow$		$\Updownarrow$
$(T_i,T_j) \in SBF$	$\Leftrightarrow$	$(T_i,T_j) \in SBD$	and	$(T_j,T_i) \in SBD$

Table 1. Basic relations between structural synchronic relations. (Note: $(T_i,T_j) \in SDS$ for $W_{ij} \Leftrightarrow (T_i,T_j) \in SLS$ for W_{ij} and $(T_j,T_i) \in SLS$ for W_{ij}).

$$T_1 = t_1 \qquad\qquad T_5 = (t_1, t_4)$$

$$T_2 = t_4 \qquad\qquad T_6 = (t_2, t_3)$$

$$T_3 = (t_1, t_2) \qquad T_7 = (t_2, t_4)$$

$$T_4 = (t_1, t_3) \qquad T_8 = (t_3, t_4)$$

But T_1 and T_2 can never be in synchronic distance relation. If $W_4 = (1\ 0\ 1\ 0)^T$, T_2 and T_4 are in distance synchronization relation only if $W_2 = (0\ 0\ 0\ 2)^T$.

4.2 On net level synchronizations

In this section some global synchronic behaviours or structures are considered. Basically they concern the maximal elements in the posets or the particular cases in which the quotient sets have a unique element.

Definitions 4.1. Let us consider a set of transitions subsets: $\theta = \{T_i \,/\, T_i \subset T\}$.

1) $<N, M_o>$ is *bounded-fair* (or *BF-synchronized*) for θ iff $(T_i, T_j) \in BF$, $\forall T_i, T_j \in \theta$.

2) The net structure N is *SBF-synchronized* for θ iff $(T_i, T_j) \in SBF$, $\forall T_i, T_j \in \theta$.   ◆

Properties 4.1.

1) $<N, M_o>$ is bounded fair for θ iff the quotient set for the BF-relation, θ/BF, has only one element.

2) N is structurally bounded fair for θ iff the quotient set for the SBF-relation, θ/SBF, has only one element.

3) If N is structurally bounded fair for θ, for all M_o, $<N, M_o>$ is bounded fair for θ.   ◆

Examples 4.2.

a) Fig. 1.1: For $\theta = \{t_i \,/\, i \in 1,2,3,4\}$ the net is not bounded-fair.

b) Fig. 1.1: For $T_1 = \{t_1, t_2, t_3\}$ and $T_2 = \{t_4\}$, the net is bounded-fair and structurally bounded fair.

c) Fig. 1.2: For $T_1 = \{t_1, t_3, t_5\}$ and $T_2 = \{t_2, t_4, t_6\}$ the net is bounded fair but not structurally bounded fair.

d) Fig. 1.2: For $T_1 = \{t_1, t_2\}$, $T_2 = \{t_3, t_4\}$ and $T_3 = \{t_5, t_6\}$ the net is structurally bounded fair.

e) Fig. 1.3: For $T_1 = \{t_1\}$, $T_2 = \{t_2, t_3\}$ and $T_3 = \{t_4\}$, the net is structurally bounded-fair.

Definitions 4.2. Let us consider a set of transitions subsets, $\theta = \{T_i \,/\, T_i \subset T\}$.

1) $<N, M_o>$ is a *BD-synchronized net* with respect to T_j for θ iff $(T_i, T_j) \in BD$, $\forall T_i \in \theta$.

2) The net structure N is *SBD-synchronized* with respect to T_j for θ iff $(T_i, T_j) \in SBD$, $\forall T_i \in \theta$.   ◆

Properties 4.2.

1) $<N,M_o>$ is a BD -synchronized net with respect to T_i in θ/BF iff T_i represents the maximal element for the BD -partial order relation.

2) N is an SBD -synchronized net with respect to T_i in θ/SBF iff T_i represents the maximal element for the SBD partial order relation.

3) If N is an SBD -synchronized net with respect to T_i in θ/SBF, $<N,M_o>$ is a BD-synchronized net with respect to T_i in θ/BF , for any initial marking, M_o . ◆

Examples 4.3. The marked net in figure 1.2 is:

a) BD -synchronized with respect any $t_i \in T$.

b) not SBD -synchronized with respect any $t_i \in T$.

c) SBD -synchronized with respect any $T_i = \{t_i, t_{i.mod.6+1}\}$.

Remark 4.1. Promptness [THIA 84] can be introduced as a particular case of BD -synchronization: A net is *prompt* with respect to T_e iff $(T\text{-}T_e, T_e) \in BD$. In other words, it is a BD -synchronized net with respect to T_e in $\theta = \{T_e, T\text{-}T_e\}$. The concept of structural promptness can be introduced in the obvious way.

Definitions 4.3. Let us consider a set of transitions subsets, $\theta = \{T_i / T_i \subset T\}$, where $W = \{W_i / W_i \in N^m, //W_i// = T_i$ and $W_{ij} = W_i - W_j \}$.

1) $<N,M_o>$ is SD-synchronized for θ iff: $(T_i, T_j) \in DS,$ $\forall T_i, T_j \in \theta$.

2) N is *structurally SD-synchronized for* θ iff: $(T_i, T_j) \in SDS,$ $\forall T_i, T_j \in \theta.$ ◆

Properties 4.3.

1) $<N,M_o>$ is SD -synchronized for θ iff $\exists W$ such that $\forall W_i \in W$, $W_i^T \cdot D$ = constant (i.e. the quotient set has an unique element).

2) N is SSD -synchronized for θ iff $\exists W$ such that $\forall W_i \in W, W_i^T \cdot R$ = constant (i.e. the quotient set has an unique element). ◆

Examples 4.4. Consider the net in figure 1.3 (see example 4.1). It is SD -synchronized (and SSD - synchronized) for $\theta = \{\{t_1, t_3\}, \{t_2, t_3\}, t_4\}$.

5. CONCLUSIONS

In this paper we have presented a unified view of several results concerning *synchrony theory* . Some of these can be found dispersed in the literature (basically, [SIFA 79], [SUZU 83], [SILV 86]).

Behavioural and structural synchronic dependences are studied by using two *relative finite delay* concepts (firing deviation bound and fairness bound) and two *linearly-based* restrictions of the above (synchronic lead and synchronic distance). Particular attention has been devoted to the algebraic characterization of *synchronic relations* (i.e. relations that characterize the existence of firing dependences or interdependences). The behavioural and structural synchronic relations appear as *equivalence relations* or, after a classification in θ, as *partial orders*. At a first glance it was suprising to discover that the bounded deviation and the lead synchronization relations were the same. Then only three different behavioural synchronic relations have been introduced, *{LS ≡ BD, BF, DS}* , and their three structural counterparts, *{SLS ≡ SBD, SBF, SDS}*.

The use of structural techniques to compute bounds for the synchronic functions seems very practical. This developments will be presented in a companion paper [SILV 87].

Finally it is worth noting that there exist many other approaches/problems in synchrony theory. Given the look of the approach adopted in this paper, a more precise but longer title might have been : "Towards a *linear-algebra-based* synchrony theory for P/T nets".

ACKNOWLEDGEMENTS

The author is indebted to J.M. Colom, J. Martínez, T. Murata and two anonymous referees. This work was supported in part by the DGA (Diputación General de Aragón) Grant # IT-2-86 and the US-Spain Joint Committee for Scientific and Technological Cooperation Grant # CCB-8409024.

REFERENCES

[ANDR 79] ANDRE C., ARMAND P., BOERI F.: Synchronic relations and applications in parallel computation. *Digital Processes ,* Vol. 5 (pp. 99-113).

[BERT 83] BERTHELOT G.: Transformations et analyse des réseaux de Petri. Application aux protocoles. *Thèse d'Etat*, Univ. P. et M. Curie, Paris, Octobre.

[BEST 84] BEST E.: Fairness and Conspiracies. *Information Processing Letters*, Vol. 18 (pp. 215-220).

[BRAU 80] BRAUER W. (ed.): *Net theory and Applications* . LNCS-84, Springer-Verlag.

[CARS 85] CARSTENSEN H., VALK R.: Infinite behaviour and fairness in Petri nets. *Advances in Petri nets -84* , LNCS 188, Springer-Verlag (pp 83-100).

[CHON 86] CHONG-YI Y.: Synchronic distances in C/E systems. *Advances in Petri nets-85*, LNCS-222, Springer-Verlag (pp. 101-121).

[GENR 80] GENRICH H., LAUTENBACH K., THIAGARAJAN P.: Elements of General net theory. In [BRAU 80], pp 21-164.

[GOLT 82] GOLTZ U., REISIG W.: Weighted synchronic distances. *Applications and Theory of Petri Nets* (C. Girault and W. Reisig, eds.), *Informatik Fachberichte* 52, Springer-Verlag (pp. 289-300).

[LEU 86] LEU D., MURATA T., SILVA M.: Maximum firing deviation in Petri Nets. *IEEE Int. Symp. on Circuits and Systems*. San José, California, May 5-7 (pp. 1008-1010).

[MART 82] MARTINEZ J., SILVA M.: A simple and fast algorithm to obtain all invariants of a generalized Petri net. *Applications and Theory of Petri Nets* (C. Girault, W. Reisig, eds.). *Informatik Fachberichte 52*, Springer-Verlag (pp. 301-310).

[MERC 86] MERCERON A.: Fair processes. *8th European Workshop on Petri Net Applications and Theory*. Oxford, July (pp. 367-387).

[MURA 85] MURATA T., WU Z.: Fair relation and modified synchronic distances. *Journal of the Franklin Institute*, vol. 320, nº 2, August (pp. 63-82).

[PETE 81] PETERSON J.L.: *Petri nets theory and the modelling of systems* . Prentice-Hall, Englewood Cliffs, New Jersey.

[REIS 85] REISIG W.: *Introduction to Petri nets* . Springer-Verlag, Berlin.

[PETR 75] PETRI C.A.: Interpretations of Net Theory. *ISF-Report 75-07, GMD,* St. Augustin.

[SIFA 79] SIFAKIS J.: Le contrôle des systèmes asynchrones: Concepts, proprietès, analyse statique. *Thèse Doct. es Sciences*, U.S.M.-I.N.P. de Grenoble, Juin.

[SILV 80] SILVA M.: Simplification des rèseaux de Petri par élimination des places implicites. *Digital Processes,* vol. 6 (pp. 245-256).

[SILV 85] SILVA M.: *Petri nets in automation and computer engineering.* Ed. AC, Madrid (in Spanish). (English translation to be published in 1987).

[SILV 86] SILVA M., MURATA T.: B-fairness and structural B-fairness in Petri nets models of concurrent systems. *Dept. AUTOMATICA, Univ. of Zaragoza, Research report 86-08,* June.

[SILV 87] SILVA M., COLOM J.M.: On the structural computation of synchronic invariants in P/T nets. To appear.

[SUZU 83] SUZUKI I., KASAMI T.: Three measures for synchronic dependence in Petri nets. *Acta Informatica* 19 (pp. 325-338).

[SUZU 85] SUZUKI I., HARNGDAR L.: Realization of global fairness by local control in live simple nets. *28th Midwest Symp. on Circuits and Systems*, Louisville, August 19-20 (pp. 245-248).

[THIA 84] THIAGARAJAN P., VOSS K.: A fresh look at Free-choice nets. *Information and Control* , vol. 62 (pp. 195-220).

The Semantics of a Net is a Net

An Exercise in General Net Theory.

Einar Smith
Wolfgang Reisig

GMD-F1
5205 St. Augustin
West-Germany

Abstract

A core issue of General Net Theory is to relate different net classes by meaning preserving mappings. To do so in a simple and elegant way, some mathematical techniques will be developed in this paper. This includes special net morphisms and equivalence relations on nets. These concepts are applied to study the relationship of (strict) high level nets to condition/event systems. Different classes of high level nets are discriminated.

Introduction

Progress in Net Theory has been achieved by studying certain *types* of nets such as CE-systems, PT-nets or several types of nets with individual tokens. Much effort was also spent in discussing distinguished net *classes* such as free choice nets or others.

Besides this *separation* of net constructs it was suggested to *integrate* net constructs by mutually relating nets. The fundamental idea of this approach is to emphasis on *mappings between nets* rather than single (types or classes of) nets. C.A. Petri introduced this approach in [P1] and denoted it as *General Net Theory* (see also [P2]). He suggested to study *net morphisms*, viz. a distinguished class of mappings between nets that respect places, transitions and the flow relations in a certain way. Net morphisms turn out as the continuous mappings w.r.t. the *net topology*, as described in [P1], [P2], [F], [GS]. Net topology, the category of nets, and classes of net morphisms constitute a challenging framework for a lot of studies.

Net Theory as well as Net Applications rely up to now only to a minor degree on net morphisms. The most prominent references to net morphisms include the definitions of processes and of refinements, and the often quoted remark, that nets of all types should in principle be conceiveable as shorthands for CE-systems.

This paper is intended as a contribution to the development of formal techniques that allow for a fruitful application of the ideas of General Net Theory. We consider some structural properties of relationships between some types of nets and suggest formal means for appropriate investigations of such relationships. As a concrete example we discuss properties of meaning preserving mappings between CE-systems and nets with individual tokens.

Part I: Some Basics of General Net Theory

1. Net Morphisms

This section contains some definitions and facts concerning net morphisms which we shall need throughout.

1.1 Mathematical Preliminaries

For a mapping $f : A \to B$ we denote the *domain* A of f by $\mathrm{dom}(f)$ and the *range* or *codomain* B by $\mathrm{ran}(f)$. If $A' \subseteq \mathrm{dom}(f)$, the set $\{f(a)|a \in A'\}$ is denoted by $f(A')$. Similarly, for $f : A \to B$ and $R \subseteq A \times A, f(R)$ stands for $\{(f(a), f(a'))|(a, a') \in R\}$. By $f|A'$ we denote the restriction of f to a subset A' of its domain. For mappings f and g with $\mathrm{ran}(f) \subseteq \mathrm{dom}(g)$, $g \circ f$ is the mapping $x \mapsto g(f(x))$ for $x \in \mathrm{dom}(f)$. The identity relation on a set A will be denoted by $\mathrm{id}_A = \{(a, a)|a \in A\}$, or sometimes simply by id, if there is no danger of confusion. For a set A let $\mathcal{P}(A) := \{B|B \subseteq A\}$ denote its powerset. For a relation R, R^* is defined as $\bigcup_{n \geq 0} R^n$.

1.2 Nets and Net Morphisms

Regarding net terminology we shall closely follow [BF]. In particular a triple $N = (S, T; F)$ is called a *net* iff

$$S \cap T = \emptyset$$
$$F \subseteq (S \times T) \cup (T \times S)$$
$$S \cup T \neq \emptyset$$
$$\mathrm{dom}(F) \cup \mathrm{ran}(F) = S \cup T.$$

$X := S \cup T$ denotes the set of *elements* of N.

In the following we assume nets N always as $N = (S, T; F)$ and automatically transpose indices and apostrophs N_1, N' etc. to its components.

(1.1) **Definition.** Let N, N' be nets. A mapping $f : X \to X'$ is a *(net) morphism*, denoted $f : N \to N'$ or $N \xrightarrow{f} N'$, iff

$$f(F \cap (S \times T)) \subseteq (F' \cap (S' \times T')) \cup \mathrm{id} \quad \text{and}$$
$$f(F \cap (T \times S)) \subseteq (F' \cap (T' \times S')) \cup \mathrm{id}.$$

This definition is equivalent to the one originally given by Petri [P1](see also e.g. [GS]). It is however significantly different from the notion of net morphism proposed in [W].

If $f : N \to N'$ and $g : N' \to N''$ are morphisms, the mapping $g \circ f : X \to X''$ clearly extends to a morphism $g \circ f : N \to N''$. $f : N \to N'$ is an *isomorphism* iff f is a bijection between X and X' and f^{-1} is a morphism from N' to N. Two nets are said to be *isomorphic* iff there is an isomorphism between them. We shall usually not distinguish between isomorphic nets. For example, when we speak of a net as being 'unique' in some sense, we always mean 'unique up to isomorphism'.

In the following we shall only consider a restricted class of morphisms:

(1.2) **Definition.** A morphism $f : N \to N'$ is a *folding* or *quotient* iff it satifies

(i) $f(S) \subseteq S'$, $f(T) \subseteq T'$, and

(ii) $F' \subseteq f(F)$

(1.3) **Remark.** This definition is slightly different from standard usage. Normally a morphism is called a folding if it satisfies (i) and a quotient if it satisfies (ii). But actually it is (1.2) which seems to conform to the intuitive meaning of 'folding'. Although 'folding' and 'quotient' are synonymous here, we shall generally prefer the term 'folding', and use 'quotient' only for the particular foldings to be introduced in section 2.2.

(1.4) **Proposition.** *A morphism $f : N \to N'$ is a folding iff $f(F) = F'$.*   □

Note that here and in the following we leave out all proofs which consist in straightforward verifications.

From (1.4) follows in particular that a folding maps S surjectively onto S', and similarly for T.

Our main concern will be foldings of CE-systems into "high level systems". These are often considered as foldings of conditions followed by a – in some sense – compatible merging of events into transition schemes. We shall follow this point of view here.

(1.5) **Definition.** A folding $f : N \to N'$ is an *S-folding* (resp. *T-folding*) iff $f|T$ (resp. $f|S$) is injective.

(1.6) **Proposition.** *Every folding $f : N \to N'$ is uniquely factorizable (decomposable) into an S-folding $f_S : N \to N''$ and a T-folding $f_T : N'' \to N'$ such that $f = f_T \circ f_S$.*   □

For a folding f we can thus speak of 'its S-factor' or 'its T-factor' meaning respectively f_S and f_T as in (1.6). Here is an example for this construct:

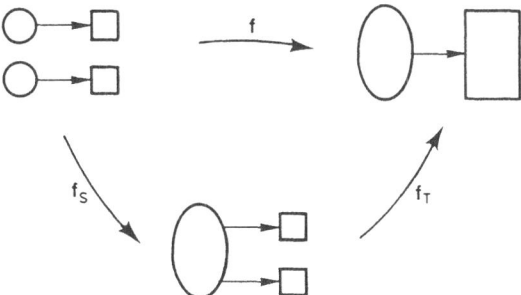

A sensible requirement for T-foldings is, that they do not fold 'too tight':

(1.7) **Definition.**

(i) A T-folding f is a *T-simplifier* iff $\forall t_1, t_2 \in T : f(t_1) = f(t_2) \Rightarrow {}^\bullet t_1 = {}^\bullet t_2 \wedge t_1^\bullet = t_2^\bullet$.

(ii) f is a *maximal T-simplifier* or a *T-simplification* iff the converse also holds, i.e. iff

$$\forall t_1, t_2 \in T : {}^\bullet t_1 = {}^\bullet t_2 \wedge t_1^\bullet = t_2^\bullet \Leftrightarrow f(t_1) = f(t_2).$$

Note that if $f : N \to N'$ is a T-simplification, then N' is *T-simple*, i.e.

$$\forall t_1, t_2 \in T' : {}^\bullet t_1 = {}^\bullet t_2 \wedge t_1^\bullet = t_2^\bullet \Rightarrow t_1 = t_2.$$

(1.8) **Proposition.** *For every net there is a unique T-simplification.* □

1.3 Special Foldings

We shall in particular be interested in foldings which explicitly express certain relationships between the flow relations in the involved nets.

(1.9) **Definition.** A folding $f : N \to N'$ is *arc-conserving* iff it maps F bijectively onto F'.

The following definitions may not seem very intuitive at the moment. They derive their motivation from the semantical properties they are related to. We refer in particular to section 3.3.

(1.10) **Definition.** A folding has *positive arc-weights* iff its T-factor is a T-simplifier.

(1.11) **Definition.** A folding has *simple arc-weights* iff it has positive arc-weights and its S-factor is arc-conserving.

Note that an S-folding f is arc-conserving iff for $s_1, s_2 \in S$, $s_1 \neq s_2$:

$$f(s_1) = f(s_2) \Rightarrow {}^\bullet s_1 \cap {}^\bullet s_2 = \emptyset = s_1^\bullet \cap s_2^\bullet.$$

(1.12) **Definition.** A folding $f : N \to N'$ has *fixed arc-weights* iff for all $t_1, t_2 \in T$ with $f(t_1) = f(t_2)$, and for all $s' \in S'$:

$$|f^{-1}(s') \cap {}^\bullet t_1| = |f^{-1}(s') \cap {}^\bullet t_2| \text{ and } |f^{-1}(s') \cap t_1^\bullet| = |f^{-1}(s') \cap t_2^\bullet|.$$

(1.13) **Theorem.** *Every folding with simple arc-weights has fixed arc-weights. Every folding with fixed arc-weights, in turn, has positive arc-weights. The converse of these implications is in general not true.* □

Again, it is fairly straightforward to verify these implications. Here are some examples and counter-examples for the different types of foldings. These examples show also that the above implications are strict: (the mappings f are unique, and not explicitly specified.)

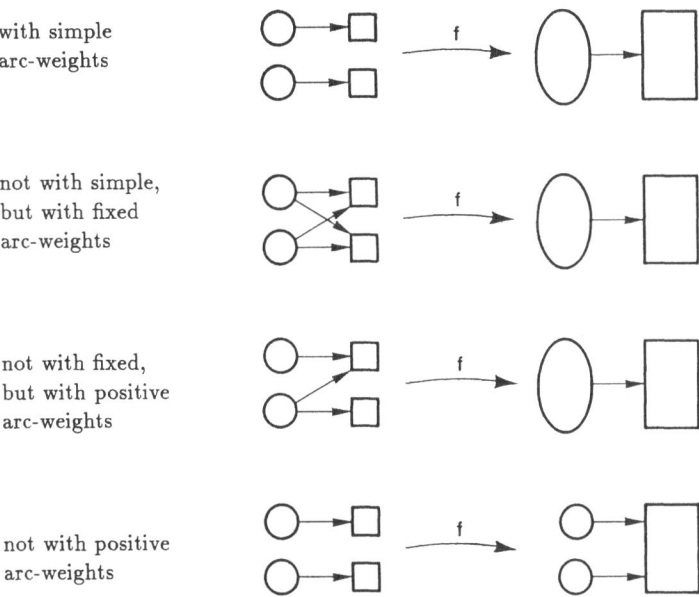

with simple
arc-weights

not with simple,
but with fixed
arc-weights

not with fixed,
but with positive
arc-weights

not with positive
arc-weights

2. Foldings and Quotient Nets

In this section we shall derive a standard representation for net-foldings. More precisely: Using basic methods from universal algebra (see e.g. [C]) we show that foldings of a net can canonically be conceived as quotient nets induced by certain equivalence relations on the set of elements of the net.

2.1 Equivalence Relations

We start with some concepts concerning equivalence relations in general. So, let X be an arbitrary set. An *equivalence relation* on X is a relation which is reflexive, symmetric and transitive. For an equivalence relation ρ on X and $x \in X$ let x^ρ stand for the equivalence class of x. For $Y \subseteq X$, $Y/\rho := \{x^\rho | x \in Y\}$ is the *quotient set of Y by ρ*. The *natural projection* associated with ρ, denoted $\mathrm{nat}(\rho)$, maps X onto X/ρ by $x \mapsto x^\rho$.

A *partition* of X is a class of subsets $A_i \subseteq X, i \in I$, where I is some index set, such that $X = \bigcup_{i \in I} A_i$ and $i \neq j \Rightarrow A_i \cap A_j = \emptyset$. A partition $\{A_i | i \in I\}$ of X canonically defines an equivalence relation ρ on X by $x \rho y :\Leftrightarrow \exists i \in I : x, y \in A_i$.

Conversely, the equivalence classes of an equivalence relation ρ form a partition $\{\,x^\rho \mid x^\rho \in X/\rho\,\}$ of X.

For two equivalence relations ρ,σ on X, the inclusion $\rho \subseteq \sigma$ means that ρ identifies less elements than σ, i.e. ρ defines a finer partition of X. Hence, in some sense we can say that X/ρ is larger than X/σ.

Assume we have an equivalence relation π on a quotient X/ρ. Sometimes we want to conceive the composed quotient $(X/\rho)/\pi$ as a simple quotient of the form X/σ. So for $x,y \in X$ we define $x\,\sigma\,y :\Leftrightarrow x^\rho\,\pi\,y^\rho$. Clearly σ is a well-defined equivalence relation on X, and we can identify $(X/\rho)/\pi$ with X/σ via the bijection $(x^\rho)^\pi \mapsto x^\sigma$. We write σ suggestively as $\rho * \pi$, so that we end up with $(X/\rho)/\pi = X/(\rho * \pi)$. Obviously $\rho \subseteq \rho * \pi$.

Conversely, let $\rho \subseteq \sigma$ be equivalence relations on X. Define an equivalence relation σ/ρ on X/ρ by $x^\rho\,\sigma/\rho\,y^\rho :\Leftrightarrow x\,\sigma\,y$. With the above we get $(X/\rho)/(\sigma/\rho) = X/\sigma$ and $\rho * (\sigma/\rho) = \sigma$.

In this way we can manipulate equivalence relations and quotients rather naturally, since after the above identification, we get that

(2.1)   'Quotients of quotients are quotients.'

2.2 Quotient Nets

After these preliminary general remarks we return to nets. Let X again denote the set of elements of a net N. If ρ is an equivalence relation on X, we want to define a net structure on X/ρ such that $\mathrm{nat}(\rho)$ becomes a folding. Consider the relation F/ρ on X/ρ given by

$$x^\rho\,F/\rho\,y^\rho :\Leftrightarrow \exists x' \in x^\rho\,\exists y' \in y^\rho : x'Fy'.$$

(2.2) **Lemma.** *The triple $(S/\rho, T/\rho; F/\rho)$ is a net iff $\rho \cap (S \times T) = \emptyset$.*

Proof. Immediate, since $S/\rho \cap T/\rho = \emptyset \Leftrightarrow \rho \cap (S \times T) = \emptyset$. $\square$

For a net N we let $\mathbf{R}(N)$ (or simply $\mathbf{R}$ if there is no danger of confusion) denote the set of equivalence relations ρ of X such that

(2.3) $\rho \cap (S \times T) = \emptyset$.

For $\rho \in \mathbf{R}(N)$ we call the net $N/\rho := (S/\rho, T/\rho; F/\rho)$ the *quotient net of N by ρ*. The collection $\{N/\rho \mid \rho \in \mathbf{R}(N)\}$ of all these quotients will be denoted by $\mathbf{Q}(N)$.

(2.4) **Theorem.** *If N is a net and $\rho \in \mathbf{R}(N)$, then $\mathrm{nat}(\rho) : N \to N/\rho$ is a folding.* $\square$

We shall use the term 'quotient' only for foldings of the form $\mathrm{nat}(\rho)$. Hence in this sense not every folding is actually a quotient. However, every folding can be *conceived* as a quotient, as the following converse of (2.4) shows.

(2.5) Theorem. *For every folding $f : N \rightarrow N'$ there exists a $\rho \in \mathbf{R}(N)$ such that $N' = N/\rho$ (up to isomorphism).*

Proof. For $x, y \in X$ let $x \rho y :\Leftrightarrow f(x) = f(y)$. Then $\rho \in \mathbf{R}(N)$. For $x^\rho \in X/\rho$ set $\overline{f}(x^\rho) := f(x)$. Now $\overline{f}$ is an isomorphism between N/ρ and N'. $\qquad\qquad\qquad\qquad\qquad\qquad\qquad\qquad\qquad$ □

(2.6) Example. Let $f : N \rightarrow N'$ be a folding where N and N' are the nets

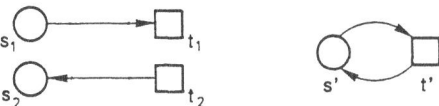

For f we immediately get $f(s_1) = f(s_2) = s'$ and $f(t_1) = f(t_2) = t'$. The equivalence relation ρ corresponding to f is then given by $s_1 \rho s_2$, and $t_1 \rho t_2$, whence $S/\rho = \{\{s_1, s_2\}\}$, $T/\rho = \{\{t_1, t_2\}\}$ and $F/\rho = \{ (\{s_1, s_2\}, \{t_1, t_2\})., (\{t_1, t_2\}, \{s_1, s_2\}) \}$.

Graphically we represent a net quotient N/ρ by enclosing the equivalence classes of ρ by dashed lines. There is no need to draw the arcs of N/ρ since they are already uniquely determined by N and ρ.

For N/ρ in the above example we get

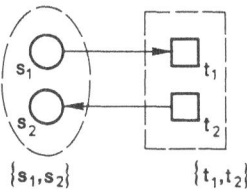

Let us use this example to illustrate a subtle but not unimportant point. If we decompose $f : N \rightarrow N'$ into its S-and T-factors (cf. (1.6)) we obtain

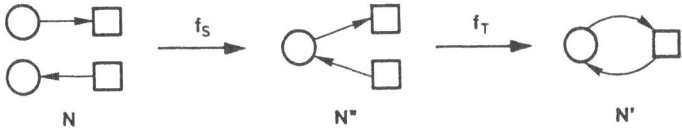

As before, we can identify N'' with a quotient of N, say N/σ:

(2.7)

In the same manner we identify N' with a quotient of N/σ, say $(N/\sigma)/\tau$:

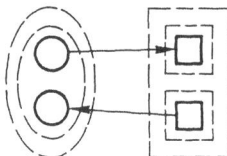

In order to identify this net with N/ρ we have to remove the innermost layer of dashed lines. Fortunately, because of (2.1) this is not problematic. Therefore we would usually also identify (2.7) with

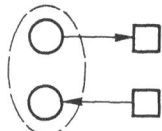

Extracting the quintessence from the above arguments, one easily arrives at the following general formulations: Let N again be an arbitrary net, and let $\rho \in \mathbf{R}(N)$. Since $\rho \cap (S \times T) = \emptyset$, ρ is the disjoint union of two equivalence relations σ on S and τ on T. We identify σ with its extension $\sigma \cup \mathrm{id}_T$ on $S \cup T$. Similarly we identify τ with the equivalence relation $\mathrm{id}_{S/\sigma} \cup \tau$ on $S/\sigma \cup T$. As an immediate consequence we obtain a version of (1.6) expressed in terms of quotients:

(2.8) $\mathrm{nat}(\rho) = \mathrm{nat}(\tau) \circ \mathrm{nat}(\sigma)$, or equivalently, $(N/\sigma)/\tau = N/\rho$

Obviously the decomposition of a $\rho \in \mathbf{R}(N)$ into an S-factor σ and a T-factor τ is again unique. This allows us to also to recast (1.10)-(1.12) in terms of equivalence relations. More precisely we define $\mathbf{R}_p(N), \mathbf{R}_f(N)$ and $\mathbf{R}_s(N)$ as the sets of those $\rho \in \mathbf{R}(N)$ where $\mathrm{nat}(\rho)$ has respectively positive, fixed and simple arc weights. (1.13) then becomes:

(2.9) **Theorem.** *Let N be a net. Then* $\mathrm{id}_X \in \mathbf{R}_s(N) \subseteq \mathbf{R}_f(N) \subseteq \mathbf{R}_p(N) \subseteq \mathbf{R}(N)$. *Furthermore, except for trivial nets, all inclusions are proper.* □

As usual we write simply $\mathbf{R}_s, \mathbf{R}_f$ and $\mathbf{R}_p$ if N can be assumed from the context.

2.3 The Lattice of Quotient Nets

In the last section we have seen how foldings of a net N correspond to equivalence relations on the elements of N. We shall therefore take a closer look at the set $\mathbf{R}(N)$ defined in section 2.2. So let N be a fixed net. We have already seen that $\sigma \leq \rho :\Leftrightarrow \sigma \subseteq \rho$ defines a partial order on $\mathbf{R}$.

(2.10) **Lemma.** *If $\rho, \sigma \in \mathbf{R}$, then also $\rho \cap \sigma$ and $(\rho \cup \sigma)^*$ are in $\mathbf{R}$.* □

In fact $\rho \cap \sigma$ is the greatest lower bound of ρ, σ in $(\mathbf{R}, \leq)$. We denote it by $\rho \sqcap \sigma$. Similarly we let $\rho \sqcup \sigma$ stand for $(\rho \cup \sigma)^*$ which is the least upper bound of ρ and σ. Hence $(\mathbf{R}, \leq)$ is a *lattice*. It even is a *complete* lattice, i.e. greatest lower bounds and least upper bounds exist for every subset of $\mathbf{R}$. In particular, $\mathbf{R}$ has a least and a greatest element.

(2.11) Lemma. $(\mathbf{R}, \leq)$ *is a complete lattice with* $\mathbf{0} := \mathrm{id}_X$ *as least and* $\mathbf{1} := (S \times S) \cup (T \times T)$ *as greatest element.* □

As in every partial order, we can define the 'immediate successor' relation in $\mathbf{R}$ by $<\cdot \ :=\ \leq -(\mathrm{id}_R \cup \leq^2)$. $\rho <\cdot\ \sigma$ means that ρ and σ define the same partition of X, except that there are precisely two equivalence classes of ρ which are merged in one class of σ, or in other words: there is no quotient of X lying strictly between X/ρ and X/σ. In particular

$$(2.12) \qquad |X/\rho| = |X/\sigma| + 1.$$

We say that $(\mathbf{R}, \leq)$ is *combinatorial* iff $\leq = (<\cdot\ )^*$. The question when $(\mathbf{R}, \leq)$ is combinatorial is easily answered.

(2.13) Proposition. $(\mathbf{R}, \leq)$ *is combinatorial iff* N *is finite.*

Proof. We only have to show " $\Rightarrow$ ": Suppose N i.e. X is infinite. If $\leq$ is combinatorial, there are $\rho_i \in \mathbf{R}, i \in \{0, \ldots, n\}$ where n is a natural number, such that $\mathbf{0} = \rho_0 <\cdot \ \cdots <\cdot\ \rho_n = \mathbf{1}$. Then there must be an i such that X/ρ_i is infinite and X/ρ_{i+1} is finite. This contradicts (2.12). □

The structure of $(\mathbf{R}, \leq)$ is now easily carried over to the set of quotients $\mathbf{Q}$ of N: If N/ρ, $N/\sigma \in \mathbf{Q}$, $N/\rho \sqcap N/\sigma$ is defined as $N/(\rho \sqcap \sigma)$, and analogously for $N/\rho \sqcup N/\sigma$. We also define $N/\rho \leq N/\sigma$ iff $\rho \leq \sigma$ in $\mathbf{R}(N)$.

(2.14) Theorem. $\mathbf{Q}(N)$ *with the relations* $\leq$ *and the operators* $\sqcup$ *and* $\sqcap$ *is a complete lattice with* N *as its least element. The greatest element of* $\mathbf{Q}(N)$ *is shaped*

If $N/\rho \leq N/\sigma$, the mapping $x^\rho \mapsto x^\sigma$ for $x \in X$ defines a folding of N/ρ to N/σ. $N/\rho <\cdot\ N/\sigma$ means that this folding cannot be decomposed into non-trivial foldings, in particular it is an S-folding or a T-folding. Now (2.13) means that all foldings of the form $x^\rho \mapsto x^\sigma$, $\rho \leq \sigma$, are decomposable into non-decomposable foldings if and only if N is finite.

In $\mathbf{Q}(N)$ we can distinguish different 'layers' corresponding to certain properties of the involved equivalence relations, which in turn stem from properties of foldings discussed in section 1.3. Let $\mathbf{Q}_p(N)$, or $\mathbf{Q}_p$, denote the set $\{N/\rho \mid \rho \in \mathbf{R}_p(N)\}$. $\mathbf{Q}_f$ and $\mathbf{Q}_s$ are defined likewise.

(2.15) Theorem. *Let* N *be a net. Then* $N \in \mathbf{Q}_s(N) \subseteq \mathbf{Q}_f(N) \subseteq \mathbf{Q}_p(N) \subseteq \mathbf{Q}(N)$. *In general all inclusions are proper.* □

Part II: The Relationship between CE-Systems and Strict High-Level Nets

3. High Level Foldings of CE-Systems

The basic step from a CE-system Σ_0 to a PrT-net Σ_1 is described in [GLT] as follows: Conditions of Σ_0 are viewed as atomic propositions with changing truthvalues. Conditions are named by instances of predicates, formed by means of individual symbols. These predicates constitute the S-elements, and the individual symbols are the tokens of Σ_1. The idea behind this is to establish in a dynamic environment a close relationship to the step from (static) propositional logic to predicate logic. One of the most important advantages of PrT–nets is their concise representation of systems. This is also one of the goals of other types of nets with individual tokens [J1] or [R2].

3.1 The Construction of High Level Foldings

In what follows, we introduce nets with individual tokens with the emphasis to get a simple and elegant relationship to CE-systems in terms of net morphisms. Furthermore we want to *remain independent of concrete inscription languages*. This is usally advantageous for analysis purposes, and the relationship to concrete types of nets with individual tokens can then easily be established.

We intend to formalize a very simple idea: Given a CE-system $\Sigma = (B_\Sigma, E_\Sigma; F_\Sigma, C_\Sigma)$, any subset $B \subseteq B_\Sigma$ of conditions may be "condensed" in order to become an S-element s_B of a "high level system" Σ'. As tokens of of s_B there may occur any conditions $b \in B$. Similarly, sets $E \subseteq E_\Sigma$ of events will serve as T-elements of Σ'.

What remains to be explained is the firing of transitions of Σ'. Σ' is intended to perform exactly the behaviour of Σ. As usual in high level nets, transitions of Σ' will fire in distinguished modes. The modes of a transition t_E will just be the events $e \in E$. And the effect of firing t_E in mode e will correspond to firing e in Σ.

In other words: the methods developed in Part I will now be extended to cover also the behavioural properties of systems.

We now give a precise representation of these ideas: Let $\Sigma = (N, C)$ be a CE-system and let $N/\rho \in \mathbf{Q}(N)$. We shall here write N as $(B_\Sigma, E_\Sigma; F_\Sigma)$ and are therefore free to use $(S, T; F)$ for N/ρ. Recall that an S-element s of N/ρ is a subset B of B_Σ. We denote this correspondence as s_B. Similarly the notion t_E for T-elements of N/ρ means that t_E is just the subset E of E_Σ .

We define a *marking* of N/ρ to be a mapping M with domain S such that for each $s_B \in S$, $M(s_B) \subseteq B$. Markings of N/ρ canonically correspond to constellations of Σ: For $k \subseteq B_\Sigma$ let $\theta_\rho(k)$ be the marking M of N/ρ defined by: $\forall s_B \in S : M(s_B) := k \cap B$.

If $\mathcal{M}$ is the class of markings of N/ρ and $K := \mathcal{P}(B_\Sigma)$ is the constellation class of Σ, then we have the

(3.1) **Lemma.** $\theta_\rho : K \to \mathcal{M}$ *is a bijection.*

Proof. Immediate, since $\{B|s_B \in S\}$ is a partition of B_Σ. □

The semantics which we shall define for N/ρ will heavily rely on this correspondence. Fundamental in this context is the

(3.2) **Definition.** A marking M of N/ρ *enables* a transition $t_E \in T$ *in a mode* $e \in E$ iff in N
$$^\bullet e \subseteq \theta_\rho^{-1}(M) \ \wedge \ e^\bullet \cap \theta_\rho^{-1}(M) = \emptyset. \qquad \Box$$

Then a *firing* of t_E in *mode* e yields the follower marking $M' := \theta_\rho((\theta_\rho^{-1}(M) - {}^\bullet e) \cup e^\bullet)$. We write this as $M[t_E(e)\rangle M'$.

In order to define the reachability relation in N/ρ we need the notion of a step:

(3.3) **Definition.** A *step* of N/ρ is a set $G \subseteq T \times E_\Sigma$ such that

(i) $(t, e) \in G \Rightarrow e \in t$

(ii) $(t_1, e_1), (t_2, e_2) \in G \Rightarrow ({}^\bullet e_1 \cup e_1^\bullet) \cap ({}^\bullet e_2 \cup e_2^\bullet) = \emptyset \qquad \Box$

A marking M of N/ρ *enables* a step G iff for all $(t, e) \in G$, M enables t in mode e. Then a *firing* of G yields as follower marking $M' = \theta_\rho(k')$, in symbols $M[G\rangle M'$, where

$$k' := (\theta_\rho^{-1}(M) - \bigcup_{(t,e) \in G} {}^\bullet e) \cup \bigcup_{(t,e) \in G} e^\bullet.$$

(3.4) **Remark.** If G is a step of N/ρ and $(t_1, e_1), (t_2, e_2) \in G$, it is perfectly possible that $t_1 = t_2$; i.e. a transition may be enabled 'concurrently with itself' (however in different modes).

We say that a marking M' is *reachable from a marking M in one step*, symbolically $M\, r\, M'$, iff there is a step G such that $M[G\rangle M'$. The *full reachability relation* on N/ρ is defined as $R := (r \cup r^{-1})^*$. Again, it is an almost literal translation of the full reachability relation R_Σ of Σ.

(3.5) **Proposition.** *For two markings M, M' of N/ρ we have $M\ R\ M'$ iff in Σ*
$$\theta_\rho^{-1}(M)\ R_\Sigma\ \theta_\rho^{-1}(M'). \qquad \Box$$

We are now ready for our main

(3.6) **Definition.** Let $\Sigma = (N, C)$ be a CE-system, let $\rho \in \mathbf{R}(N)$ and let $\mathcal{M}_C := \{\theta_\rho(c)|c \in C\}$ be the class of markings of N/ρ corresponding to the case-class of Σ. Then the system $(N/\rho, \mathcal{M}_C)$ will be called a CE-*quotient*, more precisely, the *quotient of Σ by ρ*. We denote it by Σ/ρ. $\mathbf{Q}(\Sigma)$ denotes the class $\{\Sigma/\rho|\rho \in \mathbf{R}(N)\}$ of quotients of Σ.

A CE-quotient Σ/ρ is a *strict high level system*; it is *high level* since places are marked with individual tokens, it is *strict* because the individuals are always pairwise different on each place. The term 'CE-quotient' reflects that its behaviour is defined and characterized by means of the standard semantics for CE-systems. Note in particular that we consider the full reachability relation and not only 'forward reachability', although this latter seems to be the common approach in the context of high level systems.

Here is perhaps also the right place for a general remark: It seems that the terms 'high level net' and 'high level system' are more and more used as synonyms for nets with individual tokens. Originally the term 'high level net' was reserved for real high level interpretations of nets, such as role/activity-nets or channel/agency-nets. Interpretations of nets, which can be reduced to the basic interpretation i.e. to CE-systems, were called *standard interpretations*. In section 4 we shall define the notion of *standard model*, which essentially corresponds to this approach. (To be precise, these standard models correspond to the *elementary net systems* as defined in [RT].) However, we shall in the following concede to the momentary common usage and continue to use 'high level nets' for nets with individual tokens.

3.2 Processes of CE-Quotients

One of the advantages of relying strictly on 'CE-semantics' is that also the notion of process is readily carried over from CE-systems to CE-quotients.

(3.7) Definition. Let $\Sigma = (N, C)$ be a contact-free CE-system, and let $\rho \in \mathbf{R}(N)$. A *process* of Σ/ρ is a triple $\pi/\rho := (K, \mathrm{nat}(\rho) \circ p, p)$ where $\pi = (K, p)$ is a process of Σ, as defined e.g. in [R1].

Here $\mathrm{nat}(\rho) \circ p$ labels K with elements of N/ρ. The corresponding tokens and firing modes are given by p.

Note that the mapping p from the occurence net K into the system net N is a folding in the original sense of that definition. However, since p is not necessarily surjective, it is in general not a folding or quotient in the restricted sense we use the term here.

(3.8) Example. Assume Σ and Σ/ρ are as follows:

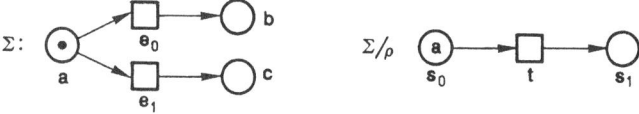

where $s_0 = \{a\}, s_1 = \{b, c\}$ and $t = \{e_0, e_1\}$. The process $\boxed{a \longrightarrow e_0 \longrightarrow b}$ of Σ is

translated into the process $\boxed{(s_0, a) \longrightarrow (t, e_0) \longrightarrow (s_1, b)}$ of Σ/ρ.

If $\pi/\rho := (K, \mathrm{nat}(\rho) \circ p, p)$ is a process of a CE-quotient Σ/ρ and $\sigma \in \mathbf{R}(N)$, then $\pi/\sigma := (K, \mathrm{nat}(\sigma) \circ p, p)$ is a process of Σ/σ. π/ρ and π/σ are identical except for the labeling functions $\mathrm{nat}(\rho) \circ p$ and $\mathrm{nat}(\sigma) \circ p$. Denoting by $\mathrm{Proc}(\Sigma/\rho)$ the set of processes of Σ/ρ for $\rho \in \mathbf{R}(N)$, we take the liberty of expressing the above as

(3.9) Proposition. *Let $\Sigma = (N, C)$ be a contact free CE-system. Let $\rho, \sigma \in \mathbf{R}(N)$. Then $\mathrm{Proc}(\Sigma/\rho) = \mathrm{Proc}(\Sigma/\sigma)$.* $\square$

Before, e.g. in (2.15) we tacitly identified a net N with its quotient N/id_X. We want to extend this identification also to CE-quotients, that is, we want $\Sigma \in \mathbf{Q}(\Sigma)$ for a CE-system Σ. There is a slight problem, however: In Σ we speak of *constellations* and *cases*, whereas in N/id_X we consider *markings*. Technically the easiest way out is to exploit the bijective correspondence between constellations and markings and make no disctinction at all. Going one step further we shall sometimes write a CE-system

(3.10) $\Sigma = (N, C)$ as (N, M_0)

where M_0 corresponds to some $c_0 \in C$ of Σ. (Recall that by definition $c \in C \Rightarrow C = [c]$.) Likewise a quotient of (N, M_0) will simply be written in the form $(N/\rho, M_0)$. Finally, note that processes of Σ are of the form (K, p) whereas processes of Σ/id_X are of the form $\pi/\mathrm{id}_X = (K, \mathrm{nat}(\mathrm{id}_X) \circ p, p)$. But this is essentially (K, p, p) so we can easily identify it with (K, p).

3.3 Distinguished Classes of High Level Foldings

In the preceeding section it was shown that each folding of a CE-system yields an equivalent high level system. We shall now discuss some subclasses of foldings that yield special types of high level systems. The distinguishing properties of these classes concern the number of tokens "flowing through" an arc when its adjacent transition fires. Again we shall without loss of generality actually consider only CE-quotients.

In a CE-system the "throughput" of an arc is constantly equal to one: Each event occurrence removes exactly one token from each precondition and adds exactly one token to each postcondition. In CE-quotients however, any integer ≥ 0 may occur as arc troughput.

As an example consider

(3.11)

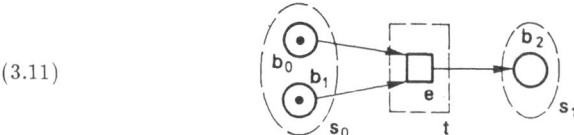

Firing t (in mode e) removes two tokens b_0 and b_1 from s_0.

The throughput of an arc may vary, depending on the firing mode of the involved transition (but fortunately not depending on the involved markings) as in

(3.12)

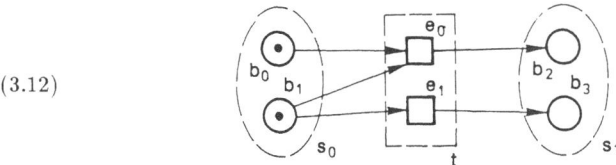

When t is fired in mode e_0, two tokens (b_0 and b_1) are removed from s_0. Firing t in mode e_1 removes however only one token from s_0 (viz. b_1).

A further consequence of the formalism of section 3.1 is that arc througputs in CE-quotients can also shrink to zero:

(3.13)

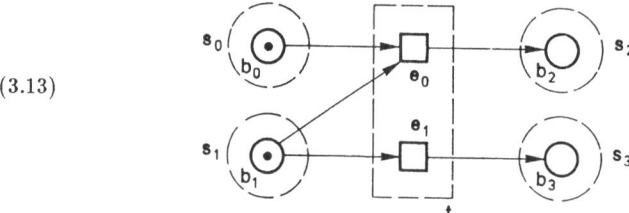

If t fires in mode e_1, the tokens of s_0 are not affected.

To prepare for formal arguments, we give a precise definition. Let $\Sigma/\rho = (N/\rho, M_0)$ be a quotient of a CE-system $\Sigma = (N, M_0)$ (cf. (3.10)). An arc (s_B, t_E) (resp. (t_E, s_B)) of N/ρ has *throughput* n (in the mode $e \in E$) iff in N $|B \cap {}^\bullet e| = n$ (resp. $|B \cap e^\bullet| = n$) for a cardinal number n.

Arcs with throughput equal to zero hint to foldings that – intuitively – do not properly express the structure of synchronisation in the underlying CE–system. It is therefore not unreasonable to exclude such foldings from consideration. This can easily be achieved by admitting only foldings with positive arc weights (see 1.10).

(3.14) **Proposition.** *The throughput of all arcs of a CE-quotient Σ/ρ are (in all modes) not smaller than one iff ρ has positive arc weights.* □

By symmetry we distinguish foldings that yield arc throughputs not greater than one. This is the case iff the S-factor σ of ρ is arc conserving. As an immediate consequence follows a characterisation of arc throughputs with the constant value one: ρ has positive arc weights and σ is arc conserving, i.e. by (1.11)

(3.15) **Proposition.** *All arc throughputs of Σ/ρ are constant with value one iff ρ has simple arc weights.* □

Of interest are also those CE-quotients where each arc has a constant (but not necessarily with the value one) throughput number independently of the mode in which the involved transition fires. (In [G] and [J2] this property is expressed by the notion of *uniformity*.)

(3.16) **Proposition** *The throughput of each arc of Σ/ρ is constant in all its modes iff ρ has fixed arc weights.* □

By (1.13) the hierarchy of quotients of N appears now as a hierarchy of the quotients of Σ, where now the distinguishing property is the semantic property of token throughput. The notions $\mathbf{Q}_p(\Sigma)$, $\mathbf{Q}_f(\Sigma)$ and $\mathbf{Q}_s(\Sigma)$ are then self-explaining.

4. Strict High Level Systems and the Standard Model

In the previous sections we have studied particular high level systems, namely quotients of CE-systems. We shall now relate this approach to existing types of high level systems such as (strict) coloured nets [J1,J2] or PrT-nets [GLT].

4.1 The Standard Model

To simplify the discussion it will be convenient to introduce yet another type of high level system as a common reference model: A fourtuple $\Sigma = (N, D, M_0, \Phi)$ will be called a *standard model*, SM for short, iff $N = (S, T; F)$ is a net and D, Φ, M_0 are mappings as explained in the following.

D associates to each $s \in S$ a set $D(s)$ of individual tokens. A *marking* of Σ is a mapping M with domain S such that $M(s) \subseteq D(s)$ for all $s \in S$. M_0 denotes a distinguished marking.

We now come to the meaning of Φ. As usual in high level systems, transitions of Σ fire in different *modes*, removing tokens from some places and adding tokens to some. The only a priori restriction we make, is that a transition in any mode may only remove tokens from its own pre-places and likewise add tokens only to its own post-places. Formally this idea can be expressed as follows: For a transition t of Σ we define its *mode space* $\Psi(t)$ as the collection of all pairs of markings (M_{pr}, M_{po}) such that

$$(4.1) \quad s \notin {}^{\bullet}t \Rightarrow M_{pr}(s) = \emptyset \text{ and } s \notin t^{\bullet} \Rightarrow M_{po}(s) = \emptyset .$$

Now Φ is defined as a mapping which associates to each transition t a subset $\Phi(t)$ of $\Psi(t)$, which we interpret as the *actual set* of modes belonging to t.

Before we define the behaviour of a SM we give an

(4.2) **Example.** Let $\Sigma/\rho = (N, M_0)$ be a CE-quotient where $N = (S, T; F)$. For $s \in S$ set $D(s) = s$ and for $t \in T$ define $\Phi(t) := \{({}^{\bullet}e, e^{\bullet}) | e \in t\}$. As discussed in (3.10) we can in the above identify e.g. ${}^{\bullet}e$ with the marking $M : s \mapsto s \cap {}^{\bullet}e$. Therefore:

(4.3)   We can easily reinterpret Σ/ρ as a SM $\Sigma' = (N, D, M_0, \Phi)$.   □

The behaviour of a SM is now defined along the usual lines: Under a marking M a transition t is *enabled* in a mode $(M_{pr}, M_{po}) \in \Phi(t)$ iff

$$\forall s \in S : M_{pr}(s) \subseteq M(s) \wedge M_{po}(s) \cap M(s) = \emptyset.$$

The forward reachability class $[M_0\rangle$ and the full reachabiltiy class $[M_0]$ can also easily be defined, so we do not give the details here.

4.2 Standard Models Conceived as CE-Quotients

We shall now show how to associate CE-quotients to SMs. Later we shall relate SMs to 'annotated high level systems'. In this way SMs can serve as a link between CE-quotients and the more prominent types of high level systems.

Let $\Sigma = (N, D, M_0, \Phi)$ be a SM with $N = (S, T; F)$ as the underlying net. We unfold (N, D, Φ) to a net $N_1 = (B_1, E_1, F_1)$ as follows

$$B_1 := \{(s, a) | s \in S, a \in D(s)\}$$

$$E_1 := \{(M_{pr}, t, M_{po}) | t \in T, (M_{pr}, M_{po}) \in \Phi(t)\}.$$

B_1 is essentially the disjoint union of the token sets $D(s)$. E_1 consists of all firing modes of Σ labelled with the transitions they belong to.

For $b := (s,a) \in B_1$ and $e := (M_{\mathrm{pr}}, t, M_{\mathrm{po}}) \in E_1$ we set $bF_1e :\Leftrightarrow a \in M_{\mathrm{pr}}(s)$ and $eF_1b :\Leftrightarrow a \in M_{\mathrm{po}}(s)$.

N_1 is an unfolding of N in the sense that we have the obvious folding $f : N_1 \to N$ given by $(s,a) \mapsto s$ and $(M_{\mathrm{pr}}, t, M_{\mathrm{po}}) \mapsto t$.

(4.4) **Proposition.** *Let $\Sigma = (B, E; F, M_0)$ be a CE-system, and let Σ/ρ be a quotient conceived as a standard model, as described in (4.3). Then the above defined unfolding, applied to Σ/ρ, yields $N_1 = (B, E; F)$ and $f = \mathrm{nat}(\rho)$, i.e. unfoldings and quotients are natural inverses of each other.* □

In the general case, the unfolding of a SM Σ cannot immediately be interpreted as a CE-system. The result of unfolding Σ may not be T-simple, and may contain S-elements which, considered as conditions, never change their truthvalue, and T-elements which, considered as events, never can occur. Upon skipping those elements, the corresponding T-simplification constitutes however in fact a CE-system. We call it *the* CE-system associated with Σ.

4.3 Distinguished High Level Models Conceived as Standard Models

We now turn to better known types of high level systems, which can be characterized as inscribed or annotated SMs. Without being exhaustive we shall look at four variations of these.

Let N be a net and D again a mapping associating a set of individual tokens with each S-element of N. We say that a mapping I is an *inscriptor* of type i $(0 \le i \le 3)$ for N iff it satisfies

Type 0: For $t \in T$, $I(t)$ is an arbitrary set, and each arc (s,t) (resp. (t,s)) is mapped to a function f_{st} (resp. f_{ts}) from $I(t)$ to the powerset of $D(s)$.

Type 1: As type 0, except that $\emptyset$ is not allowed as value of the arc-functions.

Postponing type 2 for a moment we come to

Type 3: As type 0, except that the values of the arc-functions are *elements* of the $D(s)$, not subsets.

Note that the triple (N, D, I), where I is an inscriptor of type 0, 1 or 3, can be considered as the structural part of a strict coloured net. Similarly we want inscriptors of type 2 to induce a 'PrT-like' structure on N, so we define:

I is an inscriptor of type 2 for N iff (i)-(iii) below hold:

(i) I labels each arc (s,t) (resp. (t,s)) with a set V_{st} (resp. V_{ts}) of variables ranging over $D(s)$.

(ii) For $t \in T$ all variables associated with the arcs surrounding t are distinct. We set

$$V_t := \bigcup_{s \in {}^\bullet t} V_{st} \cup \bigcup_{s \in t^\bullet} V_{ts}$$

(iii) Each $t \in T$ is inscribed with a relation R_t expressed by some formula with V_t as its set of free variables.

Let us now look at the relationship between these inscribed high level systems and SMs. In one direction this amounts to evaluating inscriptors in order to obtain firing modes for transitions. Conversely, one can ask which set of firing modes can be characterized by which inscriptors. We return to this latter question in section 4.4 below, and consider the first one here.

Assume we have a triple (N, D, I) with an inscriptor I of type $i \in \{0, \ldots, 3\}$. We must show how to evaluate I into modes sets $\Phi(t)$ for $t \in T$. The details will here be given only for the cases $i = 3$ and $i = 2$. The remaining cases can be treated similarly as the case $i = 3$.

$i = 3$:

Let $t \in T$. For $c \in I(t)$ define markings M_{pr}^c and M_{po}^c of (N, D) by

$$M_{pr}^c(s) := \begin{cases} \{f_{st}(c)\} & \text{if } s \in {}^\bullet t \\ \emptyset & \text{otherwise} \end{cases}$$

and

$$M_{po}^c(s) := \begin{cases} \{f_{ts}(c)\} & \text{if } s \in t^\bullet \\ \emptyset & \text{otherwise} \end{cases}$$

We then set for $t \in T$: $\Phi(t) := \{(M_{pr}^c, M_{po}^c) | c \in I(t)\}$.

$i = 2$:

Let $t \in T$. An *assignment* of V_t is a mapping β with domain V_t such that $\beta(V_{st})$, $\beta(V_{ts}) \subseteq D(s)$ for $s \in {}^\bullet t$ resp. $s \in t^\bullet$. For an asssignment β define a marking M_{pr}^β by

$$M_{pr}^\beta(s) := \begin{cases} \beta(V_{st}) & \text{if } s \in t^\bullet \\ \emptyset & \text{otherwise} \end{cases}$$

M_{po}^β is defined analogously. If β is an assignment of V_t, we write $\beta \models R_t$ iff β evaluates R_t to 'true'. Finally, we say that an assignment β is *feasible* (see [G]) iff $\beta|V_{st}$ and $\beta|V_{ts}$ are injective for $s \in {}^\bullet t$ and $s \in t^\bullet$. Then we set $\Phi(t) := \{(M_{pr}^\beta, M_{po}^\beta) \mid \beta \models R_t, \beta \text{ feasible}\}$.

4.4 CE-Quotients Conceived as Distinguished High level Systems

Let us now look at the other direction: Given (N, D, Φ), by what kind of inscriptors can Φ be characterized? Here we shall mainly restrict the question to CE-quotients. For the general case we only mention that Φ can always be characterized by an inscriptor of type 0:

(4.6)   For $t \in T$ set $I(t) := \Phi(t)$ and for $s \in {}^\bullet t$ define $f_{st} : I(t) \to \mathcal{P}(D(s))$ by $f_{st}((M_{pr}, M_{po})) := M_{pr}(s)$. For $s \in t^\bullet$, f_{ts} is defined analogously.

Returning now to quotients Σ/ρ of a CE-system Σ we can classify them according to how the firing modes can be characterized by inscriptors. Formally, for $i \in \{0, \ldots, 3\}$ we write $\Sigma/\rho \in \mathbf{Q}_i(\Sigma)$ iff the mode sets $\Phi(t)$ of Σ/ρ (cf. 4.2) can be described by an inscriptor of type i.

By (4.6) above we see that $\mathbf{Q}_0(\Sigma) = \mathbf{Q}(\Sigma)$, i.e. for all $\rho \in \mathbf{R}(N)$, $\Sigma/\rho \in \mathbf{Q}_0(\Sigma)$.

The following theorem summarizes the relationship between properties of the involved equivalence relations and the characterizability of the resulting CE-quotients by means of certain inscriptors.

(4.7) **Theorem.** *Let* $\Sigma = (N, C)$ *be a CE-system and let* $\rho \in \mathbf{R}(N)$. *Then the CE-quotient* Σ/ρ *is in*

$$\mathbf{Q}_0(\Sigma) \text{ iff } \rho \in \mathbf{R}(N)$$
$$\mathbf{Q}_1(\Sigma) \text{ iff } \rho \in \mathbf{R}_p(N)$$
$$\mathbf{Q}_2(\Sigma) \text{ iff } \rho \in \mathbf{R}_f(N)$$
$$\mathbf{Q}_3(\Sigma) \text{ iff } \rho \in \mathbf{R}_s(N)$$

Furthermore $\Sigma \in \mathbf{Q}_3(\Sigma) \subseteq \mathbf{Q}_2(\Sigma) \subseteq \mathbf{Q}_1(\Sigma) \subseteq \mathbf{Q}_0(\Sigma)$, *and in general all inclusions are proper.*

Proof. We prove only $\Sigma/\rho \in \mathbf{Q}_2 \Leftrightarrow \rho \in \mathbf{R}_f$. The other cases can be shown similarly (use (4.6)).

"$\Rightarrow$": Assume I is an inscriptor of type 2 for Σ/ρ. The throughput of an arc (s, t) or (t, s) is then $|V_{ts}|$ or $|V_{ts}|$, i.e. constant. By (3.16) it follows that $\rho \in \mathbf{R}_f(N)$.

"$\Leftarrow$": Let $\rho \in \mathbf{R}_f(N)$, and let (x, y) be an arc of Σ/ρ. By (3.16) there is a fixed throughput number n for (x, y). Let V_{xy} be a set of variables with $|V_{xy}| = n$. Assume that $V_{xy} \cap V_{x'y'} \neq \emptyset \Rightarrow x = x' \wedge y = y'$. For a transition t of Σ/ρ define

$$V_t := \bigcup \{V_{xy} | (x, y) \in F/\rho \, , \, t \in \{x, y\}\}.$$

We define a relation R_t in the parameters V_t by

$$R_t(V_t) :\Leftrightarrow \exists e \in t : (\forall s \in {}^{\bullet}t : V_{st} = s \cap {}^{\bullet}e \wedge \forall s \in t^{\bullet} : V_{ts} = s \cap e^{\bullet}).$$

Note that in ${}^{\bullet}t$ the dot refers to N/ρ whereas in ${}^{\bullet}e$ it refers to N. So for $t \in E_{\Sigma}/\rho$ we set $I(t) := R_t$. This concludes the definition of an inscriptor I of type 2 for Σ/ρ. □

(4.8) **Corollary.** *Let* Σ *be a CE-system. Then*
$$\mathbf{Q}_0(\Sigma) = \mathbf{Q}(\Sigma) \, , \, \mathbf{Q}_1(\Sigma) = \mathbf{Q}_p(\Sigma), \, \mathbf{Q}_2(\Sigma) = \mathbf{Q}_f(\Sigma) \text{ and } \mathbf{Q}_3(\Sigma) = \mathbf{Q}_s(\Sigma).$$ □

Conclusion

The aim of this paper is twofold: Firstly it was intended to suggest some formal techniques supporting the ideas of General Net Theory. Secondly, formally precise relationships between the semantics of high level nets and condition/event systems have been established. This was carried out, on a syntax independent level, by application of the above techniques.

It is not claimed that the morphisms of section 1 are new constructs. In fact, they are included in [GS]. A lot of similar constructs are given in [F]. Our concern was to figure out, which morphisms are suitable for certain purposes.

The problem of non-strictness, i.e. of many "identical" tokens residing coincidently on a place, has not been treated in this paper. To properly work out the relationship between strict and non-strict nets, we feel that other techniques are necessary. This will be the topic of a forthcoming paper.

References

BF: E. Best, C. Fernández: *Notations and Terminology on Petri Net Theory.* Gesellschaft für Math. und Datenverarbeitung, Bonn, Arbeitspapiere der GMD Nr. 195 (1986).

C: P.M. Cohn: *Universal Algebra.* D. Reidel Publishing Company (1981).

F: C. Fernández: *Net Topology I,II.* ISF–Reports 75.09, 76.02. GMD, St. Augustin (1975,1976).

G: H.J. Genrich: *Predicate/Transition Nets.* Proceedings of the Advanced Course on Petri Nets, Bad Honnef, West-Germany, September 1986.

GLT: H.J. Genrich, K. Lautenbach, P.S. Thiagarajan: *Elements of General Net Theory* in: W. Brauer (ed): *Net Theory and Applications.* Springer Lecture Notes in Computer Science, 84 (1980).

GS: H.J. Genrich, E. Stankiewicz–Wiechno: *A Dictionary of Some Basic Notations of Net Theory* in: W. Brauer (ed): *Net Theory and Applications.* Springer Lecture Notes in Computer Science, 84 (1980).

J1: K. Jensen: *Coloured Petri Nets and the Invariant Method.* Theoretical Computer Science 14 (1981), pp. 317–336.

J2: K. Jensen: *Coloured Nets.* Proceedings of the Advanced Course on Petri Nets, Bad Honnef, West-Germany, September 1986.

P1: C.A. Petri: *Concepts of Net Theory.* Mathematical Foundations of Computer Science, Proceedings of Symposium and Summer School, High Tatras, September 3–8, 1973. Math. Inst. of the Slovak Acad. of Science (1973), pp. 137–146.

P2: C.A. Petri: *Interpretations of Net Theory.* Interner Bericht 75-07 (2. Auflage), GMD(1976)

R1: W. Reisig: *Petri Nets, An Introduction.* Springer Verlag (1985).

R2: W. Reisig: *Petri Nets With Individual Tokens.* Theoretical Computer Science 41 (1985), pp. 185–213.

RT: G. Rozenberg, P. S. Thiagarajan: *Petri Nets: Basic Notions, Structure, Behaviour.* Springer Lecture Notes in Computer Science, 224 (1986).

W: G. Winskel: *A New Definition of Morphism on Petri Nets.* Springer Lecture Notes in Computer Science, 166 (1984).

On the mutual simulatability of different types of Petri nets

Peter H. Starke
Sektion Mathematik der Humboldt-Universitaet
DDR-1086 Berlin, PSF 1297

1. Introduction

In this paper we consider the relative power of different types of Petri nets, namely Place/Transition nets, Coloured nets, Relation nets, Predicate/Event nets, Predicate/Transition nets, Self-modifying nets and Fifo nets. We do this by investigating whether there are possibilities to simulate one type of nets by other ones or not.

It is well-known that in the bounded and finite case all these types of Petri nets describe finite-state systems, i.e. finite automata, in this sense, i.e. considered as finite-state system models, all these types of Petri nets are globally equivalent.

However, as shall be shown, in many cases we are able to refine this statement by presenting a so-called local simulation.

Definition 1.1.
A $\underline{simulation}$ of a net type A by the net type B is an algorithm which, for every marked net $N = [P,T,F,\ldots]$ of type A, constructs a net $N' = [P',T',F',\ldots]$ of type B and two mappings σ and τ, the state translation and the transition translation, respectively. The state translation σ maps the reachability set R' of N' onto the reachability set R of N, and τ is a homomorphism from T' onto T or $T \cup \{e\}$ (where e denotes the empty word) such that
(a) if $m',m'' \in R'$, $w' \in (T')^{*}$ and $m'[w'>m''$ (in N'), then
$\sigma(m')[\tau(w')>\sigma(m'')$ (in N),
(b) if $m,m^{*} \in R$, $w \in T^{*}$ and $m[w>m^{*}$ (in N), then there exist $m',m'' \in R'$, $w' \in (T')^{*}$ with $m = \sigma(m')$, $w = \tau(w')$, $m'[w'>m''$ (in N') and $m^{*} = \sigma(m'')$.
(c) if m (m', resp.) is the initial marking of N (N', resp.) then $m = \sigma(m')$.
If, moreover, σ is one-to-one, we speak of a $\underline{state\ behaviour\ simulation}$. A simulation is said to be $\underline{local}$ if the construction of N', σ and τ from N is done locally in the sense that the net N is transformed node by node (and arc by arc) using only the knowledge of the environment of this node, but without using knowledge of the whole net or its reachability set. If a transition t' of N' by τ is mapped to the empty word it is called a silent transition because its firing does not change the simulated marking, i.e. if $m'[t'>m''$ then $\sigma(m') = \sigma(m'')$.
Hence, a local simulation additionally gives a structural correspondence between N and N', which is partially invariant under net refinements. This enables us to apply net tools developed for one net type to other types of nets since errors detected in the simulating net can be translated to the simulated one.

2. Petri nets

Definition 2.1.
N = [P,T,F,V,m] is a <u>Petri net</u> (Place/Transition net) iff
(a) [P,T,F] is a net,
   i.e. P and T are finite disjoint nonempty sets and F is a subset
   of (PxT)∪(TxP) such that dom(f)∪cod(F) = P∪T,
(b) V is a mapping from F into IN^+ := {1,2,...},
(c) m_0 is a mapping from P into IN := {0,1,2...}.

Interpretation
   P is the set of (names of) places, T the set of (names of)
transitions of N. Isolated nodes (places or transitions) are
prohibited. For every arc f from F the number V(f) is its (nonzero)
multiplicity. A mapping from P into IN is called a marking of N; m
is the initial marking. To every transition t from T we assign two
markings t^- and t^+ as follows:
   $t^-(p)$:= if (p,t) ∈ F then V(p,t) else 0 end;
   $t^+(p)$:= if (t,p) ∈ F then V(t,p) else 0 end.
   For markings m,m' the sum m+m' and the relation m≤m' are
understood placewise. A transition is said to have concession at the
marking m (denoted by m[t>) iff t^- ≤ m. In this case, t can fire,
which results in the marking m' := m - t^- + t^+, written as m[t>m'.
This relation can be extended to words w from T^* inductively:
Basis: m[e>m' iff m=m',
Step: m[wt>m' iff m[w>m" and m"[t>m' for some m".
Moreover, we put
   m[w> iff m[w>m' for some m'.
Then $L(N,m_0)$:= { w | m_0[w>} is the set of all finite firing sequen-
ces of N, and $R(N,m_0)$:= { m | m_0[w>m for some w from $L(N,m_0)$} is
its reachability set.

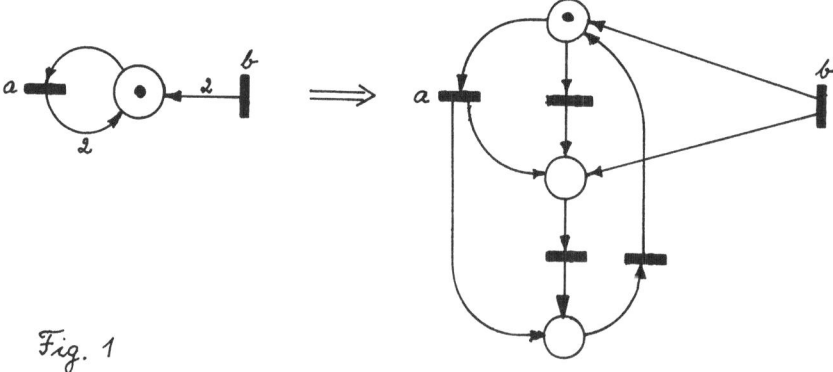

Fig. 1

Definition 2.2.
Let N = [P,T,F,V,m] be a Petri net.
   1. A place p from P is said to be <u>bounded</u> (resp. <u>safe</u>) iff
max{ m(p) | m ∈ $R(N,m_0)$ } exists (and equals 1). If all places of N
are bounded (resp. safe), the net is called bounded (resp. safe).
   2. N is an <u>ordinary</u> net iff for every arc f from F the number
V(f) equals 1.

Already HACK [5] has shown

Theorem 2.1.
There is a local simulation of all Petri nets by selfloop-free ordinary Petri nets.

The idea of the proof is to replace every place p of N with maximum input multiplicity im(p)>1 or maximum output multiplicity om(p)>1 by a ring of places and transitions (which τ maps to e) containing im(p)+om(p) places, and to distribute the multiple arcs over the places of the ring. The number of tokens in the ring equals the marking of the place p (cf. Figure 1).

A more involved construction has been given by Müller [8] which has the additional advantage to be prompt, i.e. there is a bound on the length of firing sequences of N' mapped to the empty word by τ.

Using the co-place construction (see below) one can show, more-over,

Theorem 2.2.
There is a local simulation of all bounded Petri nets by safe Petri nets.

A different construction to prove 2.2 was given by ULLRICH [12]. His idea is to split every place p with bound k>1 to places $p_0, \ldots, p_k$

such that under every reachable marking of N' exactly one of these places is marked with one token and where the token at place p_i corresponds to i tokens at p in N (cf. Figure 2). Obviously, this is a (prompt) local state behaviour simulation.

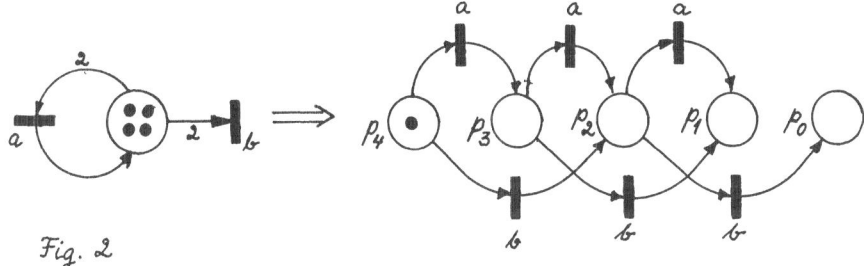

Fig. 2

Definition 2.3.
A Petri net with capacities is a pair [N,c], where N = [P,T,F,V,m_0] is a Petri net and c is a mapping from P into IN $\cup$ {∞} with $m_0 \leq c$.

Interpretation
For every place p, c(p) is the maximal number of tokens p can hold (if c(p)=∞, then this number is unbounded). The capacities restrict the fireability of transitions because the capacities must not be exceeded. The interpretations differ in the case of selfloops; we adopt the following so-called Hamburg rule:
   t has concession at m iff m[t> (in N) and $m - t^- + t^+ \leq c$.

Hence, every Petri net can be considered as a Petri net with capacities, where all the capacities are infinite.

Theorem 2.3.
There is a local state-behaviour simulation of all Petri nets with capacities by Petri nets (without capacities).

The proof is done by introducing a co-place $\bar{p}$ for every place p with $c(p) < \infty$. We put $c(p)-m_o(p)$ tokens to $\bar{p}$ in the initial marking and organize the connections of $\bar{p}$ with the transitions in such a way that at every reachable marking m it holds that

$$m(p) + m(\bar{p}) = c(p)$$

(cf. Figure 3).

Combining the above constructions we can conclude that C.A. PETRI's original model, the condition event systems which are selfloop-free ordinary Petri nets with capacity $c \equiv 1$ in our terminology, can locally simulate all bounded Petri nets.

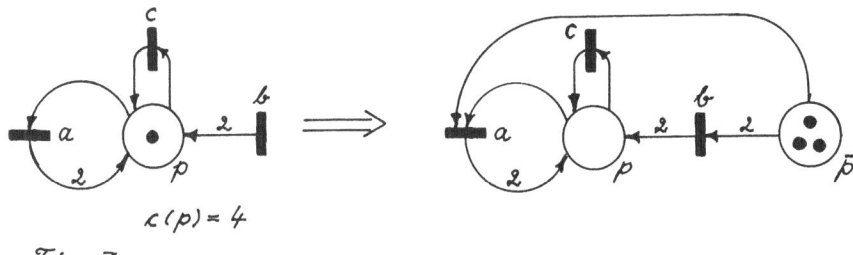

$c(p) = 4$

Fig. 3

Another 'additive' to Petri nets are the so-called <u>inhibitor arcs</u> introduced by FLYNN and AGERWALA [3]. An inhibitor arc leads from a place p to a transition t and inhibits the firing of t if the token load of p is not less than its multiplicity l. If $l > 1$, then, additionally, an ordinary arc from p to t with multiplicity less than l is allowed.

Theorem 2.4.
There is a local simulation of all bounded Petri nets with inhibitor arcs by bounded Petri nets.

By the construction displayed in Figure 3 we can assume that for every place p in N there is a co-place $\bar{p}$.
Consider an inhibitor arc of multiplicity l leading from p to the transition t. If $l > k := m_o(p) + m_o(\bar{p})$, then we can omit this arc without any change. Otherwise we split the transition t to l transitions $t_0,\ldots,t_{l-1}$ having the same connections as t. Now the inhibitor arc is replaced by a loop from $\bar{p}$ to t_i and back, where both arcs have the multiplicity $m_o(p)+m_o(\bar{p})-i$. In case that there exists an ordinary arc from p to t with multiplicity n, the transitions $t_0,\ldots,t_{n-1}$ are omitted (cf. Figure 4).

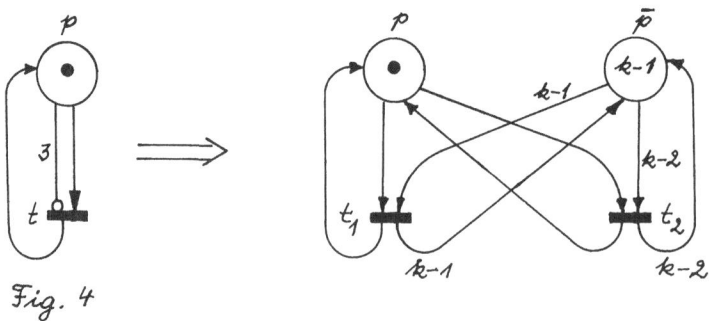

Fig. 4

As HACK [6] already has pointed out, the result cannot be sharpened, because there exits a (local) simulation of all counter machines by (unbounded) Petri nets with inhibitor arcs.

3. Self-modifying nets

Definition 3.1. (VALK [13])
$N = [P,T,F,V,m_0]$ is a self-modifying net if
  (a) $[P,T,F]$ is a net,
  (b) m_0 is a marking,
  (c) V is a mapping from F into $P \cup IN^+$, where P and IN are assumed to be disjoint.

Interpretation.
The only difference to Petri nets is that the multiplicity of an arc f can be the name p of a place. In this case, the multiplicity of the arc f changes with the marking of the place p. At the marking m the multiplicity of f equals m(p).

This net type is somewhat outside of net philosophy since there are causal relations between state-elements and transitions which are not reflected by arcs. But, carefully used, it seems to be attractive for multilevel modelling applications (cf. [1]).

Theorem 3.1.
There is a local state-behaviour simulation of all (unbounded) Petri nets with inhibitor arcs by self-modifying nets.

For convenience we consider only the case when there exists no ordinary arc in parallel with the inhibitor arc. The construction is given by Figure 5, where q is a new place only connected with t.

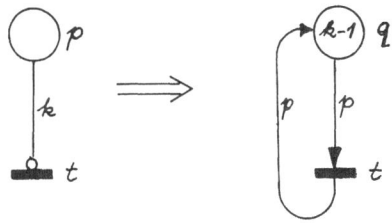

Fig. 5

Theorem 3.1 shows that we can simulate counter machines by self-modifying nets, i.e. the model has also the computational power of the TURING-machine.
On the other hand, trivially, every Petri net is a self-modifying net (without modified arcs).
In modelling real world systems the resulting self-modifying net has to be bounded. In this case the modelling power of self-modifying nets is the same as that of Petri nets:

Theorem 3.2.
There is a local state-behaviour simulation of all bounded self-modifying nets by (bounded) Petri nets.

Let N = [P,T,F,V,m$_o$] be the given bounded self-modifying net, where at most c(p) tokens can be reached at the place p. First, for every place p from P we introduce a complement place $\hat{p}$, which obtains c(p) tokens under the initial marking and which is connected with all transitions in the reverse way to p: if (p,t) ∈ F, then (t,$\hat{p}$) ∈ F' and V'(t,$\hat{p}$) = V(p,t) and if (t,p) ∈ F, then ($\hat{p}$,t) ∈ F' and V'($\hat{p}$,t) = V(t,p) (cf. Figure 6).

Obviously, for all reachable markings m' and all p from P it holds that

$$m'(p) + m'(\hat{p}) = m_o(p) + c(p)$$
and $L(N,m_o) = L(N',m_o')$.

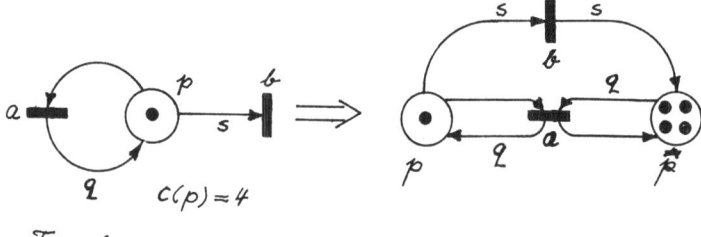

$c(p) = 4$

Fig. 6

Now we consider an arc (p,t) from F with V(p,t) = q ∈ P. Let k be the maximum token number at q and l := m$_o'$(q)+m$_o'$($\hat{q}$). We split the transition t into transitions t$_o$,...,t$_\ell$, where t$_i$ simulates the beha

viour of t at markings m' with m'(q) = i (cf. Figure 7).

4. Coloured Petri nets, Relation nets

If X is a set, every mapping m from X into IN is called a multiset over X; for x from X the integer m(x) is interpreted as the multiplicity of x in m. Hence, markings in Petri nets are multisets of places. Like for markings, we consider operations and relations for multisets componentwise. The multiset o, which for all x from X has 0 as its value, is called the empty multiset. We represent multisets as formal sums: m = $\sum_{x \in X} m(x)x$.

Definition 4.1. (JENSEN ⌊7⌋)
N = [P,T,F,C,V,m$_o$] is a <u>Coloured Petri net</u> iff
(a) [P,T,F] is a net,
(b) C is a mapping from P∪T into a system of nonempty sets,
(c) m$_o$ is a mapping which assigns a multiset over C(p) to each p from P,
(d) V is a mapping which assigns a mapping from C(t) into the system of all multisets over C(p), which is not constantly equal to the empty multiset, to every arc (p,t) or (t,p) from F.

Interpretation.
D := U{ C(t) | t ∈ T }∪U{ C(p) | p ∈ P } is the set of all colours, for p from P, C(p) is the set of colours of tokens which can be put to p (sorts of tokens) and, for t from T, C(t) is the set of all colours of the transition t (kinds of firing). A marking is a mapping which assigns a certain number of tokens of each sort to each place.

A transition t has concession at the marking m with colour c iff
(i) $c \in C(t)$,
(ii) if $(p,t) \in F$, then $V(p,t)[c] \leq m(p)$ for all p from P.
If t has concession at m with c, it can fire with c, which results in
the marking m' with
$$m'(p) := m(p) + V(t,p)[c] - V(p,t)[c],$$
where for $f \in F$, $V(f)[c]$ is to be understood as the empty multiset o.

Obviously, every Petri net is a Coloured Petri net, where $C(p) = C(t) = \{\not{c}\}$ for all p from P, t from T (multisets over a singleton correspond to numbers one-to-one).

Definition 4.2.
Let $N = [P,T,F,C,V,m_0]$ be a Coloured Petri net.
(a) N is said to be __finite__ iff the set D of all colours is finite;
(b) N is __bounded__ iff there exists a marking m covering any reachable marking.

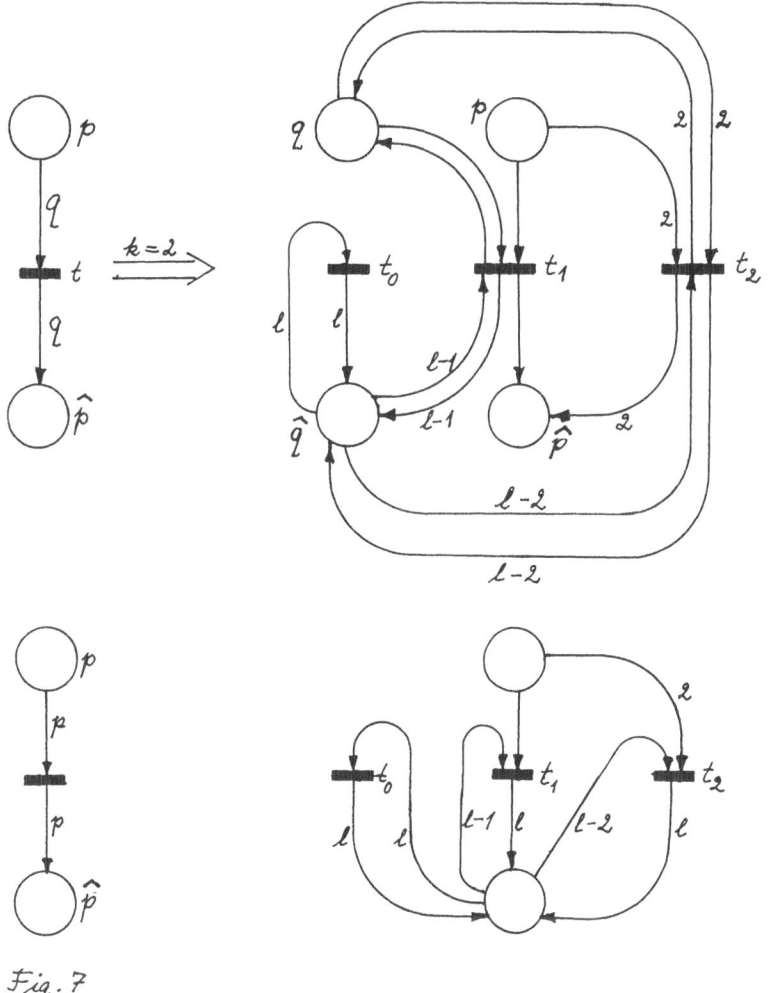

Fig. 7

Theorem 4.1.
There is a local state-behaviour simulation of all finite (bounded) Coloured Petri nets by (bounded) Petri nets.

The idea of the construction is to split the places and the transitions according to their colours, so that for any colour c from $C(p)$ we have a place $[p,c]$, and i tokens of colour c at p in N correspond to i (uncoloured) tokens at place $[p,c]$ in N'. In detail, let $N = [P,T,F,C,V,m_o]$ be a finite Coloured Petri net. Then we put for p from P, c from $C(p)$, t from T and d from $C(t)$:

$$P' := \{ [p,c] \mid p \in P \ \& \ c \in C(p) \},$$
$$T' := \{ [t,c] \mid t \in T \ \& \ c \in C(t) \},$$
$$m_o'([p,c]) := m_o(p)[c],$$
$$V'([p,c],[t,d]) := \begin{cases} (V(p,t)[c])[d], & \text{if } (p,t) \in F \ \& \ (V(p,t)[c])[d] > 0, \\ \text{not defined}, & \text{otherwise}; \end{cases}$$
$$V'([t,d],[p,c]) := \begin{cases} (V(t,p)[d])[c], & \text{if } (t,p) \in F \ \& \ (V(t,p)[d])[c] > 0, \\ \text{not defined}, & \text{otherwise}; \end{cases}$$

and F' consists of all pairs f such that $V'(f)$ is defined. All isolated nodes can be omitted. Since the value $V(f)[c]$ is not identically equal to the empty multiset, every node of N appears in N'. Obviously, N' simulates N with σ, τ where

$$\tau([t,d]) := t,$$
$$\sigma(m')(p)[c] := m'([p,c]).$$

Definition 4.3. (REISIG [9])
Let D be a nonempty set. $N = [P,T,F,V,m]$ is a <u>Relation net</u> over D iff
(a) $[P,T,F]$ is a net,
(b) m_o is a mapping which assigns a multiset over D to every place p from P,
(c) V is a mapping which to every arc f from F assigns a nonempty multiset over DxD (a multirelation).

Interpretation.
D is the set of sorts of tokens, say colours. A transition t has concession at the marking m with d from D if for all places p from Ft it holds for all d' from D

$$m(p)[d'] \geqslant V(p,t)[d,d']$$

and

$$\sum_{d' D} (\sum_{p \ Ft} V(p,t)[d,d'] + \sum_{p \ tF} V(t,p)[d,d']) > 0.$$

The second condition ensures that something happens if t fires with d. By firing t with d at m we obtain the new marking m' with

$$m'(p)[d'] := m(p)[d'] - V(p,t)[d,d'] + V(t,p)[d,d']$$

where the value 0 is assigned to all undefined terms.

Obviously, there is no essential difference between Relation nets and Coloured Petri nets. Every Relation net over D is a Coloured Petri net, where the colour function C is identically equal to D. On the other hand, every Coloured Petri net is a Relation net over the set D as set of all colours. However, we prefer Coloured Petri nets since their structure is somewhat finer.

5. Predicate/Event nets

Let $[D,OP]$ be an algebra, i.e. D is a nonempty set and OP is a set of operations (perhaps containing constants as zero-place operations)

over D. Let X be a nonempty set of variables over D and TL the term
language over X corresponding to [D,OP]. Any mapping β from X into
D is called an evaluation. It can be extended to terms from TL and
to subsets of TL canonically.

Definition 5.1. (REISIG [9])
N = [P,T,F,V,m_0] is a Predicate/Event net over [D,OP] iff
(a) [P,T,F] is a net,
(b) m_0 is a mapping from P into the powerset of D,
(c) V is a mapping which assigns to every arc f from F a nonempty
    subset of TL such that for all term1, term2 from V(f) and every
    evaluation β it holds that β(term1) # β(term2).

Interpretation.
D is a set of distinguishable tokens. Every place p represents a
predicate over D, the marking m(p) being the current extension of
this predicate. A transition t has concession at the marking m with
the evaluation β iff
    β(V(p,t)) $\subseteq$ m(p)   for all p from Ft, and,
    β(V(t,p)) $\cap$ m(p) = $\emptyset$ for all p from tF.
If t fires at m with β, then the marking m' is reached with
    m'(p) := (m(p) - β(V(p,t)))$\cup\beta$(V(t,p)),
where undefined terms amount to the empty set.

Definition 5.2.
A Predicate/Event net N=[P,T,F,V,m] over [D,OP] is said to be <u>weakly
finite</u> iff all the sets V(f) (for f from F) are finite. It is called
<u>finite</u> if D is finite.

   Obviously, every finite Predicate/Event net is weakly finite.

Theorem 5.1
There is a local state behaviour simulation of all Petri nets by
weakly finite Predicate/Event nets. There is a local simulation of
all bounded Petri nets by finite Predicate/Event nets.

   The idea is to represent i (indistinguishable) tokens at the place
p of the Petri net N by one token of type i at the place p of the
Predicate/Event net.
   Let N = [P,T,F,V,m_0] be a Petri net. If it is bounded, then let k
be the maximal number of tokens reachable at a single place. We put
$$D := \begin{cases} \{0,1,\ldots,k\}, & \text{if N is bounded,} \\ IN, & \text{otherwise,} \end{cases}$$
and
    OP := { c_i | i $\in$ D }u{ + },

where c_i is the 0-ary operation with the value i. Then we put N':=
[P,T,F',V',m_0'] over [D,OP] with m_0'(p):={m_0(p)} for p from P and
$F':=F\cup F^{-1}$. The definition of V' can be seen from Figure 8.

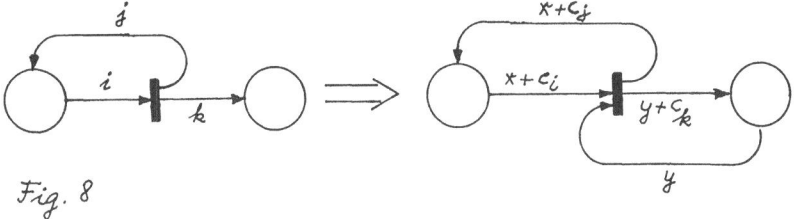

Fig. 8

Theorem 5.2.
There is a local state-behaviour simulation of all (finite) Predi-
cate/Event nets by (finite) Coloured Petri nets.

Let $N=[P,T,F,V,m]$ be a Predicate/Event net over $[D,OP]$. For t
from T by X_t we denote the set of all variables that occur in terms

from $U\{V(t,p)|p \in tF\} \cup U\{V(p,t)|p \in Ft\}$. Then $N' := [P,T,F,C',V',m'_0]$,
where
$\quad$ $C'(p) := D$ for all p from P,
$\quad$ $C'(t) := \{ \beta \mid \beta : X_t \to D \}$.

Hence, the colours of the transition t are formed by the
evaluations of the variables occurring at arcs from or to t.
Obviously, $C'(t)$ is finite if N is finite.

For any subset A of D let δ_A be the corresponding multiset, i.e.

$$\delta_A(a) := \begin{cases} 1, \text{ if } a \in A, \\ 0, \text{ else.} \end{cases}$$

Then m' is defined by
$\quad$ $m'(p) := \delta_{m_0(p)}$

and, if f is an arc from or to t, for all $\beta \in C(t)$ we put
$\quad$ $V'(f)(\beta) := \delta_{\beta(V(f))}$.

Corollary 5.3.
There is a local state-behaviour simulation of all finite Predicate/
Event nets by safe ordinary Petri nets.

6. Predicate/Transition nets

Let $[D,RL,OP]$ be a relational algebra, i.e. D is a set, RL is a
set of relations and OP a set of operations over D. Moreover, let X
be a set of variables over D and EXP the first-order language with
variables from X corresponding to $[D,RL,OP]$. Let SAT denote the set
of all satisfyable formulae from EXP, i.e. the set of all expressions exp such that there is an evaluation $\beta : X \to D$ with $Val(exp, \beta)$
= TRUE.
For any natural number n let X^n be the set of all n-tuples of
variables from X. The zero-tuple is denoted by $\varnothing$, thus, $X^0 = \{ \varnothing \}$.

Definition 6.1. (GENRICH, LAUTENBACH [4])
$N = [P,T,F,S,H,V,m]$ is a _Predicate/Transition net_ over $[D,RL,OP]$ iff
(a) $[P,T,F]$ is a net,
(b) $S: P \to IN$,
(c) $H: T \to SAT$,
(d) V is a mapping which assigns a multiset $V(f)$ over $X^{S(p)}$ to every
$\quad$ arc f from or to the place p,
(e) m_0 maps every place p to a multiset over $D^{S(p)}$, the set of all
$\quad$ S(p)-tuples of elements of D.

Interpretation.
A place p in N represents a predicate over $D^{S(p)}$ the extension of
which is given by the current marking as a multiset of S(p)-tuples of
elements of D. Hence, S(p) can be considered as the arity of the
represented predicate.

To every transition there corresponds a firing condition H(t), which is satifyable, we do not allow a transition to be blocked by its firing condition.

Let β: X ⟶ D be an evaluation. Obviously, can be extended canonically to tuples of variables (where $\beta(\varkappa)=\varkappa$) and to multisets thereof. The transition t has concession at the marking m with the evaluation β iff

    (i) Val(H(t), β) = TRUE,

    (ii) m(p) ≥ β(V(p,t)) for all p from Ft,

where again Val(exp, β) denotes the truth value of the expression exp under the evaluation β.

If t fires at m with β, the new marking m' is reached with

    m'(p) := m(p) − β(V(p,t)) + β(V(t,p)),

where undefined terms amount to the empty multiset.

Theorem 6.1.

Every Petri net is a Predicate/Transition net over the empty algebra.

Let be N = [P,T,F,V,m_o] be a Petri net. To rewrite N as a Predicate/Transition net N' over the empty algebra [∅,∅,∅] we put N' = [P,T,F,S',H',V',m'], where

    S'(p) := 0,   H'(t) := TRUE,

    V'(f) := V(f)$\varkappa$,   m'(p) := m_o(p)$\varkappa$

for all p from P, t from T and f from F. Obviously, N' is only a more sophisticated description of N.

Remark.

For every arc in a Predicate/Transition net the number of objects flowing through this arc is independent of the evaluation with which the involved transition fires. This makes it difficult to establish a local simulation of Coloured Petri nets (where the arcs have not this property) by Predicate/Transition nets.

Theorem 6.2.

There is a local state-behaviour simulation of all weakly finite Predicate/Event nets by Predicate/Transition nets.

Let N = [P,T,F,V,m_o] be a weakly finite Predicate/Event net over the algebra [D,OP]. In the first step we construct an intermediate Predicate/Event net within which the second part of the transition rule (the postcondition)

    β(V(t,p)) ∩ m(p) = ∅   for all p from tF

is always fulfilled automatically so that it need not be checked. This is done by introducing complement places in such a way that the postcondition becomes a precondition for the complement place.

For every place p we introduce the complement place $\tilde{p}$ into P, for every arc (p,t) from F we introduce the arc (t,$\tilde{p}$) and for every arc (t,p) from F the new arc ($\tilde{p}$,t) into F and put

    V(t,$\tilde{p}$) := V(p,t),   V($\tilde{p}$,t) := V(t,p), m_o($\tilde{p}$) := D−m_o(p).

Then, for all reachable markings and all p from P, it holds

    m(p) ∪ m($\tilde{p}$) = D,   m(p) ∩ m($\tilde{p}$) = ∅

and the new net has exactly the same behaviour as N.

Consider the Predicate/Transition net N' = [P,T,F,S',H',V',m'] over [D,∅,OP], where

    S'(p) := 1, m'_o(p) := $\delta_{m_o(p)}$

for all p from P. The mappings H' and V' are constructed as follows: Let t ∈ T,

    TERM(t) := U{ V(f) | f ∈ F & f incident with t }

and let φ be a one-to-one mapping from TERM(t) onto $Y_t \subseteq X$, whereby Y_t does not contain a variable occurring in a term from TERM(t).

Since N is weakly finite, the set TERM(t) is finite. We put

$$H'(t) := \bigwedge_{term \in TERM(t)} \varphi(term)=term.$$

Hence, by φ we have introduced a new variable for every term which effects the firing of t in N, and H'(t) is true iff the evaluation assigns the value of the corresponding term to this variable. Therefore, the setting

$$V'(f) := \delta_{\varphi(V(f))}$$

which assigns to every arc the multiset of those variables which correspond to terms in V(f), establishes the desired relation. It is easy to see that a firing of t in N at m has the same effect as a firing of t in N' at m'.

Theorem 6.3.
There is a local state-behaviour simulation of all Predicate/Transition nets by Coloured Petri nets.

Let $N = [P,T,F,S,H,V,m_0]$ be a Predicate/Transition net over $[D,RL,OP]$. Consider the Coloured Petri net $N' = [P,T,F,C',V',m_0]$ where

$$C'(p) := D^{S(p)},$$
$$C'(t) := \{ \beta \mid \beta : X_t \rightarrow D \ \& \ Val(H(t),\beta)=TRUE \}$$

where X_t is the set of all variables occurring in the expression H(t) or at an arc incident with t. Finally we put for every arc f incident with t and $\beta \in C(t)$

$$V'(f)[\beta] := \beta(V(f)).$$

Corollary 6.4.
There is a local state-behaviour simulation of all Predicate/Transition nets over finite algebras by Petri nets.

7. Fifo nets

Let A be a finite (nonempty) alphabet.

Definition 7.1. (FINKEL [2])
$N = [P,T,F,V,m_0]$ is a $\underline{Fifo\ net}$ over A iff
(a) [P,T,F] is a net,
(b) $m_0 : P \longrightarrow A^*$,
(c) $V : F \longrightarrow A^+ := A^* - \{e\}$.

Interpretation.
In a Fifo net every place represents a queue where processes with names from the alphabet A wait for service; thus, m(p) from A^* is the queue waiting at the place p. The service is done under the 'first in – first out' principle by the transitions of the net. To every arc f there corresponds a nonempty word V(f) over A. Let t^-, t^+ be defined as follows:

$$t^-(p) := if \ (p,t) \in F \ then \ V(p,t) \ else \ e \ end$$
$$t^+(p) := if \ (t,p) \in F \ then \ V(t,p) \ else \ e \ end$$

where e is the empty word. A transition t has concession at the
marking m: P $\rightarrow$ A* iff for all p from Ft the word V(p,t) is an
initial segment of m(p), i.e. t^{-} $\sqsubseteq$ m. If, in this case, t fires,
then a new marking m' is reached with
$$t^{-}(p)m'(p) = m(p)t^{+}(p)$$
for all p from P.

Since, for a singleton alphabet A = {a} the algebraic structures
[IN,+,0] and [A*,·,e] are isomorphic, every Petri net is a Fifo net
over a singleton and vice versa.

It is easy to see that every Fifo net can be simulated locally by
an alphabetic Fifo net, i.e. a Fifo net, where V maps F into the
alphabet A.

Since the Fifo net model is equivalent with the TURING machine,
there is no simulation of all Fifo nets by Petri nets.

Theorem 7.1.
There is a local state-behaviour simulation of all Fifo nets over the
alphabet A by Predicate/Transition nets over the algebra [A*, $\sqsubseteq$, ·],
where $\sqsubseteq$ denotes the prefix relation.

Let N = [P,T,F,V,m] be a Fifo net over A. We consider the
Predicate/Transition net N' = [P,T,F',S',H',V',m'] over [A*, $\sqsubseteq$, ·]
with

$$F' := F \cup F^{-1}, \quad S'(p) := 1, \quad m'(p) := \delta_{\{m_{0}(p)\}}$$

for all p from P. By definition, V' assigns a multiset of
(one-tuples of) variables to every arc. Here, all the values of V'
are of the form $\delta_{\{x\}}$ for x from X and the correspondence of arcs to

variables is one-to-one in the environment of each transition. By
the expression H'(t) we describe the behaviour of t in N, i.e. how t
transforms queues if it fires. Figure 9 shows how this is done.

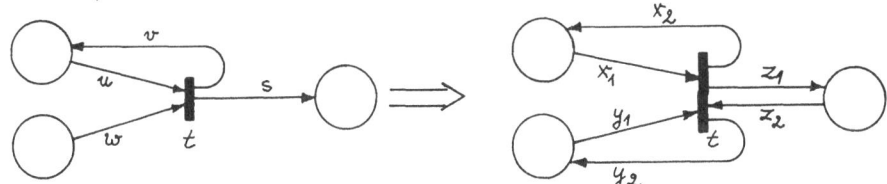

$$H(t) = u \sqsubseteq x_{1} \wedge w \sqsubseteq x_{2} \wedge ux_{2} = x_{1}v \wedge$$
$$\wedge wy_{2} = y_{1} \wedge z_{2} = z_{1}s$$

Fig. 9

Corollary 7.2.
There is a local state-behaviour simulation of all bounded Fifo nets
by safe Petri nets.

Applying the simulation of Theorem 7.1 to a bounded Fifo net we
can construct a Predicate/Transition net over the algebra of all
words of a certain fixed finite length. This algebra is finite so
that we can use Corollary 6.4.

8. Conclusions

The results underline the central role of Petri nets for net based
system modelling.

It is clear that the simulatability relation is transitive. Therefore, in the modelling of finite state systems, where only finite and bounded nets are used, all considered types of nets are locally equivalent (bisimulatable), some of them are even locally state-behaviour equivalent. This again shows the range of PETRI's basic ideas.

Moreover, we can use the above constructions to carry over notions from one type of nets (especially from Petri nets) to another one in a consistent way. Thus, e.g. a transition in a Coloured Petri net should be called live at a certain marking iff all the corresponding transitions in the corresponding Petri net are live at the corresponding marking. This additionally allows us to carry over corresponding decision procedures etc. too, i.e. to apply tools which have been developed for one type of nets to other types at least to a certain extent and for experiments.

9. References

[1] D. Corbeel, J.C. Gentina, C. Vercauter:
    Application of an extension of Petri nets to modelization and control of production processes.
    6. European Workshop on Applications and Theory of Petri nets, Espoo 1985.

[2] A. Finkel:
    About monogenous fifo Petri nets.
    3. European Workshop on Applications and Theory of Petri nets, Varenna, 1982.

[3] M. Flynn, T. Agerwala:
    Comments on capabilities, limitations and "correctness" of Petri nets.
    Computer Architecture News, Vol. 2 (1973) No. 4.

[4] H. Genrich, K. Lautenbach:
    The analysis of distributed systems by means of predicate/transition nets.
    Lect. Notes Comp. Sci. Vol. 70 (1979) 123-146.

[5] M. Hack:
    Decision Problems for Petri Nets and Vector Addition Systems.
    MIT-Project MAC, TM-59 (1975).

[6] M. Hack:
    Petri net languages.
    MIT Project MAC, Comp. Struct. Group Memo 124 (1975).

[7] K. Jensen:
    High-Level Petri nets.
    Informatik-Fachberichte Bd. 66 (1983) 166-180.

[8] H. Müller:
    Prompt and hangupfree simulations of place/transition nets by pure nets without multiple arcs.
    Petri Net Newsletter No. 15 (1983) 16-21.

[9] W. Reisig:
    Petrinetze.
    Springer-Verlag, Berlin-Heidelberg-New York, 1982.

[10] P.H. Starke:
Monogenous fifo nets and Petri nets are equivalent.
Bull. EATCS No. 21 (1983).

[11] P.H. Starke:
Petri-Netze.
VEB Deutscher Verlag der Wissenschaften, Berlin 1980.

[12] G. Ullrich:
Der Entwurf von Steuerstrukturen fuer parallele Ablaeufe mit
Hilfe von Petri-Netzen.
Universitaet Hamburg, Institut f. Informatik, IFI-HH-B-36/77
(1976).

[13] R. Valk:
Self-modifying nets.
Univ. Hamburg, Inst. f. Informatik, IFI-HH-B-34/77 (1977).

DEVELOPMENT AND APPLICATION OF PETRI NET BASED TECHNIQUES IN AUSTRALIA

F J W Symons

Telecom Australia Research Laboratories

770 Blackburn Road, Clayton, 3168, Victoria, Australia

ABSTRACT

Petri Net based techniques are currently being used in several organisations in Australia as analytical tools and models for the investigation and solution of practical problems in the specification, verification and implementation of communication and computing systems. Most of the people involved in Australia are using an extension of Petri Nets called Numerical Petri Nets (NPNs), together with a versatile computer based protocol analysis tool called PROTEAN (Protocol Emulation and Analysis).

The applications of Petri Net based techniques include the specification of Open Systems Interconnection (OSI) and Integrated Services Digital Network (ISDN) services, protocols and interfaces; the verification of OSI protocol specifications against their service specifications; the analysis of protocols and signalling systems; the generation of test sequences for testing conformance of protocol implementations; as an integral part of a systematic means of producing operational code from high level specifications of services and protocols; the study of computer architectures to enable direct execution of Petri Nets in order to take advantage of concurrency in specifications; and the development of theoretical techniques for the analysis of Petri Nets.

An overview is given of the development and application of Petri Net based techniques in Australia over the last ten years, as a tribute to the pioneering work of Dr Carl Adam Petri.

1. INTRODUCTION

This paper has been prepared as a tribute to the work of Dr Carl Adam Petri on behalf of many colleagues in Australia who are working in the areas of the development and application of Petri Net based techniques.

It is both a great pleasure and an honour to propose this deeply felt appreciation of the pioneering work of Dr Petri in leading the world towards better techniques for the specification, design, analysis and understanding of computing and communication systems. Widely available and readily accessible new types of communication and computing services and systems have the potential to enrich the lives of all people, and it is of paramount importance that techniques be developed to enable their efficient design, implementation, maintenance and enhancement. Petri Net based techniques have great potential to continue to make valuable contributions to these systems and to their universal application.

Although in Australia we are on the opposite side of the world, and about as far as it is possible to be from Dr Petri's home, many people and organisations in Australia are researching, developing and applying Petri Net based techniques to the solution of a wide range of practical problems. It is probably of interest to Dr Petri and to others in the field for me to indicate some of the current activities in Australia, and to outline the historical developments in our involvement with Petri Net based techniques.

It is of particular pleasure to me to prepare this paper, as I was one of the first in Australia to study and apply Petri Nets. It is also appropriate to send this tribute from the Telecom Australia Research Laboratories (TRL) in Melbourne, as TRL has been the focal point, launching pad, driving force and catalyst for most of the Petri Net based work performed in Australia so far.

In TRL we are very confident that Petri Net based techniques will continue to be of importance for many years to come, as they address some of the fundamental characteristics of asynchronous, concurrent communicating processes which are the foundation of modern communicating computer systems and distributed processing systems.

2. CURRENT APPLICATIONS OF PETRI NET BASED TECHNIQUES IN AUSTRALIA

By the middle of 1986, Petri Net based techniques were being studied and applied in several Australian Universities, Colleges of Advanced Education, research organisations and industrial companies. They are being used for a wide range of purposes, including the specification of Open Systems Interconnection (OSI) and Integrated Services Digital Network (ISDN) services, protocols and interfaces;

the analysis of protocols and signalling systems; the generation of test sequences for testing conformance of protocol implementations; as an integral part of a systematic means of producing operational code from high level specifications of services and protocols; the study of computer architectures to enable direct execution of Petri Nets in order to take advantage of concurrency in specifications; and the development of theoretical techniques for the analysis of Petri Nets.

Most of the Australian studies and applications have been concerned with the correct logical functioning of systems, but attention is increasingly being paid to including time in the Petri Net models so that they can be used for investigations of the performance of systems [Symons 82A, Wong 84A, Billington 85B].

Most of the people involved in Australia are using an extension of Petri Nets called Numerical Petri Nets (NPNs), together with a versatile computer based protocol analysis tool called PROTEAN (Protocol Emulation and Analysis). Both of these developments have been made by people from TRL, and a brief history of the development and application of NPNs, PROTEAN and related activities is given below.

3. INITIAL STUDIES AT THE UNIVERSITY OF ESSEX

The study and application of Petri Net based techniques in TRL has its origins in the studies I undertook at the University of Essex in England from 1975 to 1978 under a Telecom Australia Scholarship. While at the Univerity of Essex I was introduced to Petri Nets by Phillip Shaw, Michael Hills and Frank Coakley as a modelling technique which they believed had potential for application to the verification of communication protocols. (As a related study Michael Hills also stressed the need for the availability of design rules for different types of protocols, so that protocols with desired properties could be designed readily with confidence. As part of the study he could see the value in identifying canonical classes of protocols with proven properties. He was also looking for ways to derive and verify these design rules.)

Following some initial studies of Petri Nets, the idea of making particular extensions to Petri Nets to form Numerical Petri Nets arose during discussions with Phillip Shaw in May 1976. As the ideas for the extensions were triggered by attempts to use Petri Nets to model the handling of numerical data in a call establishment protocol, the name Numerical Petri Nets was chosen.

The extensions involved in Numerical Petri Nets were primarily motivated by a desire to obtain much more compact and convenient formal descriptions of complex protocols, and essentially involved the following:

- Tokens can have fields of information, or attributes;

- The tokens removed from input places on the firing of a transition are defined independently of the enabling conditions for that transition;

- The tokens added to output places on the firing of a transition are independently specified;

- An optional transition enabling condition is included, which is true or false depending on the values of data variables which may be associated with the net;

- An optional transition operation on the values of the data variables is performed when the transition fires.

The studies at the University of Essex confirmed the potential of the application of NPNs to the description and analysis of a range of different types of protocols [Symons 76A, Symons 78A]. A methodology was developed for the specification, analysis and verification of communication protocols using NPNs in conjunction with the developing CCITT Specification and Description Language, SDL; and a protocol analysis and verification software package was developed and applied to the analysis of several different Petri Net and NPN structures and to the verification by reachability analysis of a representative call establishment protocol. The protocol verification involved six different types of tokens, four different transition enabling conditions and two different "firing rules".

All the modelling and analysis studies described above were concerned only with the logical behaviour or functional performance of systems or networks. The importance was also recognised of addressing the modelling and analysis of the performance of a system in terms of throughput, delays and processing capacity required. All computer communication networks are composed of a large number of interconnected queueing systems with storage at many points. These can become extremely complex with multiple interactions or feedback between the various queueing systems, especially when mechanisms for flow control, dynamic routing and adaption to overload or unusual traffic levels, patterns and dispersion are considered. By the addition to NPNs of a timed transition, whose time to fire is a random variable, it was found possible to develop compact NPN models of a wide range of queueing systems, as described in [Symons 77A, Symons 78A, Symons 80A]. This extends the scope of the application of Petri Net based techniques to include a further class of practical problems.

Over the last few years, there has been rapidly increasing interest in many parts of the world in applying Petri Net based techniques to the specification, analysis and verification of communication protocols. This is indicated by the growing number of papers on these topics in both the annual Protocol Workshop and the annual Petri Net Workshop, and especially since 1982. On the occasion of Dr Petri's 60th birthday we are very pleased to be able to say that Australians were among the first to apply Petri Net based techniques to the specification, analysis and verification of communication protocols, as our studies reach back to 1975. In a survey [Sunshine 78A] of techniques being used for these purposes presented at the Liege Computer Networks Symposium in February 1978, the only Petri Net based studies mentioned were mine at the University of Essex.

4. STUDIES, DEVELOPMENTS AND APPLICATIONS IN AUSTRALIA

Studies in what is now called Protocol Engineering were commenced at TRL in the middle of 1979, using the University of Essex work on NPNs and protocol modelling and analysis as a starting point. The studies were integrated with several other TRL theoretical and practical investigations, as TRL has had an active interest in formal methods of specification and systematic methods of software system design and testing, including automatic implementation, since the late 1960s, as described in [Gerrand 86A]. These investigations were in support of the continuing development of Telecom Australia's customer services and networks, and were part of Telecom Australia's commitment to the development of international standards by CCITT and

ISO. As these activities developed in TRL, contact and collaboration with other groups in Australia were progressively increased.

Since 1979, several developments have been made in many aspects of protocol engineering by people in TRL and by our colleagues in other organisations, and these activities are outlined below. For most of this time the team in TRL has been led by Jonathan Billington, with significant contributions from several other people who will be identified later.

4.1 Computer Aids for the Analysis and Verification of Communication Protocols

The PROTEAN system can be used for the detection of deadlocks, livelocks, looping and other malfunctions. It can also be used to test the recovery behaviour of a protocol under error conditions or other malfunctions. Most of the analysis so far has been based on reachability analysis. Recent studies have involved language analysis, and investigations have started using invariants analysis [Billington 85A, Wheeler 85B, Wheeler 86A].

Over the last few years the facilities of the PROTEAN system have been progressively extended, and its ease of use has been improved. Significant contributions to PROTEAN have been made by the late Neil Gaylard, Michael Wilbur-Ham, John Gilmour, Jonathan Billington and Geoff Wheeler of TRL, and by Stephen Young and Sue McPherson of the University of Melbourne.

PROTEAN has a graphical display and output, and details of its applications, capabilities and planned further enhancements can be found in [Billington 85A, Wheeler 85B]. Copies of PROTEAN, on a non disclosure basis, have been made available to several Universities and research organisations in Australia in order to facilitate their studies of protocol engineering and Petri Net based techniques, and to promote national cooperation. PROTEAN has also been sold to the Jutland Telephone Company, Denmark, together with extensive manuals and detailed training courses.

4.1.1 Diagram Graphics System

As a collaborative undertaking by the Commonwealth Scientific and Industrial Research Organisation (CSIRO) and TRL, with significant funding by ICL(Australia), an interactive diagram editor system, GENIE, has been produced to enable the convenient production of SDL and NPN diagrams. Contributions were made by Graham Freeman, Pierre de Chazal and John O'Callaghan of CSIRO; and by Evelyn Swenson, Denise Hegearty, Peter Gerrand and Ron Haylock of TRL. Software has been developed by John Gilmour to convert the GENIE data files into the correct format for input to PROTEAN.

4.2 Analysis and Verification of Protocols

The PROTEAN system has been used to find and eliminate errors in Telecom Australia specifications for the T6 signalling system over digital junction signalling links and for the P1 signalling system for the connection of digital PABXs to the public networks.

PROTEAN has also been used to find errors in the CCITT D Channel ISDN single frame procedures, [Wheeler 85B]; the OSI Transport Protocol Classes 0 and 2, [Bearman 84A, Bearman 84B, Bearman 86A]; the message transfer part of the CCITT No 7 common channel signalling system; and in the packet level protocol and link access procedures of X25.

In all these analyses of protocols using PROTEAN, errors were identified in the protocol specifications; and many of the errors detected involved subtle interactions between the protocol entities, even in situations where transmission network errors such as resequencing, and loss and duplication of messages did not occur.

In all these verifications it was found possible to use NPNs to deal separately with various protocol functions in order to increase readability and understanding. The NPN definitions of these various functions, or phases, of the protocol can be combined very easily in order to form the total specification, as described in [Symons 78A, Billington 83A, Bearman 84A, Bearman 84B, Bearman 86A].

4.3 Specification and Description Techniques for Open Systems Interconnection Layers

The initial protocol modelling and analysis work concentrated on deriving NPN specifications of protocols, together with NPN specifications of the functions provided by the underlying networks and the behaviour of the end systems. An important development, as described in [Billington 83A], was to show how NPNs could be used at several different levels of abstraction, and in particular at a level of abstraction appropriate for the specification of the behaviour of Layer Services as used in the ISO/CCITT Reference Model for Open Systems Interconnection.   In these NPN specifications two major aspects are defined, namely the set of possible global sequences of the service primitives and the queueing model of a connection. Related studies using NPNs are described in [Cellary 84A].

The application of NPNs to OSI services and protocols was taken further in systems developed for international trials of the OSI Network Service and the OSI Transport Service conducted in cooperation with CSIRO, the Canadian Department of Communications and the University of Uppsala, Sweden.   For the Australian implementation, NPNs were used for the specification of the Transport Layer and Network Layer Services [O'Neill 84A], the specification of the Transport and Network Protocols and the specification of the Interfaces between layers.   NPNs were also used as implementation dependent specifications from which the software was produced.   The software for the Transport Layer was validated using protocol test sequences generated from the NPN protocol model, as described in [Ford 84A].

This practical application of NPNs for the specification and implementation of software, together with the parallel enhancement and application of PROTEAN, proved that NPNs could be used at all levels from abstract specifications of services to detailed specifications of implementations.

4.4 Verification of Protocol Specifications against Service Specifications

From the time that the concept of layer services was introduced and utilised in the OSI Reference Model, there has been a need to be able to verify a layer protocol specification against the layer service specification.   For this to be possible the relationship between the service primitives as used in service specifications and the corresponding information flow defined in the protocol specification must be defined.   Being able to use the same technique, NPNs, for both service specifications and protocol specifications opens the way for this verification task to be relatively straightforward, and the approach adopted and results achieved are described in [Billington 85A, Wheeler 85B, Wheeler 86A].

4.5   Evolution and Formal Definition of NPNs

A formal definition of NPN syntax and semantics, together with standardised notations suitable for machine recognition, has been completed, mostly by Geoff Wheeler [Wheeler 85A].   A complementary tutorial on NPNs has also been completed [Wilbur-Ham 85A]. Mainstream developments in Petri Net theory and application in the northern hemisphere have been closely monitored, and it has been found possible to align NPNs and their formal definition to a large extent with other developments in Petri Net Theory.   Self Modifying Nets and Predicate Transition Nets [Genrich 81A] have been included as subclasses of NPNs.   The formal definition and standardised notation have been retrofitted to PROTEAN which now makes use of a subclass of NPNs of wide applicability.

4.6   Industrial Applications

Under a TRL contract an Australian company, Unico, is developing implementations of four of the five OSI Transport Layer Protocols, working from NPN versions of the OSI specifications, and in particular taking advantage of the work described in [Bearman 84A, Bearman 84B, Bearman 86A].   This experience should throw further light on the value and utility of Petri Net based techniques for commercial products.

4.7   Concurrent Computer Architectures

Partly under a TRL contract, Greg Egan of the Royal Melbourne Institute of Technology has been investigating the possibilities of designing data flow architectures which can directly execute NPN specifications and take advantage of the concurrency in the specifications.

4.8   International Standards for Formal Description Techniques

Since the early 1970s, CCITT and ISO have been searching for Formal Description Techniques (FDTs) suitable for adopting as international standards for the specification of software systems, signalling systems, protocols and services.   Since 1972, TRL has been actively involved in this search [Gerrand 86A], and has made significant contributions to the international efforts.   Since 1981, TRL has been joined by CSIRO in furthering this work.   For the next generation of

FDTs, which must meet requirements not met by any currently standardised technique, Australia has proposed to both CCITT and ISO that Petri Net based techniques should be seriously considered, and NPNs in particular. We have argued that Petri Net based techniques are suitable as they are able to describe highly concurrent complex systems at several levels of detail, and are amenable to analysis and the determination of protocol properties under a range of conditions. We believe that there are other important reasons why Petri Net based techniques should be seriously considered as an international standard. These are that the same techniques can be used at several different levels of abstraction or detail, and can be used for services, protocols and interfaces of the OSI Reference Model; that Petri Net based techniques can be used very flexibly, that they have visual appeal and they offer great assistance to intuitive understanding of complex systems.

4.9   Comments and Further Activities

Looking back, we believe that much has been achieved in the development and application of Petri Net based techniques by a relatively small group of people over the last six years. In my opinion this has been due, not only to the skills, enthusiasm and dedication of the individuals involved, but also to the team management structure in TRL, to cooperation between several organisations, and to the involvement of individuals in many different aspects of protocol engineering. These aspects have ranged from theoretical studies to contributing to Telecom Australia, CCITT and ISO specifications, and to practical implementations for national and international trials. A significant level of synergy has been obtained from the blending of considerations of protocols from many different points of view, and from the approach of following ideas through to realisation.

I believe that further research is required on all the aspects of protocol engineering described above in order to develop the tools and techniques to manage effectively the increasing range, diversity and complexity of protocols. I also believe that Petri Net based techniques have the potential to be of considerable further value in all these areas, and that the best results will be obtained where there is close interaction between the people developing the theory and those whose main concern is the solution of practical problems.

There is a need to find techniques to enable us to manage complexity, and there is a complementary need to reduce the complexity of practical protocols. We need to return to the objectives of Michael Hills described above, of deriving design rules for canonical classes of protocols with guaranteed properties under specified conditions. The studies to meet these objectives should benefit from the use of Petri Net based tools, such as PROTEAN, in particular to assist in finding ways of proving the design rules.

Petri Net based techniques have much to offer these future studies on account of their ability to assist understanding and human communication, to stimulate insight into problems -- and to be one of the more enjoyable formal description techniques to work with.

5. SUMMARY

I hope that this overview of our study and application of Petri Net based techniques has shown to Dr Petri and his colleagues why several of us "down under" feel a special debt to his pioneering work, and why we believe that there are more fruits to be obtained from further study and application of Petri Net based techniques.

6. ACKNOWLEDGEMENTS

The permission of the Director Telecom Australia Research Laboratories to publish this paper is acknowledged.

Acknowledgement is given to the following people who have made contributions to the study and application of Petri Net based techniques in Australia, but whose names have not appeared in either the text or the references, being Gary Dickson, Rob Evans, Kim Nguyen, Paul Kirton, Greg Millsteed and Greg Rochlin.

7. REFERENCES

[Bearman 84A] Bearman M. Y., Wilbur-Ham M. C. and Billington J., "A Formal Specification of the OSI Class 0 Transport Protocol Using NPNs", Telecom Australia Research Laboratories Report 7736, 1984.

[Bearman 84B] Bearman M. Y., Wilbur-Ham M. C. and Billington J., "Specification and Analysis of the OSI Class 0 Transport Protocol", Proc of ICCC84, Sydney, October - November 1984, pp 597-602.

[Bearman 86A] Bearman M. Y., Wilbur-Ham M. C. and Billington J., "Analysis of the OSI Class 0 Transport Protocol", Telecom Australia Research Laboratories Report 7737, 1986.

[Billington 83A] Billington J., "Abstract Specification of the ISO Transport Service Definition using Labelled Numerical Petri Nets", Proc Third International Workshop on Protocol Specification, Testing and Verification, Ruschlikon, Switzerland, May-June 1983, pp.173-185.

[Billington 85A] Billington J., Wilbur-Ham M.C. and Bearman M.Y., "Automated Protocol Verification", Proc of the 5th International Workshop on Protocol Specification and Testing, Toulouse, June 1985, pp 2-11 to 2-22.

[Billington 85B] Billington J., "On Specifying Performance Aspects of Protocol Services", Proc International Workshop on Timed Petri Nets, Turin, July 1985.

[Cellary 84A] Cellary W., Sajkowski M. and Stroinski M., "Defining a Transport Layer using Numerical Petri Nets", Proc First International Conference on Computers and Applications, Peking, China, June 1984.

[Ford 84A] Ford W. S. and O'Neill C. J., "Design and Specification of an OSI Network Interface Protocol", Proc ICCC84, Sydney, October - November 1984, pp 591-596.

[Genrich 81A] Genrich H. and Lautenbach K., "System Modelling with High Level Petri Nets", Theoretical Computer Science 13, 1981, pp 109-136.

[Gerrand 86A] Gerrand P. H., "Experience Gained in Applying Formal Description Techniques to the Design of Complex Real Time Computing Systems", Proc First Australian Software Engineering Conference, Canberra, May 1986, pp 27-33.

[O'Neill 84A] O'Neill C. J., and Billington J., "Proposed Abstract Specification of the Network Service Definition", SDL Newsletter, No. 6, January 1984, pp 139-158.

[Sunshine 78A] Sunshine C. A., "Survey of Protocol Definition and Verification Techniques", Computer Networks Symposium, Liege, February 1978, pp F1-4.

[Symons 76A] Symons F. J. W., "Modelling and Analysis of Communication Protocols using Petri Nets", Dept of Electrical Engineering Science Telecommunications Systems Group Report No. 140, University of Essex, September 1976.

[Symons 77A] Symons F. J. W., "The Description and Definition of Queueing Systems using Numerical Petri Nets", Dept of Electrical Engineering Science Telecommunications Systems Group Report No. 143, University of Essex, March 1977.

[Symons 78A] Symons F. J. W., "Modelling and Analysis of Communication Protocols using Numerical Petri Nets", Ph. D Thesis, University of Essex, also Dept of Elec. Eng. Science Telecommunications Systems Group Report No. 152, May 1978.

[Symons 80A] Symons F. J. W., "The Description and Definition of Queueing Systems by Numerical Petri Nets", Australian Telecommunications Research, Vol. 13, No. 2, 1980, pp 20-31.

[Symons 82A]   Symons F. J. W., "The Potential of Numerical Petri Nets as a Modelling Tool for Computing Systems and Networks", Proceedings of Ninth Australian Computer Conference, Hobart, August 1982, pp 794–810.

[Wheeler 85A]   Wheeler G. R., "Numerical Petri Nets - A Definition", Telecom Australia Research Laboratories Report 7780, April 1985.

[Wheeler 85B]   Wheeler G.R., Wilbur-Ham M.C., Billington J. and Gilmour J.A., "Protocol Analysis Using Numerical Petri Nets", Proc of the 6th European Workshop on Applications and Theory of Petri Nets, Espoo, Finland, June 1985, pp 209-226.

[Wheeler 86A]   Wheeler G. R., Batten, T. J., Billington, J. and Wilbur-Ham, M. W., "A Methodology for Protocol Engineering", to be published in Proc ICCC86, Munich, September 1986.

[Wilbur-Ham 85A]   Wilbur-Ham M.C., "Numerical Petri Nets - A Guide", Telecom Australia Research Laboratories Report 7791, May 1985.

[Wong 84A]   Wong C. Y., Dillon T. S. and Forward K. E., "Analysis of Timing Aspects of Communicating Computer Systems using Timed Places Petri Nets", Proc ICCC84, Sydney, October - November 1984, pp 585-590.

QUANTITATIVE ANALYSIS OF A RESOURCE ALLOCATION
PROBLEM: A NET THEORY BASED PROPOSAL

M. Tazza

Curso de Pós-Graduação em Ciência da Computação
UNIVERSIDADE FEDERAL DO RIO GRANDE DO SUL
Caixa Postal 1501
90.001 - Porto Alegre - RS - BRAZIL

Abstract

The prevention of deadlocks is studied for a special class of resource
allocation problems. A resource decentralization algorithm identifies
safe allocation strategies. Given an initial resource availability,
each strategy is then quantitatively analysed in order to determine
the one that allows a relative maximum for the flow of processes
through a system. The structural and dynamic characteristics of the
problems suitable for analysis, as well as the proposed solution, are
described in terms of place/transition nets.

Contents

* This work was partially supported by CNPq and FAPERGS.

1   INTRODUCTION

We analyse a special class of resource allocation problems, where   an
unspecified number of processes compete for resources. The class of
problems suitable for analysis is defined by the following assumptions:
1) the processes are uniform with respect to resource demand, 2) the
demands for resources by a process are known at each step, 3) to
proceed in its activity each process seizes (takes) the necessary
resource units from a central deposit, 4) released units are returned
to the central deposit, 5) among the processes, the resources are
seized and released concurrently and in any order and 6) at the end of
a run the total number of seized units has been released by a process.

In the sequel we suppose that the reader is familiar with the   basic
structural, dynamic and terminological aspects of net theory in general
and with the place/transitions-nets (P/T-nets) in particular. As usual
the triple $(S,T;F)$ is considered as a net and the 6-tuple $N=(S,T;F,K,$
$B,Mo)$ as a P/T-net. B attaches a weight to each arc in the net and   Mo
is the initial marking of the places. The set $\bullet x(x\bullet)$ defines the
preset (postset) of an element $x \in S \cup T$ of the net. The set of
reachable markings, starting from a given marking M by occurrence   of
transitions, is designated by $[M>$. Due to the multiplicity of concepts
for the liveness of nets we define it formally. We use the notion   of
liveness for P/T-nets that requires, for each marking, the possibility
of each transition being enabled (/REIS85/).

Definition-1 : Let $N=(S,T;K,F,B,Mo)$ be a P/T-net, let $t \in T$.
i) t is called <u>live</u> iff $\forall M \in [Mo> \ \exists M' \in [M>$ such that t is M'-enabled
ii) N is called <u>live</u> iff $\forall t \in T$, t is live.

The structural and dynamic characteristics of the resource allocation
(i.e. the seize/release pattern) defined by the given assumptions   is
modelled by P/T-nets (see e.g. Fig-1).

Place r represents the central deposit of resources and its initial
marking, $Mo(r)=7$ models the initially available units. Transitions
ts_1, ts_2 and ts_3 represent <u>se</u>ize (allocation) transitions. Transitions
tr_1 and tr_2 are <u>re</u>lease transitions. The marking of place w_0, $Mo(w_0)=+$
is a reminder for an infinite number of waiting processes. This has
the same behavioural effect of an unspecified number of cyclic
processes competing for resources.

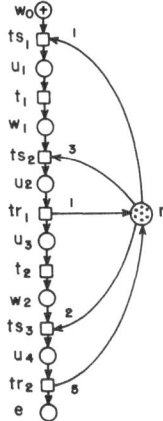

Figure 1 - P/T-net model of a resource allocation problem

The first point we are interested in is the property of liveness   in
any net that models the utilization of resources for problems in   our
class. Suppose for instance that 4 processes entered the system of
Fig-1 and are waiting in w_1. Seize transition ts_2 may now occur   and
only one process may proceed to place u_2. The only enabled   transition
at that marking is the (release) transition tr_1. The system   reaches a
situation in which three processes are waiting for resources at w_1
while the other is waiting for two units at w_2. We have only one avai<u>l</u>
able unit in r and the system has reached a dead marking. The   trivial
solution wich allows just one process in the system is not satisfactory
from the point of view of maximizing the throughput (flow of processes
through the system). Allowing more than one process in the system
without any regard to future necessities, can be dangerous for the
liveness of the system. The maximization of the flow of   processes,
while still guaranteeing the liveness of the net, represents the
second point of interest.

Lautenbach and Thiagarajan /LATH79/ presented a solution for the
liveness problem based on the strategy of guiding the net (and   hence
the system) through the safe modes of behaviour. In this sense, the
solution is optimal: not only <u>all</u> the safe allocation patterns are
allowed but also, after an initial amount of computation, no   checking
is necessary to decide to grant or to deny a particular request.
However, the class of systems considered is restricted: they   require
that all the seizures of a resource type have occurred before the
release of resources may take place.

We present a solution that guarantees the liveness of the net and allows the throughput of the system to reach a value near the theoretical maximum.

2 THE APPLICATION SCOPE

The application scope of the model we propose will be formally defined. The structural class of the resource allocation problems suitable for analysis will be defined in terms of restrictions on the flow relation F of a net (S,T;F).

Definition-1 Let (S,T;F) be a net. The triple is an __open-chain__ iff

i) $S = \{s_0, s_1, \ldots, s_n\}$ and $T = \{t_0, t_1, \ldots, t_{n-1}\}$

ii) $F = \{(s_i, t_i)|0 \leq i \leq n-1\} \cup \{(t_i, s_{i+1})|0 \leq i \leq n-1\}$

The definition above establishes a total ordering over the elements of S and T. Let $0 \leq i$ and $j \leq n$ then $s_i < s_j$ if $i < j$ and $t_i < t_j$ if $i < j$. An example of an open-chain is given in Fig.-2a.

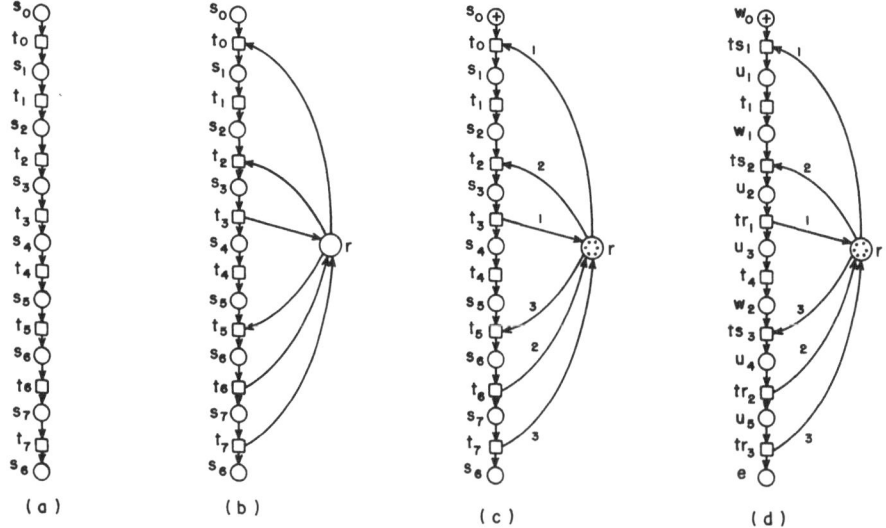

Figure 2 - a): open-chain
            b): CSCS.
            c): CQCS.
            d): CQCS with interpreted transitions and places.

Definition-2 A net $(S \cup \{r\}, T; F \cup \{(r, t_0), (t_{n-1}, r)\} \cup F')$ is a
centralized structural cycle-scheme (CSCS) iff

i) $(S,T;F)$ is an open-chain

ii) $S \cap \{r\} = \emptyset$

iii) Let $C_i = \{(r,t_i),(t_i,r)\}$, i=1,2,..., n-2

$(F' \subset \bigcup_i C_i) \wedge (|F' \cap C_i| \leq 1)$

Part one of our definition states that a subnet of a CSCS is an open-
-chain. Part two includes a new place, r, with an arc to t_0 and an arc
from t_{n-1} to r. Part three defines the flow relation between the
included r-place and the t_1, t_2, ... t_{n-2} transitions. Fig-2b shows a
CSCS. We can obviously make reference to the open-chain of a CSCS.

Definition-3 A P/T-net $(S,T;F,K,B,M_0)$ is a centralized quantitative
cycle-scheme (CQCS) iff

i) $(S,T;F)$ is a CSCS

ii) $K: S \rightarrow \mathbb{N} \cup \{i\}$ assigns a possibly unlimited (i) capacity to   each
place in S (i stands for „infinite")

iii) $B: F \rightarrow \mathbb{N} \setminus \{0\}$ attaches a weight to each arc of the net $(S,T;F)$
such that:

iii.a) the weight of any arc in the open-chain is 1, that is
$B((s_i, t_i)) = B((t_i, s_{i+1})) = 1$, i=0,1,2,...n-1

iii.b) $B((t_i, r)) \leq \sum_j B((r, t_j)) - \sum_k B((t_k, r))$ j < i, k < i

iii.c) $\sum_i B((t_i, r)) = \sum_j B((r, t_j))$

iv) $Mo: S \rightarrow \mathbb{N} \cup \{i\}$, the initial marking of the P/T-net satisfies:

iv.a) $Mo(s_0) = i$ (infinite)

iv.b) $Mo(r) \geq \sum_j B((r, t_j))$

iv.c) $Mo(s) = 0$ $\forall s \in (S \setminus \{r,s_0\})$

Fig-2c shows a CQCS based on the CSCS of Fig-2b.

The initial set of assumptions made about the structural and   dynamic
characteristics of the analisable allocation problems have been
formalized through the concept of a CQCS. In fact, $Mo(s_0)$ models   an
infinite number of uniform (with respect to resource demand) processes.
The demand for resources by a process is given by the weight of   the
arcs (r, t_i). Place r models the central deposit of resources, and its
initial marking models the initially available units. That the

resources are seized and released in any order is guaranteed by
Definition-2.iii. Definition-3.iii.b guarantees that a process   can
release a number of resources less than or equal to the number of units
taken by the process up to the moment of release. Definition-3.iii.c
says that at the end of each run all the seized units will be released
by a process.

Transitions whose preset (postset) includes place r will be   called
seize transitions (release transitions) and will be designated by   ts
(tr). Let N be a CQCS. The operator ts over N, ts(N), defines the set
Ts of seize transitions of N, Ts = {t ϵ T|(r F t)}. The operator tr
over N, tr(N), defines the set Tr of release transitions of N, Tr = {t
ϵ T|(t F r)}. Obviously Ts $\cap$ Tr = ϕ , Ts $\cup$ Tr $\subseteq$ T. Places s ϵ S in the
open-chain of a CQCS such that s ϵ •ts$_i$ will be called wait places
(w-places). Place s' ϵ S with an empty postset will be called e. All
other places in the open-chain will be called utilization places
(u-places). Fig-2d shows the same CQCS as Fig-2c with the terminology
introduced to designate transitions and places.

We have defined the application scope of the model we propose: the
concept of a quantitative cycle-scheme defines the structural and
behavioural characteristics of the resource allocation problems we
are interested in. We are ready now to deal with the problem of
guaranteeing that a dead marking will never be reached, and at the
same time, that the flow of processes through the open-chain will be
maximized.

3 GUARANTEEING THE LIVENESS

The proposed approach for guaranteeing the liveness of a CQCS is based
on the decentralization of the resources in r, which gives rise to one
(or more) decentralized quantitative cycle-schemes (DQCS). The open-
-chain of the CQCS is not affected by the decentralization nor are the
sets Ts and Tr. Place r (central deposit) will be substituted by n r-
-places r$_k$, k=1,2,...,n. The initially available units, Mo(r), will be
distributed among the n r-places, such that Σ_kMo(r$_k$)=Mo(r). From a
behavioural point of view the centralized and the decentralized cycle-
-schemes are not equivalent. Some of the behavioural modes of the CQCS
will not be possible in a DQCS. The lost modes includes those that
would originate dead markings, which is a desirable effect. Unfortunately some of the safe modes will be lost, too, as a result of the
decentralization process.

Definition-4 A P/T-net (S ∪ R,T;F,K,B,Mo) is a <u>decentralized</u>
<u>quantitative cycle-scheme</u> (DQCS) iff:

i) (S,T;F) is an open-chain

ii) R = {$r_1,r_2,\ldots, r_n$}, R∩S = ∅ , is the set of distributed r-places
such that

ii.a) $|\cdot r| = |r\cdot| = 1$, r ε R

ii.b) ∄t ε T|(t ε $\cdot r_1$) ∧ (t ε $r_2\cdot$), r_1, r_2 ε R

iii) Let I_i, i=1,2,..., n be the minimal S-invariant /LAUT 83/ that
covers place r_i ε R   and let S_i ⊂ S ∪ R be the support of I_i
(set of places covered by I_i). Then ∀s ε S ∃S_i| s ε S_i

iv) K assignes an unlimited capacity to each place in S ∪ R.

v) B: F → IN\{0} attaches a weight to the arcs in F such that

v.a) the weight of any arc in the open-chain is 1

v.b) B((r,r·)) = B((·r,r)), r ε R

vi) Mo(w_0) = i(infinite)

Mo(r) ≥ B((r,r·)), r ε R

Mo(s) = 0, ∀s ε (S \ {w_0})

The nets of Fig -3a and 3b are examples of DQC schemes. The net of
Fig-3c is not a DQCS: point iii) of our definition requires that every
place in the open-chain must be in the support of a (minimal)
S-invariant that covers a place r ε R. There is no such invariant   for
place w_1.

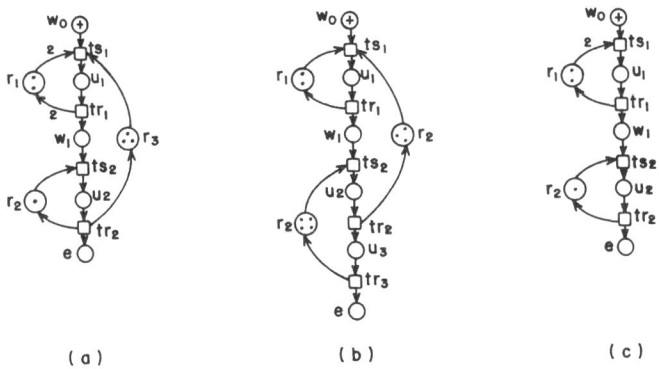

(a)   (b)   (c)

Figure 3 - a,b):DQCS
c):a net that is not a DQCS

Theorem-1 Let N= (S ∪ R, T;F,K,B,Mo) be a DQCS. N is live.

Proof: Follows from the structural characteristics and from the initial
marking of a DQCS.

We describe now an algorithm that, starting from a CQCS, yields a set
of DQCS. The decentralization algorithm considers the structural
aspects of a CQCS and the weight of the seize and release arcs. At
the same time the algorithm is presented, it will be applied to the
CQCS of Fig-1.

Algorithm-1 (decentralization of resources)

|1| (This step computes the maximal value (f) for the weight of an arc
in the DQCS).

Let $B((r, ts_i))$ and $B((tr_j, r))$, $i=1,2,...,n$ $j=1,2,...,m$ be the
weight of arcs in the CQCS.

a^+ ← max $(B((r,ts_i)))$

b^+ ← max $(B((tr_j,r)))$

f ← min (a^+,b^+)

Example (CQCS of Fig. 1): a^+ ← max(1,3,2)

$$b^+ ← max(1,5)$$

$$f ← min(3,5) (f ← 3)$$

No seize or release arc may have a weight greater that f in the DQCS's
to be originated.

|2| (For each seize and release transition in the DQCS this step
computes the number of arcs p(i) and q(j) whose weights have to be
added in order to get $B((r, ts_i))$ and $B((tr_j, r))$ in the CQCS).
$p(i)$ = min $(B((r,ts_i)), |Tr|)$
$q(j)$ = min $(B((tr_j,r)), |Ts|)$

Example: in our CQCS we have three seize transitions and two   release
transitions, $|Ts|$ = 3, $|Tr|$ = 2, i = 1,2,3, j = 1,2

$p(1)$ ← min $(B((r, ts_1)), 2)$ = 1

$p(2)$ ← min $(B((r, ts_2)), 2)$ = 2

$p(3)$ ← min $(B((r, ts_3)), 2)$ = 2

The weight of an arc (r, ts_i) in the CQCS will be "distributed" among $p(i)$ arcs in the DQCS.

$q(1) \leftarrow \underline{min}(B((tr_1,r)),3) = 1$

$q(2) \leftarrow \underline{min}(B((tr_2,r)),3) = 3$

The weight of the arc (tr_j, r) in the CQCS will be distributed among $q(j)$ arcs in the DQCS.

|3| (For each seize transition ts_i in the DQCS this step determines
    the possible values for the weight of arcs from r-places to ts_i.
    We define multisets xi such that $|xi| = p(i)$ for the possible
    values of weights. Let A_i be the set of multisets xi obtained
    for a particular transition ts_i).

    $xi = \{x_1, x_2, \ldots, x_{p(i)}\}$ such that

    a) $\sum_{x \in xi} x = B((r, ts_i))$ and

    b) $\underline{max}(x_1,x_2,\ldots, x_{p(i)}) = \underline{min}(B((r, ts_i)),f)$

Example: in step |2| we have computed the number of arcs $p(i)$ in the DQCS for each seize transition ts_i. We had $p(1) = 1$ corresponding to ts_1, $p(2)=2$ corresponding to ts_2 and $p(3)=2$ corresponding to ts_3.

For ts_1:

$p(1) = 1$, $x1 = \{x_1\}$

    a) $\sum_{x \in x1} x = B((t,ts_1)) = 1$

    b) $\underline{max}(x1) = \underline{min}(B((r,ts_1)),3)=1$

        then $x1 = \{1\}$ and $A_1 = \{\{1\}\}$

For transition ts_1 in the DQCS, $A_1 = \{\{1\}\}$ is interpreted as

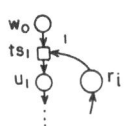

Figure 4 - Transition ts_1 in the DQCS.

For ts_2:

$p(2) = 2$, $x2 = \{x_1, x_2\}$

    a) $\sum_{x \in x2} x = B((r,ts_2)) = 3$ and

b) $\underline{\max}(x_1,x_2) = \underline{\min}(B((r,ts_2)),3) = 3$

    then $x2_1 = \{3,0\}$, $x2_2 = \{2,1\}$

    and $A_2 = \{\{3,0\},\{2,1\}\}$

The interpretation for ts_2 in the DQCS is given in Fig-5.

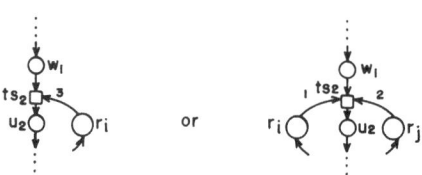

Figure 5 - Possibilities for ts_2 in the DQCS.

For ts_3:

$p(3) = 2$, $x3 = \{x_1, x_2\}$

    a) $\sum_{x\in x3}$ $x = B((r,ts_3)) = 2$ and

    b) $\underline{\max}(x_1,x_2) = \underline{\min}(B((r,ts_3)),3) = 2$

       then $x3_1 = \{2,0\}$, $x3_2 = \{1,1\}$

       and $A_3 = \{\{2,0\},\{1,1\}\}$

The interpretation for ts_3 in the DQCS:

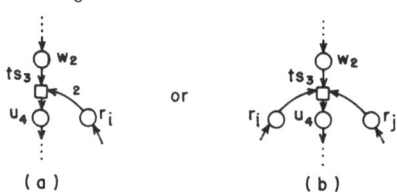

        (a)   (b)

Figure 6 - Possibilities for ts_3 in the DQCS.

|4| (For each release transition tr_j in the DQCS this step   determines
    the possible values for the weights of arcs from tr_j to r-places.
    We define multisets yj such that $|yj| = q(j)$ for the possible
    values of weights. Let B_j be the set of multisets yj obtained   for
    a particular transition tr_j).

    $yj = \{y_1, y_2, \ldots, y_{q(j)}\}$ such that

a) $\sum\limits_{y\in yj} y = B((tr_j, r))$ and

b) $\underline{max}(y_1, y_2, \ldots, y_{q(j)}) = \underline{min}(B((tr_j, r)), f)$

Example: in step |2| we have computed the number of arcs $q(j)$ in the DQCS for each release transition tr_j. We had $q(1) = 1$ for tr_1 and $q(2) = 3$ for tr_2.

For tr_1:

$q(1) = 1$, $y1 = \{y_1\}$

  a) $\sum\limits_{y\in y1} y = B((tr_1, r)) = 1$ and

  b) $\underline{max}(y1) = \underline{min}(B((tr_1, r)), 3) = 1$

     then $y1 = \{1\}$ and $B_1 = \{\{1\}\}$

The interpretation for tr_1 is:

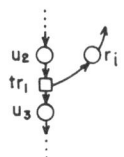

Figure 7 - Possibility for tr_1 in the DQCS.

For tr_2:

$q(2) = 3$, $y2 = \{y_1, y_2, y_3\}$ where

  a) $\sum\limits_{y\in y2} y = B((tr_2, r)) = 5$ and

  b) $\underline{max}(y2) = \underline{min}(B((tr_2, r)), 3) = 3$

     then $y2_1 = \{3,2,0\}, y2_2 = \{3,1,1\}, y2_3 = \{2,2,1\}$

     and $B2 = \{\{3,2,0\},\{3,1,1\},\{2,2,1\}\}$

The interpretation for tr_2:

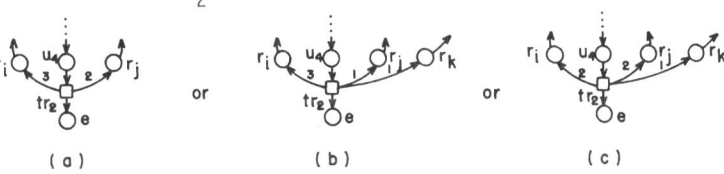

(a)   (b)   (c)

Figure 8 - Possibilities for tr_2 in the DQCS.

We carried out the decomposition of seize and release arcs. Step |5| is responsible for reassembling (seize-release) arcs with equal weights adding new r-places.

|5| (This step reassembles the seize-release arcs with equal weights)

We choose a multi-set $\mathfrak{x}$ from A_i, $i = 1,2,\ldots,$ $|Ts|$ and a multi-set $\mathfrak{y}$ from B_j, $j=1,2,\ldots,$ $|Tr|$ such that for each $x \in \mathfrak{x}$, $x > 0$ there is a $y \in \mathfrak{y}$, $y > 0$ with $x = y$.

Example: from steps |3| and |4| we have

$A_1 = \{\{1\}\}$ $B_1 = \{\{1\}\}$

$A_2 = \{\{3,0\},\{2,1\}\}$ $B_2 = \{\{3,2,0\},\{2,2,1\}, \{3,1,1\}\}$

$A_3 = \{\{2,0\},\{1,1\}\}$

The reassembling possibilities are:

$\mathfrak{x} \in A_1$ (ts_1)	$\mathfrak{x} \in A_2$ (ts_2)	$\mathfrak{x} \in A_3$ (ts_3)	$\mathfrak{y} \in B_1$ (tr_1)	$\mathfrak{y} \in B_2$ (tr_2)
{1}	{3,0}	{2,0}	{1}	{3,2,0}
{1}	{3,0}	{1,1}	{1}	{3,1,1}
{1}	{2,1}	{2,0}	{1}	{2,2,1}

The DQCS' originated through the reassembling process are represented in Fig- 9.

Given a proper initial marking, the three decentralized cycle-schemes of Fig- 9 are live. It must be noticed that while the original CQCS exhibits safe behaviour for $Mo(r) \geq 5$, the DQCS's of Fig- 9b and Fig -9c need at least 6 units. The scheme of Fig- 9d is live with 5 units.

In the next section we will analyse how to distribute the $Mo(r)$ units available in the CQCS among the n places r_k of the DQCS's in order to maximize the flow of processes through the open-chains of the DQCS's. With $Mo(r_k) \geq B((r_k, ts_i))$, $k=1,2,\ldots,n$ and $Mo(w_0) = +$ the DQCS's are live.

We introduce a further concept. A DQCS is (structurally) characterized by n cycles, one for each r-place, that share some transitions and places of the open-chain of the DQCS. Let ts_i be the transitions such

523

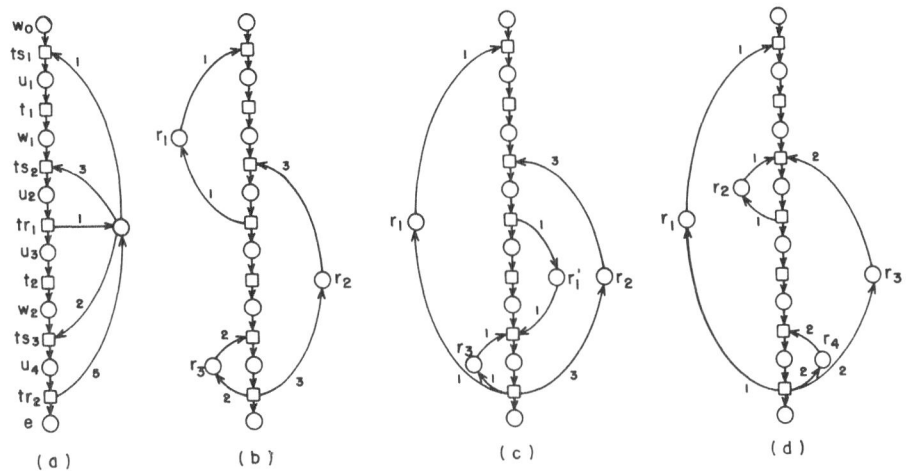

Figure 9 - a): Original CQCS (without initial marking)
         b),c),d): Corresponding decentralized schemes DQCS¹,
                   DQCS², DQCS³.

that ts_i ε $r_i \cdot$ and let tr_j be the transition such that tr_j ε $\cdot r_k$ for
a particular r_k. Each cycle together with place $\cdot ts_i$ and $tr_j \cdot$ in the
open-chain will be called a component N_k of the DQCS. Fig-10 shows
the components N_1, N_2 and N_3 of the DQCS of Fig-9b.

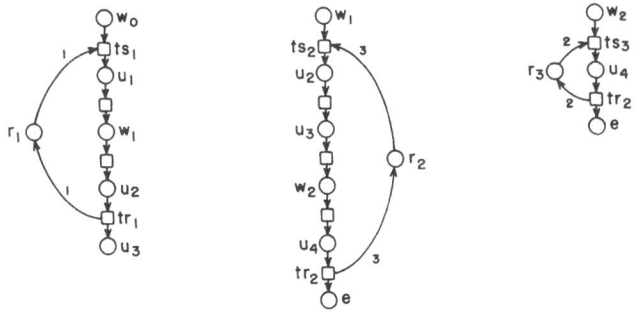

Figure 10 - Components N_1, N_2 and N_3 of DQCS¹.

In a way similar to the one used for a CQCS, we shall apply the
operators ts and tr on a component N_i of a DQCS. For example, ts(N_1)
denotes   the   seize transition ts_1 of N_1, tr(N_2) denotes   the
release transition tr_2. Obviously, tr(N_2) = tr(N3). The operator
rp (N_i) denotes the r-place of component N_i.

4 MAXIMIZING THE FLOW OF PROCESSES

In order to analyse numerically the flow of processes through the
open-chain of a DQCS, we informally introduce the concept of
utilization time Z(u) of the resources at the u-places of the scheme.
In this context a utilization time is interpreted as a 'stay time'
of the tokens in a place, as far as the transitions of the postset of
that place are concerned. A formal interpretation in terms of general
net theory is given in /RICH85/ and /TAZZ85/.

We require that a utilization time $Z(u) > 0$ be associated with each
u-place of a DQCS. If necessary, the timed DQCS will be called Z-DQCS;
otherwise we use the usual designation DQCS for a scheme with time
associated with the u-places of its open-chain.

The sum of the utilization times of a resource r_k in component N_k of a
DQCS defines the basic period Po_k of the resource ("P zero").

$$Po_k = \sum_{u \in N_k} Z(u) \qquad\qquad (eq-1)$$

The number of processes that can be concurrently served by the
component N_k is

$$m_k = \left| \frac{Mo(r_k)}{\left\lfloor B((r_k, \underline{ts}(N_k))) \right\rfloor} \right| \qquad\qquad (eq-2)$$

The flow of processes Do_k ("D zero") through the open-chain of N_k is
given by

$$Do_k = \frac{m_k}{Po_k} \qquad\qquad (eq-3)$$

(Do is measured in multiples of λ, where λ means one process per unit
of time).

Equation 3 gives the throughput of each component N_k when it is
considered as an isolated system. The decentralization algorithm
originates n components in each DQCS sharing some places and
transitions of a common open-chain. The components will interact
through the shared places and transitions and the flow through the
open-chain of the DQCS is given by

$$Do = \underline{min}(Do_1, Do_2, \ldots, Do_n) \qquad\qquad (eq-4)$$

Every component N_k with $Do_k >$ Do will be slowed down by <u>induced</u> wait times at w-places in N_k and/or at $\underline{rp}(N_k)$ until N_k reaches a value Do. The <u>index</u> <u>of</u> <u>subutilization</u> of resources of N_k is

$$U_k = 1 - \frac{Do}{Do_k} \qquad\qquad (eq-5)$$

Clearly, if all the components N_k have the same throughput then the subutilization of resources in the system is zero. To maximize the flow of processes through the open-chain of the DQCS we must distribute the available resources ($Mo(r)$) among the n r-places such that $Do_1 \simeq Do_2 \simeq \ldots \simeq Do_n$.

<u>Algorithm-2</u>: distribution of $Mo(r)$ resource units of a CQCS among n r-places of a DQCS, maximizing Do.

|1| (the initial distribution guarantees the liveness of the DQCS)

$$Mo(r_k) = B((r_k, \underline{ts}(N_k))), \quad k = 1,2,\ldots,n$$

|2| (we compute Do_k for each component N_k of the DQCS and compute the remaining available resource units R).

$$Do_k = \frac{m_k}{Po_k}, \quad k = 1,2,\ldots,n$$

$$R \leftarrow Mo(r) - \sum_k Mo(r_k)$$

|3| (we determine the slowest component N_i)
Let D be a vector such that

$$D(k) = Do_k, \quad k = 1,2,\ldots, n$$

$$i \leftarrow j \text{ such that } D(j) = \underline{min}(D(k))$$

|4| (distribution of the remaining units)

while $R \geq B((r_i, \underline{ts}(N_i)))$
do begin

$$R \leftarrow R - B((r_i, \underline{ts}(N_i)))$$

$$Mo(r_i) \leftarrow Mo(r_i) + B((r_i, \underline{ts}(N_i)))$$

$$D(i) \leftarrow \frac{Mo(r_i)/B((r_i, \underline{ts}(N_i)))}{Po_i}$$

$$i \leftarrow j \text{ such that } D(j) = \underline{min}(D(k))$$

end

|5| (there are no more sufficient available units in order to increase
the throughput of the slowest component N_i).

$$Do \leftarrow \underline{min}(D(k)), \quad k = 1, 2, \ldots, n$$

Example. Consider the three decentralized cycle-schemes of Fig-9 with
$Z(u_1) = Z(u_2) = 1z$, $Z(u_3) = 2z$ and $Z(u_4) = 1z$ (z is an arbitrary time
unit). Suppose that $Mo(r) = 14$ for the original CQCS (Fig-9a).

case a: (DQCS[1])

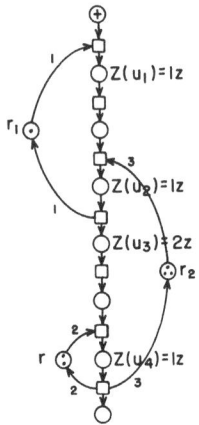

$$Po_1 = Z(u_1) + Z(u_2) = 2z$$

$$Po_2 = Z(u_2) + Z(u_3) + Z(u_4) = 4z$$

$$Po_3 = Z(u_4) = 1z$$

Figure 11: timed DQCS[1]

|1| $Mo(r_1) = 1$, $Mo(r_2) = 3$, $Mo(r_3) = 2$

|2| $Do_1 = 0,5\lambda$, $Do_2 = 0,25\lambda$, $Do_3 = 1.0\lambda$

R ← 14-6 = 8

|3| $D(k) = (0.5, 0.25, 1.0)$

i ← 2

|4| R ≥ B((r_2, ts_2))   /* 8 > 3
    R ← 8-3   /* R ← 5
    $Mo(r_2)$ ← 3+B((r_2, ts_2))   /* $Mo(r_2)$ ← 6
    D(2) ← 0.5   /* $D(k) = (0.5, 0.5, 1.0)$
    i ← 1   /* if i←2 the final result
                                                      is the same

    R ≥ B((r_1, ts_1))   /* 5 > 1
    R ← 5-1   /* R ← 4
    $Mo(r_1)$ ← 1+B((r_1, ts_1))   /* $Mo(r_1)$ ← 2

$D(1) \leftarrow 1.0$ /* $D(k)$ = (1.0, 0.5, 1.0)

$i \leftarrow 2$

$R \geq B((r_2, ts_2))$ /*4 > 3

$R \leftarrow 4-3$ /*$R \leftarrow 1$

$Mo(r_2) \leftarrow 6 + B((r_2, ts_2))$ /*$Mo(r_2) = 9$

$D(2) \leftarrow 0.75$ /*$D(k)$ = (1.0, 0.75, 1.0)

$i \leftarrow 2$

$R < B((r_2, ts_2))$ /* 1 < 3

With the allocation strategy defined by the DQCS of Fig-9b a value
Do = 0.75λ can be reached with utilization times as to Fig-11.   One
resource unit remains unused: incrementing D(1) to 1.5 will not affect
the throughput of the whole system.

case b: DQCS^2

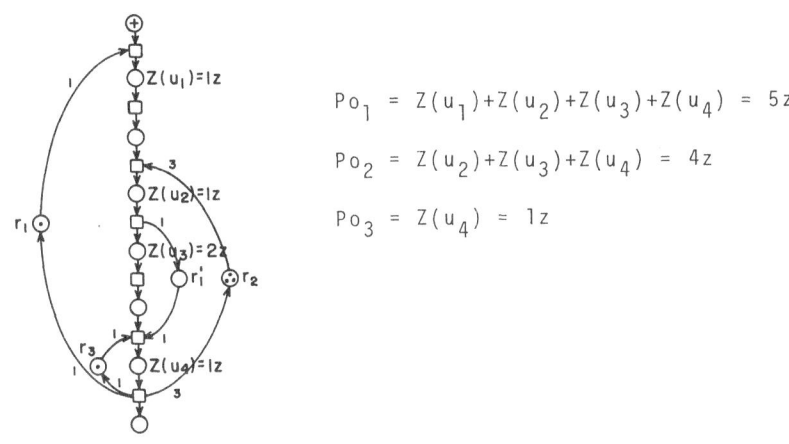

$Po_1 = Z(u_1)+Z(u_2)+Z(u_3)+Z(u_4) = 5z$

$Po_2 = Z(u_2)+Z(u_3)+Z(u_4) = 4z$

$Po_3 = Z(u_4) = 1z$

Figure 12: timed DQCS^2

|1| $Mo(r_1) = 1$, $Mo(r_2) = 3$, $Mo(r_3) = 1$

|2| $Do_1 = 0.2\lambda$, $Do_2 = 0.25\lambda$, $Do_3 = 1.0\lambda$,
    $R \leftarrow 14-5$

|3| $D(k)$ = (0.2, 0.25, 1.0)
    $i \leftarrow 1$

|4| $R \geq B((r_1, ts_1))$ /*$9 \geq 1$

    $R \leftarrow R-1$ /*$R \leftarrow 8$

    $Mo(r_1) \leftarrow 1 + B((r_1, ts_1))$ /*$Mo(r_1) = 2$

    $D(1) \leftarrow 0.4$ /*$D(k)$ = (0.4, 0.25, 1.0)

    $i \leftarrow 2$

$R \geq B((r_2, ts_2))$ /*8 $\geq$ 3
$R \leftarrow R-3$ /*R $\leftarrow$ 5
$Mo(r_2) \leftarrow 3+B((r_2,ts_2))$ /*Mo(r_2) = 6
$D(2) \leftarrow 0.5$ /*D(k)=(0.4,0.5,1.0)
$i \leftarrow 1$

$R \geq B((r_1, ts_1))$ /*5 $\geq$ 1
$R \leftarrow R-1$ /*R $\leftarrow$ 4
$Mo(r_1) \leftarrow 2 + B((r_1, ts_1))$ /*Mo(r_1) = 3
$D(1) \leftarrow 0.6$ /*D(k) = (0.6, 0.5, 1.0)
$i \leftarrow 2$
$R \geq B((r_2, ts_2))$ /*4 $\geq$ 3
$R \leftarrow R - 3$ /*R $\leftarrow$ 1
$Mo(r_2) \leftarrow 6+B((r_2,ts_2))$ /*Mo(r_2) = 9
$D(2) \leftarrow 0.75$ /*D(k)=(0.6, 0.75, 1.0)
$i \leftarrow 1$

$R \geq B((r_1, ts_1))$ /*1 $\geq$ 1
$R \leftarrow R - 1$ /*R $\leftarrow$ Q
$Mo(r_1) \leftarrow 3+B((r_1,ts_1))$ /*Mo(r_1)=4
$D(1) \leftarrow 0.8$ /*D(k)=(Q.8, 0.75, 1.0)
$i \leftarrow 2$

$R < B((r_2, ts_2))$

With this allocation strategy a value Do = 0.75λ is reached. It is
interesting to notice that the previous strategy reached the same
value with 13 resource units. This strategy however needs all the
14 units.

case c: DQCS³

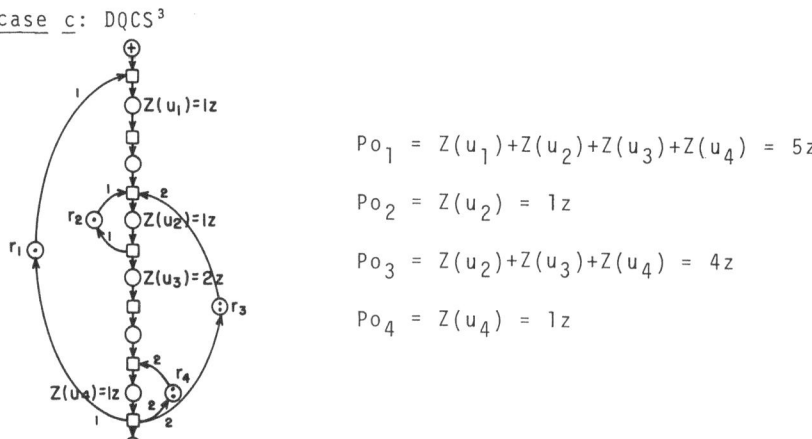

$Po_1 = Z(u_1)+Z(u_2)+Z(u_3)+Z(u_4) = 5z$

$Po_2 = Z(u_2) = 1z$

$Po_3 = Z(u_2)+Z(u_3)+Z(u_4) = 4z$

$Po_4 = Z(u_4) = 1z$

Figure 13: timed DQCS³

|1| $Mo(r_1) = 1$, $Mo(r_2) = 1$, $Mo(r_3) = 2$, $Mo(r_4) = 2$

|2| $Do_1 = 0.2\lambda$, $Do_2 = 1.0\lambda$, $Do_3 = 0.25\lambda$, $Do_4 = 1.0\lambda$
   $R \leftarrow 14 - 6$

|3| $D(k) = (0.2, 1.0, 0.25, 1.0)$
   $i \leftarrow 1$

|4| $R \geq B((r_1, ts_1))$ /*8 $\geq$ 1

   $R \leftarrow R - 1$ /*R $\leftarrow$ 7

   $Mo(r_1) \leftarrow 1+B((r_1, ts_1))$ /*$Mo(r_1)=2$
   $D(1) \leftarrow 0.4$ /*$D(k)=(0.4, 1.0, 0.25, 1.0)$
   $i \leftarrow 3$

   $R \geq B((r_3, ts_2))$ /*7 $\geq$ 2
   $R \leftarrow R - 2$ /*R $\leftarrow$ 5
   $Mo(r_3) \leftarrow 2+B((r_3, ts_2))$ /*$Mo(r_3) = 4$
   $D(3) \leftarrow 0.5$ /*$D(k)=(0.4, 1.0, 0.5, 1.0)$
   $i \leftarrow 1$

   $R \geq B((r_1, ts_1))$ /*5 $\geq$ 1
   $R \leftarrow R-1$ /*R $\leftarrow$ 4
   $Mo(r_1) \leftarrow 2+B((r_1, ts_1))$ /*$Mo(r_1)=3$
   $D(1) \leftarrow 0.6$ /*$D(k)=(0.6, 1.0, 0.5, 1.0)$
   $i \leftarrow 3$

   $R \geq B((r_3, ts_2))$ /*4 $\geq$ 2
   $R \leftarrow R-2$ /*R $\leftarrow$ 2
   $Mo(r_3) \leftarrow 4+B((r_3, ts_2))$ /*$Mo(r_3) = 6$
   $D(3) \leftarrow 0.75$ /*$D(k)=(0.6, 1.0, 0.75, 1.0)$
   $i \leftarrow 1$

   $R \geq B((r_1, ts_1))$ /*2 $\geq$ 1
   $R \leftarrow R-1$ /*R $\leftarrow$ 1
   $Mo(r_1) \leftarrow 3+B((r_1, ts_1))$ /*$Mo(r_1) = 4$
   $D(1) \leftarrow 0.8$ /*$D(k) = (0.8, 1.0, 0.75, 1.0)$

   $i \leftarrow 3$

   $R < B((r_3, ts_2))$

|5| $Do \leftarrow \underline{\min}(0.8, 1.0, 0.75, 1.0) = 0.75\lambda$

Again, with 13 of the 14 initially available resource units we reach a value 0.75λ for the flow of processes through the system shown in Fig-13.

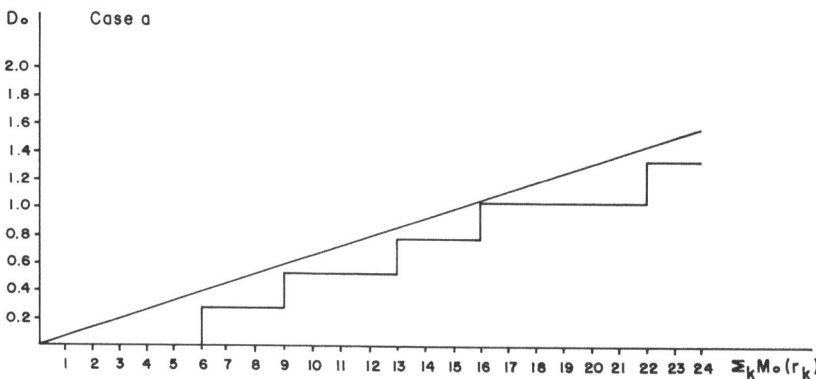

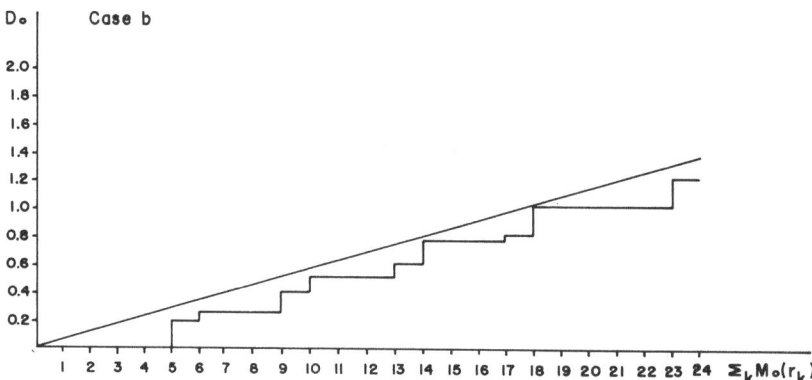

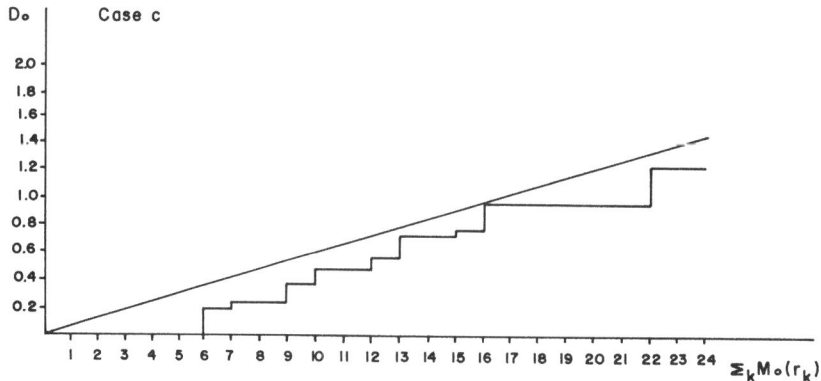

Figure 14

In Fig-14 we plot the step function throughput vs initial marking. The first conclusion is that there is no best allocation strategy: in fact, all the three may present advantages depending on the availability of resources. Suppose we have 16 available units. $DQCS^1$ and $DQCS^3$ allow a value Do = 1.0λ for the flow of processes while $DQCS^2$ allows only Do = 0.75λ. With 15 units $DQCS^1$ and $DQCS^2$ are equivalent (Do = 0.75λ) while $DQCS^3$ corresponds to Do = 0.8λ. At firts glance $DQCS^2$ would seem to be the worst strategy; it reaches Do = 0.1λ with 18 units while the other two reach the same value with 16 units. However, even $DQCS^2$ has advantages: it is the only DQCS that is live with 5 units.

The results can be used in two different ways. First, if the initial availability of resources is known, one can choose the allocation strategy that corresponds to the relative maximum for the flow of processes. Second, if a fixed value for Do is desired, we are able to select the strategy that reaches Do with a minimal number of units. Moreover, ɩnce the desidered value Do is set in the step function, the additional units needed to reach the next value can be determined by simple inspection.

5   DISCUSSION AND CONCLUSION

Resource allocation problems were analysed under two different and yet related aspects: preventing deadlocks and defining an allocation strategy that corresponds to a relative maximum for the flow of processes through the system.

Preventing deadlocks is based on the distribution of the resources of the same type. Structurally, the decentralization is a l to m mapping (one CQCS to m DQCS) where each of the originated structures defines an allocation strategy. In terms of behaviour, it is an n to m mapping, m < n, where the behavioural modes that may originate dead markings are not preserved, with the drawback.of loosing some of the safe modes. To which extent the lost modes are quantitatively relevant is still open and represents a point for future research.

The proposed solution is optimal from two points of view. First, we eliminate the need of a supervisory program to run before each seizure of resources in order to prevent deadlocks. After the execution of the decentralization algorithm the safe strategies are identified. Second, the computation of the flow of processes associated with each of the

safe strategies eliminates the need of a supervisory program for
maximizing the throughput.

One of the main results of the work is the finding that the relatively
best allocation strategy depends on the number of available resources.

Some theorems must still be formally stated and proved. One theorem
concerns the decentralization algorithm. It must be proved that
algorithm-1 originates all the decentralization possibilities with
the required structural chareacteristics (DQCS). Another theorem is
related with the claim we have made, that an allocation strategy can
be chosen so that it guarantees a value Do "near the theoretical
maximum" for the system. Intuitively, the theoretical maximum is
defined by a straight line through the point (0,0) and a point
$(\Sigma_k \text{ Mo}(r_k), \text{Do})$ which corresponds to an index of subutilization equal
zero for the resources r_k. It is obvious that each point in this line
corresponds to a value of maximal Do (i.e. zero subutilization) for a
given marking. If algorithm-1 originates all the decentralization
possibilities or if the lost safe modes in the behavioural mapping
are not quantitatively relevant, we are done. The author is convinced
of this fact, both intuitively and through exhaustive application of
the algorithm.

REFERENCES

/LATH79/ LAUTENBACH, K. and THIAGARAJAN, P.S.: Analysis of a Resource
         Allocation Problem using Petri Nets, in: Syre, J.C. (ed).
         Proc. 1st. European Conference on Parallel and Distributed
         Processing (Tolouse 1979)   260-266.

/LAUT83/ LAUTENBACH, K.: Simple Marked-graph-like Predicate Transition
         Nets. Arbeitspapiere der GMD, Nr.41 (Institut für Methodische
         Grundlagen, F1) Gesellschaft für Mathematik und Datenverarbeit
         ung mbH Bonn, july 1983.

/REIS85/ REISIG, W.: Petri nets: An Introduction. Springer-Verlag, Ber
         lin, 1985.

/RICH85/ RICHTER, G.: Clocks and their use for time modeling. In:
         Sernadas, A.; Bubenko, J: Olivé, A. (eds.): Information
         Systems: Theoretical and Formal Aspects. North-Holland,
         Amsterdam 1985, 49-66.

/TAZZ85/ TAZZA, M.: Ein netztheoretisches Modell zur quantitativen
         Analyse von Systemen (Q-Modell). Berichte der GMD, Bericht
         Nr. 149. R. Oldenbourg Verlag, Muenchen 1985.

Acknowledgments

The autor would like to thank Dr. G. Richter/GMD for his commenting
on drafts of these notes.

Existential Quantifiers in Predicate-Fact-Nets

Gerda Thieler-Mevissen

Robert-Koch-Strasse 11

6101 Rossdorf - Gundernhausen

Abstract:

It is well known that every formula of first-order predicate logic in prenex form can be represented as a strict predicate transition net (PrT-net) in which all transitions are seen to be dead. Nets wherein all predicates can carry only one specimen of an n-tuple are called strict [Ge]. In this net representation of predicate logic Skolem functions are used to get descriptions for existential quantifiers [Th].

In PrT-systems given by PrT-nets and an initial marking, the dead transitions specify the facts about that system. S-invariants are a very useful means to prove transitions of a PrT-system to be facts. In those predicate fact nets (PrF-nets) derived from S-invariants, Skolem functions do not occur.

The question arises, whether existential quantifiers do not exist in those fact nets, or whether nets carry another more natural representation for them. Sums seem to yield such an appropriate net description for existential quantifiers. But some derivation problems remain, in which sums cannot replace the Skolem functions.

1. Facts in Predicate Transition Systems

A predicate transition net $N=(S,T;F,U)$ (PrT-net) is a directed net $N=(S,T;F)$, and every element of the set of arcs F is labelled by an

n-tuple L   (n≥0),   the   components   of   L   being   elements   of U or
variables,   representing   elements of U, or functions defined on U. U
is called the universe of the net.

The elements of S are representations of first-order predicates.   All
labels of arcs incidenting with a predicate P∈S are n-tuples with the
same n, and P has to be an n-place predicate.

A PrT-net with a given initial marking M0 models a predicate
transition system (PrT-system). We will look only at nets wherein all
predicates   can carry one specimen of an n-tuple of constants at most.
In [Ge] these nets are called strict nets.

A transition is called a (first-order) fact, if it does not have
concession in any marking of the system. These dead transitions are
net   representations of invariant assertions, expressed in first-order
logic, which hold in all markings of the system.

Transitions in PrT-systems carrying variables denote a scheme which is
to be used for every possible instance of these variables.   Such an
occurrence is called here an elementary transition.

If a transition carrying variables is said to be a fact, all elementary
transitions of the transition scheme have to be dead.   So the
transition has to be a fact scheme and it represents a (finite or
infinite) conjunction of elementary facts. The fact scheme describes
a proposition in which all occurring variables are generalized [Th].

Let U, the universe of the system, be a finite set,   consisting of n
elements which shall be denoted as $u_1, \ldots, u_n$.

The   formal   sum yields all elements of U: $\sum_{u \in U} u = U$ and the set of all
pairs can be denoted as: $\sum_{p \in U} \sum_{q \in U} (p,q) = U \times U$.

An   arc labelled with a formal sum is an abreviation of a set of arcs,
each element of it labelled with exactly one summand.

Let N=(S,T;F,U) be a PrT-net and M0 the initial marking of the
respective PrT-system.   If there exists an S-invariant i of the
PrT-net, general net theory yields that the inner product of i with

the vector M0 equals the inner product of i with any follower marking M′ (cf. [GLT]):

(1.1) $i^T * M0 = i^T * M′$

With the help of (1.1) transitions of a PrT-system can be proved to be first-order facts of the system.

2. Examples of Facts with Sum Labels

This part of the paper shows examples of facts in PrT-systems. Some of their arcs may be labelled with sums. Relation (1.1) will be used to show that these transitions have to be dead.

The examples are derived from the PrT-system (fig. 1) presented in [GLT] which describes a way of consistently altering a distributed data base.

The data base consists of n copies, each of them administrated by its own manager. All managers behave in the same way and the net shows the organisation scheme describing this behaviour.

The meaning of the predicates is given by the following list:

I(u)   : u is a Database manager, who is inactive.
S(u)   : u is a Database manager, who has sent change messages.
C(u)   : u is a Database manager, who is processing change messages.
D   : no altering process is running in the Database.
H(s,r)   : the message of the sender s to the receiver r is unused.
R(s,r)   : the message of the sender s to the receiver r is waiting
           for handling.
P(s,r)   : the message of the sender s to the receiver r is processed.
A(s,r)   : the message of the sender s to the receiver r is finished.

$\sum_{\substack{r \in U \\ s \neq r}} (s,r)$ is the set of all messages which are broadcasted by one
           sender s.

The n-1 messages of one sender $s = u_K$ are enumerated by:

$$\overline{\sum_{\substack{r \in U \\ s \neq r}} (s,r)} = \sum_{j=1}^{n-1} (s,r_{sj})$$

and $r_{si} = u_i$ for $k > i$

 $r_{si} = u_{i+1}$ for $k < i$

The initial marking and the net is given as follows.

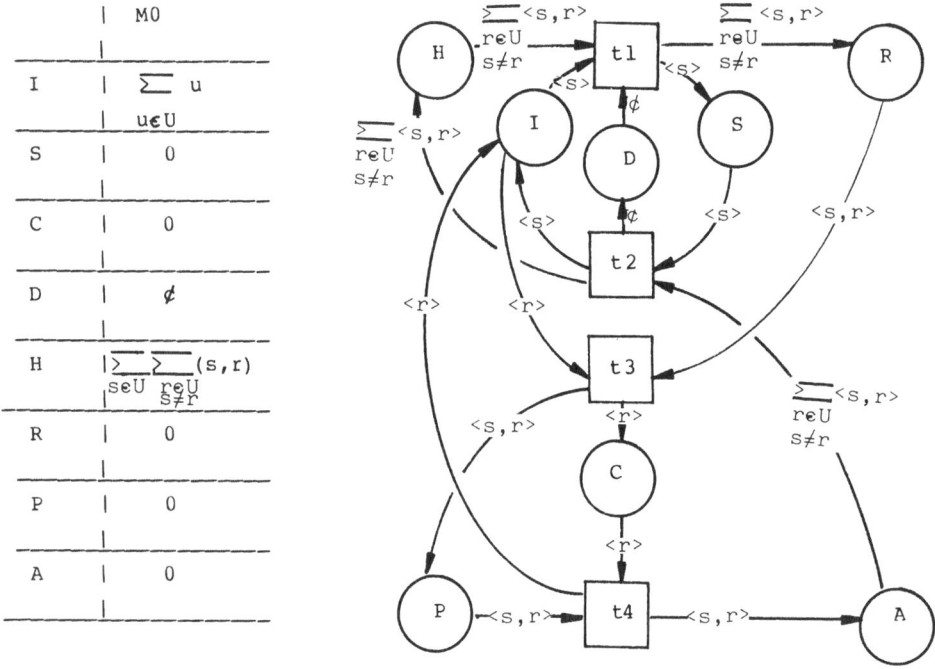

	M0
I	$\sum_{u \in U} u$
S	0
C	0
D	ϕ
H	$\sum_{s \in U} \overline{\sum_{\substack{r \in U \\ s \neq r}}} (s,r)$
R	0
P	0
A	0

Fig. 1

The modification process runs as follows:

Manager s leaves his inactive state and sends change messages to all other Database managers. In the same action the token indicating that the database is not involved in an altering process is removed (transition t1).

Every manager t leaves his inactive state and enters his processing state, takes the change messages adressed to him and handles it. All managers act independently (transition t3).

After having finished the altering of his copy manager t gets his inactive state again, and his change message is put to the finished predicate (transition t4).

When all change messages have been executed, they are removed jointly to the unused state, manager s becomes inactive again and the token on D is restored (transition t2).

Some S-invariants which are presented in the cited paper will be used to get facts of this system. The numbering of the S-invariants is taken from [GLT].

The relation (2.1) applied to the S-invariant i2 yields the following transitions to be dead. So they are facts of the system.

$$ i2^T * M' = i2^T * M0 $$

That is to say:

(2.1) $M(H) + M(R) + M(P) + M(A)$

     $= M0(H) + M0(R) + M0(P) + M0(A)$

     $= \displaystyle\sum_{s \in U} \sum_{\substack{r \in U \\ s \neq r}} (s, r)$

Among others, the facts 1, 2, and 3 (fig. 2) can be derived from relation (2.1):

Transition 1 has to be dead, because it could only have concession for a sender s, if none of the predicates H, R, P, and A contains any messages of s.

But in this case the relation (2.1) would not be fulfilled, because the sum of all markings would not contain all messages of M0(H). So transition 1 has to be a dead transition, and it shall be called fact 1.

Transition 2 has to be dead too. It has concession, when both predicates P and H contain the whole set of messages of one sender s

and the pair (s,s), the message of a sender to himself. Of course in
this case the relation (2.1) is not be fulfilled, because the sum of
M(H) and M(P) then contains pairs that are not members of MO(H).

With the same reasoning, Transition 3 has to be dead. If both
predicates R and H contain the whole set of messages of one sender r,
the sum of M(H) and M(R) contains to many summands too.

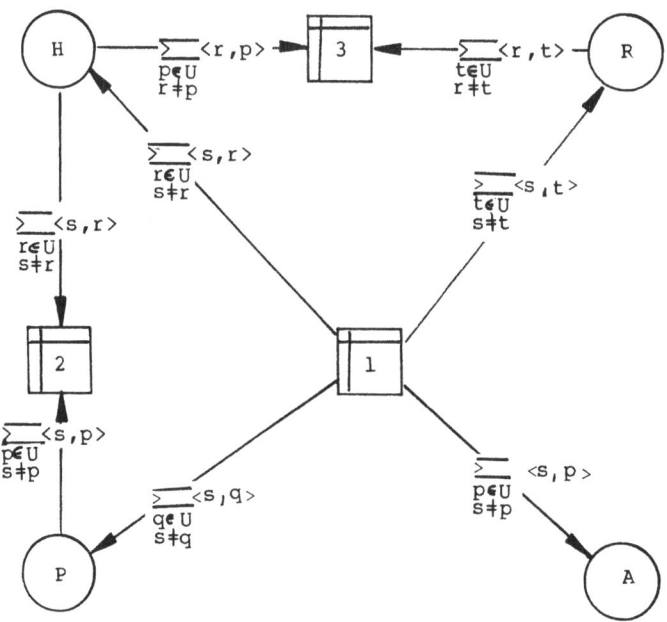

Fig. 2

The inequality of sender and receiver will always be regarded. We do
not use the predicate "=", to denote it, but we use restrictions of
the summands or inscriptions in the facts to express it.

Fact 1 is a net representation of the first order formula (see [Th]):

(F1) $\bigwedge s$ $(H(s,r_{s1})$ $\vee \ldots \vee H(s,r_{sn-1})$ $\vee A(s,p_{s1})$ $\vee \ldots \vee A(s,p_{sn-1})$
       $\vee R(s,t_{s1})$ $\vee \ldots \vee R(s,t_{sn-1})$ $\vee P(s,q_{s1})$ $\vee \ldots \vee P(s,q_{sn-1})$

(F1)   is logically equivalent to:

$$\bigwedge s \ (\bigvee r(s{\neq}r \wedge H(s,r)) \vee \bigvee p(s{\neq}p \wedge A(s,p))$$
$$\bigvee t(s{\neq}t \wedge R(s,t)) \vee \bigvee q(s{\neq}q \wedge P(s,q)))$$

and the prenex representation of fact 1 is given by:

$$\bigwedge s \bigvee r \bigvee p \bigvee t \bigvee q \ ((s{\neq}r \wedge H(s,r)) \vee (s{\neq}p \wedge A(s,p))$$
$$\vee (s{\neq}t \wedge R(s,t)) \vee (s{\neq}q \wedge P(s,q)))$$

The order of the different existential quantifiers supplied by the different sums is arbitrary, but they all follow the universal quantifier.

For every instance of the generalized variable s, another disjunction can be chosen, for which the fact holds.

Fact 2 is a net representation of (see [Th]):

(F2) $\bigwedge s \ (\neg H(s,r_{s1}) \vee \ldots \vee \neg H(s,r_{sn-1})$
$\vee \neg P(s,p_{s1}) \vee \ldots \vee \neg P(s,p_{sn-1}))$

This is logically equivalent to:

$$\bigwedge s \ (\bigvee_{\neq s} r \neg H(s,r) \vee \bigvee_{\neq s} p \neg P(s,p))$$

and the prenex representation of fact 2 is given by:

$$\bigwedge s \bigvee_{\neq s} r \bigvee_{\neq s} p \ (\neg H(s,r) \vee \neg P(s,p))$$

Again the order of the different existential quantifiers supplied by the different sums is arbitrary, but they all follow the universal quantifier.

Remark: It should be noted that reversing arcs corresponds to negation of the predicate, not of the formula.

The prenex representation of fact 3 is given by:

(F3) $\bigwedge r \bigvee_{\neq r} p \bigvee_{\neq r} t \ (\neg H(r,p) \vee \neg R(r,t))$

In the net representation of first order logic, as it is described in [Th] or in [GT] (fig 3), the propositions F1, F2, and F3 are denoted by following fact net with Skolem functions representing the existential quantifiers. The facts with Skolem functions are quoted. The differences between the two representations are very important.

1. Names of variables bound in sums are bound in the same scope as they are in the logical proposition. So for any fact these names can be chosen deliberately only respecting names of the other variables of the same fact.

On the other hand the Skolem functions a, b, c, d, p,q, r and t are global functions with names which are constants defined for the whole net. They cannot be changed for a single fact and the same symbol is not to be used for any other variable in the net (see [Th]).

2. Another important difference is given by the interpretation as a dead transition of the system. The Skolem functions are auxiliary constructs. Nothing is known about their definition. So they do not have any corresponding element in the system.

3. Skolem functions yield exactly one individual, the existence of which is stated by the expressed fact and denotes it. Sums yield a set that includes the individuals, the existence of which is expressed, but they are not denoted.

So Skolem functions may yield too little individuals, the sums yield too many.

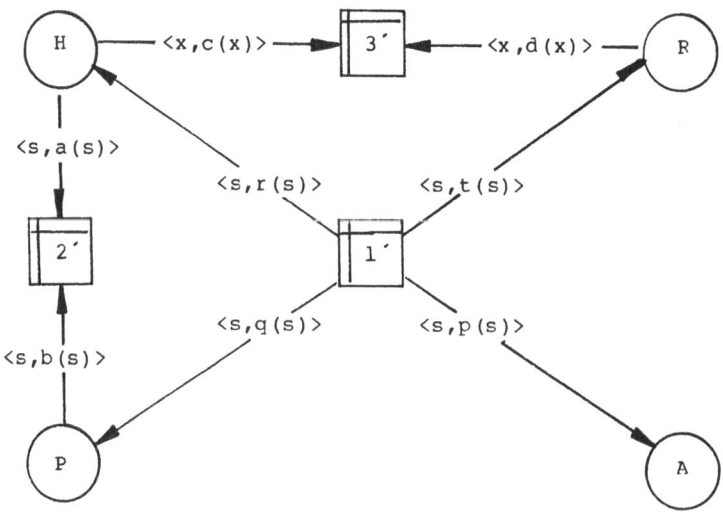

Fig. 3

S-invariant i3 together with relation (1.1) yields equation (2.2):

(2.2) $s*M(D) + ¢*M(S) = s*MO(D) + ¢*MO(S)$
$$= s*¢$$

(2.2) yields transitions 4, 5, and 6 (fig. 4) to be a facts of the system: $s \in M(S)$ implies D to be empty and $¢ \in M(D)$ implies S to be empty. So transition 4 and transition 5 cannot occur.

Transition 6 is active, if there are two different tokens sO and rO
    in the marking of S. This yields for s=sO:
$¢*M(S) \geq ¢ *(sO + rO) > ¢ * s$

So activation of transition 6 hurts relation (2.2) and 6 has to be a fact too.

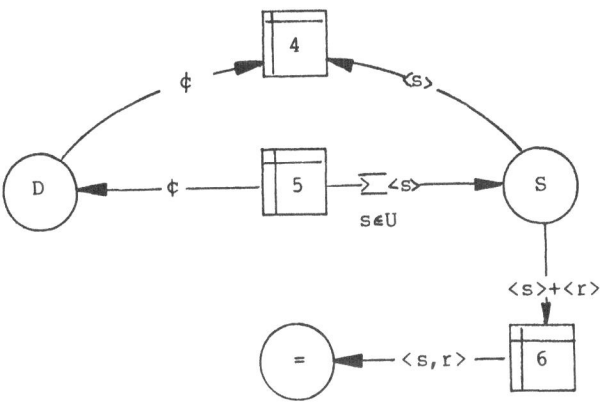

Fig. 4

Fact 4 is a net representation of the logical proposition:
(F4) $\bigwedge s \ (\neg D \lor \neg S(s))$
Fact 5 represents the formula:
(F5) $D \lor \bigvee s \ S(s)$
and the prenex representation of this fact is given by:
       $\bigvee s \ (D \lor S(s))$
Fact 6 is a net representation of the logical proposition:
(F6) $\bigwedge s \bigwedge r \ (\neg S(s) \lor \neg S(r) \lor s=r)$

The net representation of fact 5 (see Fig. 5) with a Skolem function denoting the only quantifier shows, that the representation with sums is less idealizing. The Skolem function, here the single constant s0, has to be chosen for the whole net.

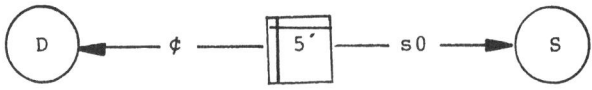

Fig. 5

Relation (1.1) applied to the S-invariant i7 yields equation (2.3):

$$(2.3) \quad s*M(H) + \sum_{\substack{p \in U \\ s \neq p}} (s,p)*M(S) = s*MO(H) + \sum_{\substack{p \in U \\ s \neq p}} (s,p)*MO(S)$$

$$= s* \sum_{s \in U} \sum_{\substack{p \in U \\ s \neq p}} (s,p)$$

(2.3) yields transitions 7,8 and 9 (fig. 6) to be facts.

$$s \in M(S) \text{ implies: } \sum_{\substack{p \in U \\ s \neq p}} (s,p) * M(S) \geq s* \sum_{\substack{p \in U \\ s \neq p}} (s,p)$$

So in this case M(H) must not contain a message of the set of the sender s and of course not the message to himself (s,s) in order not to hurt (2.3), and 7 and 8 have to be facts.

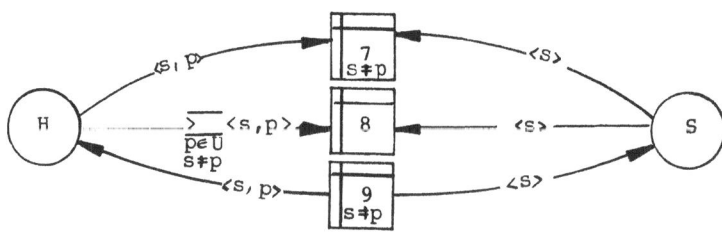

Fig. 6

On the other hand S empty yields:

$$s*M(H) + \sum_{\substack{p \in U \\ s \neq p}} (s,p)*M(S) = s*M(H)$$

$$= s* \sum_{s \in U} \sum_{\substack{p \in U \\ s \neq p}} (s,p)$$

So H contains all possible messages and 9 has to be a fact.

Fact 7 represents the proposition:

(F7) $\bigwedge s \bigwedge p \; (s \neq p \rightarrow (\neg H(s,p) \lor \neg S(s)))$

The prenex representation of fact 8 is given by:

(F8) $\bigwedge s \bigvee p \; ((s \neq p \land (\neg H(s,p) \lor \neg S(s))))$

Fact 9 represents

(F9) $\bigwedge s \bigwedge p \; ((s \neq p \rightarrow (H(s,p) \lor S(s))))$

3. Resolution on Facts with Sum Labels

The calculus of facts includes a resolution rule in order to derive new facts from given ones. Resolution is a correct interference rule, that is to say that the derived facts are logical consequences of the hypotheses. We will apply this rule directly to the facts of the last chapter.

The following denotation is used.
Let f be a fact, S an element of the preset of f, and let the label of the arc (S,f) be L. Then *f, the labelled preset of f, is given by:

$$*f = \{(S,L) \mid S \in {}^\bullet f \land L \text{ label of } (S,f)\}.$$

In the same way we define:

$$f* = \{(S,L) \mid S \in f^\bullet \land L \text{ label of } (f,S)\}.$$

Resolution Rule:

Given two facts f1 and f2, such that

(1)   a predicate P is a common element of the post-set of f1 and the pre-set of f2,

(2)   there is exactly one pair of arcs (f1,P) and (P,f2) which is labelled equally with exactly one n-tuple of constants, variables or functions (v1,...,vn).

That is to say: $f1* \cap *f2 = (P,(v1,\ldots,vn))$.

Then a new fact f3 can be derived, such that:

$$*f3 = *f1 \cup *f2 \setminus (P,(v1,\ldots,vn))$$

$$f3* = f1* \setminus (P,(v1,\ldots,vn)) \cup f2*.$$

The application of this rule will be denoted by "f3 resolution f1/f2".
(P,(v1,...,vn)) is called "the arc resolved on".

Remark 1: As sums are abreviations of sets of arcs, the arcs (f1,P)
or (P,f2) may be labelled with sums. But if they are both labelled
with a sum, these two sums must have exactly one summand
(v1,...,vn) in common.

Remark 2: In one resolution exactly on one arc (that is to say one
pair of arcs (f1,P) and (P,f2) with the same label) can be resolved.
If it would be more than one, the result is wrong, because a tautology
has been neglected.

A simple example is given by:

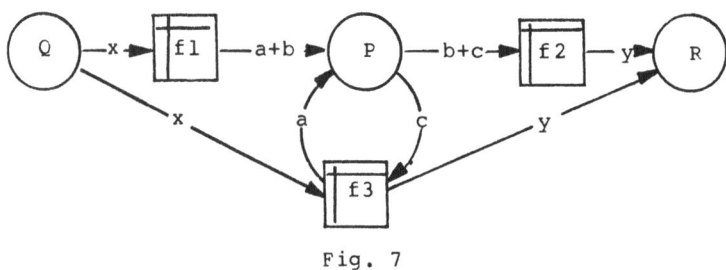

Fig. 7

In order to get the same label on a pair of arcs to resolve on, the
following rule may be applied on one or both of the facts.

Unification Rule:
The following replacements of a variable in fact are allowed:
(1)   Substitution of a variable name by another name, whereever the
      variable in the fact occurs (renaming),
or
(2)   replacement of a variable by a term (constant or function),
      whereever the variable in the fact occurs (instantiation).

If in fact f the variable x is replaced by t, this will be denoted by:
unification f: x/t

Resolution together with unification will be used to derive new facts
from those stated in the last capter.

545

Example 1: Fact *1 (fig. 8) is to be derived from facts 2 and 9.

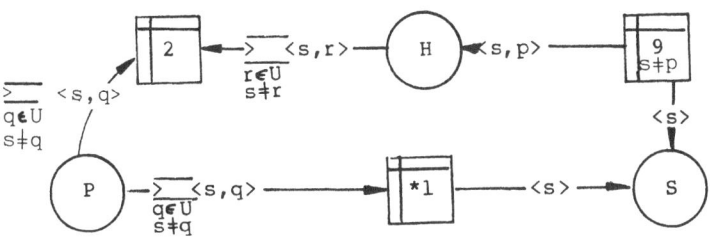

Fig. 8

Mind that for a fixed s (s=u_k with a fixed k) fact 9 denotes a scheme
of n-1 facts called 9.1... 9.n-1, such that 9.i describes a
fact-scheme which is got from fact 9 by substituting the constant u_{si}
for the variable p:

u_{si} =u_i for i<k
u_{si} =u_{i+1} for i>k

s is read as a fixed constant. The inequality s=p expressed by
leaving out the instantiation of fact 9 with s=u_k.

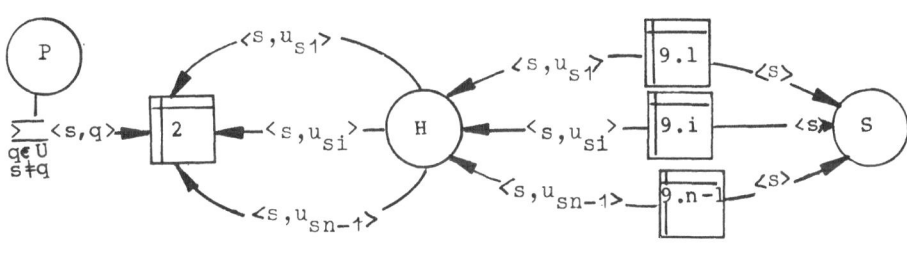

Fig. 9

Now resolution can be applied on 2 and 9.i, for any i, and the
following sequence of resolutions yields fact *1 as a derivation (fig.
10). The sum label is used again for better readability.

Derivation of *1
  2.1   resolution 2/9.1
         .
  2.i   resolution 2i-1/9.i
         .

546

2.n-2 resolution 2n-3/9.n-2

*1   resolution 2n-2/9.n-1

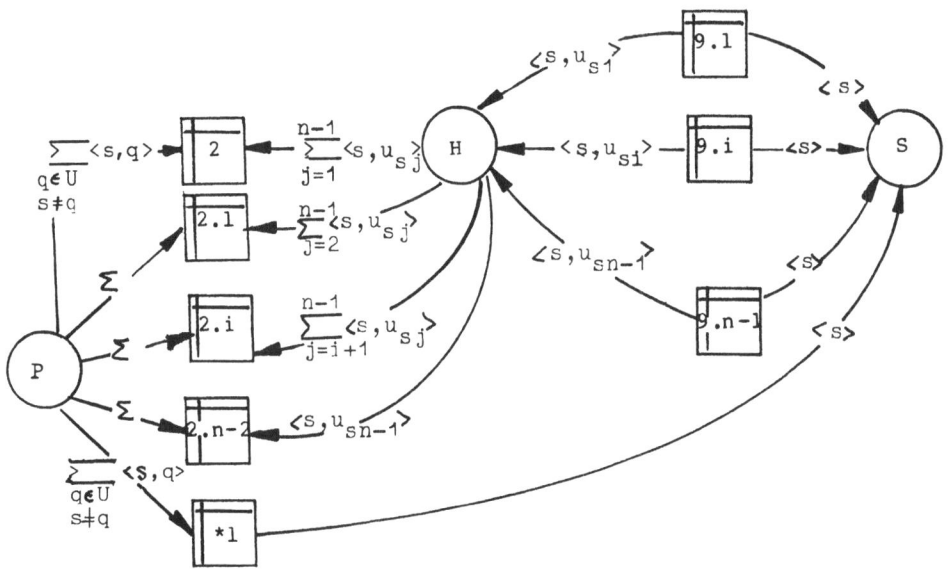

Fig. 10

This derivation can be executed for any s with the respective numbering of the rest of the set U.

The use of Skolem functions to represent F2 and F9 would simplify this derivation very much. We will show it for comparison (fig. 11). Instead of resolving all possible instances of u, we do it for every fixed s only on the one element a(s), the existence of which fact 2′ denotes.

Derivation of *1′

unification 9: p/a(s)   *1′ resolution 2′/9

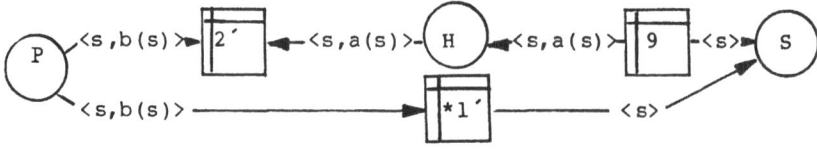

Fig. 11

The two derived facts represent the same proposition:

(F*1) $\bigwedge s \bigvee q$ (s≠q ∧ (¬P(s,q) ∨ S(s)))

But in comparison to the fact *1 with the sum label which has a direct interpretation as a dead transition in the system under view, the fact *1´ is only to be used as a representation of a logic formula.   The function a is auxiliary without a correspondence in the system.

*2 is derivable from 3 and 9 in an analogous way (fig. 12):

Derivation of *2

            Unification 9: s/r
    3.1   resolution 3/9.1

                  .

    3.i   resolution 3i-1/9.i

                  .

    3.n-2 resolution 3n-3/9.n-2
    *2   resolution 3n-2/9.n-1

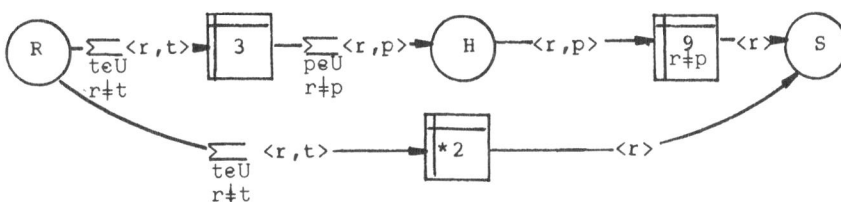

Fig. 12

Example 2: Fact *4 (fig.13) is to be derived from *1, *2 and 6 by resolution.

Derivation of *4
    *3   resolution *2/6   on S(r)
    *4   resolution *1/6   on S(s)

*4 is the net representation of the formulas:
F*4 $\bigwedge s \bigvee q \bigwedge r \bigvee t$ (¬P(s,q) ∨ ¬R(r,t) ∨ s=r)
rsp. $\bigwedge r \bigvee t \bigwedge p \bigvee q$ (¬P(s,q) ∨ ¬R(r,t) ∨ s=r)

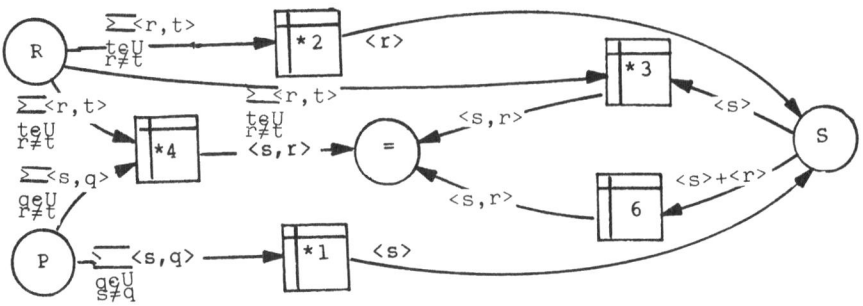

Fig. 13

The choice of an instance for q is dependent on the choice of s. So the corresponding existential quantifier has to follow the universal quantifier for s. The analogous consideration holds for r and t. But the sequence of these two pairs of quantifiers is arbitrary.

Using Skolem functions to represent the existential quantifiers in F*4 the functions, we get:

$$\bigwedge s \backslash q \bigwedge r \bigvee t \; (\neg P(s,q) \; \lor \; \neg R(r,t) \; \lor \; s=r)$$

rsp. $\quad \bigwedge s \bigwedge r \; (\neg P(s,q(s)) \;\; \lor \; \neg R(r,t(s,r)) \; \lor \; s=r)$

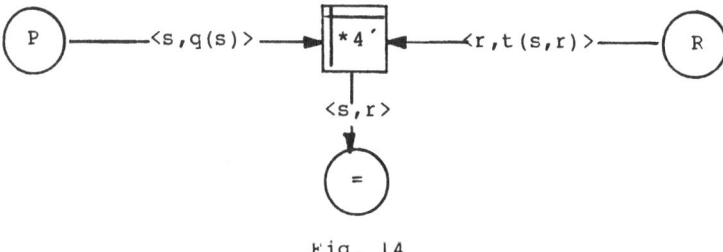

Fig. 14

Example 3: Resolutions on arcs labelled with sum may yield trivial results, which could be prevented by using Skolem-functions.

The following net (fig. 15) is a part of an example given in [GT]. We use it here only as representation of a logical formula.

The net represents the conjunction of the propositions:

E1 $\quad \bigwedge i \bigvee r \, (C(i) \rightarrow R(r))$

E2 $\quad \bigwedge x \bigwedge y \; (R(x) \land R(y) \rightarrow x=y)$

E3   C(pl)

E4   C(p2)

pl and p2 are constants.

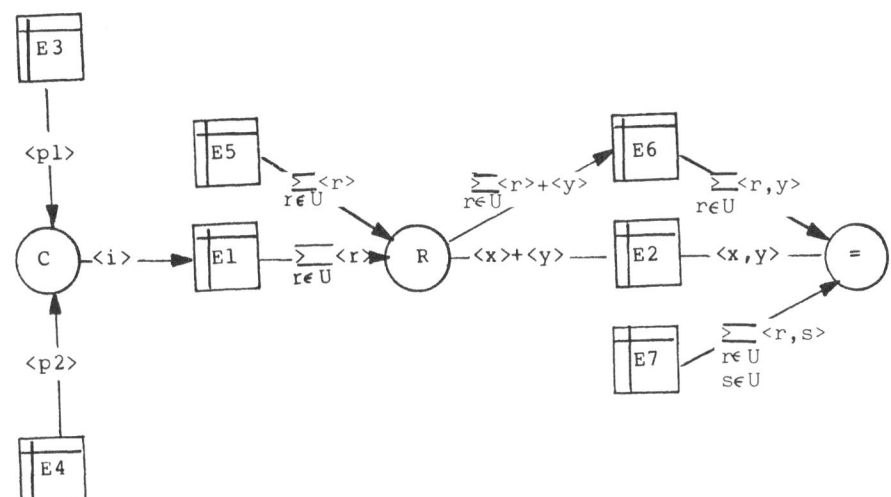

Fig. 15

E5, E6 and E7 are produced by the following derivation:

Unification El: i/pl   E5 resolution El/E3

Unification E2: x/u

        for all u∈U,

        n unifications,

        yield n transitions

                            E6 resolution E5/E2   (n-times)

Unification E6: y/u

        for all u∈U,

        n unifications,

        yield n transitions

                            E7 resolution E5/E6   (n-times)

The fact E7 states is trivial.

Using Skolem functions, the facts E5´, E6´, E7´, and E8' are   produced by the following derivation (fig. 16):

Unification El´: i/pl   E5´ resolution El´/E3

Unification E2´: x/r(pl)   E6´ resolution E5´/E2´

Unification E1´: i/p2   E8' resolution E4´/E1´

Unification E6´: y/y(p2)   E7´ resolution E8´/E6´

The fact that is stated by E7' is the wanted consequence.

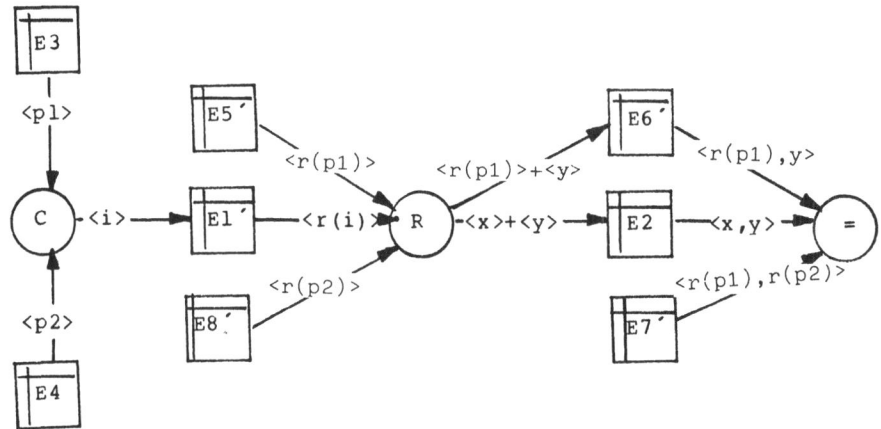

Fig. 16

These different derivations result from the fact that the sum labels denote a set of objects, which include sometimes too many elements. In these cases too many disjunctions including the trivial may be derived.

4. Resolution in Infinite Systems

The preeceeding chapters have shown the advantage of using sums in repesentations of facts of a system. Examples of direct application of resolution on those facts could be given, though sometimes the derivations become trivial. But the examples were taken from a system with a finite universe, which the individuals are taken from.

Let U be an enumerable set, $U = u_1,...,u_n,....$ and let U be the universe of a PrT-system $N^* = (S,T;F,U,M0)$. We do not consider universes, which are not enumerable, because U will be used as an index set of a sum.

Let P be an 2-ary predicate, $P \subseteq U \times U$.

The facts X1, X2 (fig.17), X3, and X4 (fig.18) denote first-order propositions in the usual way.

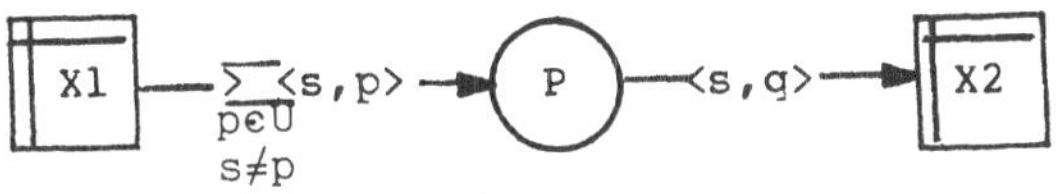

Fig. 17

$X1 \wedge X2 \qquad \bigwedge s \bigvee p\ P(s,p) \wedge \bigwedge s \bigwedge q\ \neg P(s,q)$

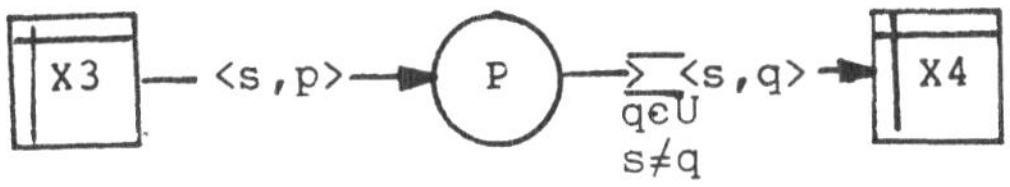

Fig. 18

$X3 \wedge X4 \qquad \bigwedge s \bigwedge p\ P(s,p) \wedge \bigwedge s \bigvee q\ \neg P(s,q)$

Both nets represent a contradictory logical formula. And in both nets the two transitions cannot be dead at the same time:

If X1 is dead, at least one pair has to be in the marking of P. So X2 has concession with that pair and is not a dead transition. If X2 is dead, no pair is allowed to be in the marking of P. So X1 has concession with the full set of pairs and is not dead.

The analogous reflections are valid for the facts X3 and X4.

As the net calculus of predicate logic is complete ([Th], it should be possible to derive the net representation of the contradiction from the nets under view. The isolated transition with empty preset and postset, so life in any marking, is this net representation of the contradiction.

The sequence:

unification X2: q/u_1	t1 resolution X1/X2
unification ti: q/u_{i+1}	ti+1 resolution X1/ti

yields a proceedure to derive the isolated fact.

But this is not a derivation, because it is not finished after a finite number of steps.

So the isolated transition is not derivable by unification and resolution, if the objects, the existence of which is proposed, are only given as elements of an infinite set, and not by their name.

As already stated, transition Xl is a fact, if there is at least one pair <s,p(s)> in the marking of the predicate P. The Skolem function denotes that element and we get the net (fig. 19), from which the isolated transition can be derived.

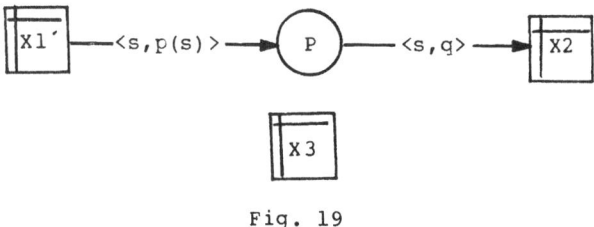

Fig. 19

Unification X2: q/p(s)
X3   Resolution   Xl/X2

From the constructive character of the fact calculus it follows that derivations in infinite universes are only possible, if the arcs to be resolved on can be denoted by names. Pure existence propositions cannot serve as hypothesis in a constructive logic.

5. Conclusions

Facts with arcs labelled with sums are a natural representation of existential quantifiers in fact nets. They have a direct interpretation as dead transitions of the respective system.

S-invariants of the sytem may yields proofs for them not to have concession in the system. The scope of the variables bound by the existential quantifier is the same as it is for the variables bound by the sum.

In systems with finite universes unification and resolution can be applied. The results of the derivations may not allways be strong enough, and Skolem functions have to be used instead. In infinite universes allways Skolem functions have to be used in order not use non constructive existential propositions as hypotheses.

6. References:

[Ge]   H.J. Genrich
        Projections of CE-Systems
        Proc. of 6th European Workshop on Application
        and Theory of Petri Nets
        Espoo, Finland 1985

[GLT]   H.J. Genrich; K. Lautenbach; P. Thiagarajan
        Elements of General Net Theory
        in: W. Brauer (Ed.)
        Net Theory and Applications
        Springer-Verlag 1980

[GT]   H.J. Genrich; G. Thieler-Mevissen
        The Calculus of Facts
        in: A. Mazurkiewicz (Ed.)
        Math. Foundations of Computer Science 1976
        Lecture Notes in Computer Science Vol. 45
        Springer-Verlag 1980

[Th]   G. Thieler-Mevissen
        The Petri Net Calculus of Predicate Logic
        Internal Report GMD-ISF 76-09

PETRI NETS FOR SEQUENCE CONSTRAINT PROPAGATION IN KNOWLEDGE BASED APPROACHES

R. Valette, H. Atabakhche

Laboratoire d'Automatique et d'Analyse des Systèmes du CNRS

7, Avenue du Colonel Roche

F - 31077 TOULOUSE CEDEX - France

ABSTRACT:

This paper deals with the possibilities of the integration of Artificial Intelligence techniques within a Petri net approach for controlling and monitoring Flexible Manufacturing Systems. Section 2 is a short presentation of knowledge representation, then the similarities between Petri nets and these techniques are described in sections 3 (static aspect) and 4 (dynamic aspect). An illustrative example is given in Section 5.

1. INTRODUCTION.

Since Petri thesis "Communication with automata" [PE 62], the model that is now known as "Petri nets", has been extensively studied and applied. During all these years, another approach for expressing causal relations has had a large impact, I mean "Artificial Intelligence". Various methods have been developed in order to represent human knowledge. Among them, two methodologies are commonly used to build expert systems: rule-based and frame-based techniques. The aim of this paper is to show that similarities can be found between Petri nets and these two approaches. As a matter of fact, Petri nets can be incorporated into these two techniques and it seems profitable to integrate these three tools, especially when the knowledge involves sequential apects and when the knowledge-based system has to interact in real-time with its environment.

2. KNOWLEDGE REPRESENTATION.

2.1. Introduction.

One of the characteristics of Artificial Intelligence is that it attempts to achieve a non procedural description of knowledge as well as the way this knowledge can be transformed. It is not surprising then, that among the two major approaches that have been developed, the first one - production rule systems - insists on knowledge transformations by means of deduction, while the second one - frame-based description - provides a concise structural representation of objects, thus privileging the static aspect.

2.2. Production rule sytems.

The kernel of a production system [HA 85] consists of:
- rules of the form IF "condition" THEN "action" where condition is a predicate involving facts contained in a knowledge base and action (or consequent part) is a set of statements modifying the veracity of the facts,
- an inference engine which is a way of searching applicable rules and in the case of conflict (when more than one rule can be applied) chooses one of them.
Nothing else, concerning the structuration of the knowledge base is assumed.

2.3. Frame-based approach.

The knowledge base can be represented by "object-oriented" languages based on "frames" [FI 85]. Frame languages provide the knowledge base builder with an easy means of describing the objects or class of objects that the system must model. Frames consist of descriptions called "slots". In many systems slots can have multiple values and a set of properties called "facets". Frames contain both declarative and procedural informations. The slots can contain values, and fire procedures when manipulating these values. They can also define relations between frames.

Very often, the properties associated with slots concern value-classes and cardinality constraints on the legal values of slots. In other words, the possible values (which can be an interval) of an object will be defined in a slot belonging to the frame associated with it. Constraint checking procedures for determining whether a slot's value-class and cardinality specifications have been respected, exist. An item can be excluded for example if the slot already has its maximum number of values or if the item is not a member of the slot's value-class. These procedures can be called by the system whenever a slot's values, value-class or cardinality specifications are changed. They can also be called by the user.

One important concept introduced by frame-based systems is that of "inheritance"; it is a way of sharing descriptive information among many frames. When an object is created, it can be linked to a preexisting one by two kinds of relations: the relation "a-kind-of" and the relation "is-a". In the first case, the new object inherits the structure and the properties of the existing one but new slots and properties can be added. In the second case, it is the instantiation of one object in a given family (the existing object). The new object will have exactly the structure and the properties defined by the family and the default values will be affected to the slots.

Finally, it must be pointed out that no standard is given for the operation of an inference mechanism on this structure.

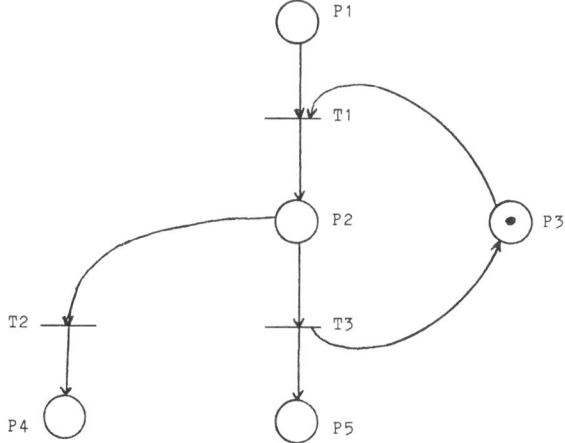

FIGURE 1: Petri net.

3. PETRI NETS FOR KNOWLEDGE REPRESENTATION.

3.1. Petri nets and production rule systems.

A parallel between rules and transitions can be drawn. With each transition, in other words, with each possible evolution of the Petri net [ZI 78], a rule is associated, giving the firing conditions (condition part of the rule) and the modifications on the marking induced by the firing of the transition (consequence part of the rule). Consequently, a Petri net expressed in the form of a production rule system is decomposed into "elementary objects" each one consisting of a transition, its input and output places. For example, the condition/event net [BF 86] in Figure 1. can be expressed by the following rules:

$$
\begin{array}{ll}
& \text{Rule 1: } \underline{\text{IF}} \text{ P1.P3.}\overline{\text{P2}} \text{ } \underline{\text{THEN}} \text{ } \overline{\text{P1}}.\overline{\text{P3}}.\text{P2} \\
(1) & \text{Rule 2: } \underline{\text{IF}} \text{ P2.}\overline{\text{P4}} \text{ } \underline{\text{THEN}} \text{ } \overline{\text{P2}}.\text{P4} \\
& \text{Rule 3: } \underline{\text{IF}} \text{ P2.}\overline{\text{P3}}.\overline{\text{P5}} \text{ } \underline{\text{THEN}} \text{ } \overline{\text{P2}}.\text{P3}.\text{P5}
\end{array}
$$

Production rule systems derived from Petri nets have special structures, but not necessarily simple ones (without variables in the rules) as in the preceding example. In fact, when predicate/transition nets [GE 79] or other high level nets are involved, the rules contain variables (predicates). The particularity of these production systems relies more on a structuration of the knowledge. A part of it (all of it if the net is not a "high level" one) has to be expressed by means of counters (places) and will be privileged. Futhermore, this kind of fact systematically appears both in the condition and in the consequent parts of the rules (see production rule 1). This is the reason why the inverse transformation although perfectly possible [ST 85] may sometimes produce nets having a very strange structure.

This characteristic of Petri nets comes from the fact that they are well-suited for describing sequences of "states"; when the fact that an object is in a new state is deduced, the fact that it was in the old state is immediately negated. On the contrary, in many rule system applications, particularly in diagnosis, the purpose of the rules is to increase knowledge about some system; in other words to add new facts to the knowledge base, and occasionnaly to delete existing facts (this is done by back-tracking procedures).

3.2. <u>Dual approach: Petri nets as frame-based systems</u>.

Since Petri nets are equivalent to production rule systems with more structuration about the knowledge, it seems natural to consider the second Artificial Intelligence technique, i.e. the frame-based one, that privileges knowledge structuration rather than knowledge transformation. Let us consider the Petri net in Figure 1., it is possible to make the dual decomposition in relation to that done in order to produce a production rule sytem. Each elementary object will be made up of a place with its input and output transitions. These objects can then be considered as frames with the following slots:
- token load of the place, (attached property: this value is zero or a positive integer less than or equal to the place capacity),
- list of the output transitions with their firing condition and marking transformation involving other frames (places),
- relation with the other frames (places) capable of altering the token load.

For example, the frame "P2" will be:
- token load: 0,
- output transition: T2 with output place P4, T3 with output places (frames) P5 and P3,
- frames capable of modifying the state: P1 and P3.

The last slot seems redundant with respect to the second one because they describe the arcs linking the transistions to the places. Nevertheless the last frame is necessary for backward chaining. It is a way of representing the input transitions of the place.

3.3. <u>Another frame-based approach for implementing Petri nets</u>.

We have seen that it was possible to make a correspondance between the transition/place duality in Petri nets and the rule/object one in the knowledge-based approach. As a matter of fact, frames present a flexible means of description and can be used to structure a knowledge base including production rules. In this context, it is possible to create a family of objects associated with each kind of object involved in Petri nets. We thus have "place-objects", "transition-objects" and in the case of high-level nets "token-objects".

For example, the basic slots of a place may be:
- token load which is a natural number in the interval "zero capacity",
- list of "token-objects" contained in the place,
- list of input "transition-objects",
- list of output "transition-objects".

It must be pointed out that if "high-level Petri nets with data structure" [SI 85] are used, the slots of the "token-objects" will be derived directly from the data structure associated with the tokens.

In conclusion, we can say that Petri nets correspond to a particular kind of frame-based approach where three specific kinds of objects have particular properties:
- places that are used for structuring "state" knowledge, such states might be states within the deduction process or estimate states of the observed system,
- transitions that correspond to deduction rules or to updating procedures of the state of the observed system,
- tokens that are objects attached to a certain "state".

Furthermore three other specific relations arc added to the two kinds of specific relations used by the inheritance mechanism ("a-kind-of" and "is-a"):
- "is-a-cause-of" which corresponds to an arc from a place to a transition,
- "can-alter" which represents an arc from a transition to a place,
- "in-the-state-of" for attaching a token to a place.

This last relation is similar to "a-kind-of" but is dynamic while all the others are static.

4. THE TOKEN PLAYER AND THE INFERENCE ENGINE.

Until now, we have shown that Petri nets could be seen as a way of structuring knowledge but nothing has been said about the inference engine. A way of implementing Petri nets in a large variety of domains and especially in manufacturing systems is the "token player" [DE 83, TH 83, MU 86]. In this section we will first describe this method, and then show the relations with the classical operations of inference engines.

4.1. Interpretation of a net.

Petri nets used for describing the monitoring of manufacturing systems always describe interacting processes. Thus, two kinds of events can be associated with the transitions during the net interpretation:
- internal events are transitions that are fired under the unique responsability of the control, they represent mere control deductions and signals sent to the environment,
- external events are transitions associated with the reception of messages from the environment, these transitions are fired as soon as the messages are received, and the control must only check if they were enabled.

4.2. The token player.

This direct implementation consists of "simulating in real-time" or "emulating" the behaviour of the interpreted Petri net by means of a token player. The structure of the token player is given in Figure 2. and can be decomposed into two behaviours.

In the first case, the token player searches the enabled transitions (regardless of the transitions associated with external events), and fires them. It goes on until the stable state is reached, i.e. a state for which no transition corresponding to an internal event can be fired. This behaviour is described in the left part of Figure 2.. Generally the search for enabled transitions is "optimized" in the sense that only transitions having at least one input place (its "trigger place") containing tokens, are examined.

In the second case, the token player is awakened by an external event (reception of a message), it searches a fireable transition associated with it, fires it and then executes the preceding procedure (first case). This corresponds to the right part of Figure 2.. If no transition associated with the external event was fireable, this means that the interaction between the modeled mechanism and its environment is erroneous and an exception handling process has to be called.

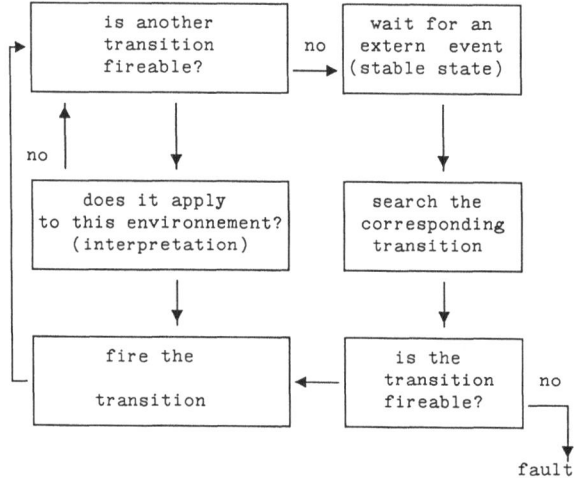

Figure 2.: The operation of the token player.

4.3. The token player as an inference mechanism.

Let us now compare the operation of the token player to that of an inference engine. The stable state corresponds to a temporary conclusion obtained by the production rule system. The reception of a message represents an information that must be transformed into a new fact by the inference engine (updating the knowledge of the estimated state of the monitored system). A rule capable of recognizing this message (a transition associated with the reception of the message) must therefore be chosen and checked to make sure it is applicable.

We can thus compare the operation of the token player provoked by an external event to "backward-chaining". In fact, backward-chaining is a backward deduction process. It consists of searching the rules the consequent part of which allows to deduce the desired conclusion, and then the rules allowing the application of the preceding rules etc... until applicable rules are found. In the case of the token player it is a very elementary backward-chaining because the result is given directly by the execution of a single rule. If an incident occurs (i.e. when no rule is found) the token player stops. Continuing in such a situation will be a real backward-chaining (which is normal because a diagnosis is necessary) to search the cause of the incident (for example a sequence of transitions that have not been fired i.e. an estimation of the state of the monitored system that is erroneous).

The operation of the token player shown in the left hand side of figure 2. corresponds to an inference engine reasonning in forward chaining. Forward chaining is the direct deduction process; the inference engine searches the rules that are applicable for the current state of the knowledge base, applies them (or a certain number of them) until the conclusion is verified. It is the case of the token player which fires all the fireable transitions from the current marking by executing the cycle given by the left hand side of figure 2., until a stable state is reached. The token player does not have any final facts because its role is to deduce all the possible facts from the information it has received to update the knowledge it has about its environment. Its goal is to reach a stable state but this is not represented in the form of a fact that must be checked. This is the reason why only forward chaining is possible in this case.

When the concept of trigger place is employed, the inference engine starts with a list of marked places, that is a sub set of facts that are known to be true, in order to deduce a sub set of appropriate rules. Rules whose purpose is to select a certain number of rules that are likely to be more important at a certain stage of the deduction are called "meta-rules". Therefore it can be said that the following meta rule is applied:
"Only the transitions corresponding to internal events
and which are outputs to a marked place, are examined".

These rules (transitions) are then examined exhaustively to see if they can be applied.

It must be pointed out that all these considerations are not altered by the utilization of the frame-based approach as far as rules (i.e. transitions) are clearly identified as specific objects.

In knowledge bases employed in sophisticated rule based systems, there may be a certainty factor associated with the facts. In other words facts whose truth is not certain can be manipulated. In a Petri net a token can be in a place or not, however a probability can be associated with the tokens [PA 85]. This concept could be employed to introduce a certainty factor and thus draw Petri net and Artificial Intelligence points of view still closer. Consequently, it will be easier to incorporate Petri nets into a knowledge-based approach and vice versa, the first one being more adapted for sequence modeling and normal operation control and monitoring, the second one being better suited for diagnosis and reconfiguration after an incident.

5. EXAMPLE.

Within the context of the monitoring of manufacturing systems, two different new approaches are being developed; one is Petri net based, the other relies on the use of Artificial Intelligence techniques [MU 86, TA 85]. In the context of an on-going project involving various research teams and supported by the G.I.E. "PROMIP", it has been decided to try integrating within a single software environment:
- a constraint-based anaysis module for job shop scheduling [ER 86],
- a knowledge-based decision module including fuzzy rules for choosing among feasible scheduling [BE 86],
- a Petri net based monitoring allowing a connection to real-time control.
The purpose of this last module is to control in real-time the execution of the production plan and to detect as soon as possible any incident and delay.

In the sequel we shall present briefly the approach concerning the Petri based monitoring and show how it can be integrated within an "Artificial Intelligence" context. Let us consider a shop where two machines "M1" and "M2" are used to fabricate parts belonging to two part routines "PR1" and "PR2". Machine "M1" can execute operations "OP1" or "OP2" and machine "M2" operation "OP3". Part routine "PR1" consists in the sequence of operation "OP1" and "OP3" and part routine "PR2" in operation "OP2".

We have chosen to represent each "sequential constraint" by a "net-object" made up of "place-objects", "transition-objects" and "token-objects". The complete synchronisation mechanism will be described by merging the transitions corresponding to the same events as it has been stated in [AL 84] and [NA 85] in order to keep the place invariants valid.

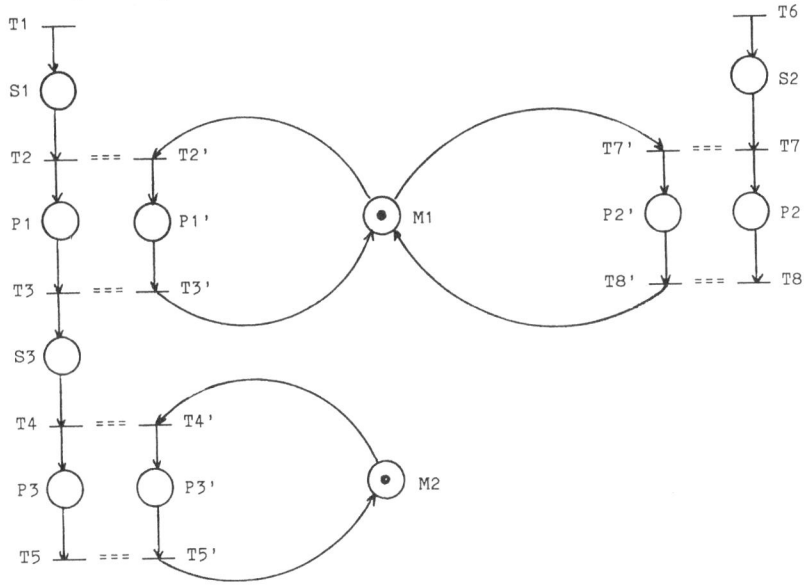

FIGURE 3: Example of a net describing monitoring operations.

Therefore, the job is modeled by means of four nets as decribed in Figure 3. Transitions T1 to T5 describe part routine "PR1", transitions T6 to T8 part routine "PR2", the net including place M1 depicts the resource allocation of machine "M1" and the net including M2 that of machine "M2". Places P1 and P1' are associated with operation "OP1", P2 and

P2' with "OP2" and P3 and P3' with "OP3". Places S1 to S3 represent intermediate stocks and the tokens are data structures giving all information that are not "state" informations about the parts and the machines.

Clearly all this synchronisation structure could have been described by means of a set of rules rather than by a set of "net-objects". However it must be pointed out that the resource allocation mechanism of machine "M1" needs four rules to be modeled (one per transition) while only one "net object" is required. Furthermore the "net object" allows an easy proof that the synchronisation mechanism is correct (by means of the place invariant) and will not be altered by any other object (because of the structured composition of the "net-objects"). This will not be the case for a pure rule-based approach with all rules at the same level.

The monitoring module has four parts:
- it has to make optimal choices concerning the ressource allocations when the schedule is not complete (some operations remain unordered),
- it has to update the informations concerning the state of the manufacturing shop from messages sent by the local control, after checking the consistency of the message content in relation to the present state,
- it has to check the consistency of the plan in relation to the present state of the shop,
- it has to make a correct diagnosis in case of incident.

The first one will be realized by a call. from the token player, to a specific rule based system, each time a conflict between two transitions is detected. The description of all the "token objects" present in the input places of the conflicting transitions will be passed as an argument (i.e. the description of the input stock of the machine). When the control returns to the token player, the argument will be the name of a "transition object" to be fired and of a "token object" to be affected to the color variable attached to the input arc of the transition. The kinds of rules that can be found in this specific rule based system are:
- choose the most urgent part,
- choose the part with the earliest "earliest starting time",

- choose the part with the earliest "latest starting time",

- ...

In this case, the inference engine of the rule based system and the token player cooperate but remain relatively autonomous. The cooperation is simply facilitated by the use of a common language (LISP) and a common structuration of the knowledge by means of frames.

The second one corresponds mainly to the normal operation of the token player, at each reception of a message, it checks if it corresponds to an enabled transition and fires it. This can be complemented by temporal verifications. With each place a maximum delay is associated (watch dog), if a token remains longer in this place the exception process is called. In this manner it can be detected, for example, that a tool is worn and that the machining operation cannot procede without a tool change.

The third aspect of the monitoring module is relatively similar to the preceding one. The token player will check if the token load of some places (corresponding to machine input stocks) does not exceed their capacity. By means of a calendar it will also be possible to associate watch dogs to the places representing stocks in order to detect immediately when the latest starting time of a given operation has arrived.

The last apect of the monitoring module is the most interesting. As it has been mentionned previously, diagnosis are generally based on backward chaining whereas the token player operates by means of forward chaining. Let us consider an example. Suppose that the marking of the net is such that places P1 and P1' are marked (operation "OP1" is on) and that a message "end of operation "OP2"" is received. At the begining of the diagnosis we face the situation where two "possible" markings correspond to two "possible" situations:

- the last message is erroneous,

- the last message is true and consequently the present marking of the net is erroneous; this implies that an error has occured previously; either transition T3' should have been fired, or transition T2' has been fired instead of transition T7'.

The diagnosis will consist in analysing the possibility and the certainty of each of these alternatives by means of rules and messages sent to the local controllers in order to check all the informations. The interesting point is that these rules will operate directly on the net structure and that sometimes the "transition object" will be used by the diagnosis inference engine. For example, the deduction that if, for the monitoring module, operation "OP2" is not on (there is no token in place P2') it is because either the decision of begining it (transition T7' fired) has not been taken or the signal "end of "OP2"" have been received (transistion T8' fired) is done by considering the corresponding transitions as rules.

Consequently, this aspect of the monitoring module requires actually an integration of the Petri net aspect in the Artificial Intelligence aspect. A first prototype of a monitoring module is under development on an UNIX system. The language LISP (dialect LE_LISP) has been chosen. For the moment, no standard inference engine has been chosen because the purpose of this first prototype is to analyse how the Petri net structure can be utilized by a rule based diagnosis mechanism.

6. REFERENCES.

AL 84 P. Alanche, K. Benzakour, F. Dollé, F. Gillet, P. Rodrigues, R. Valette: "PSI a Petri net based simulator for flexible manufacturing systems", Lecture Notes in Computer Science 188, 1984, p.1-14.
BE 86 Bensana E., Corrège M., Bel G., Dubois D.: "An expert system approach to industrial job-shop scheduling", IEEE International Conference on Robotics and Automation, San Francisco, april 1986.
BF 86 E. Best, C. Fernandez: "Notations and Terminology on Petri Net Theory", Arbeitspapiere der GMD 195, Jan. 1986.
DE 83 H. Demmou, M. Courvoisier, E. Thuriot, R. Valette: "A new synchronization scheme for microprocessor based real-time control system", IECON'83, San Francisco, nov. 1983, p.237-242.
ER 86 Erschler J., Esquirol P.: "Decision-aid in job shop scheduling: a knowledge based approach", IEEE International Conference on Robotics and Automation, San Francisco, april 1986.

FI 85 R.   Fikes,   T.   Kehler: "The role of frame-based representation in reasonning", Communications of the ACM, vol. 28, n9, p.904-920 sept. 85.

FO 83 M. Fox, B. Allen, S. Smith, G. Strohm: "ISIS: a constraint-directed approach to job shop scheduling", Technical   Report   CMU-RI-TR-83-8,   Carnegie-Mellon University, june 1983.

GE 79 H.J.   Genrich,   K.   Lautenbach: "The analysis of distributed systems by means of Predicate/transition nets", Semantics of concurrent computation Evian 1979, p.123-146.

HA 85 F. Hayes-Roth: "Rule based systems", Communications of the ACM, vol. 28, n9, p.921-932, sept. 85.

LA 82   Laurière: "Représentation   et   utilisation   des connaissances", TSI, vol. 1, n 1 et 2, 1982.

LA 84 Laurent: "La structure de contrôle dans les systèmes experts", TSI, vol. 3, n 3, 1984.

MU 86 Murata T., Komoda N., Matsumoto K., Haruna K.: "A Petri net based controller for flexible and maintainable sequence control and its applications in factory automation", IEEE trans. on Industrial Electronics, vol. IE-33, n1, feb. 1986.

NA 85 Y. Narahari, N. Viswanadham: "On the invariants of coloured Petri nets", 6th European Workshop on Applications and Theory of Petri nets, Espoo, Finland, 1985.

PA 85 Pagnoni A.:   "Stochastic invariance in predicate-transition nets", 6th European Workshop on Applications and Theory of Petri nets, Espoo, Finland, 1985.

PE 62 C.A. Petri: "Kommunikation mit Automaten", Ph.D. Thesis, University of Bonn,   1962,   also "Communication with automata" supplement 1, Technical Report RADC-TR-65-377 1966.

SI 85   Sibertin-Blanc: "High-level   Petri   nets with data structure", 6th European Workshop on Applications and Theory of Petri nets, Espoo, Finland, 1985.

ST 85 R. Steinmetz, S. Theissen: "Integration of Petri nets into a tool for consistency checking of expert systems with rule-based knowledge representation", 6th European Workshop on Applications and Theory of Petri Nets, Espoo, Finland, 1985.

TA 85 Tashiro T., Komoda N., Tsushima I., Matsumoto K.: "Advanced software for constraint combinational control of discrete event systems - rule-based control software for factory   automation",   Compint   85   Computer   aided technologies, Montréal, sept. 1985.

TH 83 E. Thuriot, R. Valette, M. Courvoisier: "Implementation of a centralized synchronization concept for production systems", IEEE Real-Time Symposium, Arlington, USA, dec. 1983.

ZI 78 Zisman M.D.: "Use of production systems for modelling asynchronous concurrent processes", in Pattern Directed Inference Systems, p.53-68, D. Waterman and F. Hayes-Roth, Academic Press Inc., ISBN 0-12-737550-3, 1978.

EXTENSION AND INTENSION OF ACTIONS

Rüdiger Valk

Fachbereich Informatik, Universität Hamburg

Rothenbaumchaussee 67-69, D-2000 Hamburg 13

ABSTRACT   The principle of extensionality in net theory is examined from a new point of view. Including side conditions it becomes possible to describe different intensions of an action.

Key words: extension, intension, condition/event-system, side conditions.

CONTENTS

1. Introduction

'Event', 'transition' or 'action' are central notions in net theory. Let us start with an example (after [Peterson 81]). Some task A (see Fig. 1.1) has to be executed first on machine M_1 and then on machine M_2 or M_3 . Machines M_1 and M_2 are operated by operator O_1 whereas O_2 operates either M_1 or M_3 .

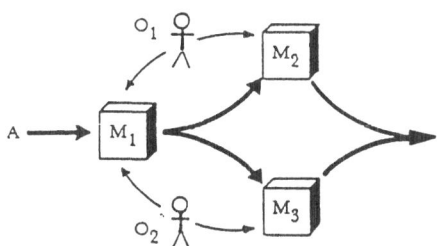

Fig. 1.1: Task A for three machines and two operators

In Fig. 1.2(a) a global state transition of some action α for this example is given, changing task A from "not executed" into "A in M_1 with O_2". At the same time, M_1 changes from "M_1 idle" to "M_1 busy" etc. Action α may be characterized by the conditions which are changing (and only by these!). This leads to the representation by a transition α in Fig. 1.2(b) connecting corresponding conditions.

a)

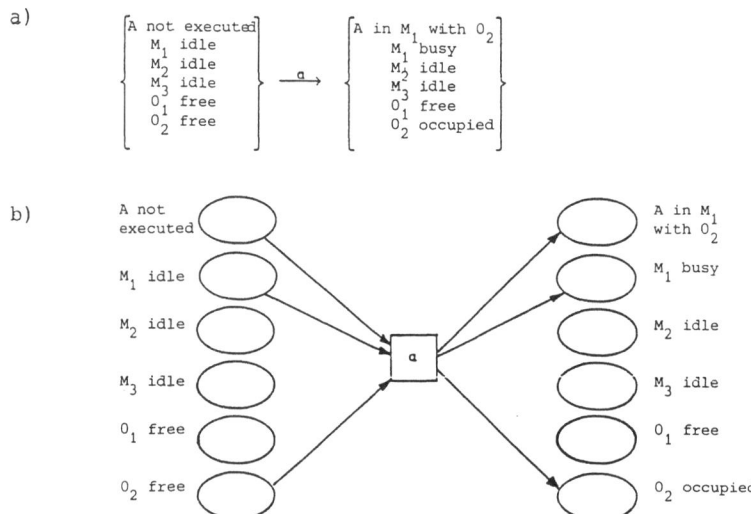

b)

Fig. 1.2: State change and transition for action α

Fig. 1.3 shows a net representation for the example including action α .

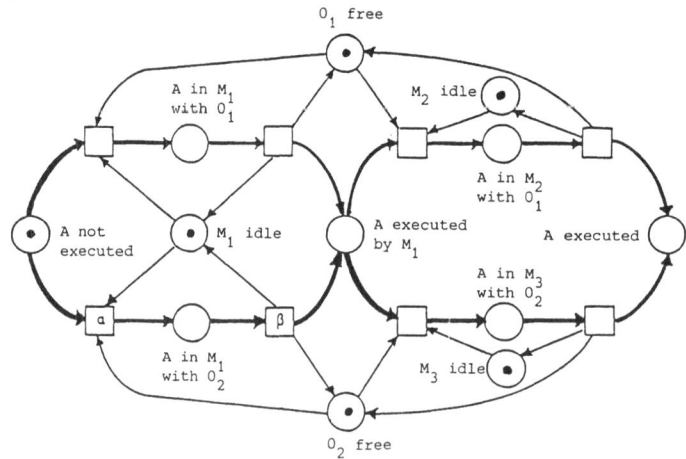

Fig. 1.3: Net for example of Fig. 1.1

The example illustrates the "principle of extensionality for actions":

"Actions are represented by the change of conditions they affect". This principle leads to the definition of transition rules in all kinds of nets. It is of general importance in computer science since it allows

a) abstraction from internal implementations of actions,
b) reduction to affected conditions only and
c) concurrency of actions if their affected conditions are disjoint.

To our knowledge this principle was first formulated by C.A. Petri in 1978: "It means that a repeatable event is fully characterized by the extension of the change in conditions effected by each of its occurrences [Petri 78]". In some different form it was called the "Extensionality Principle of Elementary Changes" in [Petri 86], where we read: "A point x which denotes an interaction between signals ... can be precisely defined by the beginning and ending of states in its vicinity, and no state can begin and end at the same point x ".

To prepare the discussion below we should recall the general usage of the notion "extensionality". In [Carnap 42] we find: "Logical connections which have truth-tables are often called truth-function, because the truth-value of a full sentence depends merely upon the truth-values of the components. Following Russel, we call connections of this kind and their connectives extensional and a full sentence of such a connection extensional with respect to the components". In natural languages the extension of a predicate denotes the class of objects, making this predicate true. On the other hand, the intension of a sentence or predicate denotes its meaning or causal interdependence. For instance, the predicates "organism having a heart" and "organism having kidneys" may have the same extension (if the corresponding classes are identical), but certainly have different intensions (example from [Quine 61]). The "thesis of extensionality" says that (from a logical point of view) different intensions of a predicate with the same extension are a matter of taste but not objective [Carnap 55]. For instance, the formal implication "if p then q" does in general not reflect the intension of the causal dependence "q since p". In axiomatic set theory the axiom of extensionality states that two sets x and y are equal if they contain the same elements:

$$\forall z: (z \in x \Leftrightarrow z \in y) \Rightarrow x = y$$

2. Safe S/T-systems

The extensional nature of transition rules in Petri nets was formalized by the "principle of extensionality" in case/transition
models and condition/event-systems ([Genrich/Lautenbach/Thiagarajan 80]
page 26). For reasons discussed in section 3 and 4 condition/event-
systems do not permit side-conditions. Such side conditions may be very
useful, however, in high level modelling of systems. If, for instance,
in the example of Fig. 1.3 transitions α and ß are considered as one
abstract action, the condition "M_1 idle" becomes a side condition with
the possible meaning "M_1 available".

a)

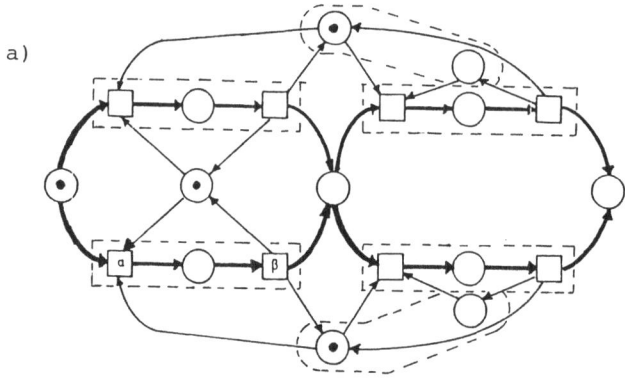

b)

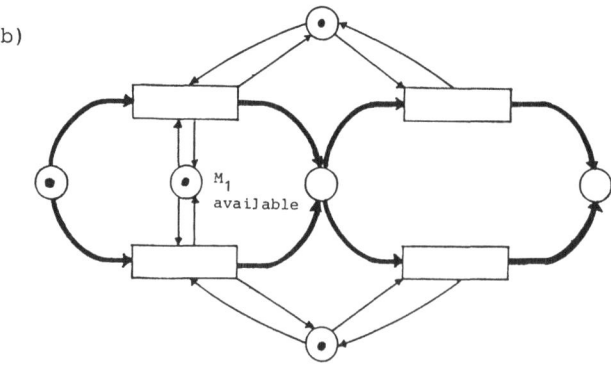

Fig. 2.1: Abstraction of the net in Fig. 1.3

Such an abstraction is shown in Fig. 2.1. Note that no transition is
enabled when the net of Fig. 2.1(b) is interpreted as condition/event-
system in the sense of [Reisig 85], for instance!. Condition/event-

systems also contain other features such as the forward and backward reachability class, which are not used in every context. Therefore, we consider condition/event-systems here as place/transition systems with capacity one on all places. To avoid confusion we will call them "safe S/T-systems".

(2.1) Definition:

A quadrupel $\Sigma = (S,T,F,m_o)$ is called a <u>safe S/T-system</u> if

a) S and T are finite and disjoint sets of <u>places</u> and <u>transitions</u>, respectively.

b) $F \subset (S \times T) \cup (T \times S)$ is the <u>flow relation</u>.

c) $m_o \subset S$ is the <u>initial case</u>.

We adopt the transition rule introduced in [Jantzen/Valk 80] and used in many papers, also in [Jessen/Valk 86]. By this rule the transition of Fig. 2.2(a) is enabled. It could be interpreted as an abstraction of the refinement of Fig. 2.2(b), where the side condition might be a reusable resource.

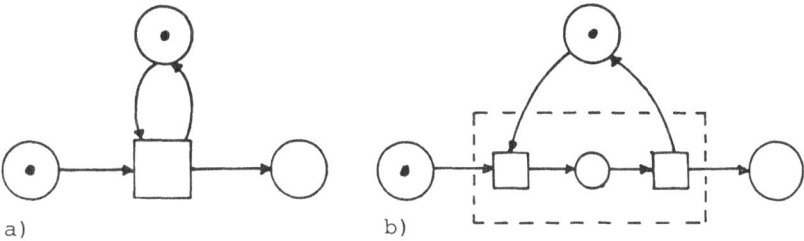

a)   b)

Fig. 2.2: Transition with side condition in a) and refinement in b)

(2.2) Definition:

For an element $x \in S \cup T$ of a safe S/T-system $\Sigma = (S,T,F,m_o)$ we define $^{\bullet}x := \{y \mid (y,x) \in F\}$ and $x^{\bullet} := \{y \mid (x,y) \in F\}$.

A transition $t \in T$ is said <u>to be enabled</u> in a case $m \subset S$ (or t <u>has concession</u> in m) (formally: $m \overset{t}{\to}$) if $^{\bullet}t \subset m$ and $(t^{\bullet} \smallsetminus {}^{\bullet}t) \cap m = \emptyset$. If t is enabled in m then $m' := (m \smallsetminus {}^{\bullet}t) \cup t^{\bullet}$ is the <u>follower case</u> of m under t (formally: $m \overset{t}{\to} m'$).

(2.3) Definition:

a) By R(m) we denote the smallest set of cases such that $m \in R(m)$ and with $m_1 \in R(m)$, $m_1 \overset{t}{\to} m_2$ also $m_2 \in R(m)$. R(m) is the <u>set of reachable cases from m</u> or the <u>reachability set of m</u> . $R(\Sigma) :=$ $R(m_o)$ is the reachability set of a given safe S/T-system $\Sigma = (S,T,F,m_o)$.

b)   For a safe S/T-system $\Sigma = (S,T,F,m_o)$ the graph $G(\Sigma) := (R(\Sigma),E)$ with nodes $R(\Sigma)$ and labelled edges $E = \{(m_1,t,m_2), \mid m_1 \overset{t}{\to} m_2\}$ is called the <u>reachability graph</u> of Σ .

(2.4) Definition:

A safe S/T-system $\Sigma = (S,T,F,m_o)$ is called

a)   <u>simple</u> (or T-simple) if $\,{}^\bullet t_1 = {}^\bullet t_2\,$ and $\,t_1{}^\bullet = t_2{}^\bullet\,$ imply $\,t_1 = t_2$ for all $t_1,t_2 \in T$,

b)   <u>pure</u> if $(s,t) \in F$ implies $(t,s) \notin F$, i.e. there are no <u>side conditions</u> or <u>side places</u> $s \in {}^\bullet t \cap t^\bullet$.

3. The principle of extensionality

Usually, the behaviour of a dynamic system is specified by means of a state space and the transition rules which determine the set of possible future states, given a present state. In our approach, a state is described by means of those conditions which hold concurrently in that state. Such a set of conditions holding while nothing changes is called a case or constellation [Genrich/Lautenbach/Thiagarajan 80]. This specification of system behaviour can be obtained by observation of an existing system or by the construction of a new one. It is represented by graph, which is called "case/transition model".

(3.1) Definition:

A <u>case/transition model</u> is a quadrupel $M = (B,C,T,r)$ where

a) $B = \{b_1,\ldots,b_n\}$ is a finite set of <u>conditions</u>,

b) $C \subset P(B) := \{A \mid ACB\}$ is the set of possible <u>cases</u>,

c) T is a finite set of <u>transitions</u>,

d) $r \subset C \times T \times C$ is the <u>relation of case change</u>.

(3.2) Definition:

A case/transition model $M = (B,C,T,r)$ is usually represented by a directed graph with set of nodes C and labelled edges r . M is said to be <u>reachable</u> from $c_o \in C$ if for every $c \in C$ there is a path from c_o to c .

In the following definition to every transition $t \in T$ of a case/ transition model M we associate a set $pre(t)$ of pre-conditions, a set $side(t)$ of side-conditions and a set $post(t)$ of post-conditions together with some axioms.

(3.3) Definition:

A case/transition model $M = (B,C,T,r)$ satisfies the <u>principle of extensionality</u>, if there are mappings

$$\text{pre}: T \to P(B)$$
$$\text{side}: T \to P(B) \quad \text{and}$$
$$\text{post}: T \to P(B)$$

having the following properties:

(E1) $\forall\ (c_1,t,c_2) \in r : \text{pre}(t) = c_1 - c_2 \wedge \text{post}(t) = c_2 - c_1 \wedge \text{side}(t) \subset c_1 \cap c_2$

(E2) $\forall\ c_1 \in C\ \forall t \in T : \text{pre}(t) \cup \text{side}(t) \subset c_1 \wedge \text{post}(t) \cap c_1 = \emptyset \Rightarrow \exists\ c_2 :$
$(c_1,t,c_2) \in r$

(E3) $\forall\ t_1,t_2 \in T: (\text{pre}(t_1) = \text{pre}(t_2) \wedge \text{side}(t_1) = \text{side}(t_2)$
$\wedge\ \text{post}(t_1) = \text{post}(t_2)) \Rightarrow t_1 = t_2$

If, in addition, $\text{side}(t) = \emptyset$ for all $t \in T$, then we say that M satisfies the <u>principle of extensionality without side conditions</u>.

Axiom (E1) expresses that the changes of conditions are equal for all occurrences of t . By (E2) a transition occurs if pre- and side-conditions are satisfied, but no post-condition holds. (E3) characterizes each transition by its "extension": $(\text{pre}(t), \text{side}(t), \text{post}(t))$. Hence this axiom is similar to the axioms of extensionality of set theory as given at the end of section 1.

A given case/transition model can satisfy the principle of extensionality in different ways. The case/transition model in Fig. 3.1(b) satis-

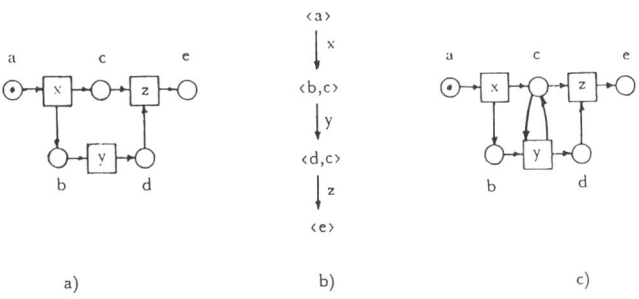

a)   b)   c)

Fig. 3.1: Two safe S/T-systems having the same reachability graph

fies the principle of extensionality, either by defining $\text{side}(y) = \emptyset$ or by taking $\text{side}(y) = \{c\}$. These possibilities are shown by the safe

S/T-systems in Fig. 3.1(a) and (c), resp. If nothing is known but the case/transition model, the choice between (a) and (c) is arbitrary. The conflict could be solved by learning something of the intension, namely whether c and y are interacting or not. Therefore, we call this an "intensional difference". Such intensional differences are avoided if we restrict our attention on the principle of extensionality without side condition (see. Def. 3.3). However, the reader should observe that conflicts by the intensional difference also exist in real modelling and that it might be desirable to have a formal way to express this difference. For instance, in Fig. 2.1(b) it depends on the intension of modelling whether "M_1 available" is included or not. (We obtain a different situation, however, if we take concurrency into consideration). As suggested in Fig. 3.1, there is a strong relation between case/transition models and safe S/T-systems. If in the following theorem a case/transition model and a reachability graph are stated to be equal, then this is supposed to be true for their corresponding graphs.

(3.4) Theorem

a)   The reachability graph $G(\Sigma)$ of a simple safe S/T-system $\Sigma = (S,T,F,m_o)$ is a case/transition model M , satisfying the principle of extensionality and reachable from m_o . If Σ is pure, then M is without side conditions.

b)   If M= (B,C,T,r) is a case/transition model, which satisfies the principle of extensionality and is reachable from some case c_o , then there is a simple safe S/T-system Σ having a reachability graph identical with M . If M is without side conditions, then Σ is pure.

Proof

a)   Let $G(\Sigma) = (R(\Sigma),E)$ be the reachability graph of a given simple S/T-system $\Sigma = (S,T,F,m_o)$. Then M= $(S,R(\Sigma),T,E)$ is a case/transition model reachable from m_o .

Now we define for all $t \in T$:
$$side(t) := \{s \in S \mid (s,t) \in F \wedge (t,s) \in F\}$$
$$pre(t) := {}^{\bullet}t - side(t)$$
$$post(t) := t^{\bullet} - side(t)$$

These functions satisfy the axioms (E1),...,(E3):

<u>Axiom E1:</u>   By the definition of a follower case (Def. 2.2) for $(m_1,t,m_2) \in E$ it follows $pre(t) = m_1-m_2$, $post(t) = m_2-m_1$ and $side(t) \subset m_1 \cap m_2$.

Axiom E2:   If pre(t) ∪ side(t) ⊆ m₁ , then t has concession, since
•t= pre(t) ∪ side(t) . Then there is some m₂ such that (m₁,t,m₂) ∈ E .

Axiom E3:   If pre(t₁)= pre(t₂) and side(t₁)= side(t₂) and post(t₁)
= post(t₂) , then •t₁= pre(t₁) ∪ side(t₁) = •t₂ and t₁•= post(t₁)
∪ side(t₁) = t₂• . Since Σ is simple, it follows t₁ = t₂ .

If Σ is pure, then side(t) = ∅ for all t∈T .

b)   Now let M= (B,C,T,r) be a case/transition model, which satisfies
the principle of extensionality and is reachable from c₀ . Then we
define the following safe S/T-system Σ= (S,T',F,c₀) by:

    S := B
    T':= {t∈T | ∃ c₁,c₂ : (c₁,t,c₂) ∈ r}
    F := {(b,t) | b ∉ pre(t) ∪ side(t)} ∪ {(t,b) | b ∈ post(t) ∪ side(t)}

Each transition t∈T' appears in some edge (c₁,t,c₂) ∈ r of M .
From (E1) we therefore know pre(t)= c₁-c₂ , post(t)= c₂-c₁ and
side(t) ⊆ c₁ ∩ c₂ . Hence, pre(t), side(t) and post(t) are disjoint
sets. We now prove that Σ is simple. Suppose that t₁ and t₂ are
transitions from T' with •t₁ = •t₂ and t₁• = t₂• . By the definition
of F this means pre(t₁) ∪ side(t₁) = pre(t₂) ∪ side(t₂) and post(t₁)
∪ side(t₁) = post(t₂) ∪ side(t₂) .

We now prove pre(t₁) ⊆ pre(t₂) . Suppose b ∈ pre(t₁) but b∉ pre(t₂) .
Then by b ∈ pre(t₁) ∪ side(t₁) = pre(t₂) ∪ side(t₂) we conclude
b ∈ side(t₂) and b ∈ side(t₂) ∪ post(t₂) = side(t₁) ∪ post(t₁) .
By b ∈ pre(t₁) ∩ (side(t₁) ∪ post(t₁)) we obtain a contradiction, hence
pre(t₁) ⊆ pre(t₂) .

In a similar way one can show pre(t₁) ⊃ pre(t₂) , side(t₁) = side(t₂)
and post(t₁) = post(t₂) . By axiom (E3) we therefore obtain t₁ = t₂
and Σ is simple.

    If M is without side conditions, then side(t) = ∅ for all t∈T .
Then, since pre(t) and post(t) are disjoint, {(b,t),(t,b)} ⊆ F is
impossible and Σ is pure.

    Now it remains to prove that M is the reachability graph of Σ .
By definition the initial case of Σ is c₀ and all cases are reachable
from c₀ . We have to prove that for all t∈T' and m₁ ∈ R(Σ) we have
c₁ ⟶ᵗ c₂ iff (c₁,t,c₂) ∈ r . In fact, from c₁ ⟶ᵗ c₂ we have •t ⊆ c₁
and t•∖•t ∩ c₁ = ∅ . Hence, by pre(t) ∪ side(t) ⊆ c₁ and post(t)
∩ c₁ = ∅ with (E2), we conclude ∃ c₂' : (c₁,t,c₂') ∈ r . But then it
is necessary that c₂' = c₂ . On the other hand, if (c₁,t,c₂) ∈ r ,

then $\mathrm{pre}(t) = c_1 - c_2$, $\mathrm{side}(t) \subset c_1 \cap c_2$ and $\mathrm{post}(t) = c_2 - c_1$. Hence, $\cdot t = \mathrm{pre}(t) \cup \mathrm{side}(t) \subset c_1$ and $(t\cdot\searrow\cdot t) = \mathrm{post}(t)$ and $(t\cdot\searrow\cdot t) \cap c_1 = \emptyset$. Then by definition 2.2 we obtain $c_1 \overset{t}{\to} c_2$.

□

4. Applications

The proof of part (b) of the theorem is constructive and can be used to find a safe S/T-system from a specification, given by a case/ transition model.

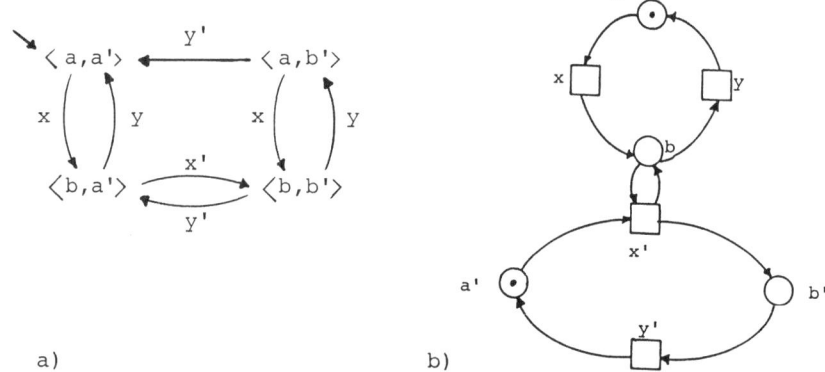

a)   b)

Fig. 4.1: A safe S/T-system in (b) constructed from the case/
           transition model in (a)

Fig. 4.1(a) shows a case/transition model satisfying the principle of extensionality. Note that $\mathrm{side}(x') = \emptyset$ is impossible! The resulting safe S/T-system is shown in Fig. 4.1(b).

In Fig. 4.2 both directions of the theorem are used several times. In the first two steps the (dead) transition x is lost (x has no exten- sion!). In the next step, the permanent condition b is omitted (this could be added in general to the construction of the theorem). The last two steps are characterized by choosing $\mathrm{side}(z) = f$ or $\mathrm{side}(z) = \emptyset$.

In [Carnap 55] the following experiment is discussed. An English speaking linguist is studying the semantics of vocabulary of the German person Karl. He observes
                    Pferd = horse
whereas a second linguist notes
                    Pferd = horse or unicorn
Since unicorns do not exist, both linguists associate the same extension with Karl's word "Pferd". With respect to really existing objects, no

581

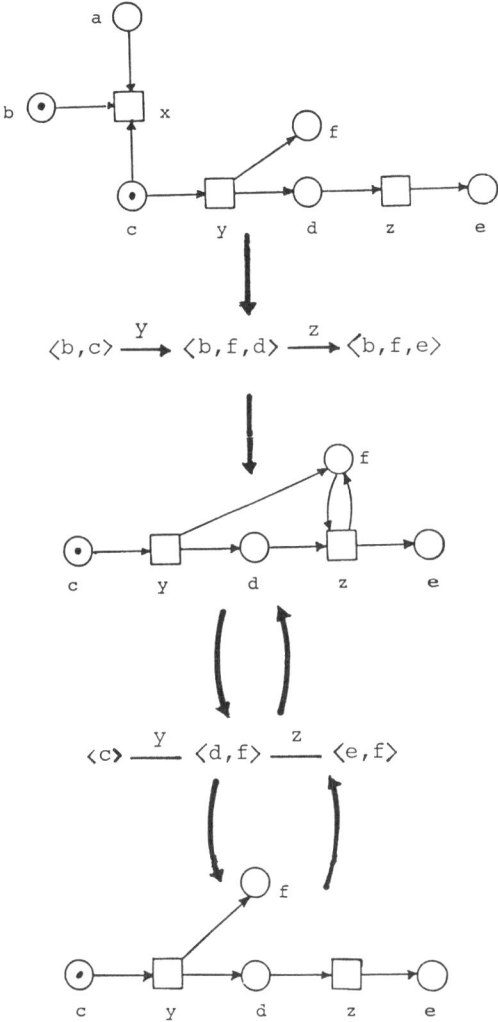

Fig. 4.2: Iterated application of the theorem

distinction can be made between the two observations. If, however, modal
sentences are used (e.g. "it is possible that"), than the two cases
could be distinguished. This experiment is similar to the situation of
action x in Fig. 4.2. S/T-systems with or without transition x have the
same extension, since x is dead. However, if we interpret the first S/T-
system as a scheme for possible actions (depending on different initial

cases), then the situation changes. Also here modal logic could be adequate.

Our last example shows that the theorem can be used to design systems from specifications [Jessen/Valk 86]. Fig. 4.3(a) shows two independent sequential systems with reachability graph in Fig. 4.3(b).

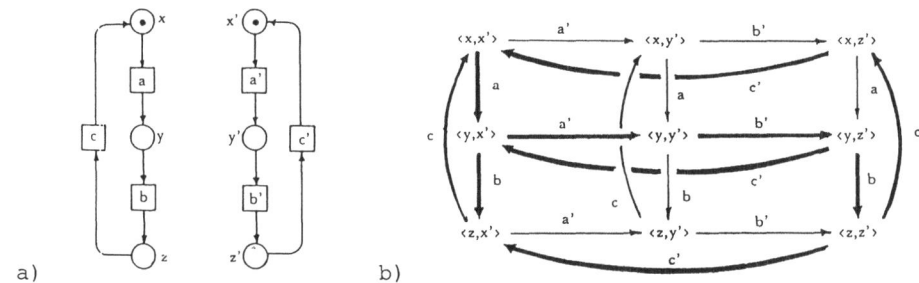

a)   b)

Fig. 4.3: Two independent sequential systems

Now we want to change the system in such a way that z and z' hold in mutual exclusion. Therefore, we cut off the case <z,z'> in Fig. 4.3(b) and apply our theorem. However, this modified case/transition model does not satisfy the principle of extensionality!

A solution seems to be possible if x' and y' are allowed to become side conditions. But then b has to be replaced by b_1 and b_2 with side(b_1) = x' and side(b_2) = y' . In a similar way, b' is replaced by b_1' and b_2' . Then we obtain a case/transition model satisfying the principle of extensionality. The resulting safe S/T-system is shown in Fig. 4.4.

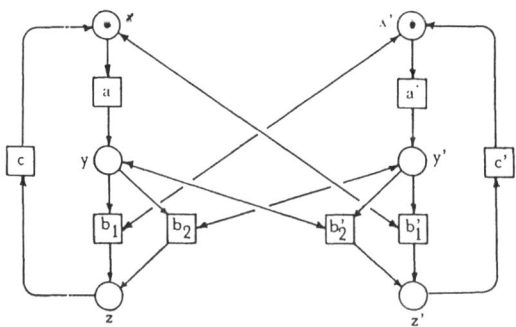

Fig. 4.4: Solution with side conditions

An alternative way to satisfy the principle of extensionality consists in explicitly introducing the condition   f:= "neither z nor z'" ,
as done in Fig. 4.5.

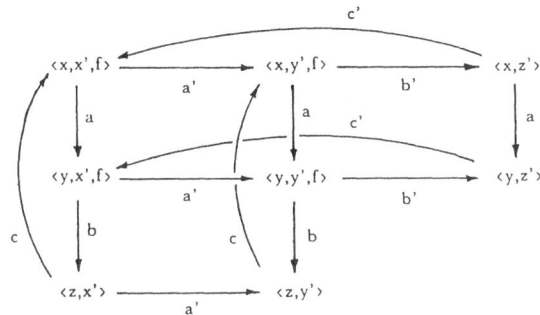

Fig. 4.5:   Case/transition model with mutual exclusion

Applying now the theorem, we obtain the classical solution of
Fig. 4.6. Obviously, this solution is simpler. This may be a hint that

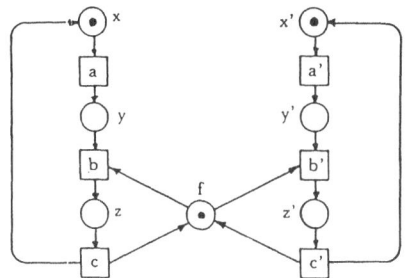

Fig. 4.6:   Solution without side conditions

satisfying the principle of extensionality without side conditions
should be preferred, if possible. However, there are also cases where
side conditions lead to conceptual simpler solutions.

Note that the solution of Fig. 4.4 is "transition-oriented" whereas the solution of Fig. 4.6 is "place-oriented". Hence, we find here
again some kind of duality, which is so typical for general net theory.

5. References

[Brauer 80]   Brauer W. (ed.): Net Theory and Applications, Lecture
              Notes in Computer Science No. 84, Springer, Berlin 1980
[Carnap 42]   Carnap, R.: Introduction to Semantics and Formalization
              of Logic, Harvard Univ, Press, Cambridge, Mass., 1942

[Carnap 55]   Carnap, R.: Meaning and Synonymy in Natural Languages,
                 in: Philosophical Studies 6 (1955)

[Genrich/Lautenbach/Thiagarajan 80]
                 Genrich, H.J., Lautenbach, K., Thiagarajan, P.S.:
                 Elements of General Net Theory, in [Brauer 80]

[Jantzen/Valk 80]   Jantzen, M., Valk, R.: Formal Properties of Place/
                 Transition Nets, in [Brauer 80]

[Jessen/Valk 86]   Jessen, E., Valk, R.: Rechensysteme, Springer,
                 Berlin, 1986

[Peterson 81]   Peterson, J.L.: Petri Net Theory and the Modelling of
                 Systems, Prentice Hall, Englewood Cliffs (1981)

[Petri 78]   Petri, C.A.: Concurrency as a Basis of Systems Thinking,
                 St. Augustin: Gesellschaft für Mathematik und Datenver-
                 arbeitung Bonn, Interner Bericht, ISF-78-06 (1978);
                 also: Jensen, F.V., Mayoh, B.H., Moller, K.K. (eds.),
                 Proc. 5th Scandinavian Logic Symposium, Universitets-
                 forlag, Aalborg, 1979

[Petri 86]   Petri, C.A.: Concurrency Theory, Proc. Advanced Course
                 on Petri Nets, Bad Honnef, 1986

[Quine 61]   Quine, W,V,: Two Dogmas of Empiricism, in: From a logi-
                 cal point of view, 2nd. rev. ed., Cambridge, Mass., 1961

[Reisig 85]   Reisig, W.: Petri Nets, Springer, Berlin 1985

INTERFACE AS A BASIC CONCEPT FOR
SYSTEMS SPECIFICATION AND VERIFICATION

Klaus Voss

Gesellschaft für Mathematik und Datenverarbeitung (GMD)

5205 St. Augustin

Federal Republic of Germany

Abstract

To specify or to verify a system means to prescribe or to analyze its connections with the environment into which it shall be or in which it is embedded. From this point of view, its internal construction and its internal behaviour are relevant only as far as they determine its communication and cooperation with the outside. Hence, both specification and verification are concerned with a part of the system only, namely its interface with its environment. As a basis to deal with such partially determined systems, an appropriate means to represent them and a suitable equivalence notion are needed. This paper discusses the use of strictly labelled net systems for the representation of system interfaces and the notion of interface equivalence. The latter notion is a generalization of bisimulation and is intended to preserve the most important properties of sytems at the interface with their environment. It can constitute a basis for deriving concepts of compatibility to be used in answering the question whether and to what extent a realized system fulfills a given specification.

1. Introduction

In 1980, C.A. Petri launched a project in GMD which was entirely devoted to investigating and clarifying the notion of interface and its role in system modelling and analysis. The recognition of 'interface' as a system notion of central importance dates back to 1962 when C.A. Petri investigated questions concerning the communication between modules via common border events, called "Kommunikationsformen" (forms of communication) in his dissertation [Pe62]. His in-

tention was to find formulas for describing precisely the communication between a
net and its environment, and to establish a partial order among the forms of communication of different modules characterizing their relative 'determinedness' (see
chapter 4 of [Pe62]).

This paper tries to sketch one of the results of the project mentioned above. Chapter 2 is concerned with the representation of partially determined or partially visible systems by strictly labelled marked nets. It introduces the notion of interface and its relevance for system specification and verification. Related to this
context, chapter 3 defines an equivalence relation between partially determined systems and shows that the most important system properties are preserved under this
so-called interface equivalence. Chapter 4 suggests ways how to apply these concepts
to the problems of specification and verification and how to define such a partial
order as mentioned above. All theorems are stated here without proofs which can be
found in [Vo86].

2. Partially determined systems and the notion of interface

The goal of this section is to present a means to deal with models of systems which
we have only partial knowledge of. Our approach will be to treat the system to be
investigated as a subsystem of a more comprehensive system. To avoid confusion, a
model of the subsystem from now on will be called a module. If the module does not
yet exist and has to be specified, we are often satisfied in prescribing its
cooperation with its neighbours and we are not concerned with its detailed inner
structure and with the invisible part of its behaviour. If the module already exists, we often are unable to look into it and we can experience only its appearance
and behaviour on its border to the outside. In both cases, the module is regarded
not in isolation but in connection with its environment (neighbours, outside), and
the interest is directed towards the coordination and communication between them
[Pe62]. In order to be able to delimit the environment of a module there must be a
decision on the real system to be modelled (the object system), on its borderlines
and on the perspective and the level of abstraction adopted for the model. Thus, for
the rest of this section, we assume that the model of the object system includes
both the module under investigation and its (pre-given) environment.

The basis of this investigation is the General Net Theory [Pe77b]. Our means to
model real systems are elementary net systems (EN systems) $\Sigma = (N, M^0)$, as defined in
[RT86], where $N = (S, T; F)$ is a finite net and M^0 its initial case (marking). The
S-elements are called conditions and the T-elements are called events. If the events
of a subset u of T are concurrently enabled at a case M and if their occurrences
lead to a case M', we write $M[u>M'$. For an EN system, its state space is defined to

be the set of cases thus reachable from M^0, i.e., the forward case class $[M^0>$. In addition, we require that all net systems are simple and that every event (out of T) has an occurrence at a case in $[M^0>$. If, as will be done frequently in the sequel, we consider the full case class $[M^0]$ as the state space of an EN system, then we arrive at condition/event systems (CE systems). Finally, we always assume that net systems are contact-free. The following naming convention is used throughout this paper: whenever a net is considered as a system, i.e. with a case class, its name contains a 'Σ', and whenever a net or a system is considered with a labelling function its name contains a '$*$'.

Suppose that we have decided on the part of the real world (object system) to be considered and on the level of modelling resolution to be adopted. Suppose further that we have a found an elementary net system model $\Sigma = (N, M^0)$ of the object system which contains a sub-model $\Sigma B = (B, MB^0)$, the module to be investigated. Of course we demand that B should be a subnet of N. Recall that $B = (SB, TB; FB)$ is a subnet of $N = (S, T; F)$ iff $SB \subseteq S$, $TB \subseteq T$ and $FB = F \cap (SB \times TB \cup TB \times SB)$. Concerning the initial case MB^0 of B, we choose $MB^0 = M^0/B$, denoting M^0 restricted to the conditions of B. What then is the 'environment' of B?

The first basic assumption adopted for describing the connection of a module B and its environment Z is the following:

A set of distinguished nodes is supposed to be shared by B and Z. Any communication or synchronization between B and Z takes place via these shared nodes.

The shared nodes shall separate the module from its environment. Hence we arrive at the following definition which is illustrated schematically in figure 1:

Definition 1. Let $N = (S, T; F)$ be a simple net and $B = (SB, TB; FB)$ be a subnet of N. Let $Q \subseteq SB \cup TB$ be a set of distinguished nodes of B such that $Z = (SZ, TZ; FZ)$ with $SZ = S - (SB - Q)$, $TZ = T - (TB - Q)$ and $FZ = F \cap (SZ \times TZ \cup TZ \times SZ)$ is also a subnet of N.
Then Q is called the interface of B and Z, and
    Z is called the environment of B with respect to Q (in N),
iff
(i)   there exists no path from an element of B to an element of Z not containing an element of Q and
(ii) $F \cap (Q \times Q) = \emptyset$.   □

Condition (i) says that every directed path from B to Z and every directed path from Z to B contains an interface element, and condition (ii) says that there is no direct connection between interface elements.

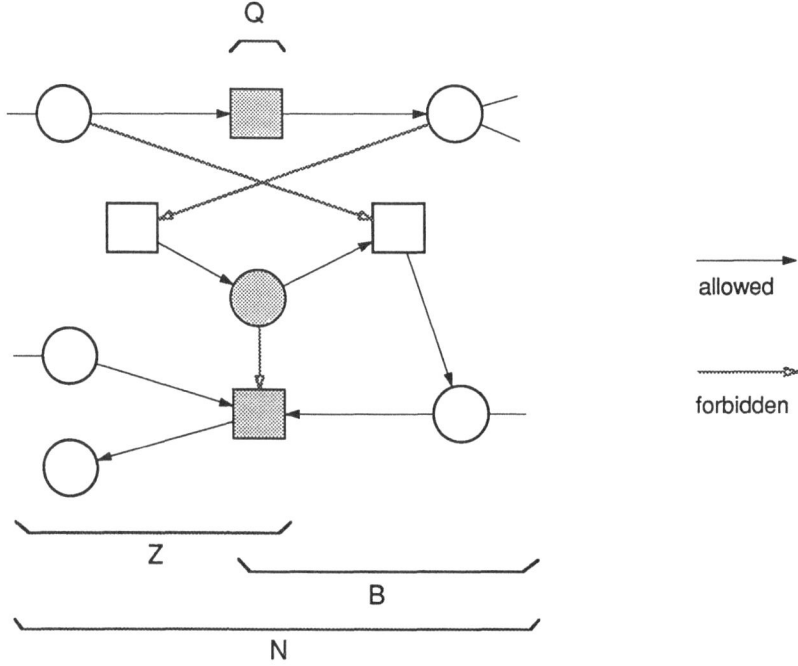

Q

allowed

forbidden

Z

B

N

Figure 1. A sample net N and its components
        B (module), Z (environment) and Q (interface)

The interface Q defines a decomposition of N in two subnets B and Z such that
SB ∪ SZ = S, TB ∪ TZ = T and (SB ∩ SZ) ∪ (TB ∩ TZ) = Q. Moreover, an orientation of
the interface is determined such that, for each interface element, every arc
attached to it is connected with a node either of B - Q or of Z - Q.

In the following, a crucial point will be to distinguish that part of a net N which
we have knowledge of, from that part of N which is not known to us. In the literature we often encounter the concept of an external 'observer' who is able to see the
interface between a module and its environment (being parts of a composite system)
but to whom the rest of the system is hidden. Adopting the terminology from literature, the nodes of the interface Q (the 'observable' part of N) are called external
and the nodes belonging to N - Q (the 'hidden' part of N) are called internal. Thus
we have sets of external events, Te = Q ∩ T, of external conditions, Se = Q ∩ S, of
internal   events,   Ti = T - Q = T - Te,   and   of   internal   conditions,
Si = S - Q = S - Se.

Returning to the dynamics of the system Σ, we assign the initial marking M^0/Z to Z and thus arrive at the environment subsystem $\Sigma Z = (Z, M^0/Z)$ of Σ. Fusing ΣB and ΣZ by identifying their shared nodes Q and by taking the union $M^0/B \cup M^0/Z$ as the initial case then yields Σ again.

When investigating the cooperation between ΣB and ΣZ we obviously are interested in that part of the system behaviour that takes place or leaves traces on the external nodes. A precise description of this 'interface behaviour' is needed. When starting with the behaviour of the composite system Σ, our attention is focussed on its 'projections' on Q.

A first step to come to a description of the interface behaviour of a system Σ is to discriminate external from internal nodes by labelling them appropriately. The elements of Q, the external nodes, all get different labels from an alphabet Y, whereas all remaining internal nodes get an identical label $\not\subset$, which is not an element of Y.

Definition 2. Let $N = (S, T; F)$ be a simple net with an interface $Q \subseteq S \cup T$. Let Y denote an alphabet and $\not\subset \notin Y$.
The labelling function L: $S \cup T \longrightarrow Y \cup \{\not\subset\}$ is called (Q-)strict iff L/Q is an injection into Y and L(x)=$\not\subset$ for all x $\in$ (S$\cup$T)-Q.
A labelled net N$*$ = (N, Q, L) with a Q-strict labelling function L is called a strictly labelled net.
For the sake of simplicity we then can choose Y to be Q, such that each interface element is labelled with its own identity. $\square$

An extensive discussion of the reasons why a strict labelling is appropriate in the context of the present investigation can be found in [Vo86]. These reasons are briefly compared with those guiding other theories concerned with 'observable behaviour' or 'languages produced by systems' and related equivalence notions like [Mi80, HM80, HBR81, Pa81, BR83]. It also discusses equivalence concepts developed within Net Theory like [An82a, An82b, DDPS83, Vo83, YE83, Po86a, Po86b, NT84, Be86]. Compared with these approaches, the refinement to strict labelling is, as we argue in [Vo86], a sensible restriction in our context. In one aspect, however, our definition is a generalization in that it permits to label not only events but also conditions.

3. Interface equivalence

Now we can tackle the central problem of what constitutes the behaviour of a system on an interface. Consider two strictly labelled nets N$*_n$ = (N_n, Q_n, L_n) with interfaces Q_n, n=1,2. As said already, Q_n are chosen as the label alphabets. N$*_1$ and N$*_2$

are called <u>interface-bijective</u> if there exists a bijection between the external events and a bijection between the external conditions. Under this assumption, matters can be simplified by <u>identifying</u> the external nodes of the systems, $Q:=Q_1=Q_2$. This implies, of course, that the labelling functions L_1 and L_2 have the same range $Q \cup \{\cent\}$. If we now proceed from nets to systems and if then we define an equivalence relation in the set of all strictly labelled systems $\Sigma*$ with the same interface Q, the answer to the central question could be: the interface behaviour of such a system is the equivalence class to which it belongs.

This answer has to be substantiated still in order to get more precise and more practicable. Indeed, it converts the initial problem into the task to choose an appropriate equivalence notion for this purpose. This choice depends mainly on the selection of those properties of the systems which are esteemed essential for the interface behaviour and which therefore should be invariants of the equivalence relation. In view of our goals, roughly speaking, the desired invariants should be the enlogic and the synchronic properties of the interface. To be more specific, the equivalence should preserve liveness and synchronic distances of the external events. Moreover, it should be a (natural) generalization of the equivalence relation that is commonly used for condition/event systems. The latter reads as follows (cf. [Re85a], paragraph 2.4):

<u>Definition 3</u>. The condition/event systems $\Sigma_1 = (N_1, M^0_1)$ and $\Sigma_2 = (N_2, M^0_2)$, with $N_n = (S_n, T_n; F_n)$, $n=1,2$, are <u>gh-equivalent</u> (or <u>equivalent</u> for short), written $\Sigma_1 \sim gh\sim \Sigma_2$, iff
$g: [M^0_1] \longrightarrow [M^0_2]$ and $h: T_1 \longrightarrow T_2$ are bijections such that for all markings $M^1_1, M^2_1 \in [M^0_1]$ and for all event sets $u \subseteq T_1$:
$M^1_1 [u> M^2_1 \Longleftrightarrow g(M^1_1) [h\blacksquare(u)> g(M^2_1)$
(where $h\blacksquare$ is defined by $h\blacksquare(u) = \{h(t) | t \in u\}$). □

In the above definition 3, $h\blacksquare$ is the bijection between the power sets of T_1 and T_2 induced by h. In the same way also a labelling function $L: S \cup T \longrightarrow Y \cup \{\cent\}$ can be extended to a function $L\blacksquare$ of the power set of $S \cup T$ into the power set of $Y \cup \{\cent\}$. In the sequel, the distinguishing $\blacksquare$ will be omitted. Note that the label of the empty step and the label of the empty case is always $\{\cent\}$.

Our preferred equivalence relation is a generalization of the notions of observational equivalence [Mi80] and bisimulation [Pa81]. In order to meet the requirements stated above, it shall be formulated for condition/event systems (rather than elementary net systems). The main reason for this is that a synchronic measure is defined only for systems with their <u>full</u> case class as will be discussed at the end of this chapter.

For the rest of this section 3 (with exception of the last three paragraphs) we will consider simple nets $N_n = (S_n, T_n; F_n)$, n=1,2, for which interfaces $Q_n \subseteq S_n \cup T_n$ and labelling functions $L_n: S_n \cup T_n \longrightarrow Q_n \cup \{\cent\}$ are defined as above. This leads to the strictly labelled nets $N*_n = (N_n, Q_n, L_n)$. On the other hand, we want to attribute intial cases M^0_n to N_n such that the resulting systems $\Sigma_n = (N_n, M^0_n)$ are (contact-free) condition/event systems. Thus, we finally arrive at strictly labelled CE systems $\Sigma*_n = (N_n, Q_n, L_n, M^0_n)$.

<u>Definition 4.</u> Let $\Sigma*_n = (N_n, Q_n, L_n, M^0_n)$ be strictly labelled CE systems as defined above, n=1,2, such that $N*_1$ and $N*_2$ are interface-bijective ($Q: = Q_1 = Q_2$).
Let $u_n \subseteq T_n$ (u_n may be empty).
Then the relation $R \subseteq [M^0_1] \times [M^0_2]$ is called a <u>simulation</u> of $\Sigma*_1$ by $\Sigma*_2$, and we say that $\Sigma*_2$ <u>simulates</u> $\Sigma*_1$ with respect to R, iff
(i) $(M^0_1, M^0_2) \in R$,
(ii) $(M_1, M_2) \in R$ ===>
       a) For every step $M_1[u_1 > M'_1$ (in $\Sigma*_1$) there exists $M_2[u_2 > M'_2$ (in $\Sigma*_2$):
           $(M'_1, M'_2) \in R$ $\wedge$ $L_1(M'_1) = L_2(M'_2)$ $\wedge$ $L_1(u_1) = L_2(u_2)$, and
       b) For every $M'_1[u_1 > M_1$ (in $\Sigma*_1$) there exists $M'_2[u_2 > M_2$ (in $\Sigma*_2$):
           $(M'_1, M'_2) \in R$ $\wedge$ $L_1(M'_1) = L_2(M'_2)$ $\wedge$ $L_1(u_1) = L_2(u_2)$.
The relation R is called a <u>bisimulation</u> between $\Sigma*_1$ and $\Sigma*_2$ iff $\Sigma*_1$ simulates $\Sigma*_2$ and $\Sigma*_2$ simulates $\Sigma*_1$ (both with respect to R).
$\Sigma*_1$ and $\Sigma*_2$ are called <u>interface equivalent</u>, written $\Sigma*_1 \sim if\sim \Sigma*_2$, iff there exists a bisimulation relation R between them.   □

<u>Proposition 5.</u> Let $\Sigma*_n$, n=1,2, be as above. If R is a bisimulation between $\Sigma*_1$ and $\Sigma*_2$, then
(i)   For all $M_1 \in [M^0_1]$ there exists $M_2 \in [M^0_2]$: $(M_1, M_2) \in R$,
      and vice versa.
(ii) If $(M_1, M_2) \in R$, $(M'_1, M_2) \in R$ and $M_1[u_1 > M'_1$ then $L_1(u_1) = \{\cent\}$,
      and vice versa.   □

It can be shown easily that $\sim if\sim$ is an equivalence relation. The interface behaviour of a labelled system is characterized by the $\sim if\sim$-equivalence class to which it belongs. To be more specific, the interface behaviour is represented - according to definition 4 - by sequences of (forward and backward) steps when projected on the interface nodes.

The subject of behaviour description is one of the most essential topics in any system theory. It has been pursued in great depth and detail in, e.g., [Pe77a, GR83, Be84, NT84, BFP85, Re85b, RT86] in the context of Net Theory. Not trying to join the debate we just remark that, for our purposes, sequences of event occurrences would not suffice as we want to distinguish true concurrency from arbitrary interleaving,

whereas the granularity of processes is estimated as unnecessarily high in our context.

A few examples shall provide a flavour of the interface equivalence notion
(figure 2). For discussing these examples, a few simplifying conventions shall be
adopted: In net pictures, the identifier of a node is always placed inside and its
label is placed outside of it. For each net system of figure 2, all conditions are
supposed to be labelled with ϕ and this ϕ will always be omitted in the picture. In
the following text, whenever a set consists of only one element, say e, we often
write e instead of {e}.

- The systems $\Sigma *_1$ and $\Sigma *_7$ are <u>not</u> interface equivalent (if-equivalent) because condition (ii) a) or b) of definition 4 is not satisfied.
- $\Sigma *_1$ and $\Sigma *_3$ are <u>not</u> if-equivalent because any relation $R_{13} \subseteq [M^0{}_1] \times [M^0{}_3]$ would
  have to assume either $(a, f) \in R_{13}$ or $(b, f) \in R_{13}$ or both. In the first case there ex-
  ists no forward step and in the second case there exists no backward step from f
  (in $\Sigma *_3$) which corresponds to the β-labelled step 1 from a to b (in $\Sigma *_1$), contradicting (ii) a) and b). In the third case, we have a contradiction to
  proposition 5 (ii) as the label of step 1 equals β and not ϕ.
- The systems $\Sigma *_1$ and $\Sigma *_2$ are also <u>not</u> if-equivalent. Arguing symmetrically to the
  example above, there either exists no backward step from e (in $\Sigma *_2$) corresponding
  to the β-labelled backward step 1 from b to a (in $\Sigma *_1$), or there exists no M_1 (in
  $\Sigma *_1$) such that $(M_1, e) \in R_{12}$.
- Clearly, neither $\Sigma *_1$ nor $\Sigma *_2$ are if-equivalent to $\Sigma *_4$. Moreover, $\Sigma *_3$ is <u>not</u> if-
  equivalent to $\Sigma *_4$. This is true because we first observe that $(f, i) \in R$ for every
  feasible bisimulation relation R. Then step 4 (in $\Sigma *_3$) has, as counterpart in
  $\Sigma *_4$, either 6 or the empty step yielding either $(g, j) \in R$ or $(g, i) \in R$. Both is impossible because the β-labelled step 5 in $\Sigma *_3$ has no equivalent step in $\Sigma *_4$.
- $\Sigma *_3$ is <u>not</u> if-equivalent to $\Sigma *_5$ because there exists no bisimulation relation
  which could relate case q (in $\Sigma *_5$) to a reachable case $M_3 \in [M^0{}_3]$ (in $\Sigma *_3$).
- Similar arguments show that $\Sigma *_5$ is <u>not</u> if-equivalent to $\Sigma *_6$.
  {(m,r),(n,s),(p,t),(q,u),(q,v)} defines a bisimulation between them.
- Also $\Sigma *_9$ is <u>not</u> if-equivalent to $\Sigma *_{10}$. In fact, the relation
  R = {(({a,c},{e,g,i}), ({b,c},{f,h,i}), ({a,d},{e,h,j}), ({b,d},{f,g,j})} is a
  simulation of $\Sigma *_{10}$ by $\Sigma *_9$, whereas R^{-1} is <u>not</u> a simulation of $\Sigma *_9$ by $\Sigma *_{10}$ because
  the step {1,2} in $\Sigma *_9$ has no counterpart in $\Sigma *_{10}$. Roughly speaking, 'arbitrary
  interleaving' can be simulated by 'concurrency', but not vice versa.
- However, $\Sigma *_1 \sim if\sim \Sigma *_8$ by the relation {(a,{w,y}),(a,{w,z}),(b,{x,y}),(b,{x,z})}.
  According to this example, a β-labelled (external) event can be replaced by a
  step consisting of an external event with label β and one or more internal
  events, hence with label ϕ.
- $\Sigma *_1 \sim if\sim \Sigma *_{11}$ is obvious.

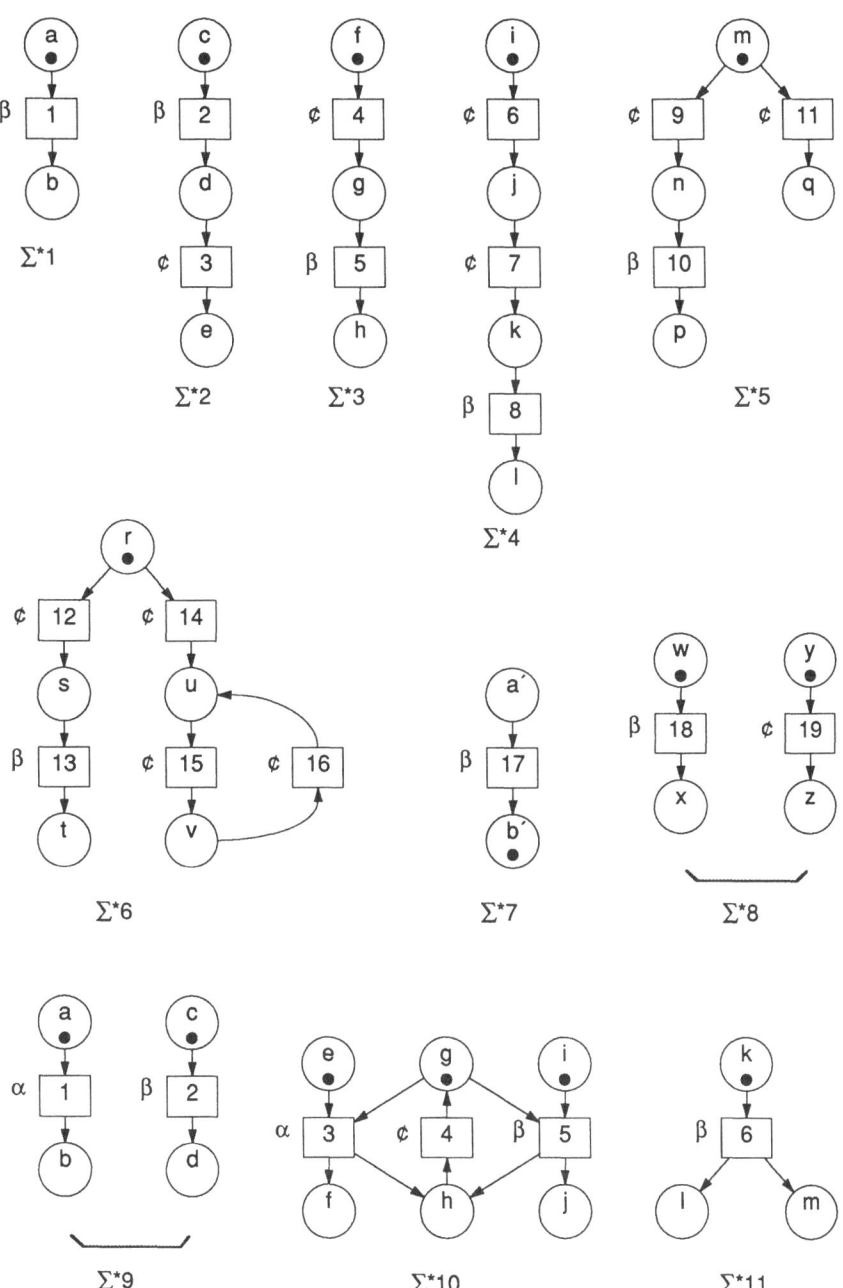

<u>Figure 2</u>. Examples of strictly labelled CE systems
(All conditions are supposed to be labelled with ¢)

In the following, some of the most crucial properties of the interface equivalence shall be enumerated. It must be emphasized that the notions of liveness or synchronic distance shall be applied to __external__ events only. As long as we are dealing only with the behaviour of systems on an interface, we cannot expect to gain much knowledge about occurrences of internal events. Bearing this in mind, we adopt the usual definitions of a __live__ event and the __synchronic distance__ of two sets of events from the literature, e.g., [RT86], def. 11.4., and [Re85a], 4.1, def. (a)-(f).

Proposition 6. Let $\Sigma\ast_n$, n=1,2, be two interface-bijective systems as above. As usual, Se:=Se$_1$=Se$_2$ and Te:=Te$_1$=Te$_2$.
If $\Sigma\ast_1 \sim$if$\sim \Sigma\ast_2$, then for all t$\in$Te:
t is live in $\Sigma\ast_1$ <===> t is live in $\Sigma\ast_2$. $\Box$

Next we compare the interface equivalence with the gh-equivalence of definition 3. Not surprisingly, the two notions coincide if the interfaces consist of exactly __all__ events, because then we have at once a bijection between them. If there are also internal events then if-equivalence is weaker than gh-equivalence because one cannot say anything definite about the internal events. If, however, the interfaces include all events and possibly some conditions, then gh-equivalence is weaker than if-equivalence because it does not relate single conditions but only cases to each other.

Proposition 7. Let $\Sigma\ast_n = (N_n, M^0{}_n, Q_n, L_n)$, n=1,2, be two interface-bijective strictly labelled systems and $\Sigma_n = (N_n, M^0{}_n)$ the underlying unlabelled CE systems, where $N_n = (S_n, T_n; F_n)$. As usual, Q:=Q$_1$=Q$_2$.
a) If $Q \subseteq T_1 \cap T_2$, then: $\Sigma_1 \sim$gh$\sim \Sigma_2$ ===> $\Sigma\ast_1 \sim$if$\sim \Sigma\ast_2$.
b) If $Q \supseteq T_1 \cup T_2$, then: $\Sigma\ast_1 \sim$if$\sim \Sigma\ast_2$ ===> $\Sigma_1 \sim$gh$\sim \Sigma_2$.
c) If $Q = T_1 = T_2$, then: $\Sigma_1 \sim$gh$\sim \Sigma_2$ <===> $\Sigma\ast_1 \sim$if$\sim \Sigma\ast_2$. $\Box$

As an example for a), consider the systems $\Sigma\ast_1$ and $\Sigma\lambda_8$ of figure 2 which are if-equivalent. The event 19 of $\Sigma\ast_8$ is internal, hence $Q \subset T_8$. And obviously there does not exist a bijection g between T_1 and T_8. An example for b) can be derived from the systems $\Sigma\ast_1$ and $\Sigma\ast_{11}$ of figure 2 assuming now that conditions b (in $\Sigma\ast_1$) and l and m (in $\Sigma\ast_{11}$) are external, carrying different labels. Thus $Q_1 \supset T_1$ and $Q_{11} \supset T_{11}$. Relating case a to c and case b to {l,m} and relating event 1 to event 6 yields bijections which determine a gh-equivalence between $\Sigma\ast_1$ and $\Sigma\ast_{11}$. However, as the label of b is different from the label of {l,m}, the two systems are not if-equivalent.

Proposition 8. Let $\Sigma\ast_n$, n=1,2, be as above and $\Sigma\ast_1 \sim$if$\sim \Sigma\ast_2$.
Let U_n = {u | there exists M$\in$[M$^0{}_n$]: M enables u}.
a) If $Q = S_1 \cup T_1 = S_2 \cup T_2$, then there exist bijections
   $S_1 \longleftrightarrow S_2$ and $T_1 \longleftrightarrow T_2$.

b) If $Q = T_1 = T_2$, then there exist bijections
   $[M^0{}_1] \longleftrightarrow [M^0{}_2]$ and $T_1 \longleftrightarrow T_2$.

c) If $Q = S_1 = S_2$, then there exist bijections
   $S_1 \longleftrightarrow S_2$ and $U_1 \longleftrightarrow U_2$.   □

Finally, we are prepared now to state that interface equivalence preserves the
synchronic distances between those interface events which are bijectively related,
i.e., identified with each other. And this assertion is extended in the usual way to
sets of external events. If $\Sigma*$ is a strictly labelled CE system, then SD shall
denote the synchronic distance function defined for pairs of event subsets (cf.
[Re85a], p. 47 f).

Proposition 9. Let $\Sigma*_n$ be two interface-bijective systems as above, and Te:=Te$_1$=Te$_2$.
If $\Sigma*_1$ ~if~ $\Sigma*_2$,   then:
For all T',T" $\subseteq$ Te: SD$_1$ (T',T") = SD$_2$ (T',T").   □

As could be observed from the examples in figure 2, the notion of interface
equivalence is very strong. Its justification lies, first, in its property of
preserving liveness and synchronic distance of the interface events. Secondly, it
could serve as a reference for validating any weaker notion. When looking at other
equivalence notions defined in the literature and when trying to compare them with
each other and with interface equivalence, the first necessary step is to formulate
them in one common language. As many of these notions consider only sequences of
event occurrences and not concurrent occurrences in describing 'observable behaviour', it is also desirable for a sensible comparison to generalize the notions to
capture concurrency. Pursuing this goal, in [Po86a] a number of well known
equivalence notions have been translated in terms of Net Theory and have been generalized to include concurrency of the interface behaviour. The result of the comparison in [Po86a] is reported below.

One property is common to all equivalence notions investigated in [Po86a]: they and
their 'concurrent counterparts' all take into account only the class of forward
reachable cases of labelled systems. This may be motivated by the observation that
(most) real effects cannot be reverted. Indeed, the advantage of considering the
full case class lies mainly in the possibility for backward reasoning and not in the
- doubtful - expectation that the systems should be able to really proceed backwards.

Accordingly, in the rest of this section we recall one weaker version called forward
interface equivalence, and in the following section we show how a notion of compatibility could be derived from the interface equivalence. Both concepts are felt to be
of high practical significance. Thus the following definition is proposed.

Let $\Sigma\star_n = (N_n, M^0{}_n, Q_n, L_n)$, n=1,2, be interface-bijective strictly labelled elementary
net systems whose state spaces are the forward case classes $[M^0{}_n>$, but which
otherwise meet the assumptions stated at the beginning of this chapter. The forward
(bi-)simulation is then a relation $R \subseteq [M^0{}_1> \times [M^0{}_2>$ which is defined according to
definition 4, but omitting condition (ii) b). This yields the notion of forward interface equivalence (fif-equivalence).

Obviously, bisimulation implies forward bisimulation, and interface equivalence is
stronger than forward interface equivalence. Liveness of the external events is
preserved by forward equivalence [Vo86]. What is lost by weakening the definition in
this way, shall be characterized by the use of three properties (which are interrelated). First, a direct comparison with the gh-equivalence of CE systems is no
longer possible because the latter one relies on the full case class. Secondly, by
the same reason, one cannot say anything about synchronic distances when requiring
them to define a metric in net systems. Thirdly, there appears to be an asymmetry in
treating internal events. This can be demonstrated by means of figure 2. While both
$\Sigma\star_2$ and $\Sigma\star_3$ are not interface equivalent to $\Sigma\star_1$, it turns out that $\Sigma\star_2$ is forward
interface equivalent (fif-equivalent) to $\Sigma\star_1$ but that $\Sigma\star_3$ is not.

Using the examples of figure 2, a further difference between the if- and the fif-
equivalence shall be pointed at.
- $\Sigma\star_3$ and $\Sigma\star_4$ are fif-equivalent (but not if-equivalent) because
  $\{(f,i),(f,j),(g,k),(h,l)\}$ defines a bisimulation relation between them.
- $\Sigma\star_5$ and $\Sigma\star_8$ are fif-equivalent (but not if-equivalent) because
  $\{(m,r),(n,s),(p,t),(q,u),(q,v)\}$ is a bisimulation relation between them.
- However, $\Sigma\star_3$ and $\Sigma\star_5$ are neither fif-equivalent nor if-equivalent.
Among others, these examples show that any single internal move (a case consisting
only of internal conditions) can be replaced equivalently by an - even infinite -
non-branching sequence of internal moves. However, it makes a difference whether, at
a case, there is only one possibility to proceed (like in $\Sigma\star_3$) or whether a choice
exists to continue by a sequence consisting of hidden events only (like in the right
part of $\Sigma\star_5$ or of $\Sigma\star_8$).

A comparison of several equivalence notions has been executed in [Po86a]. Adding the
results of [Vo86] concerning interface equivalence yields the following list in
which each notion is strictly stronger than those notions appearing later in the
list:
- behaviour equivalence [An82a, An82b],
- forward interface equivalence [Vo83, Vo86],
- bisimulation equivalence [Pa81],

- observation equivalence [Mi80],
- exhibited-behaviour equivalence [DDPS84],
- failure equivalence [HBR81, Br83],
- string equivalence (based on event sequences).

The main differences between the above mentioned equivalence notions have been discussed and demonstrated by examples in [Po86a] and in [Vo86].

4. Interface equivalence in specification and verification

The final goal of this paper is to sketch the application of the concepts of interface and interface equivalence to the area of system design and analysis. This can only be a brief introduction because a thorough discussion of this subject would exceed the scope of this contribution. Therefore we will simplify questions as far as feasible and we will concentrate on the introduction of the most crucial notions and concepts.

First, for every model discussed throughout this chapter, we assume a fixed level of detail. Thus we are not concerned with questions of model refinement or abstraction. Secondly, when designing a module B, we must know the environment into which it shall be embedded. For our purpose it is sufficient to stick to a fixed environment Z of module B. However, a deeper investigation of the entire (iterative) design process would have to consider the are mutual influences between the module and its environment.

We first mention some steps to be done and some assumptions to be made in designing a system. We always start with a fixed environment Z which is modelled by a net system, i.e., a net with an initial case. A subset Q of nodes is distinguished as the intended interface of Z to the module to be built. To <u>design</u> the module B means to define a net system B, in which a subset Q' of nodes constitutes the interface to the environment. As stated previously, a bijection has to be defined between the interface Q of the environment and the interface Q' of the module and, to simplify matters, we assume that the respective interface elements are identified. This is demonstrated for q_1, q_2, q_3 in figure 3. By fusing Z and B via this identification and by taking the union of the initial cases of Z and B as initial case, we arrive at a <u>system</u> which is designated as Σ or more clearly as $\Sigma(B,Q,Z)$. (We always assume that B, Z and Σ meet all essential preconditions which constitute the basis of our investigation.)

Having constructed the system $\Sigma(B,Q,Z)$, the next aim is to describe its behaviour on the interface Q between B and Z, called <u>interface behaviour</u>. To this end, all elements of Q, the external nodes, are labelled with different labels from an alphabet

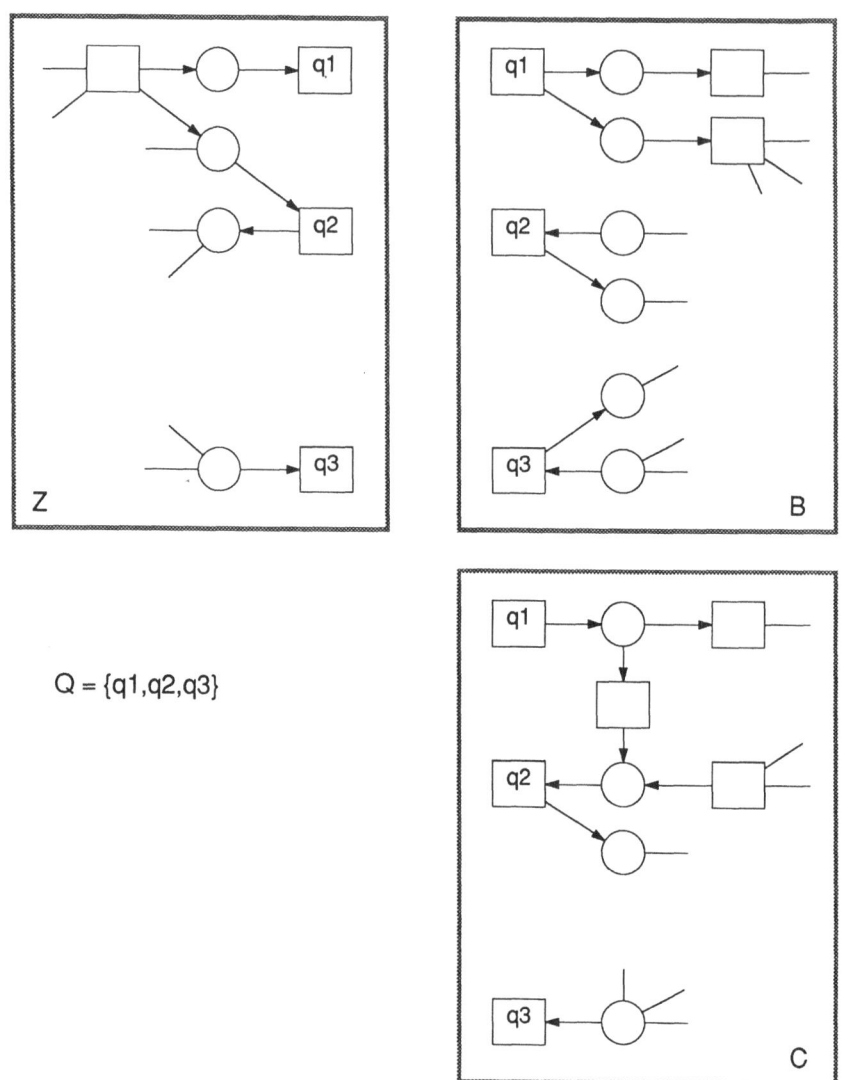

Figure 3. Modules B and C fitting to environment Z at interface Q

Y and all remaining internal nodes get the label ¢, which is not an element of Y. This operation transforms $\Sigma(B,Q,Z)$ into a strictly labelled elementary net system $\Sigma(B,Q,Z)*$ whose behaviour on Q can be represented by the sequences of steps involving interface elements.

Under the assumption of a given interface Q and a given environment Z, a specification of a module is the description of the intended behaviour of the module on Q

when the module is fused with the environment Z at Q (yielding a composite system
$\Sigma(B,Q,Z)*$). We demand that a specification should determine the module in mind 'up
to fif-equivalence'. That means that a specification defines a whole ~fif~-class of
modules each of which exhibits the specified behaviour. Thus, a specification may be
represented as a labelled system $\Sigma(B,Q,Z)*$ which is constructed by fusing Z and an
arbitrary module B of the specified module class at the interface Q and labelling it
as usual. To state it differently, two modules B and C meet the same specification
(w.r.t. Q and Z) if and only if $\Sigma(B,Q,Z)*$ ~fif~ $\Sigma(C,Q,Z)*$.
(Presenting a sample module of the ~fif~-class is merely one - maybe even impractical - way out of several possible ways to state a specification. In general it is
not advisable to prescribe rigidly just one method for expressing a specification.
Practical reasons rather demand for a variety and even a mixture of different means
to be permitted for this purpose. As an example, it would be desirable to define a
specification as a combination of positive and negative statements on the behaviour,
i.e., as requirements and prohibitions. The exploration of appropriate specification
languages in this context constitutes a promising goal for future work.)

Normally, the design of a module C has to meet a given specification, say $\Sigma(B,Q,Z)*$.
To verify if a module C satisfies a specification means to check if it belongs to
the specified ~fif~-class, i.e., by examining whether $\Sigma(C,Q,Z)*$ ~fif~ $\Sigma(B,Q,Z)*$
holds or not. In other words, the result of the verification is positive only if B
and C can be exchanged in the given environment without changing the interface
behaviour. In this case, C could be called fully compatible (at the interface Q to
environment Z) with the specification or with B.

Interface equivalence is quite a strong notion. It discriminates modules with only
'slightly different' interface behaviour. In practice, the related notions of verification and of compatibility may sometimes be too strong. Take as an example a
specification which demands, at a particular case M, the concurrent occurrences of
the interface events q_1 and q_2. Let B denote a module that exactly fulfills the
specification, i.e., for which $\Sigma(B,Q,Z)*$ exhibits exactly the intended interface
behaviour. Now consider a module C, satisfying precisely all requirements of the
specification with the exception that q_1 and q_2 are enabled always in sequential
order, first q_1 then q_2, at M. The environment Z is prepared, by the specification,
to accept concurrent occurrences of q_1 and q_2, but if connected to C, experiences
only sequential occurrences of them. In many practical applications, C would be
welcomed as satisfying the specification at least 'weakly'. One reason for demanding
a weaker notion stems from the fact that specifications often are incomplete in the
sense of not determining exactly, i.e. not more and not less than, the desired
behaviour. Rather they establish a framework leaving room for choosing among
alternatives and for imposing sensible restrictions in the actual design.

Following this idea, we arrive at a problem which again C.A. Petri has observed and investigated already in his dissertation [Pe62]. Assume that we start with an 'experimenter' who is able to perform experiments with a module by trying to execute common actions with it and observing whether they are successful or not. Thus we see the experimenter as a given (sub-)system with a fixed set of its nodes, distinguished as constituting the interface with an unknown module, communicating with the module via shared events. Thus the experimenter and the observed module together form a composite system, and the results of the experiments can be recorded as the behaviour of the composite system on the interface. Assume that another module, fitting to the same interface, is to replace the first one. The problem then is to describe and compare the forms of communication that the experimenter experiences at that interface when communicating with different modules. Obviously, the desired weaker notion cannot be a symmetric relation, but rather has to be a (partial) order among the modules. To illustrate the point, the subsequent example - in a slightly modified version - is taken from [Pe62], p.78.

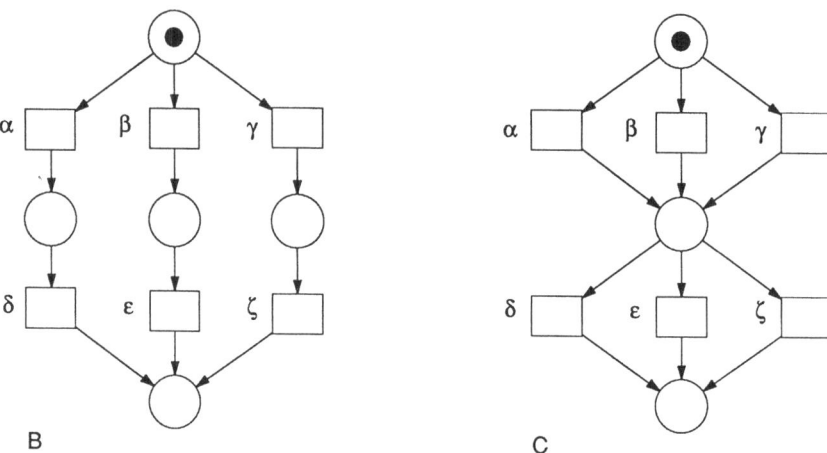

<u>Figure 4</u>. Upward compatible modules

Figure 4 presents two candidate modules B and C whose interface behaviour in connection with an experimenter Z is to be examined. All events are supposed to be external, i.e. belonging to the interface, whereas all conditions are internal (labels ¢ omitted). Clearly, the potential behaviour of B is more restricted than that of C. We are looking for a partial order ⊆ relation among modules with the same interface elements, such that C ⊆ B. Inspecting definition 4 suggests at once one sensible solution.

Definition 10. Let an environment Z and an interface Q be given.   Let B and C denote
two modules both fitting to Q.
C ⊆ B iff
$\Sigma(C,Q,Z)*$ simulates $\Sigma(B,Q,Z)*$.   □

One expected property of this simulation is that all synchronic distances between
interface transitions of $\Sigma(B,Q,Z)*$ are less   than   or   equal   to   the   corresponding
synchronic distances of $\Sigma(C,Q,Z)*$.   And of course, simulation preserves liveness. As
an example,   refer to the systems $\Sigma*_9$ and $\Sigma*_{10}$ of figure 2 or to the modules B and C
of figure 4.

If we define simulation for elementary net systems with the forward case classes as
their state spaces and with the properties required at the beginning   of   chapter 2,
then   we get the same definition as in [Be86],   chapter 3,   applied to safe net systems. As shown in [Be86],   also this simulation preserves the liveness of the interface events. But again, no statement is possible concerning synchronic distances.

The partial order notion in definition 10 depends on Z and Q. If we require C ⊆ B to
hold   for   all   Z   fitting   to   the   given   Q,   then   we   get   a   more comprehensive
characterization of the interface behaviour of   modules   (see   [La85]).   Under   this
assumption,   we   would   say   that   module   C is upward compatible with module B with
respect to interface Q.   We do not address here the inherent difficulties of such   a
definition, ·but merely mention it as a promising subject for further investigation.

5. Conclusion

The areas of system modelling,   specification, design, verification and analysis can
be treated all within one common conceptual framework.   This has to be   based   on   a
concept   of   equivalence   of   partially determined or partially known systems.   This
paper presents and motivates the notion of interface equivalence   on   the   basis   of
General   Net   Theory and it demonstrates its appropriateness for the above mentioned
issue.   In accordance with most other approaches we assume a module and its environment which communicate through a set of common elements which we call interface.   We
suppose that the environment is fixed and the interface is given.   Then the only information   on the module can be derived from the observed behaviour of the module on
the   interface   to   its   environment.   Thus   interface   behaviour   can   be   used   as
representing
- the specification of a module still to be built,
- the result of analyzing an existent module,
- the object of verification, comparing a specification with an existent module.

For representing partially determined systems, marked Petri nets are used whose nodes carry labels which denote the kind of 'observation' attributed to that nodes. The notion of interface equivalence deviates in a number of essential points from other similar notions:
- Not only transitions (events) but also local states (conditions) may contribute to the interface behaviour.
- The nets are labelled in a strict manner such that they can be derived from an underlying unlabelled net by selecting a subset of nodes to form the interface.
- Interface behaviour is not restricted to consist of sequences of event occurrences only, but may include concurrency as well.
- Not only the future but also the past of behaviour patterns are considered in defining the interface equivalence. Thus a lot of undesired unsymmetries and deficiencies of the notion can be avoided.

The ultimate justification to present this new notion of equivalence lies in the fact that it preserves exactly the most essential properties: the enlogic and synchronic structure of partially determined systems. Not surprisingly, this equivalence is very strong. But even in cases where a weaker version is preferable, one may draw profit from considering very consciously and carefully every omission and alleviation of the original strong concept.

The way how to establish a practice oriented method of system specification and analysis on this basis could only be outlined in this paper. More results have to be elaborated in the future which will enable the definition of net based specification languages and the development of computer assisted tools for the design and the verification of distributed concurrent systems.

Acknowledgements

The author is very much indebted to a number of colleagues at GMD for fruitful discussions and to two referees for many valuable hints for improving the paper.

References

[An82a] Andre, C.: Behaviour of a Place-Transition Net on a Subset of Transitions. Informatik-Fachberichte 52, Springer-Verlag, Berlin, 1982, 131-135.
[An82b] Andre, C.: Use of Behaviour Equivalence in Place-Transition Net Analysis. Informatik-Fachberichte 52, Springer-Verlag, Berlin, 1982, 241-250.
[Be84]   Best, E.: Concurrent Behaviour: Sequences, Processes and Axioms. Arbeitspapiere der GMD Nr. 118, GMD, St. Augustin, Nov. 1984.
[Be86]   Best, E.: Structure Theory of Petri Nets: the Free Choice Hiatus. Advanced Course on Petri Nets 1986. To appear.

[BFP85] Best, E., Fernandez, C., Plünnecke, H.: Concurrent Systems and Processes.
         Final Report on the Foundational Part of the Project BEGRUND, GMD-Studien
         Nr. 104, GMD, St. Augustin, March 1985.

[BR83]   Brookes, S.D., Rounds, W.C.: Behavioural Equivalence Relations Induced by
         Programming Logic. Lecture Notes in Computer Science 154, Springer-Verlag,
         Berlin, 1983, 97-108.

[Br83]   Brookes, S.D.: On the Relationship of CCS and CSP. Proc. ICALP '83, Lecture
         Notes in Computer Science, Springer-Verlag, Berlin, 1983.

[DDPS83] De Cindio, F., De Michelis, G., Pomello, L., Simone, C.: Equivalence No-
         tions for Concurrent Systems. Informatik-Fachberichte 66, Springer-Verlag,
         Berlin, 1983, 29-39.

[DDPS84] De Cindio, F., De Michelis, G., Pomello, L., Simone, C.: Exhibited-
         Behaviour Equivalence as the basis for Concurrent Systems Design.
         Universita Degli Studi di Milano, Istituto di Cibernetica, Internal Report,
         Feb. 1984.

[GR83]   Goltz, U., Reisig, W.: The Non-sequential Behaviour of Petri Nets. Infor-
         mation and Control, Vol. 57, Nos. 2-3, May/June 1983, 125-147.

[HBR81] Hoare, C.A.R., Brookes, S.D., Roscoe, A.W.: A Theory of Communicating
         Sequential Processes. Technical Report PRG-16, Oxford Univ. Comp. Lab.,
         Programming Research Group, 1981.

[HM80]   Hennessy, M., Milner, R.: On Observing Nondeterminism and Concurrency. Lec-
         ture Notes in Computer Science 85, Springer-Verlag, Berlin, 1980.

[La85]   Larsen, K.G: A Context Dependent Equivalence Between Processes. Lecture
         Notes in Computer Science 194, Springer-Verlag, Berlin, 1985, 373-382.

[Mi80]   Milner, R.: A Calculus of Communicating Systems. Lecture Notes in Computer
         Science 92, Springer-Verlag, Berlin, 1980.

[NT84]   Nielsen, M., Thiagarajan, P.S.: Degrees of Non-determinism and Concurrency:
         A Petri Net View. Report DAIMI PB-180, Aarhus Univ., Oct. 1984.

[Pa81]   Park, D.M.: Concurrency and Automata on Finite Sequences. Report, Computer
         Science Dept., Univ. of Warwick, 1981.

[Pe62]   Petri, C.A.: Kommunikation mit Automaten. Schriften des
         Rheinisch-Westfälischen Instituts für Instrumentelle Mathematik an der
         Universität Bonn, Nr. 2, Bonn, 1962.

[Pe77a] Petri, C.A.: Non-sequential Processes. GMD-ISF Report 77-05, GMD,
         St. Augustin, 1977.

[Pe77b] Petri, C.A.: General Net Theory. In Shaw, B. (ed.): Computing Systems
         Design: Proc. of the Joint IBM Univ. of Newcastle upon Tyne Seminar. Univ.
         of Newcastle upon Tyne, 1977.

[Po86a] Pomello, L.: Some Equivalence Notions for Concurrent Systems. Lecture Notes
         in Computer Science 222, Springer-Verlag, Berlin, 1986, 381-400.

[Po86b] Pomello, L.: Observing Net Behaviour. This volume.

[Re85a] Reisig, W.: Petri Nets - An Introduction. EATCS Monographs on Theoretical
         Computer Science, Vol. 4, Springer-Verlag, Berlin, 1985.
[Re85b] Reisig, W.: On the Semantics of Petri Nets. IFIP Working Conference "The
         Role of Abstract Models in Information Processing", Wien, Jan. 1985.
[RT86]   Rozenberg, G., Thiagarajan, P.S.: Petri Nets: Basic Notions, Structure,
         Behaviour. Lecture Notes in Computer Science 224, Springer-Verlag, Berlin,
         1986, 585-668.
[Vo83]   Voss, K.: On the Notion of Interface in Condition/ Event-Systems.
         Informatik-Fachberichte 66, Springer-Verlag, Berlin, 1983, 278-291.
[Vo86]   Voss, K.: System Specification with Labelled Nets and the Notion of Inter-
         face Equivalence. Arbeitspapiere der GMD 211, GMD, St. Augustin, June 1986.
[YE83]   Yoeli, M., Etzion, T.: Behavioral Equivalence of Concurrent Systems.
         Informatik-Fachberichte 66, Springer-Verlag, Berlin, 1983, 292-305.

SPECIFICATION AND VERIFICATION OF ASYNCHRONOUS CIRCUITS

USING MARKED GRAPHS

Michael Yoeli
Computer Science Department
Technion - Israel Institute of Technology
Haifa 32000, Israel

ABSTRACT

This paper deals with the behavioral specification and description of
digital circuits operating asynchronously. In particular, it is shown
how the behavior of a composite circuit may be derived from the behavioral description of its components. The paper combines the theory of
trace structures developed by M. Rem and J.L.A. van de Snepsheut with
a suitable extension of the theory of marked graphs. It consequently
achieves a considerable simplification of the way composite circuits
are modeled and verified.

1. INTRODUCTION

This paper deals with the behavioral specification and description of digital
circuits and devices, operating asynchronously. In particular, we show how the
behavior of a composite circuit or system may be derived from the behavioral description of its components. This method of "behavioral composition" may evidently be
applied to the verification of complex digital systems, obtained by a suitable
composition of basic building blocks. By "verification" we mean a formal proof,
showing that a composite system satisfies a given behavioral specification.

The elegant theory of communicating processes, due to C.A.R. Hoare [Ho85], has
been suitably extended and applied to the study of asynchronous (VLSI) circuits by
M. Rem [Re83] and J.L.A. van de Snepsheut [Sn85]. This paper combines the approach
of [Re83] and [Sn85] with a suitable extension of the theory of marked graphs
[CHEP71], [GL73]; by doing so, it achieves a considerable simplification of the way,
composite circuits are modeled and verified. This simplification is related to the
fact that Petri nets (marked graphs are a particular class of such nets) have the
capability of suitably capturing the concept of "true concurrency" [Pe80], whereas
most algebraic theories of communicating processes [Ho85], [Sn85], [Mi80] replace
true concurrency by "interleaving".

Another feature of this paper is its restriction to deterministic systems.
This avoids the complications involved in properly formalizing the behavior of non-
deterministic systems (cf. [Mi80], [Ho85], [YE83]). Indeed, by using marked graphs
to model asynchronous circuits, we further restrict our considerations to a particular
class of deterministic circuits. However, this class plays an important role in
connection with self-timed systems [Se80].

The approach developed in this paper can be extended from marked graphs to Petri nets in general. Such an extension, which preserves the advantage of suitably capturing the concept of true concurrency rather than replacing it by interleaving, enables us to model arbitrary asynchronous circuits, including non-deterministic ones (cf. [Yo85]).

2. BEHAVIOR OF DIGITAL CIRCUITS - INFORMAL DISCUSSION

In this section we introduce the concept of (input-output) behavior of digital circuits and devices in an informal way.

As illustrative example we consider a well-known sequential circuit, namely the (Muller) C-element [Se80], which plays an important role in the design of self-timed systems. The circuit has two binary inputs A, B and one binary output Z. We denote by A↑ (A↓) a change of A from 0 to 1 (from 1 to 0). This notation is used in the state diagram of Fig. 2.1, which may be considered as a way of specifying the input-output behavior of the C-element. Two of the states shown in Fig. 2.1 are unstable. When the circuit is in an unstable state, its output has the tendency to change and will actually change, provided the inputs remain unchanged. Generally, the behavior specification of a circuit or device will also make some assumptions about the "admissible" behavior of the environment. Namely, we assume that the "environment" is provided by some other component, designed in such a way as to only produce admissible inputs to the circuit in question. This approach is particularly appropriate in connection with the design of self-timed systems [Se80].

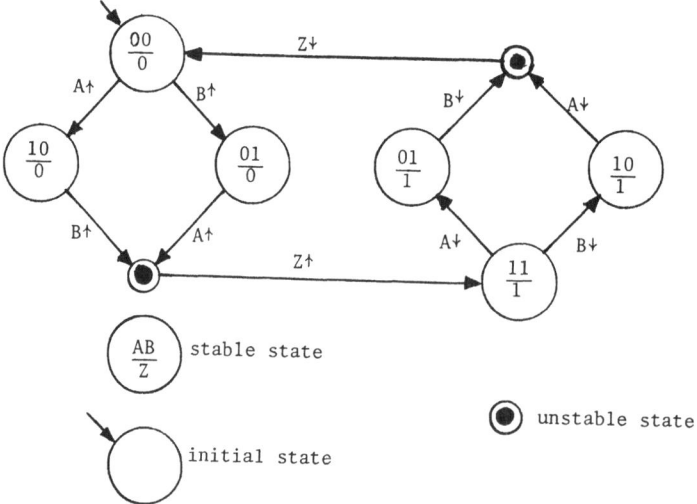

Figure 2.1 - State diagram of C-element

In the case of the C-element we assume that the inputs will not change as long as the C-element is in an unstable state. The state diagram of Fig. 2.1 incorporates additional restrictions on the behavior of the environment. All the assumptions about the environment are motivated by the way the C-element is used as a component within a larger self-timed system (See Section 6).

The asynchronous circuit shown at the gate-level in Fig. 2.2 may be considered to be an implementation of the C-element, as specified in Fig. 2.1. The analysis of the circuit of Fig. 2.2 is rather straightforward. In general, however, the detection and classification of racing phenomena in gate networks offer considerable difficult-ies (cf. [BY79]).

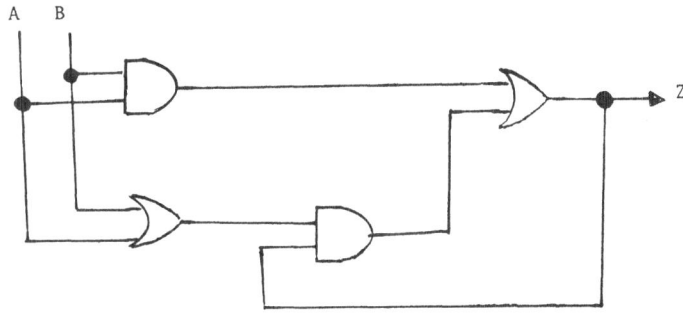

Figure 2.2 - A gate-level implementation of the C-element

Further difficulties arise in connection with a precise race analysis of MOS circuits, represented at the switch-level (cf. [LN84], [YB85], [BY85]). However, in this paper we shall not be concerned with the problems involved in deriving a gate-level or switch-level MOS (VLSI) implementation from a given behavioral specification. The diagram of Fig. 2.1 may be viewed as a finite automaton FA over the (event) alphabet

$$E = \{A\uparrow, A\downarrow, B\uparrow, B\downarrow, Z\uparrow, Z\downarrow\}.$$

If all the states of this automaton are viewed as accepting states, the language defined by FA becomes

$$L(FA) = pref((A\uparrow \$ B\uparrow); Z\uparrow; (A\downarrow \$ B\downarrow); Z\downarrow)^*$$

where ';', '$', and 'pref' denote catenation, shuffle, and prefix-closure, respectiv-ely. Thus, $A\uparrow \$ B\uparrow = \{(A\uparrow; B\uparrow), (B\uparrow; A\uparrow)\}$. The prefix-closure pref L of any language L over the alphabet Σ is defined by

$$pref\ L = \{x \in \Sigma^* \mid \exists y \in \Sigma^* : (x;y) \in L\}.$$

The language L(FA) over the event alphabet E may be chosen as another way of
specifying the behavior of the C-element (and its environment).

In the so-called "2-cycle signaling scheme" [Se80] both high-going and low-going
voltage transitions are viewed as (input or output) signals. Consequently, let us
denote by lower-case letters (a,b,z) any change of the corresponding binary value
(A,B,Z). We call the alphabet $\Sigma = \{a,b,z\}$ the _signal_ alphabet of the C-element.
The sequential behavior of the C-element is the language over Σ consisting of all
feasible, finite sequences of signals (i.e. elements of Σ). Evidently the sequential
behavior of the C-element is given by the language

$$S(CE) = pref((a \ \$ \ b);z)^*$$

This is the language derived from L(FA) by replacing A↑ and A↓ by a, B↑ and
B↓ by b, and Z↑ and Z↓ by z. Conversely, L(FA) can be reconstructed provided
the sequential behavior of the C-element and its initial state (i.e. A = B = Z = 0)
are given.

Note that the ordered pair <E,L(FA)> specifies a _process_ [Ho85] or _trace_
structure [Re83], [Se85], with E as its alphabet and L(FA) as its set of traces.
The ordered pair $<\Sigma,S(CE)>$ may be viewed similarly. This viewpoint, similarly to
our approach so far, replaces "true concurrency" by "interleaving", as mentioned in
the introduction. Indeed, up till now we have disregarded the possibility that both
inputs of the C-element may change simultaneously. We now introduce the concept of
"parallel behavior", which does take simultaneous input or output changes (signals)
into account.

The (finite) _parallel behavior_ of a digital circuit is defined as the set of
all feasible finite sequences of both single and multiple input and output changes.
The parallel behavior of the C-element becomes

$$pref[((a \ \$ \ b) \cup \{\{a,b\}\});z]^*$$

This expression represents a (regular) language over the alphabet $\{a,b,z,\{a,b\}\}$.
In the sequel we show how marked graphs may be applied to the precise and concise
definition of the parallel behavior of digital circuits and devices.

The "infinite" behavior of a digital circuit (cf. [YE83]) is also of some interest,
particularly in connection with studies of "fair" behavior. However, in this paper
we restrict our consideration to finite behaviors.

Two discrete systems which coincide in their sequential behavior, may differ in
their parallel behavior. Relevant concepts of "behavioral equivalence" are introduced
and compared in [YE83], [RV83], [CMPS83]. These studies apply Petri nets to the
modelling of discrete systems, but do not deal with the problem of "behavioral
composition". By restricting the device model to marked graphs (rather than Petri
nets in general) we considerably simplify the problems of "behavioral equivalence"
and "behavioral composition".

3. MARKED GRAPHS

Marked (directed) graphs [CHEP71] form a particular class of Petri nets [Pe81]. In this section we summarize the basic concepts required for this paper.

A marked graph is a triple G = <V,E,M> where

(1) <V,E> is a finite, directed graph with V as its set of vertices or nodes, and E its set of edges. Let e ∈ E be an edge from node u to node v. We say that the edge e is an outedge of node u and an inedge of node v. Self-loops (i.e, u=v) and multiple edges are admitted.

(2) M is a function from E into ω, the set of nonnegative integers. M(e) is the marking of e.

It is customary to indicate the marking M(e) = k by placing k tokens (black dots) on edge e in the graphical representation of <V,E>. Alternatively, edge e may be labeled by k. If k = 0, the label is usually omitted. Examples of marked graphs are shown in Fig. 3.1. The marked graph GCE of Fig. 3.1(a) "represents" the behavioral specification of the C-element. This statement will be made precise in the sequel.

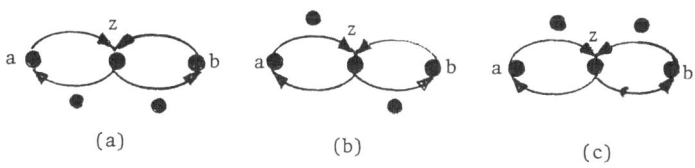

(a)   (b)   (c)

Figure 3.1 - (a) Marked graph GCE; (b) The marking obtained by
firing node a in (a); (c) The marking obtained by
firing node b in (b).

An edge e of the marked graph G = <V,E,M> is marked, iff M(e) > 0. A node v of G is firable iff all its inedges are marked. The firing of a firable node v consists of decreasing the marking of all its inedges by 1, and of increasing the marking of all its outedges by 1. Evidently, if node v has indegree 0, v is always firable. In the marked graph of Fig. 3.1(a), nodes a and b are both firable. The firing of the node a yields the marking shown in Fig. 3.1(b). Node b is still firable in Fig. 3.1(b), and its firing yields Fig. 3.1(c). In Fig. 3.1(c) only node z is firable; the outcome of firing z is the "initial" marking shown in Fig. 3.1(a).

Let M' be the marking obtained by firing node v in G = <V,E,M>. This is denoted by M $\overset{v}{\rightarrow}$ M'. This notation is extended to finite strings w ∈ V* in the usual way. Namely, (1) $M \overset{\lambda}{\to} M$, where λ denotes the empty string; (2) $M \overset{vw}{\to} M'$, where $v \in V$ and $w \in V^*$, iff $M \overset{v}{\to} M''$ and $M'' \overset{w}{\to} M'$ for some intermediate marking M''.

Let $G = <V,E,M>$ be a marked graph. A finite string $w \in V^*$ is a <u>firing sequence</u> of G iff $M \overset{w}{\to} M'$ for some marking M'. We denote by $L(G)$ the <u>L-language</u> of G, i.e. the set of all the firing sequences of G. For the marked graph GCE of Fig. 3.1(a) one easily verifies that

$$L(GCE) = pref((a \$ b);z)^*$$

i.e. $L(GCE)$ coincides with the sequential behavior $S(CE)$ of the C-element (see Section 2).

In recent years, various authors have been concerned with "multiple-firing" sequences of Petri nets [YE83], [RV83], [CMPS83]. Following [YE83], we now introduce the relevant concepts, as they apply to marked graphs.

Let $U = \{u_1,\ldots,u_k\}$, $k \geq 1$, be a set of nodes of $G = <V,E,M>$. An inedge (outedge) of U is an inedge (outedge) of any of the u_i's. U is <u>firable</u> in G iff each $u_i \in U$ is firable, i.e. all the inedges of U are marked. The outcome of <u>concurrently firing</u> U is the marking M' obtained from M by decreasing the marking of each inedge of U by 1, and increasing the marking of each outedge of U by 1. We again use the notation $M \overset{U}{\to} M'$. Let w denote any permutation of $u_1 u_2 \ldots u_k$. One easily verifies that w is a firing sequence of G and that $M \overset{w}{\to} M'$. Similarly as before, we extend the notation $M \overset{U}{\to} M'$ to finite strings (of nonempty subsets of V). Let $\tilde{V} = 2^V - \{\emptyset\}$. We define the π-language [YE83] (or <u>subset language</u> [RV83]) of G to be

$$\pi(G) = \{w \in \tilde{V}^* \mid \exists M': M \overset{w}{\to} M'\}.$$

In the marked graph GCE of Fig. 3.1(a), nodes a and b are concurrently firable, i.e. the subset $U = \{a,b\}$ of V is firable. The outcome of firing U is the marking shown in Fig. 3.1(c). Furthermore, $\pi(GCE)$ evidently correspond to the parallel behavior of the C-element (see Section 2).

The following proposition relates the L-languages and π-languages of marked graphs. We omit its rather simple proof (cf. [RV83]).

<u>Proposition 1</u>: Let G and G' be marked graphs. Then

$$\pi(G) = \pi(G') \quad \text{iff} \quad L(G) = L(G').$$

This "nice" property of marked graphs is not shared by other Petri nets (see [YE83], [RV83]).

Given a marked graph $G = <V,E,M>$ without self-loops, its L-language $L(G)$ can be characterized by means of "Δ-languages" introduced in [YG78] (see also [Yo82]).

Let $w \in V^*$ and $v \in V$. Following [Ho85] we denote by $w \downarrow v$ the number of occurrences of the symbol v in the string w. Let e be an edge of G from node u to node v. With e we associate its Δ-language $\Delta(e)$ defined as follows:

$$\Delta(e) = \{w \in V^* \mid \forall x \in \text{pref}\{w\}: (x \downarrow v) - (x \downarrow u) \leqslant M(e)\}.$$

The L-language of G is the intersection of such Δ-languages. Indeed we have

Proposition 2: Let $G = \langle V,E,M \rangle$ be a marked graph without self-loops. Then

$$L(G) = \bigcap_{e \in E} \Delta(e).$$

This proposition follows immediately from the results of Petri net languages derived in [YG78] (see also [Ha76], [Yo82]).

4. LABELED MARKED GRAPHS

The nodes of marked graphs, as considered so far, represented observable (input or output) changes. We now wish to extend our model by incorporating nodes which represent internal (hidden) changes of the circuit in question. Such nodes will be labeled by λ, where λ represents the empty string. In the sequel it will also be convenient to represent an observable change (or signal) by a vertex label, rather than the vertex name, as we did so far. These considerations lead to the following definition.

A labeled marked graph (LMG) is a triple $\Gamma = \langle G,\Sigma,\eta \rangle$, where $G = \langle V,E,M \rangle$ is a marked graph, Σ is a finite alphabet (of signals) and η is a mapping $\eta: V \rightarrow \Sigma \cup \{\lambda\}$, satisfying the conditions

(1) $\Sigma \subseteq \eta(V)$; (2) $\eta(v) = \eta(v')$ and $v \neq v' \Rightarrow \eta(v) = \lambda$.

We refer to $\eta(v)$ as the label of node v.

Conditions (1) and (2) assure that every letter in Σ appears exactly once as a node label of G. The mapping η is extended to finite strings of nodes in the obvious way. The language $L(\Gamma)$ is defined accordingly:

$$L(\Gamma) = \{\eta(w) \mid w \in L(G)\}.$$

Next, we extend the mapping η to any nonempty subset U of V as follows (cf. [YE83]).

$$\eta(U) = \text{if } \{\eta(v) \mid v \in U\} = \{\lambda\} \text{ then } \Lambda \text{ else } \{\eta(v) \mid v \in V \text{ and } \eta(v) \in \Sigma\},$$

where Λ denotes the empty multiple-firing sequence. The language $\pi(\Gamma)$ is defined by

$$\pi(\Gamma) = \{\eta(W) \mid W \in \pi(G)\}.$$

Two LMGs Γ and Γ' will be called equivalent (notation: $\Gamma \cong \Gamma'$) iff $\pi(\Gamma) = \pi(\Gamma')$.

One easily verifies that Proposition 1 may be extended to LMGs (cf. [RV83]).

Proposition 3: Let Γ and Γ' be LMGs. Then

$$\pi(\Gamma) = \pi(\Gamma') \quad \text{iff} \quad L(\Gamma) = L(\Gamma').$$

Thus, $\Gamma \cong \Gamma'$ iff $L(\Gamma) = L(\Gamma')$.

5. COMPOSITIONS OF LABELED MARKED GRAPHS

Two digital circuits may be interconnected in the obvious way: an output of one circuit is connected to an input of the other circuit. In this section we wish to model two types of such an interconnection. In the first type the interconnection also appears as an output of the composite circuit. In the second type the interconnection is not externally observable.

Interconnection of both types are assumed to be delay-free. Interconnections which may introduce delays will be discussed in Section 8.

Interconnections of the first type will be modeled by the "composition" of LMGs, to be defined next. Let Γ_1 and Γ_2 be two node-disjoint LMGs. Their composition $\Gamma_1 \| \Gamma_2$ is obtained from their union by joining into a single node each pair of matched nodes, i.e. two nodes having the same label $\sigma \neq \lambda$.

We shall consider two LMGs Γ_1 and Γ_2 to be equal iff Γ_2 can be obtained from Γ_1 by a change of node names (and edge names). Hence node (and edge) names are omitted in the graphical representation of LMGs. An example of the composition of LMGs is shown in Fig. 5.1. It is rather evident that the composition of LMGs is commutative and associative.

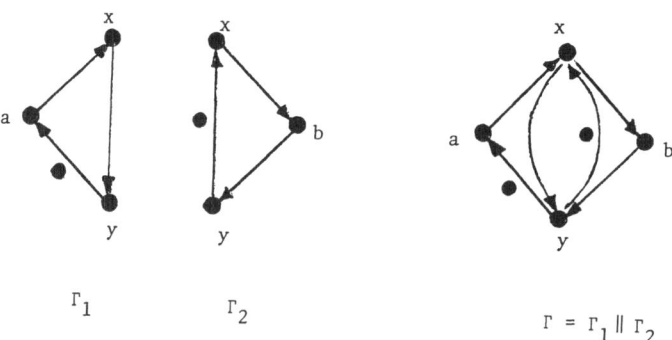

Figure 5.1 - Example of the composition of LMGs

Let w be a string of symbols and Σ a finite alphabet. The restriction $w{\uparrow}\Sigma$ is obtained from w by omitting all symbols not in Σ. This definition is extended to L↑Σ for any set L of strings in the evident way. Part (a) of the following proposition relates the composition of LMGs to the concurrency operator (∥) in [Ho85], as well as to the weave operator (w) in [Sn85].

Proposition 4: Let $\Gamma_i = <G_i, \Sigma_i, \eta_i>$, i = 1,2, be LMGs. Then

(a) $L(\Gamma_1 \| \Gamma_2) = \{w \in (\Sigma_1 \cup \Sigma_2)^* | w \uparrow \Sigma_1 \in L(\Gamma_1) \text{ and } w \uparrow \Sigma_2 \in L(\Gamma_2)\}$

(b) $L(\Gamma_1 \| \Gamma_2) = [L(\Gamma_1) \$ (\Sigma_2 - \Sigma_1)^*] \cap [L(\Gamma_2) \$ (\Sigma_1 - \Sigma_2)^*]$,

where '$' denotes the shuffle operator.

This proposition is easily proven. We omit the details.

We now proceed to model interconnections of the second type. Let $\Gamma = <G, \Sigma, \eta>$ be an LMG, and $\Sigma' \subseteq \Sigma$. The reduction $\Gamma \setminus \Sigma'$ is obtained from Γ by replacing all node labels in Σ' by λ. Clearly $L(\Gamma \setminus \Sigma') = L(\Gamma) \uparrow (\Sigma - \Sigma')$. The merge $\Gamma_1 \underline{\|} \Gamma_2$ of two LMGs $\Gamma_i = <G_i, \Sigma_i, \eta_i>$, i = 1,2, is defined by

$$\Gamma_1 \underline{\|} \Gamma_2 = (\Gamma_1 \| \Gamma_2) \setminus (\Sigma_1 \cap \Sigma_2).$$

The merge operation ($\underline{\|}$) is commutative, but, in general, not associative. As to $L(\Gamma_1 \underline{\|} \Gamma_2)$ we evidently have

$$L(\Gamma_1 \underline{\|} \Gamma_2) = L(\Gamma_1 \| \Gamma_2) \uparrow (\Sigma_1 \Delta \Sigma_2)$$

where $\Sigma_1 \Delta \Sigma_2$ denotes the symmetric difference of Σ_1 and Σ_2, i.e.

$$\Sigma_1 \Delta \Sigma_2 = (\Sigma_1 - \Sigma_2) \cup (\Sigma_2 - \Sigma_1) = (\Sigma_1 \cup \Sigma_2) - (\Sigma_1 \cap \Sigma_2).$$

Thus, the merge operator ($\underline{\|}$) corresponds to the blend operator (b) in [Sn85].

6. ILLUSTRATIVE EXAMPLES

In this section we illustrate some applications of the concepts introduced so far. For this purpose it will be convenient to introduce a notation for "cyclic" LMGs. A cyclic LMG consists of a (directed) cycle of length $n \geqslant 2$, containing a single token. We denote by $*[a_1, a_2, \ldots, a_n]$ a cyclic LMG of length n, with nodes labeled consecutively $a_1, a_2, \ldots, a_n$; only the edge from a_n to a_1 is marked (by a single token). For example, the LMG Γ_1 of Fig. 5.1 is denoted by $\Gamma_1 = *[a, x, y]$.

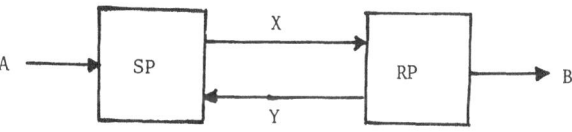

Figure 6.1 - Control part of a simple protocol

Example 6.1 - A Simple Protocol

Fig. 6.1 shows the control part of a simple communication protocol. The component SP is part of the sending process, RP is part of the receiving process. Signals on A, X, and B indicate the availability of a new message on associated message channels. A signal on Y is a request for the next message. We adopt the "2-cycle signaling scheme" [Se80]: a signal on e.g. A, denoted by a, is any change of the binary value of A (i.e. both A↑ and A↓ are signals on A). Initially, $A = X = Y = B = 0$. The component SP of Fig. 6.1 is represented by the LMG $GSP := *[a,x,y]$ and the component RP by $GRP := *[x,b,y]$. The LMGs GSP and GRP coincide with the LMGs Γ_1 and Γ_2 of Fig. 5.1, respectively. It is of interest to consider Fig. 6.1 as a composite circuit, with input A and output B. If we assume the interconnections X and Y to be delay-free, the overall circuit of Fig. 6.1 may be represented by $GSP \parallel GRP$. By definition

$$GSP \parallel GRP = (GSP \parallel GRP) \backslash \{x,y\} = \Gamma \backslash \{x,y\},$$

where Γ is the LMG $\Gamma = \Gamma_1 \parallel \Gamma_2$ shown in Fig. 5.1. In the next section we prove formally the rather evident equivalence

$$GSP \parallel GRP \cong *[a,b]. \tag{EQ1}$$

In Section 8 we show that this equivalence remains valid, even if the interconnections X and Y are assumed to introduce delays.

Example 6.2 - The Control Structure PAR2

We consider a parallel control structure PAR2 which initiates and supervises the concurrent operation of two devices. The outside connections are shown in Fig. 6.2, and the representing LMG (GPAR2) in Fig. 6.3.

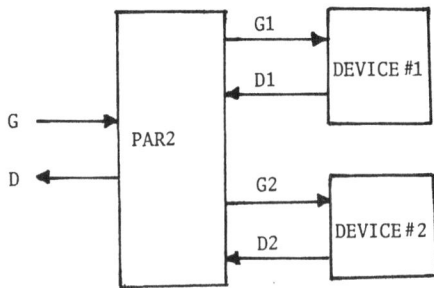

Figure 6.2 - Parallel control structure PAR2

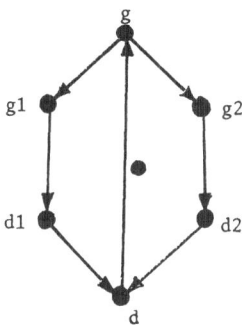

Figure 6.3 - The LMG GPAR2 representing the control
structure PAR2 of Fig. 6.2

We again adhere to the 2-cycle signalling scheme. Initially all inputs and outputs
are equal and the structure is idle. Upon the arrival of a go signal g(G↑ or G↓),
the control structure becomes active and issues signals g1 and g2. These signals
initiate the operation of the corresponding devices. Each device issues, upon
completion of its operation, the corresponding completion or done signal (d1 or d2).
The control structure awaits the arrival of both completion signals d1 and d2,
whereupon it produces the output signal d and becomes idle again. Fig. 6.4 shows
a possible decomposition of the control structure PAR2.

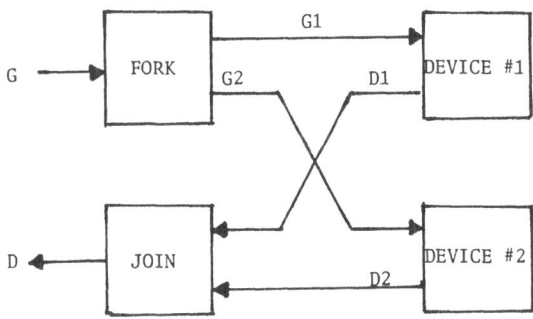

Figure 6.4 - Decomposition of the control structure PAR2

An LMG (GFORK) representing the FORK component is shown in Fig. 6.5. The FORK
component may be implemented simply by connecting the input terminal G to the output
terminals G1 and G2. The JOIN component is represented by GCE(d1,d2;d), i.e. the
marked graph GCE of Fig. 3.1(a), with nodes a, b, z labeled d1, d2, d respectively.
The devices #1 and #2 can be represented by *[g1,d1] and *[g2,d2], respectively.

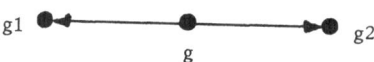

Figure 6.5 - The LMG GFORK representing the FORK component
of Fig. 6.4

Finally, the behavior of the environment connected to input G and output D may be
represented by *[g,d]. Formally the decomposition of Fig. 6.4 can be verified by
proving the equivalence

$$GPAR2 \cong GFORK\|GCE(d1,d2;d)\| *[g1,d1]\| *[g2,d2]\| *[g,d]. \qquad (EQ2)$$

In the next section we derive simplification techniques by means of which the above
equivalence is easily proven.

7. SOME SIMPLIFICATION TECHNIQUES

In this section we derive techniques for the "simplification" of LMGs. By
simplification of an LMG Γ we mean the removal of edges or nodes in such a way that
the resultant LMG is equivalent to Γ. An elementary LMG consists of two nodes and
a single edge between the two nodes. We denote the following elementary LMG

$$a \circ \xrightarrow{k} \circ b$$

by [a(k)b]. In the sequel, Γ (with or without subscripts or superscripts) will always
denote an LMG.

The following proposition is a special case of the net reduction rules in [BRV80].
Proposition 5: Let

$$\Gamma_1 = [a_1(k_1)a_2]\| \ldots \| [a_{n-1}(k_{n-1})a_n], \quad n \geq 2$$

and

$$\Gamma = \Gamma_1\| [a_1(k)a_n]\| \Gamma_2,$$

where $k \geq k_1 + k_2 + \ldots + k_{n-1}$. Then

$$\Gamma \cong \Gamma_1\| \Gamma_2.$$

Proof: Assume that the firing of any node of Γ changes the values of k and k_i
($1 \leq i < n$) into k' and k_i'. We show that $k' \geq \Sigma k_i'$, i.e. the inequality $k \geq \Sigma k_i$
is preserved. Indeed, if node a_1 is fired, we have $k' = k+1$, and $\Sigma k_i' = \Sigma k_i +1$.
Similarly, the firing of node a_n yields $k' = k-1$ and $\Sigma k_i' = \Sigma k_i -1$. The firing
of any other node preserves the values of k and Σk_i. It follows that any firing
sequence applied to Γ will preserve the inequality $k \geq \Sigma k_i$, particularly $k \geq k_{n-1}$.

Now, the removal of the edge $[a_1(k)a_n]$ from Γ could only influence the firability of node a_n. However, in view of the preservation of the inequality $k \geqslant k_{n-1}$, the edge $[a_1(k)a_n]$ can be removed without affecting the firability of a_n. Hence, $L(\Gamma) = L(\Gamma_1 \| \Gamma_2)$.

Formulated somewhat informally, Proposition 5 asserts that an edge of an LMG from node a_1 to node a_n with $k \geqslant 0$ tokens is redundant, provided there exists a directed path $a_1, a_2, \ldots, a_n$ with token count $k' \leqslant k$. The following proposition is evident (cf. [BRV80]).

<u>Proposition 6</u>: Let Γ contain a marked self-loop (cycle of length 1). Let Γ' be the LMG obtained from Γ by removing the marked self-loop. Then $\Gamma \cong \Gamma'$.

An important simplification technique oconsists of removing nodes labeled λ, as illustrated in Fig. 7.1. Generally the elimination of a node v labeled λ

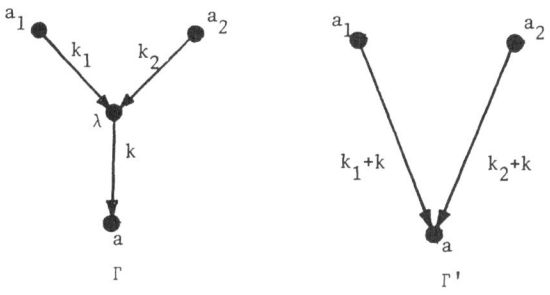

Figure 7.1 - LMG Γ' obtained from LMG Γ by λ-elimination

consists of removing node v together with all its inedges and outedges. Each pair consisting of an inedge $[a_1(k_1)v]$ and an outedge $[v(k_2)a_2]$ is replaced by a single edge $[a_1(k_1+k_2)a_2]$.

One easily verifies the following.

<u>Proposition 7</u>: Let Γ be an LMG and Γ' the LMG obtained from Γ by applying the λ-elimination step as specified above. . Then $\Gamma' \cong \Gamma$.
A few examples will illustrate the applicability of the preceding simplification rules.

<u>Example 6.1</u> (continued)

We wish to prove (EQ1), stated in Section 6. We have $GSP = \Gamma_1$, $GRP = \Gamma_2$, $GSP \| GRP = \Gamma$ where Γ_1, Γ_2 and $\Gamma = \Gamma_1 \| \Gamma_2$ are shown in Fig. 5.1. Proposition 5 is applicable to the LMG Γ of Fig. 5.1: namely the two edges between nodes x and y are redundant. Thus, $GSP \| GRP = \Gamma \cong *[a,x,b,y]$. Consequently, $GSP \| GRP \cong *[a,x,b,y] \setminus \{x,y\}$. Applying λ-elimination (Proposition 7), we obtain

618

$$*[a,x,b,y]\setminus \{x,y\} \cong *[a,b],$$

thus proving the equivalence

$$\text{GSP} \parallel \text{GRP} \cong *[a,b].\tag{EQ1}$$

Example 6.2 (continued)

We now indicate the proof of (EQ2), stated in Section 6. The LMG represented by the right-hand side of (EQ2) is shown in Fig. 7.2. By applying Proposition 5 one easily verifies that all the curved edges in Fig. 7.2 are redundant. Since the remaining LMG coincides with GPAR2 of Fig. 6.3, the equivalence (EQ2) is proven.

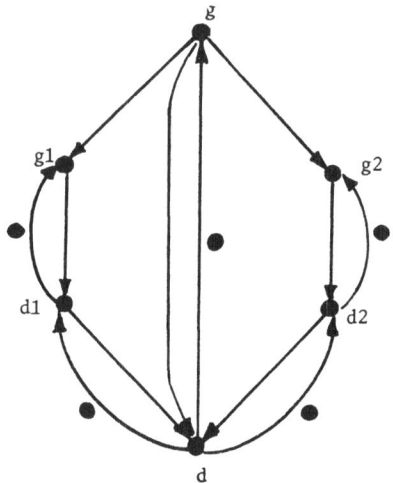

Figure 7.2 - An LMG equivalent to GPAR2 (curved edges are redundant)

Example 7.1

A signal buffer B_k, $k \geqslant 1$, will output a signal z in response to each signal a, received as input. The buffer B_k is capable of storing up to k input signals. An LMG GBk = GBk(a,z), specifying the behavior of B_k, is shown in Fig. 7.3(a). If buffers B_k and B_m are connected in cascade, a buffer B_{k+m} is obtained.

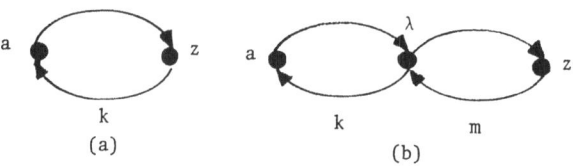

(a)   (b)

Figure 7.3 - (a) GBk(a,z)
              (b) GBk(a,y) $\parallel$ GBm(y,z)

Indeed, we have

Proposition 8:   GBk(a,y) $\parallel$ GBm(y,z) $\cong$ GBn(a,z),   where   n = k+m.

Proof: The LMG GBk(a,y) $\parallel$ GBm(y,z)   is shown in Fig. 7.3(b).   By applying λ-elimination (Proposition 7) and Proposition 6, GBn is obtained.

The language L(GBk) can be characterized by applying Proposition 2.   We thus obtain:

$$L(GBk) = \{w \in \{a,z\}^* \mid \forall x \in \text{pref}\{w\}: 0 \leqslant (x{\downarrow}a) - (x{\downarrow}z) \leqslant k\}.$$

Thus, B_k corresponds to (k,0)SYNC(a,z) in [Sn85].

8.   COMMUNICATION GRAPHS

In this section we model interconnections of circuits which are not necessarily delay-free.   We shall refer to such interconnections as channels (cf. [Ho85]). Signals transmitted over channels are communication signals.   A signal which is output onto channel c is denoted c! and an input signal received over channel c is denoted c?.

A communication graph (CG) is an LMG $\Gamma = <G,\Sigma,\eta>$, whose signal alphabet Σ is partitioned into a set of action signals and a set of communication signals.

Consider a circuit consisting of two components connected by channels.   We model the components by CGs and the overall circuit by their "connection" defined below. Our concept of connection of CGs corresponds to the "agglutinate of directed trace structures" in [Sn85].

Let $\Gamma_i = <G_i,\Sigma_i,\eta_i>$, i = 1,2, be two CGs.   If either x! $\in \Sigma_1$ and x? $\in \Sigma_2$ or x! $\in \Sigma_2$ and x? $\in \Sigma_1$, we refer to x! and x? as communication signals between Γ_1 and Γ_2.   We denote by $\text{cosig}(\Gamma_1,\Gamma_2)$ the set of all such signals.   The connection $\Gamma_1 \bot \Gamma_2$ is defined by $\Gamma_1 \bot \Gamma_2 = (\Gamma_1 | \Gamma_2) \backslash \text{cosig}(\Gamma_1,\Gamma_2)$ where

$$\Gamma_1 | \Gamma_2 := \Gamma_1 \parallel \Gamma_2 \parallel \{[x!(0)x?] \mid x! \in \text{cosig}(\Gamma_1,\Gamma_2)\}.$$

As illustration of this concept, let Γ_1 = *[a,x!,y?] and Γ_2 = *[x?,b,y!].   The CGs Γ_1 and Γ_2 may be considered to represent the SP component of Fig. 6.1 and its RP component, respectively.   In this representation, the links X and Y of Fig. 6.1 are assumed to be channels.   The CG $\Gamma_1 | \Gamma_2$ is shown in Fig. 8.1.   By the above definition of connection, $\Gamma_1 \bot \Gamma_2 = (\Gamma_1 | \Gamma_2) \backslash \{x!,x?,y!,y?\}$.   One easily verifies that $\Gamma_1 \bot \Gamma_2 \cong$ *[a,b].   Indeed, by Proposition 5, $\Gamma_1 | \Gamma_2 \cong$ *[a,x!,x?,b,y!,y?].   Applying λ-elimination, we obtain

$$\Gamma_1 | \Gamma_2 \cong \text{*[a,x!,x?,b,y!,y?]} \backslash \{x!,x?,y!,y?\} \cong \text{*[,a,b]}.$$

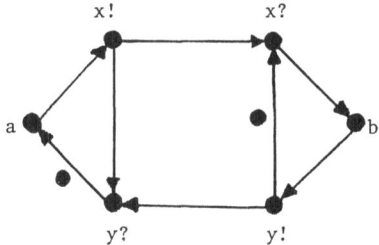

Figure 8.1 - Communication graph Γ

It follows that the interconnection $\Gamma_1 \bot \Gamma_2$ is "delay-insensitive" (cf. [Sn85]). This concept will now be formulated precisely.

Let Γ be a CG. We denote by $\hat{\Gamma}$ the LMG obtained from Γ by replacing each communication signal (x! or x?) by the corresponding channel symbol (x). An interconnection $\Gamma_1 \bot \Gamma_2$ is <u>delay-insensitive</u> iff $\Gamma_1 \bot \Gamma_2 \cong \hat{\Gamma}_1 \parallel \hat{\Gamma}_2$.

In the above example, where Γ_1 = *[a,x!,y?] and Γ_2 = *[x?,b,y!], we have $\hat{\Gamma}_1$ = *[a,x,y] = GSP (see Example 6.1) and $\hat{\Gamma}_2$ = *[x,b,y] = GRP. Since GSP $\parallel$ GRP $\cong$ *[a,b] we indeed have $\Gamma_1 \bot \Gamma_2 \cong \hat{\Gamma}_1 \parallel \hat{\Gamma}_2$.

The connection of two CGs is not necessarily delay-insensitive. A simple example is the following.

<u>Example 8.1</u> Let Γ_1 = *[a,x!] and Γ_2 = *[b,x?]. We have aa $\in \Gamma_1 \bot \Gamma_2$ but aa $\notin \hat{\Gamma}_1 \parallel \hat{\Gamma}_2$. Thus, the connection $\Gamma_1 \bot \Gamma_2$ is not delay-insensitive. Consider now the CGs Γ_3 = *[a,x!,y?] and Γ_4 = *[b,y!,x?]. One easily verifies that

$$\Gamma_3 \bot \Gamma_4 \cong \hat{\Gamma}_3 \parallel \hat{\Gamma}_4 \cong \hat{\Gamma}_1 \parallel \hat{\Gamma}_2.$$

The above example illustrates a method of "extending" a given connection $\Gamma_1 \bot \Gamma_2$ which is not delay-insensitive into a connection $\Gamma_3 \bot \Gamma_4$ which <u>is</u> delay-insensitive and satisfies the condition $\hat{\Gamma}_3 \parallel \hat{\Gamma}_4 \cong \hat{\Gamma}_1 \parallel \hat{\Gamma}_2$. The extension is obtained by "incorporating" additional channels, in order to provide a suitable "handshaking" protocol.

However, the precise formulation of these concepts and the discussion of methods for designing delay-insensitive connections are beyond the scope of this paper (cf. [Sn85]).

CONCLUSIONS

This paper has demonstrated the applicability of labeled marked graphs to the specification, composition, and verification of asynchronous circuits, particularly delay-insensitive (speed-independent, self-timed) circuits.

The following are some further recommended research topics.

(a) Incorporation of structural considerations (e.g. liveness, boundedness).

(b) By considering labeled Petri nets, rather than marked graphs, and their composi- tions, the approach of this paper can be extended to the modeling of arbitrary asynchronous circuits, particularly circuits which behave non-deterministically (cf. [Yo85]).

(c) The approach developed in this paper may be combined with related approaches dealing with communicating systems which involve message-passing (cf. [PY84], [YP86]).

REFERENCES

LNCS   = Lecture Notes in Computer Science, Springer-Verlag.

[BRV80]   Berthelot, G., Roucairol, G. and Valk, R., "Reduction of Nets and Parallel Programs", *Advanced Course on General Net Theory*, W. Brauer, ed., LNCS 84, 1980.

[BY79]   Brzozowski, J.A., and Yoeli, M., "On a Ternary Model of Gate Networks", *IEEE Trans. Computers*, C-28, March 1979, 178-184.

[BY85]   Brzozowski, J.A., and Yoeli, M., "Combinational Static CMOS Networks", Research Report CS-85-42, Department of Computer Science, University of Waterloo, Ontario, Canada, December 1985.

[CMPS83] De Cindio, F., De Michelis, G., Pomello, L. and Simone, C., "Equivalent Notions for Concurrent Systems", *Applications and Theory of Petri Nets*, A. Pagnoni and G. Rozenberg, eds., Informatik-Fachberichte 66, Springer-Verlag, 1983.

[CHEP71] Commoner, F., Holt, A.W., Even, S., and Pnueli, A., "Marked Directed Graphs", *J. Comp. and Syst. Sciences 5*, 1971, 511-523.

[GL73]   Genrich, H., and Lautenbach, K., "Synchronizationsgraphen", *Acta Informatica 2*, 1973, 143-161.

[Ha76]   Hack, M., "Petri Net Languages", Technical Report 159, Lab. for Computer Science, Mass. Inst. Tech., March 1976.

[Ho85]   Hoare, C.A.R., *Communicating Sequential Processes*, Prentice Hall, 1985.

[LN84]   Lengauer, T., and Näher, S., "An Analysis of Ternary Simulation as a Tool for Race Detection in Digital MOS Circuits", *VLSI Algorithms and Architect- ures*, International Workshop, Amalfi, Italy, 1984.

[Mi80]   Milner, R., *A Calculus of Communicating Systems*, LNCS 92, 1980.

[Pe80]   Petri, C.A., "Concurrency", *Advanced Course on General Net Theory*,
         W. Brauer, ed., LNCS 84, 1980.

[Pe81]   Peterson, J., *Petri Net Theory and the Modeling of Systems*, Prentice-Hall,
         1981,

[PY84]   Pehrson, B., and Yoeli, M., "A Communication System Net Model for Specifica-
         tion and Verification of Distributed Systems", *Proc. 4th Int. Conf. on
         Protocol Spec., Verif., and Testing*, Skytop, 1984, North-Holland Publ.

[Re83]   Rem, M., "Partially Ordered Computations with Applications to VLSI Design",
         Foundations of Computer Science IV, Pt.2, MC-Tract 159, 1-44, J.W. de Bakker,
         J. van Leeuwen, eds., Mathematical Center, Amsterdam, 1983.

[RV83]   Rozenberg, G., and Verraedt, R., "Subset Languages of Petri Nets, Part I",
         Theor. Comp. Sci. 26, 1983, 301-326.

[Se80]   Seitz, C.L., "System Timing", in: C. Mead and L. Conway, *Introduction to
         VLSI Systems*, 218-262, Addison-Wesley, 1980.

[Sn85]   Van de Snepscheut, J.L.A., *Trace Theory and VLSI Design*, LNCS 200, 1985.

[YB85]   Yoeli, M., and Brzozowski, J.A., "A Mathematical Model of Digital CMOS Net-
         works", *1985 Canadian Conference on VLSI*, Toronto, November 1985, 117-120.

[YE83]   Yoeli, M., and Etzion, T., "Behavioral Equivalence of Concurrent Systems",
         Applications and Theory of Petri Nets, 292-305, A. Pagnoni and G. Rozenberg,
         eds., Informatik-Fachberichte 66, Springer-Verlag, 1983.

[YG78]   Yoeli, M., and Ginzburg, A., "Petri Net Languages and their Applications",
         Research Report CS-78-45, Department of Computer Science, University of
         Waterloo, Ontario, Canada, November 1978.

[Yo82]   Yoeli, M., "Synthesis of Concurrent Systems", *Application and Theory of
         Petri Nets*, 183-186, C. Girault and W. Reisig, eds., Informatik-Fachberichte
         52, Springer-Verlag, 1982.

[Yo85]   Yoeli, M., "A Net Model of Communicating Systems", Technical Report #372,
         Department of Computer Science, Technion-IIT, Haifa, Israel, July 1985.

[YP86]   Yoeli, M., and Pehrson, B., "Behavior-Preserving Reductions of Communicat-
         ing System Nets", Technical Report #1, Swedish Institute of Computer
         Science, Kista, Sweden, February 1986.